BASICS OI

The Sun continually ejects matter sonic plasma: the solar wind, which engulfs the Earth and the other planets, shaping their environments.

Basics of the Solar Wind presents a modern introduction to the subject, starting with basic principles and including the latest advances from space exploration and theory. The book discusses the structure of the solar interior and atmosphere, the production of the solar wind, and its perturbations. It addresses the basic physics of the objects of the Solar System, from dust to comets and planets, and their interaction with the solar wind. The final sections explore the subject from an astrophysical point of view, including the interaction with the interstellar medium, cosmic rays and winds from other stars. The book contains a historical survey and a short introduction to plasma physics.

This volume is the first to present a comprehensive basic coverage of this subject. The topics are discussed at various levels of difficulty, by including qualitative as well as quantitative treatments and emphasising physical processes rather than mathematics or observation. This book will appeal to students and researchers in physics, astronomy, space physics and engineering, geophysics and atmospheric sciences.

Cambridge Atmospheric and Space Science Series
Editors: J. T. Houghton, M. J. Rycroft and A. J. Dessler

This series of upper-level texts and research monographs covers the physics and chemistry of different regions of the Earth's atmosphere, from the troposphere and stratosphere, up through the ionosphere and magnetosphere and out to the interplanetary medium.

NICOLE MEYER-VERNET is an astrophysicist at the Observatoire de Paris and a Research Director with the Centre National de la Recherche Scientifique (CNRS), Paris. She has been involved in pioneering studies with a variety of spacecraft including the Voyagers and Ulysses, for which she received an Académie des Sciences prize and several NASA Group Achievement Awards.

Cambridge Atmospheric and Space Science Series

EDITORS

Alexander J. Dessler
John T. Houghton
Michael J. Rycroft

TITLES IN PRINT IN THIS SERIES

M. H. Rees
Physics and chemistry of the upper atmosphere

R. Daley
Atmosphere data analysis

J. R. Garratt
The atmospheric boundary layer

J. K. Hargreaves
The solar–terrestrial environment

S. Sazhin
Whistler-mode waves in a hot plasma

S. P. Gary
Theory of space plasma microinstabilities

M. Walt
Introduction to geomagnetically trapped radiation

T. I. Gombosi
Gaskinetic theory

B. A. Kagan
Ocean–atmosphere interaction and climate modelling

I. N. James
Introduction to circulating atmospheres

J. C. King and J. Turner
Antarctic meteorology and climatology

J. F. Lemaire and K. I. Gringauz
The Earth's plasmasphere

D. Hastings and H. Garrett
Spacecraft–environment interactions

T. E. Cravens
Physics of solar system plasmas

J. Green
Atmospheric dynamics

G. E. Thomas and K. Stamnes
Radiative transfer in the atmosphere and ocean

T. I. Gombosi
Physics of space environment

R. W. Schunk and A. F. Nagy
Ionospheres: Physics, plasma physics, and chemistry

I. G. Enting
Inverse problems in atmospheric constituent transport

R. D. Hunsucker and J. K. Hargreaves
The high-latitude ionosphere and its effects on radio propagation

R. W. Schunk and Andrew F. Nagy
Ionospheres

M. C. Serreze and R. G. Barry
The Arctic Climate System

BASICS OF THE SOLAR WIND

NICOLE MEYER-VERNET
Observatoire de Paris and
Centre National de la Recherche Scientifique

CAMBRIDGE
UNIVERSITY PRESS

CAMBRIDGE UNIVERSITY PRESS
Cambridge, New York, Melbourne, Madrid, Cape Town,
Singapore, São Paulo, Delhi, Mexico City

Cambridge University Press
The Edinburgh Building, Cambridge CB2 8RU, UK

Published in the United States of America by
Cambridge University Press, New York

www.cambridge.org
Information on this title: www.cambridge.org/9781107407459

© N. Meyer-Vernet 2007

This publication is in copyright. Subject to statutory exception
and to the provisions of relevant collective licensing agreements,
no reproduction of any part may take place without the written
permission of Cambridge University Press.

First published 2007
First paperback edition 2012

A catalogue record for this publication is available from the British Library

ISBN 978-0-521-81420-1 Hardback
ISBN 978-1-107-40745-9 Paperback

Cambridge University Press has no responsibility for the persistence or
accuracy of URLs for external or third-party internet websites referred to in
this publication, and does not guarantee that any content on such websites is,
or will remain, accurate or appropriate. Information regarding prices, travel
timetables, and other factual information given in this work is correct at
the time of first printing but Cambridge University Press does not guarantee
the accuracy of such information thereafter.

To my sons François and Alain Meyer,

*and to the memory of my father, Jean Vernet,
for the intellectual enrichment he gave me*

Contents

Preface		*page* xiii
1	**The wind from the Sun: an introduction**	1
1.1	A brief history of ideas	1
	1.1.1 Intermittent particle beams?	2
	1.1.2 Permanent solar corpuscular emission?	4
	1.1.3 The modern solar wind	6
1.2	Looking at the Sun	8
	1.2.1 Basic solar properties	9
	1.2.2 The solar spectrum	10
	1.2.3 The solar disc	13
	1.2.4 Sunspots, magnetic fields and the solar cycle	15
	1.2.5 Around the Sun: chromosphere and corona	18
1.3	Observing the solar wind	24
	1.3.1 Observing near the ecliptic	24
	1.3.2 Exploring the third dimension with Ulysses	28
	1.3.3 A simplified three-dimensional picture	33
	References	37
2	**Tool kit for space plasma physics**	41
2.1	What is a plasma?	42
	2.1.1 Gaseous plasma	44
	2.1.2 Quasi-neutrality	44
	2.1.3 Collisions of charged particles	48
	2.1.4 Plasma oscillations	54
	2.1.5 Non-classical plasmas	56
	2.1.6 Summary	57
2.2	Dynamics of a charged particle	58
	2.2.1 The key role of the magnetic field	58
	2.2.2 Basic charge motion in constant and uniform fields	59
	2.2.3 Non-uniform magnetic field	62
	2.2.4 Adiabatic invariants	65
	2.2.5 Summary	66
2.3	Many particles: from kinetics to magnetohydrodynamics	66
	2.3.1 Elements of plasma kinetics	66

		2.3.2	First-aid kit for space plasma fluids	72
		2.3.3	Elements of magnetohydrodynamics	85
		2.3.4	Waves and instabilities	96
		2.3.5	Summary	100
	2.4	Basic tools for ionisation		101
		2.4.1	Energy of ionisation and the size of the hydrogen atom	101
		2.4.2	Ionisation by compressing or heating	102
		2.4.3	Radiative ionisation and recombination	103
		2.4.4	Non-radiative ionisation and recombination	105
	2.5	Problems		107
		2.5.1	Linear Debye shielding in a non-equilibrium plasma	107
		2.5.2	Mean free path in a plasma	108
		2.5.3	Particles trapped in a planetary magnetic field	108
		2.5.4	Filtration of particles in the absence of equilibrium	109
		2.5.5	Freezing of magnetic field lines	110
		2.5.6	Alfvén wave	110
		2.5.7	Why is the solar wind ionised?	110
	References			110
3	**Anatomy of the Sun**			**113**
	3.1	An (almost) ordinary star		113
		3.1.1	Hydrostatic equilibrium of a large ball of plasma	114
		3.1.2	Luminosity	116
		3.1.3	Energy source and timescales	118
		3.1.4	The mass of a normal star	121
	3.2	Structure and dynamics		123
		3.2.1	Modelling the solar interior	124
		3.2.2	Convective instability	125
		3.2.3	Convective energy transfer	128
		3.2.4	The quiet photosphere	132
		3.2.5	Solar rotation	135
	3.3	Some guesses on solar magnetism		137
		3.3.1	Elements of dynamo theory	138
		3.3.2	Solar kinematic dynamos	142
		3.3.3	Concentrating and expelling the magnetic field	145
		3.3.4	Lorentz force restriction on dynamo action	148
		3.3.5	Elementary physics of magnetic flux tubes	149
		3.3.6	Surface magnetic field	154
	3.4	Problems		158
		3.4.1	Conductive heat transfer in the solar interior	158
		3.4.2	Timescale for radiative transport	158
		3.4.3	Solar differential rotation	158
		3.4.4	Twisted magnetic flux tube	159
		3.4.5	The heat flux blocked by sunspots	159
	References			160

Contents

4 The outer solar atmosphere — **165**
- 4.1 From the photosphere to the corona — 166
  - 4.1.1 The atmosphere in one dimension — 166
  - 4.1.2 One more dimension — 168
  - 4.1.3 Three dimensions in space — 169
  - 4.1.4 ...and one dimension in time — 169
  - 4.1.5 A (tentative) look at the solar jungle — 172
- 4.2 Force balance and magnetic structures — 174
  - 4.2.1 Forces — 175
  - 4.2.2 Force-free magnetic field — 177
  - 4.2.3 Magnetic helicity — 181
  - 4.2.4 Inferences on magnetic structure in the low corona — 185
- 4.3 Energy balance — 186
  - 4.3.1 Radiative losses — 186
  - 4.3.2 Radiative and conductive timescales — 187
  - 4.3.3 Temperature structure — 188
- 4.4 Some prominent species — 190
  - 4.4.1 Spicules — 190
  - 4.4.2 Magnetic loops — 191
  - 4.4.3 Prominences — 193
- 4.5 Time variability — 194
  - 4.5.1 Empirical facts — 194
  - 4.5.2 Hints from physics — 197
  - 4.5.3 Further difficult questions — 200
- 4.6 Coronal heating: boojums at work? — 203
  - 4.6.1 The energy budget and how to balance it — 204
  - 4.6.2 Heating through reconnection events — 205
  - 4.6.3 Heating by waves — 206
  - 4.6.4 Filtration of a non-Maxwellian velocity distribution — 209
- 4.7 Hydrostatic instability of the corona — 214
  - 4.7.1 Simplified picture of a static atmosphere — 214
  - 4.7.2 Magnetic field effects — 215
- 4.8 Problems — 217
  - 4.8.1 Elementary temperature profile — 217
  - 4.8.2 Helicity of a string wrapped around a doughnut — 217
  - 4.8.3 A static solar atmosphere? — 218
- References — 218

5 How does the solar wind blow? — **223**
- 5.1 The basic problem — 225
  - 5.1.1 The solar wind on the back of an envelope — 225
  - 5.1.2 Nasty questions, or why it is complicated — 227
- 5.2 Simple fluid theory — 228
  - 5.2.1 The isothermal approximation — 228
  - 5.2.2 Breeze, wind or accretion? — 232
- 5.3 Letting the temperature vary — 237

	5.3.1	Energy balance	237
	5.3.2	Polytrope approximation	239
	5.3.3	Changing the geometry	246
	5.3.4	Further pushing or heating the wind	247
	5.3.5	What about viscosity?	249
5.4	A mixture of fluids		250
	5.4.1	Simple balance equations	251
	5.4.2	Observed proton and electron temperatures	253
	5.4.3	The role of collisions	254
	5.4.4	Heat flux	256
	5.4.5	The electric field	257
	5.4.6	Fluid picture balance sheet and refinements	261
5.5	Kinetic descriptions		262
	5.5.1	Some notations	262
	5.5.2	Observed proton and electron velocity distributions	263
	5.5.3	Non-collisional electron heat flux	267
	5.5.4	Exospheric models	268
	5.5.5	Kinetic models with collisions and wave–particle interactions	273
5.6	Building a 'full' theory?		274
	5.6.1	More and better observations (beware of hidden assumptions)	274
	5.6.2	Difficult theoretical questions	275
5.7	Problems		277
	5.7.1	Transonic flows in ducts: the de Laval nozzle	277
	5.7.2	The hysteresis cycle of an isothermal flow	279
	5.7.3	Spherical accretion by a star: the Bondi problem	280
	5.7.4	A wind with polytrope protons and electrons	281
	5.7.5	Playing with the kappa distribution	282
	5.7.6	'Temperature' or 'temperatures'?	283
	5.7.7	Non-collisional heat flux	284
	5.7.8	An imaginary wind with charges of equal masses	285
	References		286

6 Structure and perturbations — 291

6.1	Basic large-scale magnetic field		291
	6.1.1	Parker's spiral	291
	6.1.2	Basic heliospheric current sheet and other currents	296
	6.1.3	Magnetic field effects on the wind	299
6.2	Three-dimensional structure during the solar cycle		300
	6.2.1	Warped heliospheric current sheet	301
	6.2.2	Observed large-scale structure	301
	6.2.3	Connecting the Sun and the solar wind, or: where do the fast and slow winds come from?	305
6.3	Major perturbations		308
	6.3.1	Interaction between the fast and slow winds	308

	6.3.2	Coronal mass ejections in the solar wind	309
	6.3.3	Associated shocks	311
6.4	Waves and turbulence		315
	6.4.1	Waves	315
	6.4.2	Turbulence	318
6.5	Minor constituents		326
	6.5.1	Abundances: from the Universe to the solar wind	326
	6.5.2	Helium and heavier solar wind ions	327
	6.5.3	Pick-up ions	328
6.6	Problems		329
	6.6.1	Parker's spiral	329
	6.6.2	Heliospheric currents	329
	6.6.3	Coplanarity in MHD shocks	330
	6.6.4	Kraichnan's spectrum in magnetofluid turbulence	330
References			330

7 Bodies in the wind: dust, asteroids, planets and comets   **335**

7.1	Bodies in the wind		336
	7.1.1	Various bodies	336
	7.1.2	Mass distribution	338
	7.1.3	Mass versus size	341
	7.1.4	Atmospheres and how they are ionised	344
	7.1.5	Planetary magnetic fields and ionospheric conductivity	347
7.2	Basics of the interaction		348
	7.2.1	Properties and spatial scales of the flow	348
	7.2.2	Being small: electrostatic charging and wakes	352
	7.2.3	Being large: the importance of conductivity	358
	7.2.4	Large objects with a conducting atmosphere	362
	7.2.5	Large magnetised objects	365
	7.2.6	Bow shocks	368
	7.2.7	Not being constant: sputtering and evaporation	371
7.3	The magnetospheric engine		372
	7.3.1	Basic structure	375
	7.3.2	Energy, coupling and timescales	378
	7.3.3	Storms, substorms and auroras	385
7.4	Physics of heliospheric dust grains		390
	7.4.1	Forces	390
	7.4.2	Evaporation	394
7.5	Comets		394
	7.5.1	Producing an atmosphere	397
	7.5.2	Ionising the atmosphere	400
	7.5.3	Pick-up of cometary ions	401
	7.5.4	Magnetic pile-up	403
	7.5.5	The plasma tail	404
	7.5.6	X-ray emission	406
	7.5.7	The dust tail	408

7.6	Problems		409
	7.6.1	Electrostatic charging in space	409
	7.6.2	Magnetic pile-up	409
	7.6.3	Chapman–Ferraro layer	410
	7.6.4	Interaction of the solar wind with Venus and Mars	411
	7.6.5	Ring current	411
	7.6.6	Does Vesta have a magnetosphere?	412
	7.6.7	Gas–dust drag in a comet: another nozzle problem	412
References			413
8 The solar wind in the Universe			**419**
8.1	The frontier of the heliosphere		419
	8.1.1	The Local Cloud	420
	8.1.2	Basics of the interaction	421
	8.1.3	The size of the solar wind bubble	424
8.2	Cosmic rays		425
	8.2.1	Cosmic rays observed near Earth	426
	8.2.2	Rudiments of the acceleration of particles	430
	8.2.3	Modulation of galactic cosmic rays by solar activity	436
	8.2.4	'Anomalous cosmic rays'	439
8.3	Examples of winds in the Universe		440
	8.3.1	Some basic physical processes in mass outflows	441
	8.3.2	Some empirical results on stellar winds	443
	8.3.3	The efficiency of the wind engine	445
8.4	Problems		448
	8.4.1	Energy density of cosmic rays	448
	8.4.2	Power law distribution of accelerated particles	448
	8.4.3	The size of an astrosphere	448
	8.4.4	Instability of a star's atmosphere produced by radiation pressure	448
References			449
Appendix			451
Index			457

Preface

Why chase the wind?
J. Cocteau, *Antigone*

For science-fiction writers and some space engineers, the 'wind from the Sun'[1] is a wind of photons – the light we see, whose pressure might allow solar sailing and drive space windjammers through the solar system. Yet the Sun blows another kind of wind, made of material particles, whose importance is considerable since it bathes the whole Solar System and shapes all planetary environments.

This wind has many faces. To the layman, it sounds rather mysterious, being made of a strange medium, a plasma: the fourth state of matter. Not only do its tempests affect our everyday technology by disrupting communications and power stations, but it drives two bewildering sky displays: comet tails and auroras. To the space scientist, in contrast, the solar wind is a close companion, and the challenge is to explore and tame a jungle where his or her instruments reveal a strange fauna. The plasma physicist is delighted to find there a number of stunning surprises and extreme properties which are virtually impossible to simulate in the laboratory. And to the astronomer who is trying to understand how cosmic bodies – from planets and comets to stars and galaxies – eject particles into space, it is the only stellar wind that can be studied in detail.

The solar wind has been explored *in situ* by numerous space probes, from inside Mercury's orbit to far beyond the distance of Neptune, and, quite recently, at virtually all heliocentric latitudes. The last decade has seen an explosion in the volume of data, and the solar wind is now measured in almost embarrassing detail. Yet, from the beginning of modern physics to the present epoch, its origin has motivated – and still motivates – much debate.

This book explores the physics involved, from the solar origin, to the frontier of the Solar System. The object of the game is to retrieve (in a quantitative, albeit approximate, way) the basic properties from first principles, within the limits of our incomplete understanding, keeping in mind that Nature always turns out to be subtler than we had imagined.

This book is intended for scientists, for technical workers involved in space missions, for science students and teachers, and more generally for those who enjoy the application of basic physics to a realm unattainable in Earth's laboratories. The emphasis is aimed at physical intuition rather than mathematical

---

[1] Clarke, Arthur C. 1972, *The Wind from the Sun*, London, Victor Gollancz.

rigour. The calculations only require a basic background in physics and mathematics and assume no prior knowledge of plasma physics, for which a first-aid kit is given in Chapter 2.

This subject has hideously complicated aspects, and I had to make gross simplifications in order to avoid the fundamental ideas being lost in a morass of details. Resisting the temptation of replacing basic understanding by classification, detailed mathematics, and/or computer modelling, creates a dilemma: how be useful to non-specialists, without angering the specialists too much. I therefore made no attempt to be comprehensive, either in the topics, or in the references. Instead, I have tried to follow Victor Weisskopf, who used to say at the start of a course: 'I will not cover the subject, I will try to uncover part of it.'[2] In the same spirit, the references are meant to help the readers, not to give credit to the authors. Unless otherwise stated, units are SI.

Although I take full responsibility for the errors that have crept in,[3] I should not give the impression that this book was written by me alone. Many people of diverse languages and cultures have contributed, either personally or through their writings. It is impossible to acknowledge all of them properly and to give credit to the scientists whose viewpoints influenced me. I offer my warm thanks to the generous friends and colleagues who have scrutinised sections of the manuscript and provided suggestions for improvements, or contributed in other ways, especially to Jean-Louis Steinberg (who got me started on space research), Alex Dessler (who got me started on this book), Marcia Neugebauer, Ludwik Celnikier, Joseph Lemaire, Marco Velli, Serge Koutchmy, Jorge Sanchez-Almeida, Pascal Démoulin, Dominique Bockelée, Karine Briand, Darrell Strobel, Rosine Lallement, Guillaume Aulanier, Françoise Launay, Milan and Antonella Maksimovic and Danielle Briot. I owe much to my friends and colleagues of the Observatoire de Paris at Meudon for the warm environment and numerous discussions, and to several outstanding former graduate students for their insightful and stimulating questions. This work would not have been possible without the kind help of the efficient staffs of the library at Meudon (Observatoire de Paris) and of the laboratory LESIA (CNRS and Universities Paris 6 and 7). The students who endured my lectures on the solar wind at the University Paris 11, and on astrophysical plasmas at the Observatoire de Paris (and the Universities Paris 6, 7 and 11) have contributed in no small way too. Thanks to all of them! Last but not least, I am very grateful to my family and my friends for their encouragement and help. Special thanks are due to my son François Meyer for the drawings he made to illustrate this book.

Nicole Meyer-Vernet
CNRS
Observatoire de Paris (Meudon, France)

---

[2] Weisskopf, V. F. 1989, *The Privilege of Being a Physicist*, New York, W. H. Freeman, p. 32.

[3] I encourage readers to send me typographical or other errors at nicole.meyer@obspm.fr. I intend to post an updated list of errors at www.lesia.obspm.fr/~meyer/BSW.html.

1
The wind from the Sun: an introduction

'First accumulate a mass of Facts: and *then* construct a Theory.'
That, I believe, is the true Scientific Method. I sat up, rubbed my eyes, and began to accumulate Facts.
Lewis Carroll, *Sylvie and Bruno*

Not only does the Sun radiate the light we see – and that we do not see – but it also continually ejects into space 1 million tonnes of hydrogen per second. This wind is minute by astronomical standards; it carries a very small fraction of the solar energy output, and compared to the violent explosions pervading the universe it blows rather gently. Yet it has amazing effects on the solar surroundings. It blows a huge bubble of supersonic plasma – the heliosphere – which engulfs the planets and a host of smaller bodies, shaping their environments. It also conveys perturbations that can be seen in our daily life.

The object of this chapter is twofold. To give a concise historical account of the key ideas and observations that made our modern view of the solar wind emerge; and introduce the main properties of the Sun and of its wind, and their interpretation in terms of basic physics. The latter goal requires some tools of plasma physics, and will be developed in the rest of the book.

1.1 A brief history of ideas

The idea that planets are not moving in a vacuum is very old. In some sense our modern view of a solar wind filling interplanetary space has replaced the Aristotelian quintessence, the impalpable pneuma of the Stoic philosophers, and the swirling 'sky' introduced two thousand years later by Descartes. In some sense only, as there is a major difference: the solar wind is made of normal matter whose behaviour is – to some extent – understood, even though this matter is in a special state, a *plasma*, having unusual properties as we will see in Chapter 2.

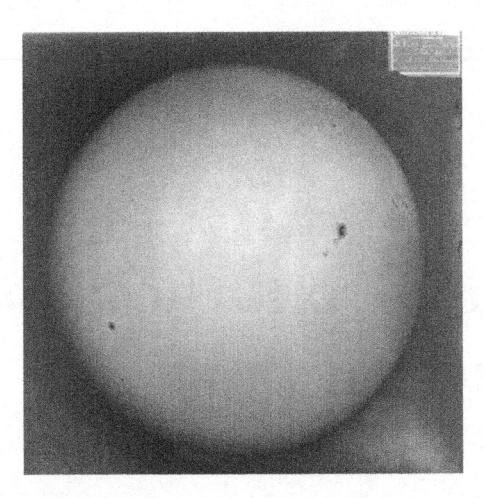

Figure 1.1 An early photograph of the Sun showing sunspots, made by Jules Janssen at the Meudon Observatory on 22 June 1894. (Observatoire de Paris, Meudon heritage collections.)

Ironically enough, the solar wind contains vortices – as did Descartes' sky and also the luminiferous ether imagined later by Maxwell – and even though those vortices have nothing in common with their ethereal ancestors, they are barely better understood. And not only does the solar wind transmit sound and light as did the ancient ether, but it also carries a host of waves that Maxwell could not have dreamt of.

Even though the idea is an ancient one, most of the solar wind story took place over little more than a century. At the end of the nineteenth century, only a couple of far-seeing scientists had imagined that a solar wind might exist. At the beginning of the twenty-first century, hordes of space probes have explored the solar wind and it is honoured with a secure place in astronomy textbooks.

Since there are eminent accounts of how this concept emerged and developed (see for example [9], [25]), it is not my intention here to trace a detailed history. I shall only give a few hints[1] as to how the ideas evolved to fit reasonably well into the logical structure of modern physics and astronomy.

1.1.1 Intermittent particle beams?

When did the story begin? Perhaps around the middle of the nineteenth century, when the British amateur astronomer Richard Carrington, who was drawing sunspots (see Fig. 1.1) from a projected image of the Sun,[2] suddenly saw two patches of peculiarly intense light appear and fade within 5 minutes in the largest sunspot group visible [7], [5]. Carrington had witnessed what we now call a

---

[1] Partly taken from Meyer-Vernet, N. 2005, *Bul. Inst. d'Astron. et de Géophys. Georges Lemaître* (Université Catholique de Louvain, Louvain-La-Neuve, Belgium).

[2] By far the safest way for amateurs to observe the Sun.

A brief history of ideas

Figure 1.2 Auroral display seen from the ground in Lapland. (Photo courtesy of C. Molinier.) For hints on auroral physics see Section 7.3.

solar flare: a giant explosive energy release on the Sun – and a very strong one. Some time later, the magnetic field at Earth was strongly perturbed (Fig. 7.22), large disturbances appeared in telegraph systems, and intense auroras spread over much of the world (see Fig. 1.2). The connection between geomagnetic perturbations and auroras was already known, and Carrington suggested that both phenomena might be due to the special event he had seen on the Sun; but he was careful to point out that his single observation did not imply a cause-and-effect relationship. Carrington was correct; we will see later that solar flares are sometimes, although not always, followed by geomagnetic disturbances and auroras; this happens when the Sun releases a massive cloud of gas that reaches the Earth's environment and perturbs it.

Carrington was not the first to suspect the Sun of producing auroras and magnetic effects on Earth, as a correlation between the number of sunspots and geomagnetic disturbances had already been noted. The cause, however, was unclear. And the aurora was not better understood, even though its electric and magnetic nature had been identified a century before – a major improvement over the ingenious scheme based on the firing of dry gases proposed by Aristotle. To summarise the state of auroral physics around the mid 1880s (see [10]): 'The scientific theories ... are more abstruse than the popular ones, but equally fail ...', as the Norwegian scientist Sophus Tromholt put it with splendid irreverence.

All that pointed to a connection between the Sun and terrestrial magnetic disturbances, and the idea was taken seriously by some physicists near the end of the nineteenth century. Assuming 'that the Sun is powerfully electrified, and repels similarly electrified molecules with a force of some moderate number of times the gravitation of the molecules to the Sun', George Fitzgerald suggested that [12]: 'matter starting from the Sun with the explosive velocities we know possible there, and subjected to an acceleration of several times solar gravitation, could reach the Earth in a couple of days'. In other words, the Earth was bombarded by intermittent beams of charged particles coming from the Sun

Figure 1.3 Birkeland on two of his fronts [4]. Left: with his *terella* apparatus – a magnetised sphere subjected to a beam of electrons in a vacuum chamber. Right: with some of his instruments for aurora detection.

and accelerated by an electrostatic field, just as an electrode in a giant vacuum tube.[3]

1.1.2 Permanent solar corpuscular emission?

In this context, an essential step was taken by the Norwegian physicist Kristian Birkeland, who worked in the closing years of the nineteenth century and the opening years of the twentieth (Fig. 1.3). These were fabulous times for a physicist: X-rays and radioactive decay were just being discovered, J. J. Thomson was unveiling the electron, Hendrik Lorentz was developing the electron theory and building steps on the route which led Einstein to change our vision of the Universe and Max Planck was explaining the spectrum of radiation, among other major accomplishments. Applied science was rising, too: the first aeroplane was close to being born, and Guglielmo Marconi made the first long-distance radio transmission, an engineering performance which led to the discovery of the Earth's ionosphere.

Birkeland worked on three fronts: theory, laboratory experiments with a model Earth and observation [11]. Not only did he develop the ideas put forward by Fitzgerald and others, but in order to test them he organised several polar expeditions and made the largest geomagnetic survey up to that time [4]. He also put forward a number of ingenious ideas that stand up well today, and above all, he submitted a crucial point: since auroral and geomagnetic activity was produced by solar particles and was virtually permanent, the inescapable conclusion was that the Earth environment was bombarded in permanence by 'rays of electric corpuscles emitted by the Sun'.

---

[3]This was written 5 years before J. J. Thomson's 1897 paper on 'cathode rays' (*Phil. Mag.* **44** 293), and showed remarkable insight. The solar wind is made of charged particles – protons and electrons – and we will see later in this book that indeed the heliospheric electric field pushes the protons outwards with a force of a few times the solar gravitational attraction.

A brief history of ideas 5

Figure 1.4 An old drawing of Donati's comet, shown over Paris on 5 October 1858. (From A. Guillemin, 1877, *Le Ciel*, Paris, Hachette.) A modern view of cometary physics may be found in Section 7.5.

Put in modern terms, Birkeland was suggesting that the Sun emits a continuous flux of charged particles filling up interplanetary space: nearly our modern solar wind. This was a great change of perspective from the picture of the Sun emitting sporadic beams separated by a vacuum. However, many of these ideas were far ahead of the time, some were incorrect, and above all, the revered Lord Kelvin submitted impressive arguments showing that the Sun could not produce geomagnetic disturbances.[4] As a result, Birkeland's work was largely ignored by the scientific community.

These ideas lay in obscurity for many years. And when the *solar corpuscular radiation* (as it was called) resurfaced – albeit on independent grounds – to explain geomagnetic activity, it was once again in the form of occasional beams emitted by the Sun by some exotic process in a (slightly dusty) vacuum.

This remained the leading view until the middle of the twentieth century, when the concept of a continuous solar emission was to re-emerge through an entirely separate line of work. Comets have two classes of tails, one class nearly straight, made of gas, the other curving away, made of dust (Fig. 1.4). The

---

[4]On 30 November 1892, he concludes in his Presidential Address to the Royal Society: 'It seems as if we may also be forced to conclude that the supposed connection between magnetic storms and sun-spots is unreal, and that the seeming agreement between the periods has been a mere coincidence' (Kelvin, Thompson, W. 1892, *Proc. Roy. Soc. London A* **52** 300).

curved shape of dust tails is produced by solar radiation pressure and gravity acting on the dust grains. But the gaseous tails raised an intriguing problem: they were observed to always point straight away from the Sun (with a slight aberration angle)[5] and to exhibit irregularities that appeared to be accelerated away from the Sun. What caused these properties? The current explanation in terms of solar radiation pressure acting on the cometary gas failed by several orders of magnitude.

This problem was brilliantly solved by the German physicist Ludwig Biermann in the early 1950s: solar photons could not do the job, but perhaps solar corpuscular radiation would? Biermann thus developed a model for the interaction of cometary particles with those coming from the Sun, that neatly explained the comets' gaseous tails *if* they were subjected to a permanent flux of charged particles coming from the Sun (see [3] and references therein).

Although Biermann's original arguments regarding the interaction process between particles are now known to be incorrect, this was again a crucial change of perspective from the current view. Since comets' orbits pass at all heliolatitudes, the inescapable conclusion was that the Sun was emitting particles in *all* directions at *all* times. Half a century after Birkeland's work, the concept of a continuous solar corpuscular emission was resurfacing, with stronger observational and theoretical support.

But at the same time a different conclusion was reached by the English physicist Sydney Chapman through a completely different path. The outer atmosphere of the Sun – called the *corona* (see Fig. 1.12) – was known to be very hot. Chapman, who had pioneered the calculation of the kinetic properties of gases, found that this hot ionised atmosphere conducted heat so well that it should remain hot out to very large distances. As a result, particles have such large thermal speeds even far from the Sun that they can go very far against its gravitational attraction; this makes the density decline very slowly, so slowly that the solar atmosphere should extend well beyond the Earth's orbit [6]. In other words, the Earth was to be immersed in the *static* atmosphere of the Sun.

1.1.3 The modern solar wind

How could the ubiquitous solar corpuscular flux found by Biermann coexist with this static solar atmosphere? Both are plasmas and, as we will see later, the coexistence of plasmas having such different bulk velocities has very nasty consequences. The great achievement made by Eugene Parker in 1958 was to realise that [24]: 'however unlikely it seemed, the only possibility was that Biermann and Chapman were talking about the same thing'. So Biermann's continuous flux of solar particles was just Chapman's extended solar atmosphere expanding away in space as a supersonic flow. This comes about because this atmosphere is so hot, even far away from the Sun, that neither the solar gravitational attraction nor the pressure of the tenuous interstellar medium can confine it. At

---

[5]With hindsight, the aberration angle of comet plasma tails, determined by the relative speeds of the radially moving solar wind plasma and the comet, yielded a correct solar wind speed of about 400 km s^{-1}. For details, see Brandt, J. C. 1970, cited in Chapter 5.

A brief history of ideas 7

Figure 1.5 Mariner 2, the spacecraft that served to identify the solar wind, and the first one to have reached a planet other than the Earth. (Image by National Aeronautics and Space Administration (NASA).)

last the modern solar wind concept was born. Parker's theory was not only an eminently elegant solution which brought together existing evidences, but it made numerous testable predictions: in particular the wind was to flow at several hundreds of kilometres per second radially away from the Sun. Irony of ironies, these ideas were so novel that the paper presenting them [23] encountered difficulty in being published in the eminent *Astrophysical Journal*, on the grounds that the author was not familiar with the subject [25]. The referees of the *Astrophysical Journal* were not the only scientists to be displeased by Parker's theory, and a hot debate followed as to whether or not the Sun was capable of emitting a supersonic wind.

Observation was needed to settle the debate. But measuring the solar wind was a heroic challenge in those years when space technology was just springing up. It took four Russian missions and seven American ones – most of which failed due to problems of launching – to get an unambiguous result. The most successful of the Russian spacecraft, Lunik II, launched in 1959 (only 2 years after the first Russian Sputnik went up and the United States launched their first satellite in reply) detected a flux of positive ions; however, this observation was not entirely conclusive because the direction of the particles' velocity was unknown. The ultimate proof came in 1962 from the American spacecraft Mariner 2 [22], which was en route for Venus after having miraculously survived an impressive series of failures (Fig. 1.5). As Marcia Neugebauer superbly puts it [20]: 'We had data! Lots of it! There was no longer any uncertainty about the existence and general properties of the solar wind.'

So ended the first age of the solar wind story. With these results Parker's ideas rose to prominence, and within a few years, in spite of some dissenting voices, the solar wind concept acquired the respected state of a physical reality. Even now, decades later, Parker's ideas serve as a reference for a large part of what is understood on the subject. However, the very elegance of this theory

Figure 1.6   Cartoon published in the French journal *Le Grelot* (March 1873).

masked a number of fundamental difficulties that we will examine later, and of which an historical account may be found in [18]. As we will see in this book, the story has many other chapters and there is still a long way ahead.

1.2   Looking at the Sun

> My Sun, the golden garden of your hair
> Has begun to flame
> And the fire has spread over our corn field
>
> Already the green ears are parched
> Pressed by the presence of your breath
> And the last drop of their sweet is wrung for them
>
> Strike us with the rain of your arrows,
> Open to us the door of your eyes,
> Oh Sun, source of beneficent light.
>   *(Quecha poem to the Sun[6])*

To us the Sun is no longer a god, and we do not have to feed it with hearts and blood to keep it moving across the sky, as did the Aztecs. Modern scientists (Fig. 1.6) worship it by building beautifully instrumented observatories scattered over the Earth's surface, far underground and in space, to analyse its radiation.

---

[6]Ferguson, D. 2000, *Tales of the Plumed Serpent*, London, Collins & Brown.

Table 1.1 *Basic solar properties*

Mean distance to the Earth	$d_\oplus = 1.5 \times 10^{11}$ m
Mass	$M_\odot = 2. \times 10^{30}$ kg
Radius	$R_\odot = 7. \times 10^{8}$ m
Luminosity	$L_\odot = 3.84 \times 10^{26}$ W

The last quarter of the twentieth century saw the vertiginous growth of the scope and quality of solar data. A number of new techniques have emerged, making a revolution in the achievements of ground-based telescopes, while several beautifully instrumented spacecraft have been launched, enabling one to study wavelengths inaccessible from the ground. Among others, the satellites Yohkoh (launched in 1991), SoHO (launched in 1995), TRACE (launched in 1998) and RHESSI (launched in 2002) are currently in operation and devoted to solar observations.

We outline below some characteristics of the Sun and make a short survey of what its radiation tells us. We shall try to complete and unify this impressionistic picture later, equipped with the tools of plasma physics. A very accessible survey of solar physics is [13]; general accounts aiming at physical processes are given in [17], [14], [33].

1.2.1 Basic solar properties

Some basic properties of the Sun are listed in Table 1.1. Although these are rounded figures, they are exact to better than 1%, and may be considered as nearly constant by human standards; indeed, as we shall see later, many years will pass before changes in solar properties may oblige human beings to transfer the Earth to a more convenient star (see [2] and references therein).

The solar distance is called the astronomical unit (AU); it is used as a basic unit in the Solar System and beyond. So is the solar mass, which is negligibly altered by the solar wind ejection. Indeed, the solar wind pours out in space roughly 10^9 kg s^{-1}, which amounts to only $10^{-4} M_\odot$ over the Sun's age of a few 10^9 years. Note that the wind is not the only source of solar mass loss; the mass–energy equivalence tells us that the luminosity $L_\odot$ – the energy lost by the Sun per second via electromagnetic waves – yields a mass loss of $L_\odot/c^2 = 4.3 \times 10^9$ kg s^{-1}; this amounts to about four times the mass carried away by the solar wind, and thus barely alters the Sun's mass either; we will return later to the solar energy source.

The Sun's radius is that of the visible disc, which is almost perfectly round and whose diameter is a little more than half a degree as seen from Earth; it is sharply defined because virtually all the light we receive from the Sun originates in a thin layer a few hundred kilometres thick: the *photosphere*, where – going outwards – matter changes rapidly from completely opaque to almost completely transparent, letting radiation escape freely into space.

At the mean distance $d_\oplus$ of the Earth (but outside its atmosphere), the flux received from the Sun in the form of electromagnetic radiation by a surface perpendicular to the rays of sunlight is

$$S = L_\odot / (4\pi d_\oplus^2) \tag{1.1}$$
$$= 1.37 \times 10^3 \ \mathrm{W\,m^{-2}}$$

which is called the *solar constant*. One square metre of the Earth's surface receives, however, much less: around 200 W in average, partly because only half the surface is sunlit and the radiation does not arrive at right angles, and partly because some of the incident energy is absorbed in the atmosphere. This energy sustains virtually all life on Earth; it may also harm it, as the Quecha poem reminds us. Despite its name, the solar constant is more variable than the other basic solar properties; it may vary by up to a few thousandths over several days and even more over long periods; it shows in fact variability at virtually all timescales, thereby raising a strong interest in climatology circles [15].

The solar luminosity enables one to derive one more basic property. If the Sun were radiating as a blackbody, that is if the Sun's disc emitted thermal radiation of temperature T_{eff}, the luminosity would be given by Stefan–Boltzmann law as

$$L_\odot = \sigma_S T_{eff}^4 \times 4\pi R_\odot^2 \tag{1.2}$$

where $\sigma_S = 5.67 \times 10^{-8}$ is the Stefan–Boltzmann constant in SI units. Putting the figures given in Table 1.1 into (1.2), we deduce the *effective temperature* of the Sun

$$T_{eff} = 5800 \ \mathrm{K}. \tag{1.3}$$

This is about 20 times hotter than the temperature at the Earth's surface, and it is this temperature difference that sustains life there. The effective temperature T_{eff} would be the actual temperature of the emitting outward layer of the Sun – the photosphere – if the radiation were thermal. We will see later that the actual photospheric temperature is close to this value.

1.2.2 The solar spectrum

The solar radiation has been studied at virtually all wavelengths, from gamma rays on the short wavelength side of the spectrum, to radio waves on the long wavelength side. How close to thermal is it?

Spectral distribution

Figure 1.7 shows the spectral distribution S_λ of the energy received from the Sun at Earth (outside the Earth's atmosphere) per unit wavelength per unit sunlit surface area perpendicular to the Sun's direction per unit time, at wavelengths ranging from 10^{-13} m to a few metres; in this range the intensity spans over

Looking at the Sun 11

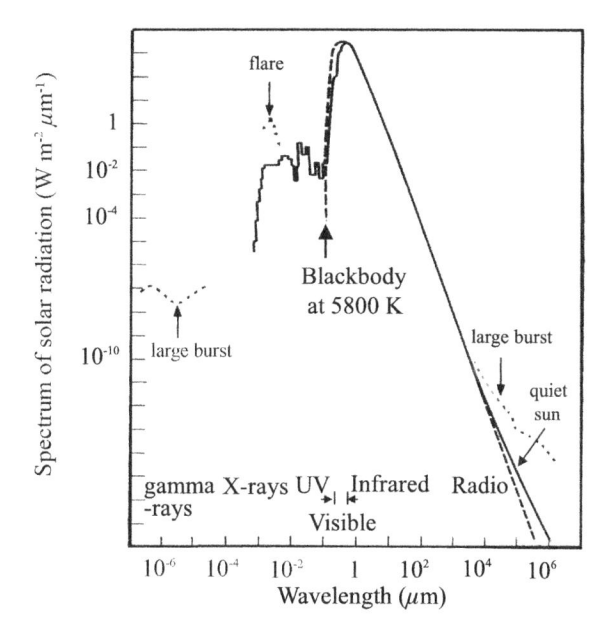

Figure 1.7 The measured spectrum of solar radiation (continuous line, with the contribution of bursts dotted), compared with the blackbody spectrum (1.4) (dashed).

23 orders of magnitude. The visible and near-infrared spectrum, which makes only a minute part of the wavelength range, takes the lion's share of the energy. The solar constant S introduced in (1.1) is the integral of S_λ over wavelengths.

On this measured spectrum we have superimposed (dashed line) the spectrum of thermal radiation:

$$S_{T\lambda} = \pi B_\lambda R_\odot^2 / d_\oplus^2 \qquad (1.4)$$

where

$$B_\lambda = \frac{2hc^2}{\lambda^5} \frac{1}{e^{hc/\lambda k_B T} - 1} \qquad (1.5)$$

is the Planck emissivity at wavelength λ (per unit wavelength per unit surface area perpendicular to the Sun's direction per unit solid angle per unit time), calculated with $T = T_{\mathit{eff}}$ given by (1.3).[7]

One sees that the continuous and dashed lines are rather close to each other in the near and middle ultraviolet (UV), and nearly indistinguishable in the visible and in the infrared. This means that the solar spectrum is close to thermal in that range which is the one contributing most to the total flux. This gives some sense to the value of T_{eff} we derived from Stefan–Boltzmann's law.

---
[7]Beware that the spectra in Fig. 1.7 are plotted in $W\,m^{-2}\,\mu m^{-1}$, i.e. not in SI units.

In contrast, one sees in Fig. 1.7 that the UV radiation below about 0.1 μm greatly exceeds that of a 5800 K blackbody; instead of falling almost exponentially the level remains of the order of 3×10^{-2} W m^{-2}μm^{-1}. This UV emission below 0.1 μm is of great importance. Integrated over wavelengths it amounts only to about 3×10^{-3} W m^{-2}, less than 10^{-5}th of the total solar flux. Yet, this radiation has a crucial property: at $\lambda = 10^{-7}$ m the frequency is $f = c/\lambda = 3 \times 10^{15}$ Hz, so that the characteristic energy hf of a photon is about 12 eV. We will see later that this is close to the binding energy of the hydrogen atom, which sets the order of magnitude of the energy needed to strip atoms of their outer electron. It is this part of the solar spectrum that is responsible for most ionised environments in the inner solar system, from planetary ionospheres (Section 7.1) to comets (Section 7.5).

At these short wavelengths, the spectral level is far from constant and becomes increasingly variable at shorter wavelengths. The largest variations can reach several orders of magnitude and occur during flares – those giant eruptions that Carrington described for the first time, and to which we will return in Section 4.5. We have sketched this variable component as a dotted line.

One sees in Fig. 1.7 that the long wavelength part of the spectrum behaves very differently. In the radio range, one can distinguish two superimposed components. The first one is a smooth spectrum – the so-called *quiet sun* radiation, which continues smoothly from the infrared with an amplitude deviating more and more from the 5800 K thermal spectrum as the wavelength increases; near $\lambda \simeq 1$ m, it would correspond to an effective temperature about 200 times hotter, that is of the order of 10^6 K. The second component is made of intermittent bursts whose amplitude may exceed the quiet level by several orders of magnitude, and whose spectrum does not even have a small resemblance to Planck's law; it is sketched as a dotted line.

Spectral lines

On the continuous spectrum shown in Fig. 1.7 are superimposed a huge number of spectral lines, which do not appear clearly on the figure because its resolution is too low; spectral lines even dominate the ultraviolet range. Millions of lines are observed in the solar spectrum, of which only a very small fraction have been analysed with the sophisticated tools of atomic and molecular physics. Since no two chemical elements have the same spectrum, spectral lines act as bar codes which enable spectroscopists to determine the chemical composition of the outer layers of the Sun, in addition to a number of physical properties. Practically all known chemical elements have been detected on the Sun. Hydrogen is by far the most abundant, about 92% in number density; the rest is mostly helium, the other elements making up only about 0.1% of the total. On the whole, and apart from a few significant exceptions, this is rather similar to the rest of the universe. We will return to this point in Section 6.5.

These lines still reveal more information. As particles are moving, the lines are Doppler-shifted, which enables one to derive the particle velocities. The random speeds produce a random frequency shift which broadens the lines,

Looking at the Sun

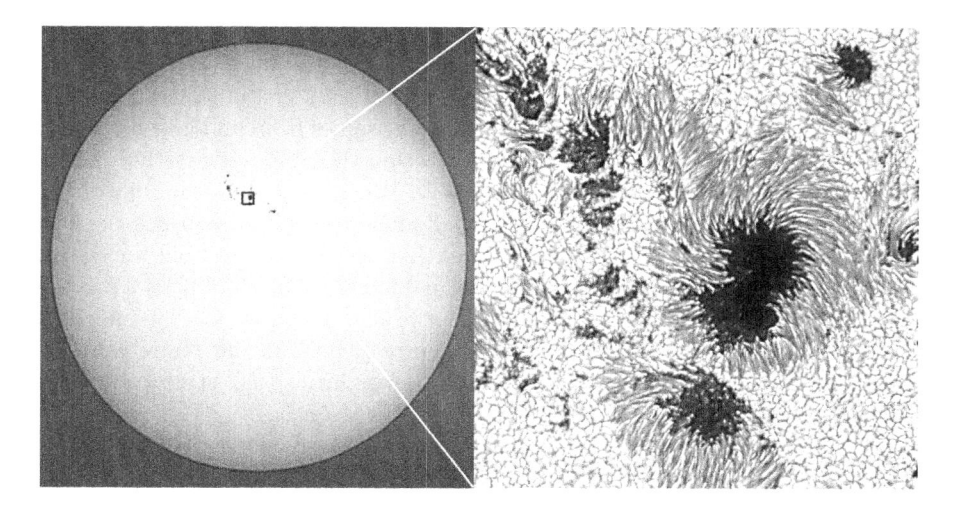

Figure 1.8 A modern photograph of the solar disc in white light, taken with the Swedish 1-metre Solar Telescope on La Palma, Canary Islands (left panel) on 15 July 2002, and a detailed view of a part of the group of sunspots visible near the centre of the disc (right panel). (Photographs courtesy of Royal Swedish Academy of Sciences.)

so that the line shapes reveal both the particle thermal speeds (that is the temperature) and the turbulent motions of the medium. A further quantity can be derived: the bulk velocity of the medium, which produces a net shift of the lines.

The spectral lines still reveal more. As the medium is magnetised, the Zeeman effect modifies the lines, which may enable one to derive the strength and direction of the magnetic field. As we shall see below, the magnetic field plays a major role in solar properties; in fact it is a key to understanding most phenomena occurring in the solar atmosphere.

1.2.3 The solar disc

Figure 1.8 is a modern image of the solar disc in white light, showing a group of sunspots (left panel) and an enlarged view of this group (right panel). Apart from sunspots, the figure shows two main features:

- the whole disc (left panel) appears darker near the edge than near the centre,

- at small scales (right panel), the space outside sunspots is covered with a granular pattern.

Both features reveal important properties of the solar surface. Let us consider in turn the disc as a whole, and its small-scale structure.

The disc as a whole

Why does the disc appear darker near the edge? To understand this *limb darkening*, we must consider where the observed light comes from and how it travels. The observed light comes from some depth within the Sun, a depth that is limited by absorption of light along its trajectory. When we look at the edge of the disc, light is travelling towards us at a small angle to the solar surface, so that it encounters more absorbing solar material than when it comes from the centre of the disc, in which case it is travelling normal to the surface. Hence we can see deeper into the Sun when we look at the centre than when we look at the edge. Therefore, the greater brightness near the centre means that deeper regions are hotter. This is not surprising; as radiation escapes into space, energy is lost so that the temperature decreases outwards.

From the variation in brightness from centre to limb, one can calculate that the temperature decreases outwards by more than 2000 K over a distance of a few hundred kilometres, where light absorption by the solar material varies from almost complete to negligible.

Why does light absorption by the solar material decrease so sharply outwards? This is because the density drops sharply; furthermore we have seen that the temperature also falls off, so that the ionisation changes. But why does the density drop so sharply? The answer lies in two simple properties of the solar material. First, it behaves as a perfect gas. We have seen that it is essentially made of hydrogen atoms, whose mass is about the proton's mass m_p; hence the pressure P and mass density ρ are related by the perfect gas law

$$P = \rho k_B T / m_p. \tag{1.6}$$

Second, the gas is close to hydrostatic equilibrium in the solar gravitational field. Thus at distance r from the Sun's centre, the outward pressure force per unit volume dP/dr balances the gravitational attraction per unit volume ρg; writing $dP/dr = \rho g$ and substituting (1.6), we deduce

$$dP/P = (m_p g / k_B T) \, dr.$$

Integrating over r in a small range of distance above the solar surface over which g and T do not vary much, we find

$$P \propto \exp(-r/H) \tag{1.7}$$
$$H = k_B T / m_p g. \tag{1.8}$$

Since the Sun's gravitational acceleration at distance $r \simeq R_\odot$ is

$$g = M_\odot G / R_\odot^2, \tag{1.9}$$

equations (1.7) and (1.8) show that the pressure and mass density decrease outwards with the scale height

$$H = \frac{k_B T R_\odot^2}{m_p M_\odot G}. \tag{1.10}$$

Looking at the Sun 15

With the parameters in Table 1.1 and $T = T_{\mathit{eff}}$ (given in (1.3)), this yields $H = 1.8 \times 10^5$ m.[8] Hence over a distance of a few hundred kilometres the density falls by a large factor; this corresponds to the width of the photosphere. We shall study the photosphere in more detail in Section 3.3.

Small-scale structure

Let us now examine the solar disc at small scales (right panel of Fig. 1.8, which includes a large sunspot region). Outside sunspots, it is covered with bright irregular 'granules' of different sizes, more or less polygonal in shape, separated by darker lines. These granules have a maximum size of about 2000 km, and detailed observation shows that they seem to have a life of their own: they appear, fragment or explode, living typically for a few minutes. What is their origin? They are believed to be the uppermost signature of convective motions arising below the photosphere. As we said, the Sun is hotter in the interior, and, somewhat like water in a kettle heated from below, the solar material develops convective patterns. Spectroscopy shows that the bright central regions of granules are moving upwards, while the darker edges move downwards, with speeds of the order of 1 km s^{-1}. Granules can be thought of as fountains where hot solar material is rising near the centre, and then flows horizontally towards the edges; in doing so the material cools by radiating to outer space so that upon arriving at the edges it is cooler, and thus appears darker than the central part of the granules. Having cooled, the material is denser and sinks back into the Sun. The upward flow in the central region of the granules compensates for this disappearing cold material.

Granulation is not the only structure presumably convective in origin observed at the photosphere. It is the most conspicuous of what may be a continuum of convective flow scales. In particular, a large-scale velocity pattern – the so-called *supergranulation* – is observed, albeit with little if any intensity contrast. These large cellular structures, of scale $\sim (2\text{--}3) \times 10^4$ km and living for about a day, are identified as horizontal flows moving at about 0.5 km s^{-1} towards the periphery of the cells, with much slower vertical upflows at the centre of the structures and downflows at their boundaries.

Convection plays a major role in the Sun's behaviour: on a large scale, it transports heat to the photosphere from the hotter layers below; it also drives small-scale motions of the solar material. We will return to this phenomenon in Chapter 3, and shall see that it plays an important role in the generation and structure of the magnetic field.

1.2.4 Sunspots, magnetic fields and the solar cycle

Sunspots

Let us now examine sunspots – the kind of structure Carrington was observing when he witnessed a solar flare. Sunspots can be observed with simple tools;

---

[8] We will see in Section 3.2 that the scale height is slightly smaller because the presence of a small quantity of atoms heavier than hydrogen increases the mean mass per particle.

in some cases they are big enough to be seen by the unaided eye.[9] Sunspots have a remarkable property: their number varies in a nearly cyclic way, with a period of about 11 years. They generally occur in groups on the solar disc. Figure 1.8 (right panel) shows a detailed view of such a group,[10] where individual spots appear as a dark *umbra* about 10^4 km across surrounded by a filamentary penumbra.

Some sunspots may last for weeks; this has enabled early observers to follow their motion and to find that the Sun appears to rotate about its axis once in about 27 days (as seen from Earth). Detailed observation shows that the Sun does not rotate as a solid body: its outer layers rotate more slowly in polar regions than at the equator. Correcting for the Earth's motion which changes the Sun's apparent rotation speed when it is observed from Earth, one finds that in an inertial frame, the Sun's outer layers rotate once in 25 days near the equator and only about once in 30 days near 60° solar latitude.

Sunspots appear dark because they are colder than the normal photosphere; since radiation varies as the fourth power of temperature (1.2), they radiate much less than their surroundings. What makes them cold? An answer was suggested in 1941 by Biermann. We have said that the normal photosphere is heated by convective motions that mix the surface and the hotter layers below. However, sunspots have a special property: they are permeated with an intense magnetic field, and as we will see later, this magnetic field holds matter so tightly that it can inhibit convective motions, thereby preventing heat from reaching them.

To summarise, sunspots exhibit three basic properties:

- they are colder than their surroundings by about 2000 K,

- their magnetic field is strong: about 0.3 T (nearly vertical at the centre); they often appear in pairs with opposite magnetic field polarities,

- their number and location on the solar disc follow regular laws, with a cycle of 11 years.

The opposite magnetic field directions observed in nearby sunspots are illustrated in Fig. 1.9; this structure is not surprising since from Maxwell equations the total magnetic flux through the solar surface must remain zero. We will return to sunspot physics in Section 3.3.

The sunspot cycle is illustrated in Fig. 1.10, which shows the number of sunspots visible on the Sun as a function of time.[11] Sunspots are not the only solar feature observed to follow such a cycle. They are the most conspicuous of a variety of structures having virtually all scales, which are governed by the magnetic field and often follow the same 11-year cycle.

---

[9] A warning for absent-minded readers: the Sun is by far the most dangerous astronomical object to observe; it should never be looked at, either directly or through an optical device, without an appropriate filter.

[10] Details on the structure of these sunspots may be found in Scharmer, G. B. *et al.* 2002, *Nature* **420** 151.

[11] Computed with the international sunspot number convention, till daily observations of sunspots were started at the Zurich observatory.

Looking at the Sun

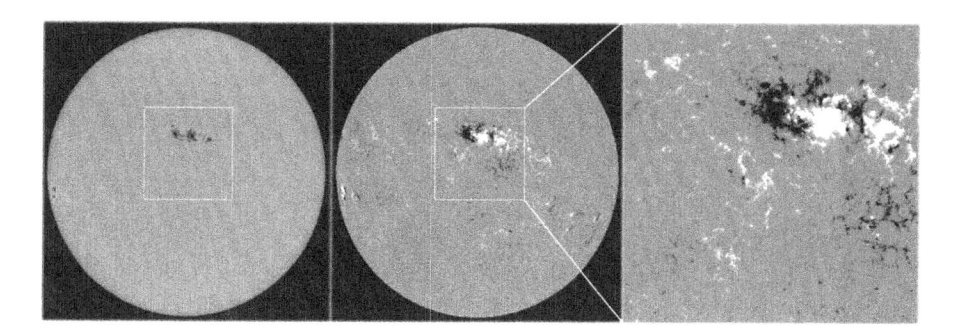

Figure 1.9 Magnetogram of the solar disc (middle panel) taken from the SoHO spacecraft on 16 July 2002; white represents upward magnetic field along the line of sight, black downward. The left panel is a white-light image of the solar surface taken a few minutes before, showing a few sunspots; the right panel is an enlarged magnetogram. (Images from SoHO/MDI, European Space Agency (ESA) and NASA, Stanford-Lockheed Institute for Space Research.)

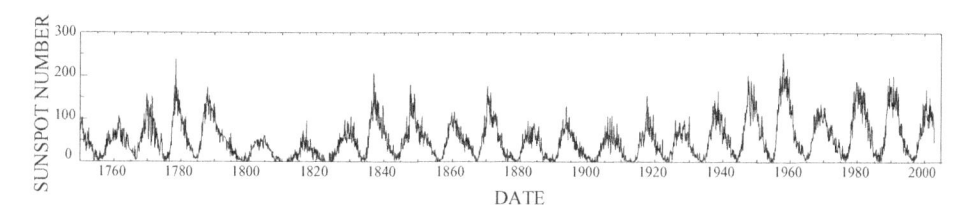

Figure 1.10 Monthly mean sunspot numbers from 1750 to 2003.

Magnetic field on the Sun

What does the photospheric magnetic field look like far away from sunspots? On a large scale and very roughly, it can be thought of as produced by a giant bar magnet whose polarity reverses every 11 years (so that the true period is 22 years). On the whole, this yields opposite average fields of the order of perhaps $(1-5) \times 10^{-4}$ T near the solar poles at the minimum of the cycle, when the solar magnetic dipole is roughly aligned with the rotation axis and the number of sunspots is minimum. This dipolar field gradually weakens and reverses its direction near the maximum of the cycle, when the number of sunspots is maximum; at that time the large-scale field is no longer dipole-like and has a complex multipolar structure.

However, observation at smaller scales reveals a more complicated pattern (Figs. 1.9 and 1.11) and shows a remarkable phenomenon: the magnetic field at the photosphere has an intermittent structure, showing concentrations as small as perhaps 100 km across,[12] where the field can be as high as about 0.15 T and is changing rapidly; since these thin structures cover only a very small fraction of

---

[12] Or less, since the best angular resolution achieved at present is not less than 0.1 arc–sec, and detecting smaller features requires extremely subtle techniques.

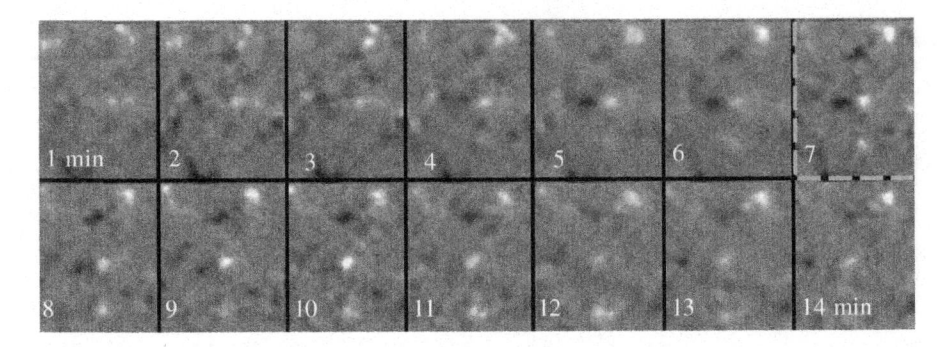

Figure 1.11 A time sequence of magnetograms of a small area on the solar surface, taken with the Swedish telescope at La Palma; the resolution is roughly 200 km on the solar surface. White represents upward vertical magnetic field, black downward. The displayed area on the Sun is approximately 7000 × 9000 km^2 and the magnetograms are taken at intervals of 1 min. (From [29].)

the solar surface, the average field is much smaller. Furthermore, one observes nearly everywhere a kind of magnetic carpet of fields having opposite vertical directions (Fig. 1.11), so that in 'quiet' regions the mean magnetic field may be much smaller than the average absolute value [29]. An important observational clue is that the small field concentrations generally evolve at the same timescale as convective flows, tending to accumulate at the edge of the convective cells.

What causes the solar magnetic field and its cycle? What determines the scale of its structures and their relation to one another? A hint may be obtained by comparing the energies of the magnetic field and of the particle thermal motions. The thermal energy per particle at temperature T is $3k_BT/2$; with a particle number density of about $n \simeq 10^{23}$ particles/m^3 at $T \simeq 5000$ K in the photosphere, the density of thermal energy is $3nk_BT/2 \sim 10^4$ J m^{-3}. In small magnetic structures having a strong magnetic field $B \sim 0.15$ T, the magnetic energy density $B^2/2\mu_0 \sim 10^4$ J m^{-3}, so that both types of energy are in equilibrium. In sunspots, whose magnetic field is much stronger (and temperature weaker), the magnetic field dominates and tends to hold matter, whereas in normal regions of smaller magnetic field, matter should drive the field. We shall put this rough argument on a sounder footing in Section 3.3 and will try to unveil part of the enigma of solar magnetic fields, using the tools of plasma physics. However, more than three centuries after the time of Galileo, when sunspots were first observed with a telescope, their structure is still not fully understood.

1.2.5 Around the Sun: chromosphere and corona

Chromosphere

Were the density to continue to decrease outwards at the rate estimated above, the solar atmosphere would not extend very far away ... and there would be no solar wind. But a striking phenomenon occurs above the photosphere: instead of continuing to decrease outwards, the temperature starts to increase; over an altitude of 2000 km or so the temperature rises to 10^4 K or so. This relatively high temperature enables atoms to be excited to energy levels from which they emit spectral lines which should not be observed otherwise. In particular the red H_α line – the less energetic of the Balmer series of hydrogen – contributes to the bright purplish-red crescent seen during total solar eclipses that is at the origin of the name of this region: the *chromosphere*.

Corona

At the top of the chromosphere a still more striking phenomenon occurs: the temperature T jumps by two orders of magnitude; the density ρ decreases by roughly the same factor to prevent too large changes in the pressure, which is proportional to ρT; above this *transition region* the temperature profile flattens and remains of the order of 10^6 K over a few solar radii. This hot atmosphere is called the *corona* – the Sun's crown; not only is it very extended because the scale height increases with temperature, but it is at the origin of the solar wind.

How can the corona be hotter than the photosphere? Normally, heat should flow from the hot corona to the cold photosphere, not the opposite; how then does the Sun manage to heat the corona? This question has worried several generations of physicists who have attacked it on many fronts, but nobody has yet come out with a satisfying answer. The fundamental gaps are so important that: 'we cannot state at the present time why the Sun is obliged by the basic laws of physics to produce the heliosphere', as Eugene Parker put it not long ago [26]. We will discuss this question in Section 4.6.

We have seen that the density decreases exponentially outwards with a small scale height at the photosphere; this density decrease continues in a gentler way in the chromosphere, with a somewhat greater scale height – due to the greater temperature (cf. (1.10)); and a further density decrease by two orders of magnitude occurs in the transition region. Hence the corona is a very tenuous medium that does not radiate much.

Visible coronal radiation

The brightness of the brighter inner part of the corona in visible radiation is indeed only 10^{-6}th of the solar disc's brightness. This faint radiation is produced by scattering of sunlight in the Earth's atmosphere and barely reaches 10^{-4}th of the sky luminosity; hence, the corona cannot be seen from Earth under normal conditions. However, by a happy accident, the Sun and the Moon are seen from the Earth with virtually the same angular size, so that the Moon occults precisely the whole solar disc during total eclipses. Since this occultation occurs above the Earth's atmosphere, the sky brightness is also lowered, so that the corona can be seen on these occasions, even by the unaided eye. Total eclipses are rare, and other techniques have been developed to study the corona and the chromosphere, in particular by devising artificial

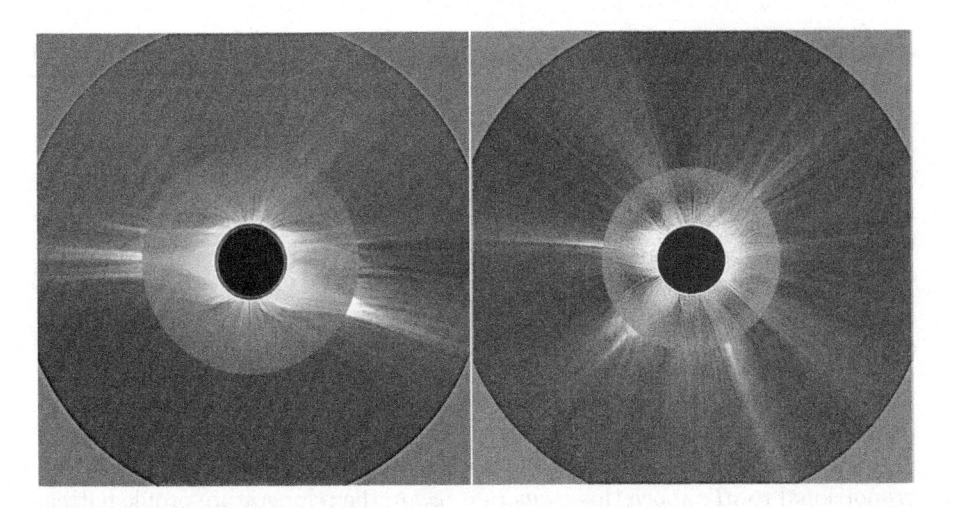

Figure 1.12 The solar corona after solar activity minimum (left panel) and near solar activity maximum (right panel). The images are composites of eclipse photographs taken respectively in Guadeloupe on 26 February 1998 (left, obs. C. Viladrich) and in Angola on 21 June 2001 (right, obs. J. Mouette), and nearly simultaneously from the LASCO-C2 coronagraph on the spacecraft SoHO (ESA and NASA). (Composites by Institut d'Astrophysique de Paris – CNRS; by courtesy of S. Koutchmy.)

eclipses inside optical instruments. Figure 1.12 is a composite obtained with both techniques, at different epochs of solar activity. The inner part shows the visible appearance of the corona observed from Earth during an eclipse; the outer part is obtained by making an artificial eclipse aboard a spacecraft on the same days.[13]

These images raise a number of questions. In the first place, what produces the observed radiation?

This question can be addressed, up to a point, in a very simple way. At $T \simeq 10^6$ K, the mean kinetic energy of a particle, $3k_BT/2$, amounts to 130 eV – roughly ten times more than the binding energy of the hydrogen atom. Hence not only do ambient electrons have enough energy for ionising hydrogen atoms by collisions and knocking out the outer electron of heavier atoms, but they can also knock out a large part of the more strongly bound inner electrons. The corona is thus a mixture of ions, including several-times ionised ones, and free electrons: a plasma. These free electrons are subjected to the solar radiation, are accelerated by the wave electric field, and in turn radiate at the same frequency; this is called *Thomson scattering*. Thomson scattering of sunlight is responsible for the visible radiation of the corona at altitudes up to a few solar radii.

From analysis of this radiation, one can deduce the density of electrons in the corona. Let us perform a simple estimate. The lower part of the corona is

---

[13] An opaque disc is put into the telescope to mask out the bright central emitting region.

close to hydrostatic equilibrium, hence the density is expected to fall roughly in an exponential way with a scale height given by (1.10); H is in fact twice larger because the medium is essentially made of protons and electrons, so that the mean mass per particle is $\mu \simeq (m_p + m_e)/2 \simeq m_p/2$ instead of m_p; we will return to this point later. With $T \simeq 10^6$ K, we find $H \simeq 0.1 \times R_\odot$. With such a small scale height, most of the scattered radiation comes from the electrons lying low in the corona, at distances from the Sun's centre close to $R_\odot$, so that the flux of solar radiation they receive is about $L_\odot/4\pi R_\odot^2$.

The ratio of the power radiated by one electron to the incident flux of radiation is given by Thomson's cross-section

$$\sigma_T = 8\pi r_e^2/3 \tag{1.11}$$

where the so-called classical electron radius is

$$r_e = \frac{e^2}{4\pi\epsilon_0 m_e c^2} = 2.8 \times 10^{-15} \text{ m} \tag{1.12}$$

so that each electron radiates per second:

$$\sigma_T \times L_\odot/4\pi R_\odot^2 \simeq L_\odot r_e^2/R_\odot^2 \quad \text{(radiation of one electron)} \tag{1.13}$$

from (1.11). We have seen that the coronal brightness at low altitudes is about 10^{-6}th of the solar value. Therefore the total number of scattering electrons is about

$$N \sim 10^{-6} \times R_\odot^2/r_e^2 \sim 6 \times 10^{40}.$$

A large fraction of this number lies close to the base of the corona, thus within a thin shell of surface $4\pi R_\odot^2$ and width H; we deduce the mean electron number density near the base of the corona

$$n_e \sim \frac{N}{4\pi R_\odot^2 H} \sim 10^{14} \text{ m}^{-3}.$$

We will see in Section 4.1 that the actual electron density there is indeed of this order of magnitude.

Coronal structure

The appearance of the corona in Fig. 1.12 raises another question. Near solar activity minimum, one sees bright complex structures at small and mid latitudes, with bright streamers extending more or less radially above them, whereas the polar regions appear rather uniform. In contrast, near solar activity maximum bright structures are seen all around the Sun.

What is the origin of these structures and of their variation with solar activity?

The observed emission is produced by particles, which we have seen to be electrically charged. Charged particles act as tracers of magnetic field lines,

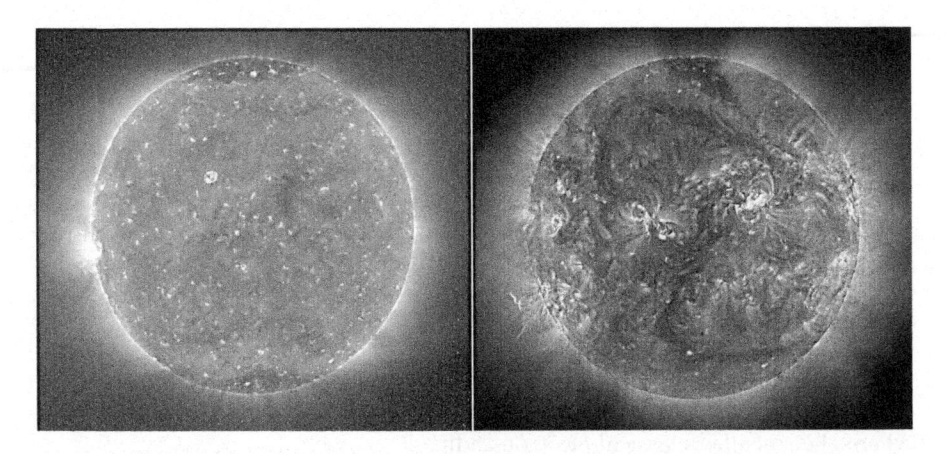

Figure 1.13 Images of the Sun from the Extreme Ultraviolet Imaging Telescope (at 195 Å) on the spacecraft SoHO on respectively 5 May 1996, when solar activity was minimum (left panel), and on 2 November 2001, when solar activity was maximum (right panel). (Images by SoHO/EIT consortium, ESA and NASA.)

somewhat as do iron filings sprinkled on a sheet of paper placed near a magnet. Indeed, as we will see in Section 2.3, the plasma and the magnetic field are intimately linked together, and one may ask whether the magnetic field might drive the plasma, somewhat as a magnet drives iron filings. A hint can be obtained by comparing the energy densities. We have seen that at the photosphere, the thermal energy of the particles is normally greater than the magnetic energy, except in the localised regions where the magnetic field is very strong. However, the fast density decrease with altitude makes the particle energy fall more rapidly than does the magnetic energy. In the upper chromosphere and the lower corona, therefore, magnetic energy tends to dominate, so that the magnetic field is expected to drive the plasma. Hence the changes in the plasma structures are driven by changes in the solar magnetic field during the sunspot cycle.

Coronal emission of ultraviolet and X-rays

We saw that the radiation responsible for the white light images of the corona shown in Fig. 1.12 is mainly produced by scattering of solar radiation by free electrons. It is interesting to compare these images to images taken in the X-ray or the extreme ultraviolet range.[14] Figure 1.13 shows such images, obtained in a narrow spectral band in the extreme ultraviolet when solar activity was respectively minimum (left panel) and maximum (right panel).

How is this emission produced? We have seen that the coronal temperature is so high that heavy elements are stripped of many of their electrons. The

---

[14] These images must be taken from space since radiation at X-ray and extreme ultraviolet wavelengths is absorbed by the upper atmosphere of the Earth.

Looking at the Sun

Figure 1.14 Image taken by the Transition Region and Coronal Explorer (TRACE) in April 2001 in a narrow spectral band around 171 Å, containing emission lines of eight- and nine-times ionised iron atoms that reveal structures having a temperature of about 10^6 K. (Image by Stanford-Lockheed Institute for Space Research and NASA.)

collisions with ambient electrons that ionise these elements can also knock their bound electrons into higher energy levels from which they radiate, producing spectral lines. Since the mean energy of electrons is $3k_BT/2$ and that of photons of wavelength λ is hc/λ, the plasma at $T \simeq 10^6$ K (or more) should radiate strongly at wavelengths of the order of $\lambda \simeq 2hc/3k_BT \simeq 10^{-8}$ m (or less), which corresponds to the extreme ultraviolet or the X-ray range. Furthermore, the emission is proportional to the number of emitting ions and to the number of exciting electrons, so that it increases as the density squared, thereby increasing the contrast of the structures.

Another reason for the contrast of those images taken in a narrow spectral band is that different spectral lines are emitted by different ions, which in turn are abundant in different temperature ranges. Hence a given spectral line reveals plasma at a particular temperature. For example, the spectral line in which the images of Fig. 1.13 have been obtained is emitted by iron atoms that are ionised 11 times and are abundant around 1.5×10^6 K; at lower temperatures, ambient electrons have not enough energy to produce such a high degree of ionisation, whereas at higher temperatures, they tend to produce a higher degree of ionisation.

Figure 1.13 shows two extreme types of large-scale structures: bright and dark. Bright regions appear highly structured and are especially numerous at activity maximum. As on images in the visible range, these bright structures outline the lines of force of the magnetic field. These lines emerge from the Sun and close back to its surface, forming large arches or loops, which are generally located above sunspot groups. Figure 1.14 is an image of such loops; one sees that they have a fibril or thread-like structure. On the other hand, dark regions appear more uniform (although they do contain small-scale structures), and they radiate less essentially because they are less dense. Near solar activity minimum these dark *coronal holes* are located around the solar poles; over each coronal hole the large-scale vertical field is observed to have a uniform direction that is opposite in the north and south polar caps. Around activity maximum, when the predominant magnetic field polarity around the poles is being reversed, the polar coronal holes disappear but smaller ones appear throughout the disc.

On these basic structures are superimposed a host of time-varying phenomena having virtually all scales from the solar radius down to the smallest observable scale. What produces these features? We shall tackle this intriguing question in Chapter 4.

1.3 Observing the solar wind

> 2ND WITCH I'll give thee a wind.
> 1ST WITCH Th'art kind.
> 3RD WITCH And I another.
> 1ST WITCH I myself have all the other,
>   And the very ports they blow,
>   All the quarters that they know ...
>   W. Shakespeare, *Macbeth*

A solar astronomer is like a child in a toy museum, who can look but not touch (even via instruments) – until some bold and insightful space agency sends a probe to the Sun (Fig. 5.15). This involves the difficult art of determining where the radiation comes from and how it has been altered by propagation. The solar wind physicist has an easier task: he or she can make direct measurements with space probes that analyse *in situ* the solar wind particles and fields. Since the beginning of the space age, several tens of spacecraft have explored the heliosphere at virtually all latitudes up to the outskirts of the solar system, and returned a host of data (Fig. 1.15). Strangely enough, they have found that there is not a single wind but several – albeit one is more basic than the others.

1.3.1 Observing near the ecliptic

Some subtleties of space exploration

Most space probes lie close to the *ecliptic* – the plane in which the Earth and most of the planets orbit the Sun. There is a simple reason for that. A spacecraft leaving the Earth starts with a velocity vector equal to the Earth's orbital velocity plus that provided by the launcher; since the Earth's velocity is about 30 km s^{-1} and lies in the ecliptic plane, one must give to the spacecraft a velocity perpendicular to the ecliptic of at least this amount to put it into an orbit angled far from this plane; this is outside the capabilities of existing rocket technology (Fig. 1.16). As a result, there is an armada of space probes exploring the solar wind near the ecliptic.

Observing the solar wind 25

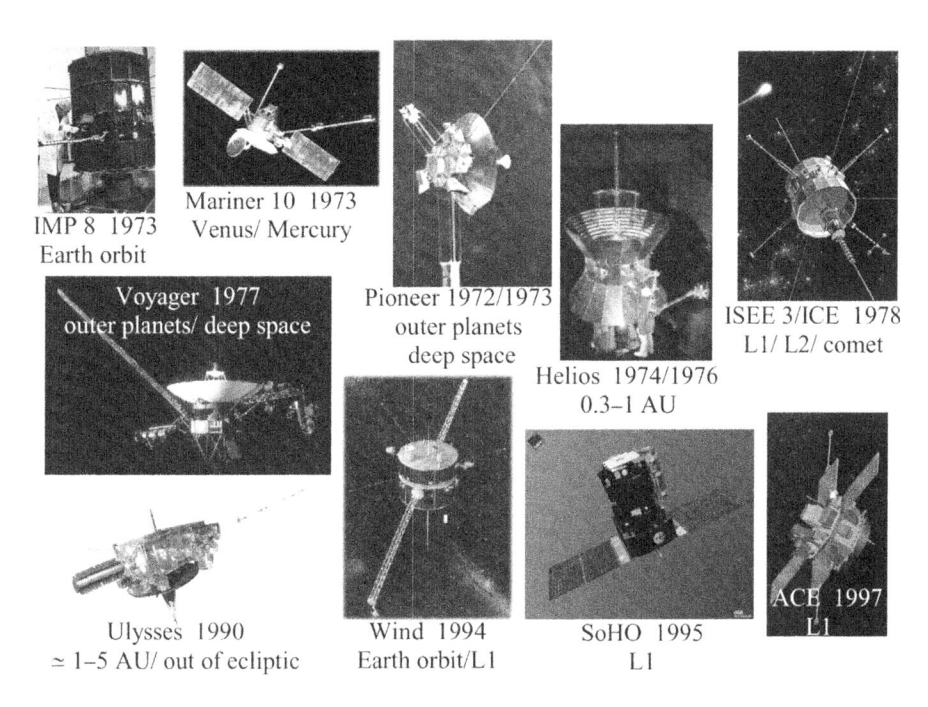

Figure 1.15 Some notable spacecraft that have explored the solar wind, with their dates of launch, and their main targets. (Images by NASA and ESA.)

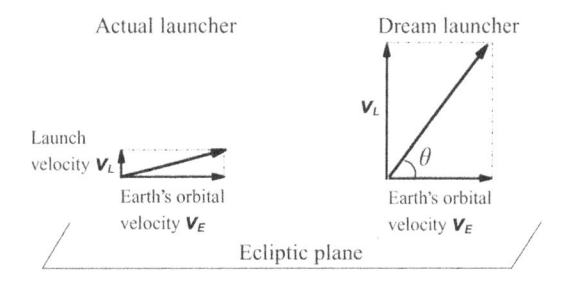

Figure 1.16 The difficulty of sending a spacecraft outside the ecliptic: because the Earth speed v_E is much greater than the launch speed v_L, the spacecraft velocity $\mathbf{v}_L + \mathbf{v}_E$ in the solar frame makes a very small angle to the ecliptic.

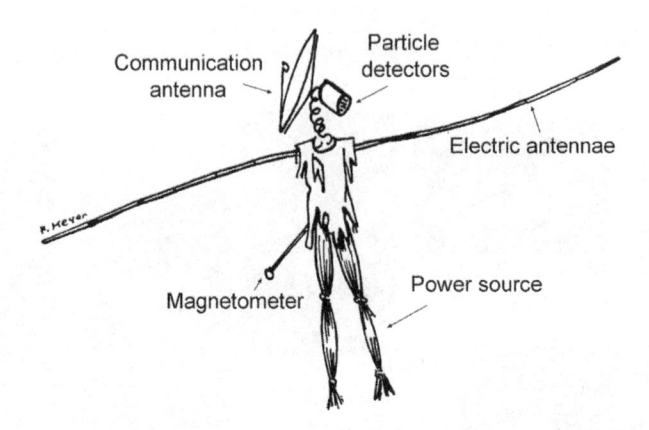

Figure 1.17 Simplified 'strawman payload' of the space agencies: the minimum required for solar wind *in situ* observation. (Drawing by F. Meyer.)

At the extremes of distances, the spacecraft Helios 1 and 2, launched respectively in 1974 and 1976, have approached the Sun up to about 0.3 AU, whereas Pioneer 10 and 11 (launched respectively in 1972 and 1973) and Voyager 1 and 2 (launched in 1977) are aiming at the outskirts of the Solar System. A horde of spacecraft are or have been watching the solar wind impinging on the Earth at 1 AU from the Sun. For doing so, they lie between the Sun and the Earth, at the right position for the Sun's and the Earth's gravitational attractions to combine to make the spacecraft orbit the Sun at the same angular speed as does the Earth.[15]

The solar wind, as the corona, is essentially made up of electrons and protons plus a small proportion of heavier ions, and it carries a magnetic field. As we shall see in Section 2.3, particles and fields are intimately coupled in plasmas, so that in order to explore them, space probes should carry at least a particle detector, a magnetometer and an electric antenna measuring waves, in addition to power and communication resources and to the necessary software (Fig. 1.17); most spacecraft generally carry additional instruments.

Several winds

Figure 1.18 shows typical measurements of key solar wind parameters: the mean velocity of protons (top panel) and the mean number density of electrons[16] (middle panel) obtained by the spacecraft WIND (operated by NASA [16]) in June 1995, close to solar activity minimum.[17]

---

[15]The so-called L1 Lagrange point, located at about 250 Earth's radii from the Earth towards the Sun.

[16]Because protons and electrons are the main constituents and carry opposite charges, their number densities and mean velocities must be roughly equal to keep the plasma electrically neutral.

[17]The mean velocity of protons (roughly equal to their mean radial velocity) and the density of electrons are plotted as (about) 10 minutes averages of data from respectively the ion

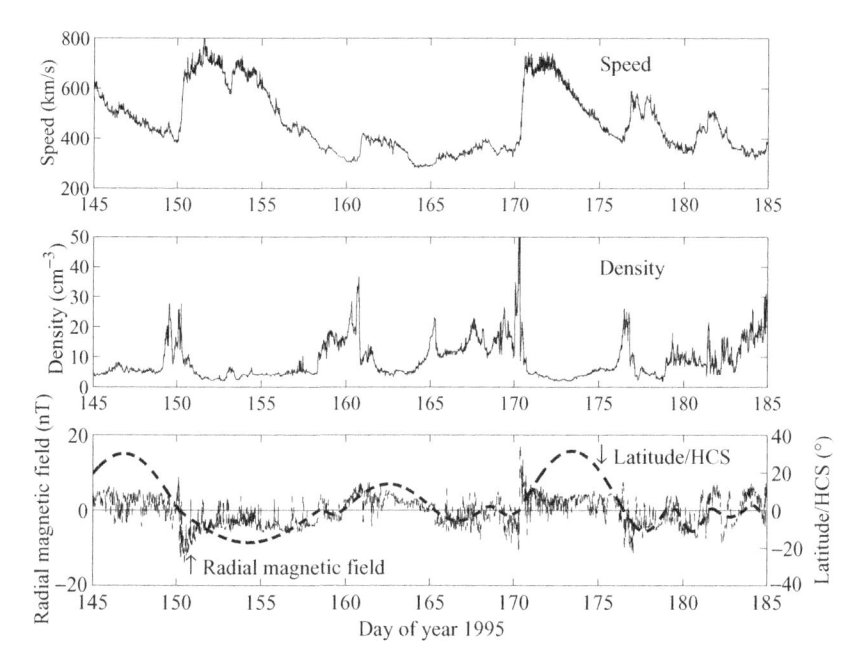

Figure 1.18 Mean proton velocity (top panel) and mean electron density (middle panel) measured on the spacecraft WIND in June 1995, at 1 AU from the Sun in the ecliptic. The bottom panel shows the radial component (Sun-centred) of the magnetic field and the latitude of WIND with respect to the heliospheric current sheet (HCS), that we will discuss in Sections 1.3.3 and 6.2. (Data courtesy of C. Salem.)

What does Fig. 1.18 tell us? In addition to small fluctuations at short timescales, the speed and the density exhibit a characteristic pattern: the speed varies by up to a factor of two or more and the density varies by up to a factor of ten, nearly simultaneously at intervals of about a week. The abrupt speed increases are followed by slow decreases; the density peaks when the speed changes abruptly, and (apart from the density peaks) the density is low (and uniform) when the speed is high and vice versa. This pattern of alternate fast and slow streams is repeated – with some modification – at the rotation period of the Sun, and is accompanied by changes in the magnetic field and in other properties. One sees in the bottom panel of Fig. 1.18 that the sign of the radial component of the magnetic field changes as a new fast stream is encountered, and remains constant within it.

---

electrostatic analyser (Lin, R. P. *et al.* 1995, *Space Sci. Rev.* **71** 125) and the thermal noise (Meyer-Vernet, N. *et al.* 1998, in *Measurement Techniques in Space Plasmas: Fields, Geophys. Monogr. Ser.* **103**, ed. R. F. Pfaff, *et al.*, Washington, DC, American Geophysical Union, p. 205) part of the WAVES receiver (Bougeret, J.-L. *et al.* 1995, *Space Sci. Rev.* **71** 231), which are the best instruments for measuring these respective properties on WIND (Salem, C. *et al.* 2003, *Ap. J.* **585** 1147).

This pattern has been observed by all spacecraft orbiting near the ecliptic at moderate distances from the Sun. It suggests the existence of several wind states: one slow/dense/structured and one fast/tenuous/uniform, in addition to transient and intermediate stages where the wind properties are still different (Section 6.3).

What is the origin of these states and of the changes between them? To understand this, we must realize that all these observations were performed near the ecliptic plane. This plane is very peculiar because the solar spin axis makes an angle of only 7.25° with the normal to the ecliptic. Furthermore near solar activity minimum, the dipolar component of the solar magnetic field is also nearly parallel to the spin axis, but not exactly so (making an angle generally larger than 7.25°). As the Sun spins, rotation of the Sun's dipolar magnetic pattern places an in-ecliptic observer alternatively in two opposite magnetic hemispheres. In contrast, a spacecraft sufficiently far from the ecliptic would remain in the same solar magnetic hemisphere as the Sun spins.

1.3.2 Exploring the third dimension with Ulysses

> O frati, dissi, che per cento milia
> Perigli siete gicenti all'occidente,
> A questa tanto picciola vigilia
> De' vostri sensi, ch'è del rimanente
> Non vogliate negar l'esperienza,
> Diretro al sol, del mondo senza gente.
> *La Divina Commedia di Dante Alighieri,*
> *Canto XXVI*[18]

So did Ulysses encourage his sailors in Dante's version of the Odyssey. Less famous is the record of a round-table discussion that gathered a lot of distinguished scientists in 1959, at a symposium on the exploration of space [28]:

> *Mr Hibbs*: I should like to ask whether there is a particular importance in performing experiments out of the plane of the ecliptic.

The fortitude of the scientists and engineers who made this dream come true in spite of the difficulties and sent the spacecraft Ulysses (Fig. 1.19) where no probe had ever flown is reminiscent of the mythical Greek warrior.

When the idea of an out-of-ecliptic mission arose, nobody knew how to realise it, and only in the 1970s did the idea appear technically feasible. The American and European space agencies then proposed a joint package of two spacecraft that were to be launched in 1983, and to sweep towards opposite sides of the ecliptic plane – using Jupiter's assist[19] – in order to pass nearly

---

[18] Tell me brothers, would you,/who braved a hundred thousand perils/to go ever farther to west,/now be loath to roam a realm/which reaches to the Sun/and harbours not a single living soul? (Trans. L. M. Celnikier.)

[19] Gravity assist from a planet is like playing billiards with a spacecraft, using the planetary gravitational field to deviate the spacecraft, as the edges of a billiard table deviate the ball.

Observing the solar wind

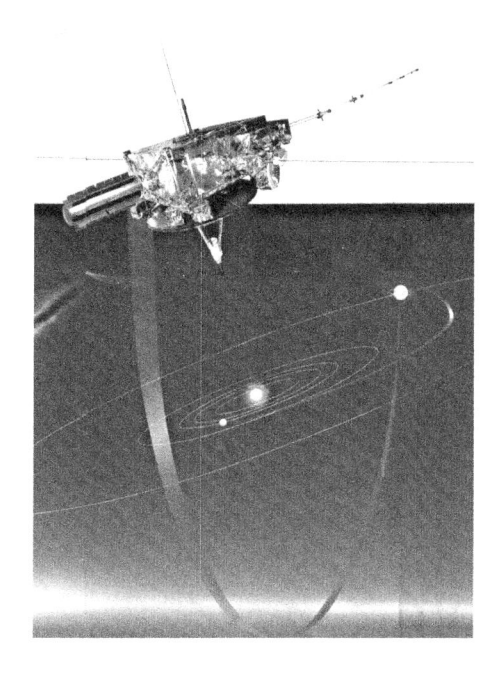

Figure 1.19 An artist's view of Ulysses, the first and only spacecraft to have reached high heliocentric latitudes (ESA and NASA).

simultaneously over opposite solar poles and to achieve a stereoscopic view of the solar wind [31]. But in the early 1980s, the National Aeronautical and Space Administration (NASA) decided to cancel the US spacecraft because of financial and technical difficulties. The project was reduced to a single spacecraft, to be built by the European Space Agency (ESA), launched by NASA with the Space Shuttle, and equipped with European and American instruments. In late 1983, however, the mission had still to wait: the spacecraft was ready but the launcher was not. And 1986 saw a catastrophic event: the Space Shuttle Challenger blew up, a few months before the planned launch of Ulysses, once more delaying the mission.

The launch took place at last in October 1990 – close to solar activity maximum. In 1991 the probe travelled in the ecliptic towards Jupiter. In February 1992 it swung around Jupiter into an elliptic orbit inclined by $80°$ to the ecliptic (Fig. 1.20). It then travelled into the Sun's southern hemisphere, passed over the south polar region in late 1994, crossed the ecliptic plane at 1.3 AU from the Sun, and passed over the north polar region in 1995 – near solar activity

---

Since the planet is moving, the spacecraft's speed with respect to the Sun changes upon reflection, even though the reflection is elastic in the frame of the planet; in this way, the spacecraft can not only change direction, but also gain or lose kinetic energy at the expense of the planet (see Gurzadyan, G. A. 2002, *Space Dynamics*, New York, Taylor & Francis). We will encounter a similar effect in Section 2.1, with the electromagnetic fields replacing gravitation, and charged particles replacing spacecraft.

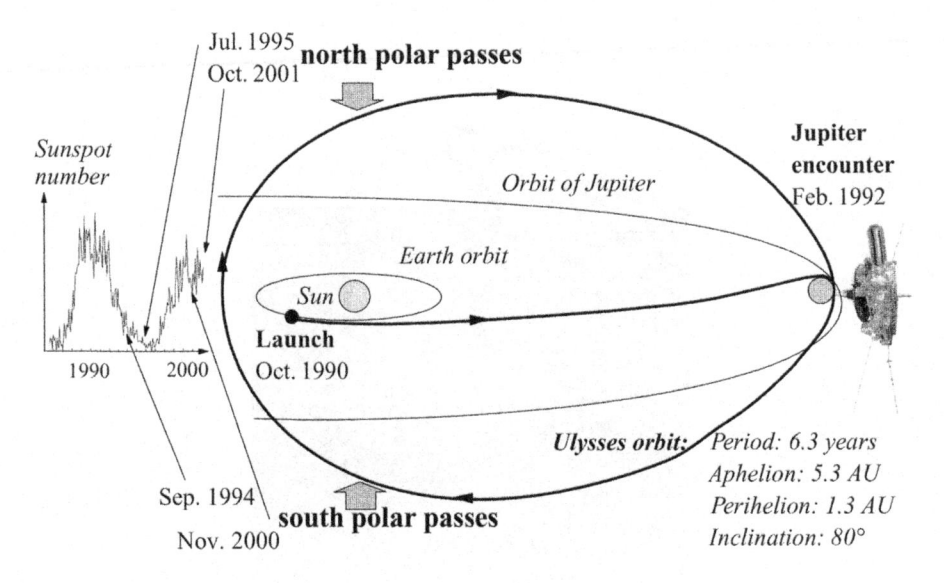

Figure 1.20 Sketch of Ulysses' trajectory. The dates of the first two passages over the solar poles are indicated on the left, along with solar activity. The orbital period is such that the fast pole-to-pole transits, covered in less than a year on the perihelion side, take place alternatively near sunspot minimum and maximum.

minimum. The second orbit took it over the polar regions once more in 2000–2001, this time near solar activity maximum. And a third orbit will take it again over the polar regions in 2007, near solar activity minimum. Ulysses carries a panel of instruments measuring the charged and neutral particles over a wide range of energies, the dust grains, the magnetic field and waves, including X- and gamma rays [32].

The orbit is especially suitable for studying the heliosphere. The orbital period is nearly half a solar activity cycle, and the pole-to-pole transit near perihelion takes less than a year – a time-span during which solar activity and distance do not change much. Hence at each passage along this part of the orbit, which take place alternately near solar activity minimum and maximum, Ulysses measures how the solar wind varies with heliocentric latitude, other parameters being roughly constant. On the other hand, the distance, latitude and solar activity vary simultaneously during the aphelion phase, when the spacecraft is moving less rapidly.

The solar wind in three dimensions

What did Ulysses find?

First of all, did Ulysses observe the recurrent pattern of two wind states observed by in-ecliptic spacecraft? Figure 1.21 shows the mean proton velocity (top panel) and the electron density (middle panel) measured by Ulysses in

Observing the solar wind 31

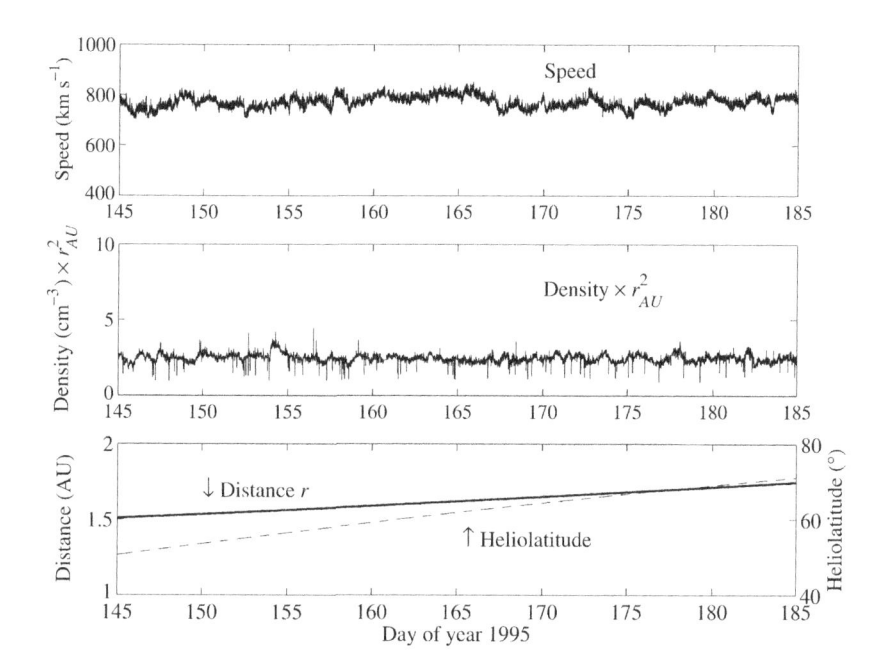

Figure 1.21 Mean proton velocity (top panel) and mean electron density (middle panel) measured on Ulysses in June 1995. The density is normalised to 1 AU by multiplying by Ulysses' distance (in AU) squared. The radial magnetic field (not shown) has a constant (positive) sign. The bottom panel shows the spacecraft heliocentric distance (thick line) and latitude (dashed line) in solar co-ordinates. (Data courtesy of K. Issautier.)

June 1995, the same month as the WIND observations plotted on Fig. 1.18 were acquired.[20] Figure 1.21 does not show any trace of the two-state wind pattern observed by in-ecliptic spacecraft. The speed remains close to $750 \, \text{km s}^{-1}$, while the density (normalised to the distance of 1 AU) is also roughly constant at $2.5 \, \text{cm}^{-3}$, as expected from conservation of particles expanding radially at a constant speed. Note that the density scale in Fig. 1.21 is enlarged by a factor of five with respect to that in Fig. 1.18, to emphasise the small variations. In June 1995, Ulysses was at about 1.6 AU from the Sun and 65° heliolatitude, i.e. very far from the equatorial plane of the solar magnetic dipole.

What did it observe elsewhere?

This is best summarised in Fig. 1.22, which shows how the speed changes with latitude and with solar activity. On this figure, the speed is plotted in

---

[20]The speed is from the SWOOPS particle electrostatic analyser (Bame, S. J. et al. 1992, Astron. Astrophys. Suppl. Ser. **92** 237), and the density from thermal noise analysis of data from the URAP instrument (Stone, R. G. et al. 1992, Astron. Astrophys. Suppl. Ser. **92** 291), that are the best ways of measuring these respective properties on Ulysses (Issautier, K. et al. 1999, J. Geophys. Res. **104** 6691).

32   *The wind from the Sun: an introduction*

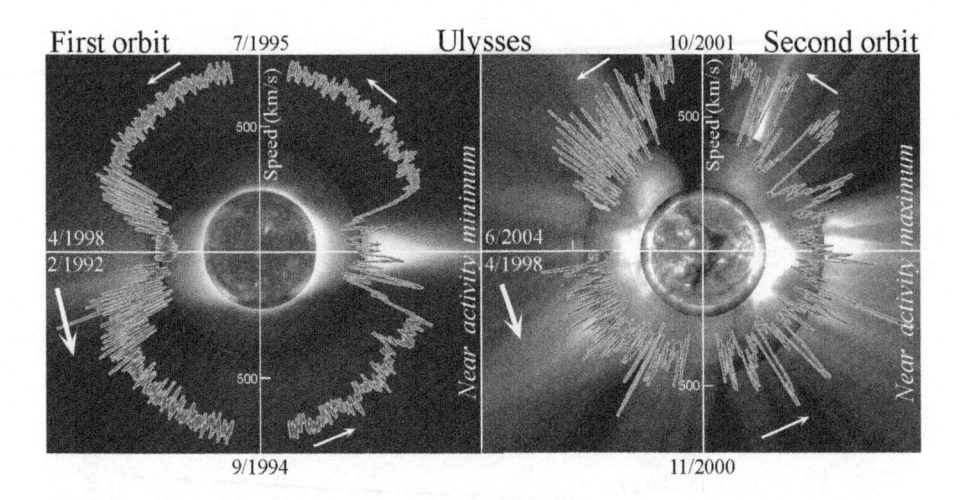

Figure 1.22 Solar wind speed as a function of heliocentric latitude, plotted in polar co-ordinates during Ulysses' first two orbits. The data are plotted over solar images obtained on 17 August 1996 (near activity minimum; left) and 7 December 2000 (near activity maximum; right), with SoHO EIT(195 Å), and LASCO instruments and the Mauna Loa K coronameter (0.7–0.95 µm). (Adapted from [19].)

polar co-ordinates, i.e. the distance from the centre is proportional to the speed at each latitude. The left and right panels show the first and second orbits, which took place respectively around solar activity minimum and maximum. The data are superimposed on images of the corona typical of these periods of solar activity. In both panels, time starts on the left-hand side (southwards) and progresses counter-clockwise along the orbit as indicated by the arrows and the dates.

Consider first the structure near solar activity minimum (left-hand panel of Fig. 1.22). The coronal image on which the data are superimposed shows the simple structure we have already seen in the left-hand panel of Fig. 1.12, with dark coronal holes on the polar caps and bright streamers extending outwards near the equatorial plane. The solar wind structure reflects this simplicity: the speed is nearly constant at all latitudes except in a narrow band of $\pm 20°$ around the equator where the speed pattern resembles the two-state structure seen by near-ecliptic spacecraft. This simplicity is shared by the other properties. The sign of the radial component of the magnetic field remains constant within each hemisphere (except for some brief reversals), being outwards in the north and inwards in the south, except in a narrow equatorial band where polarities are alternating.

Around solar activity minimum, therefore, the heliosphere has an outstandingly simple structure, essentially made of a quiet, fast and tenuous wind where the radial component of the magnetic field has a constant sign (opposite in opposite hemispheres). Comparison with solar observations shows that this fast

Table 1.2 Occurrence and variability of the fast and slow winds

	Fast wind	Slow wind
At activity minimum	Ubiquitous outside equator	Only near equator
At activity maximum	Occurs as narrow streams	Ubiquitous outside streams
Variability	Low	Large

Table 1.3 Basic properties of the fast and slow winds

	Speed v (m s^{-1})	Electron density n (m^{-3})	Mass loss through a sphere $\left[\rho v \times 4\pi r^2\right]_{1\text{AU}}$ (kg s^{-1})	Ram pressure $\left[\rho v^2\right]_{1\text{AU}}$ (Pa)
Fast	7.5×10^5	2.5×10^6	10^9	2.6×10^{-9}
Slow	4×10^5	7×10^6	1.5×10^9	2.1×10^{-9}

wind arises from the inactive solar regions, especially the large coronal holes surrounding each pole, where the magnetic field has a constant polarity (opposite at opposite poles). The pattern of fast and slow winds recurring at the solar rotation period is restricted to a narrow latitude band surrounding the solar equatorial plane, that is akin to the equatorial region where bright streamers are observed in the corona. This picture is consistent with observations by near-ecliptic spacecraft, which are located at low latitudes.

Consider now the structure near activity maximum (right-hand panel of Fig. 1.22). The superimposed coronal image shows the complex picture typical of solar activity maximum, which we have already seen in the right-hand panel of Fig. 1.12, with bright streamers extending radially all around the Sun. The solar wind structure reflects this complexity, with alternating fast and slow streams of small scale – observed at all latitudes, in addition to transients. This complex structure is shared by the magnetic field, whose polarity alternates, and by other properties. Near activity maximum, therefore, the pattern of alternating fast and slow streams is observed at all latitudes, with, however, somewhat smaller speeds and scales than near activity minimum.

We have summarised in Tables 1.2 and 1.3 the basic properties of these wind states (see [21], [19]).[21]

1.3.3 A simplified three-dimensional picture

These Ulysses observations bear out – albeit with some modifications – a simple picture of the heliosphere near solar activity minimum, that had already been

---

[21] The values of ρv and ρv^2 are slightly greater than $nm_p v$ and $nm_p v^2$ respectively because there is a small proportion of helium.

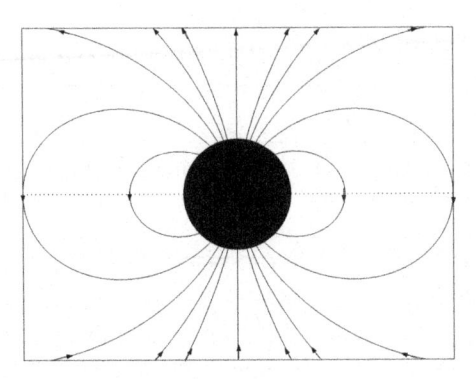

Figure 1.23 Magnetic field lines of a magnetic dipole aligned with the vertical axis. The black disc sketches the region (here the solar interior) where the currents producing the magnetic field are flowing.

hinted at from remote-sensing observations, and from data of previous spacecraft that had gone slightly outside the ecliptic.

Dipolar magnetic field

To begin with, let us assume that the Sun behaves as a huge magnetic dipole, producing a magnetic field as sketched on Fig. 1.23. The actual solar magnetic field is more complicated, but the dipole is a reasonable starting approximation because the farther out from the Sun, the smaller are the effects of the solar non-dipolar components. Indeed, outside the region containing electric currents, Maxwell's equations

$$\nabla \times \mathbf{B} = \mu_0 \mathbf{J} \qquad (1.14)$$
$$\nabla \cdot \mathbf{B} = 0$$

with $\mathbf{J} = 0$, imply $\mathbf{B} = -\nabla \psi$, with the Laplace equation $\nabla^2 \psi = 0$; hence the magnetic field can be developed in a multipolar series where the term of order n decreases with distance as $r^{-(n+2)}$ (e.g. [8]). The dipolar term corresponds to $n = 1$ and becomes dominant with increasing distance.

With the components of the dipolar magnetic field in the radial (B_r) and latitudinal (B_θ) directions (Appendix), the field lines are solutions of the differential equation $dr/(rd\theta) = B_r/B_\theta = -2\sin\theta/\cos\theta$, where r and θ are respectively the distance and latitude in a spherical co-ordinate system whose symmetry axis (vertical) is aligned with the magnetic dipole axis. Putting $u = \cos\theta$, this yields $dr/r = 2du/u$, whose integration gives $r \propto u^2 = \cos^2\theta$. Hence, a field line crossing the magnetic equatorial plane ($\cos\theta = 1$) at heliocentric distance L is given by $r = L\cos^2\theta$.

In this geometry, all field lines are closed; each one leaves the surface, extends outwards to a maximum distance L in the magnetic equatorial plane and then returns towards the surface on the other side (Fig. 1.23).

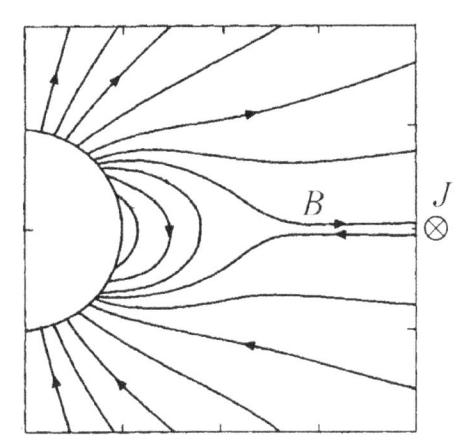

Figure 1.24 Simple model of how a dipolar magnetic field imposed at the solar surface is modified by a wind flowing along the field lines. At large distances the oppositely directed magnetic fields imply an electric current flowing in a thin sheet perpendicular to the figure, a picture that will be refined in Section 6.2. (Adapted from [27].)

A dipole plus a wind

This nicely simple picture, however, concerns a magnetic dipole in a vacuum, which the Sun is certainly not; the electric charges of the plasma around the Sun produce currents that change the magnetic field according to Maxwell's equations (1.14). Furthermore this plasma manages to flow outwards, with an energy density greater than that of the magnetic field, as we will see in Chapters 5 and 6. How does this modify the picture? We have already said (and shall explain in Section 2.3) that the plasma and the magnetic field lines are strongly tied together, so that the plasma can move along the field lines but not across them, just as beads on a necklace. Near the poles, where the magnetic lines extend nearly radially, the outgoing wind can flow unimpeded along them. In the equatorial regions, however, the situation is different because an outgoing wind would have to flow perpendicularly to the dipolar field lines (Fig. 1.23).

What happens in that case? Since the wind energy density exceeds that of the magnetic field, the wind pushes out the field, drawing the field lines outwards to the extent that they become nearly parallel to the equator and no longer return to the surface. In this way, the outgoing wind can flow along the field lines everywhere. This is sketched in Fig. 1.24, which shows a simplified solution of this problem [27]. Because the lines parallel to the equatorial plane come from opposite ends of the dipole, they represent magnetic fields having opposite directions. Hence, at large distances, the magnetic field direction changes abruptly at the equator. This implies that a thin sheet of current flows along the magnetic equatorial plane in a direction normal to the figure; in three

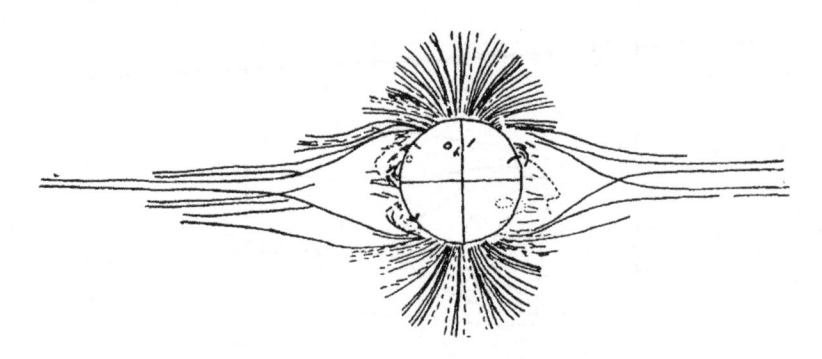

Figure 1.25 Drawing of the appearance of the corona at the 30 June 1954 eclipse, near solar activity minimum. (Adapted from [30].)

dimensions, it forms an annular current sheet that separates the magnetic fields originating at opposite poles.

Although Fig. 1.24 refers to a highly idealised situation, the real corona might sometimes resemble it, as suggested by Fig. 1.25, which is an old drawing of the visual appearance of the corona during a solar eclipse [30]. One must be cautious in interpreting such visual appearances since they result from different structures lying along the line of sight of the observer, and projection effects can be misleading. However, the figure suggests a current sheet extending into space near the solar magnetic equatorial plane, that separates regions of opposite magnetic polarities. We will see in Section 6.2 that this sheet is not a simple plane, having a complex warped shape which has been likened to the skirt of a spinning ballerina [1].

How does this picture change with solar activity? During a few years near minimum, when the solar magnetic field is not far from that of a dipole making a small angle with the rotation axis, the current sheet is a (slightly warped) plane making a small angle with the solar equator. As solar activity rises, the solar magnetic field becomes more complicated, making the sheet warp and thread its way towards polar regions, finally sometimes breaking up near activity maximum, when the large-scale solar magnetic field becomes rather disorganised. As solar activity decreases, the large-scale magnetic field reorganises itself towards a dipolar structure whose axis is again close to the spin axis, but with a direction opposite to the one in the previous cycle.

How is the wind velocity related to this magnetic structure? Naively, one expects that the flow of the wind will be unimpeded and stationary everywhere except near the current sheet where complex geometry-dependent effects should occur. In this sense, this simple picture, conceived in the mid 1970s, gives a straightforward explanation for the basic geometric structure of the heliosphere, sketched in Fig. 1.26. Returning to the WIND observations of Fig. 1.18, we can now relate them to the position of the spacecraft with respect to the heliospheric current sheet, which is shown in the bottom panel. We can see that the sign of the radial component of the magnetic field is the same as that of the latitude

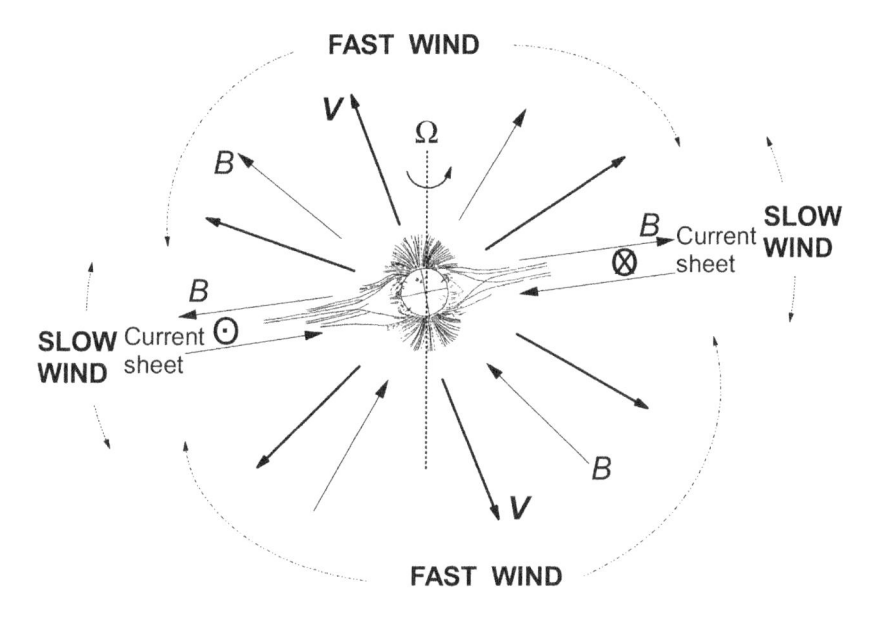

Figure 1.26 Simplified picture of the large-scale structure of the solar wind near sunspot minimum, when the solar magnetic dipole makes a small angle with the spin axis (dotted line). The velocity and field lines are sketched in bold and thin lines respectively. The magnetic polarity is the one that existed when the WIND and Ulysses observations shown in Figs. 1.18 and 1.22 were acquired; this polarity reverses every 11 years.

with respect to the current sheet, as expected from the sketch in Fig. 1.26. As the Sun spins, the WIND spacecraft in the ecliptic lies alternately above and below the heliospheric current sheet. At each crossing of the sheet, the direction of the magnetic field changes and a slow and variable wind is observed. This picture is also consistent with observations from Ulysses, which spends most of its trajectory far from the current sheet near solar minimum, and measures a fast stationary wind with a constant magnetic polarity in each solar hemisphere except at low latitudes.

We shall revisit this grossly simplified picture later. Before doing so, we have to introduce some basic plasma physics.

References

[1] Alfvén, H. 1977, Electric currents in cosmic plasmas, *Rev. Geophys. Space Phys.* **15** 271.

[2] Badescu, V. and R. B. Cathcart 2006, Use of class A and class C stellar engines to control Sun movement in the Galaxy, *Acta Astronautica* **58** 119.

[3] Biermann, L. 1953, Physical processes in comet tails and their relation to solar activity, *Mém. Soc. Roy. Sci. Liège (Ser. 4)* **13** 291.

[4] Birkeland, K. R. 1908, 1913, *The Norwegian Aurora Polaris Expedition 1902–1903*, vols. 1 and 2, Christiania, Norway, H. Aschehoug & Co.

[5] Carrington, R. C. 1860, Description of a singular appearance seen on the Sun on September 1, 1859, *Mont. Not. Roy. Soc.* **20** 13.

[6] Chapman, S. 1957, Notes on the solar corona and the terrestrial ionosphere, *Smithsonian Contr. Astrophys.* **2** 1.

[7] Clerke, A. M. 1887, *A Popular History of Astronomy during the 19th Century*, Edinburgh, Adam & Charles Black.

[8] Dennery, P. and A. Krzywicki 1967, *Mathematics for Physicists*, New York, Harper & Row.

[9] Dessler, A. J. 1967, Solar wind and interplanetary magnetic field, *Rev. Geophys.* **5** 1.

[10] Eather, R. H. 1980, *Majestic Lights*, Washington DC, American Geophysical Union.

[11] Egeland, A. and W. J. Burke 2005, *Kristian Birkeland: The First Space Scientist*, New York, Springer.

[12] Fitzgerald, G. F. 1892, Sunspots and magnetic storms, *The Electrician* **30** 48.

[13] Foukal, P. V. 2004, *Solar Astrophysics*, New York, Wiley.

[14] Giovanelli, R.G. 1984, *Secrets of the Sun*, Cambridge University Press.

[15] Haigh, J. D. *et al.* eds. 2004, *The Sun, Solar Analogs and the Climate*, New York, Springer.

[16] Harten, R. and K. Clark 1995, The design features of the GGS Wind and Polar spacecraft, *Space Sci. Rev.* **71** 23.

[17] Kippenhahn, R. 1994, *Discovering the Secrets of the Sun*, New York, Wiley.

[18] Lemaire, J. and V. Pierrard 2001, Kinetic models of solar and polar winds, *Astrophys. Space Sci.* **277** 169.

[19] McComas, D. J. *et al.* 2003, The three-dimensional solar wind around solar maximum, *Geophys. Res. Lett.* **30** 1517.

[20] Neugebauer, M. 1997, Pioneers of space physics: a career in the solar wind, *J. Geophys. Res.* **102** 26887.

[21] Neugebauer, M. 2001, in *The Heliosphere Near Solar Minimum*, ed. A. Balogh *et al.*, New York, Springer, p. 43.

[22] Neugebauer, M. and C. W. Snyder 1962, The mission of Mariner II: preliminary observations, *Science* **138** 1095.

[23] Parker, E. N. 1958, Dynamics of the interplanetary gas and magnetic fields, *Ap. J.* **128** 664.

[24] Parker, E. N. 1997, in *Cosmic Winds*, ed. J. R. Jokipii *et al.*, Tucson AZ, University of Arizona Press, p. 3.

[25] Parker, E. N. 2001, in *The Century of Space Science*, ed. J. A. M. Bleeker *et al.*, New York, Kluwer, p. 225.

[26] Parker, E. N. 2001, A critical review of sun–space physics, *Astrophys. Space Sci.* **277** 1.

References

[27] Pneuman, G. W. and R. A. Kopp 1971, Gas–magnetic field interactions in the solar corona, *Solar Phys.* **18** 258.

[28] Simpson, J. A. *et al.* 1959, Round-table discussion, *J. Geophys. Res.* **64** 1691.

[29] Title, A. 2000, Magnetic fields below, on and above the solar surface, *Phil. Trans. Roy. Soc. London A* **358** 657.

[30] Vsekhsvjatsky, S. K. 1963, in *The Solar Corona*, ed. J. W. Evans, London, Academic Press, p. 271.

[31] Wenzel, K.-P. 1980, The scientific objectives of the international solar polar mission, *Phil. Trans. Roy. Soc. London A* **297** 565.

[32] Wenzel, K.-P. 1992, The Ulysses mission, *Astron. Astrophys. Suppl. Ser.* **92** 207.

[33] Zirker, J. B. 2002, *Journey from the Center of the Sun*, Princeton University Press.

2

Tool kit for space plasma physics

Most of the Universe is made of plasma. And yet, plasmas are very rare on the Earth, where solids, liquids and gases – the three primary states of matter – are ubiquitous (Fig. 2.1). These states are the result of a competition between thermal energy and intermolecular forces. In solids, the latter win, maintaining the atoms and/or molecules at nearly fixed positions, whereas thermal energy merely produces vibrations around these positions [9]. In gases on the contrary, thermal energy wins, making the particles almost completely free. Liquids are in between: the intermolecular forces are sufficiently strong to resist compression, but sufficiently weak to enable deformation and flow; it is not surprising that this intermediate state is less well understood than the other two [19].

Common experience and elementary physics tell us that we may transform a solid into a liquid by heating it; this weakens the bonds between molecules so that they may move slightly, enabling matter to change shape. This requires an amount of energy per molecule somewhat smaller than the binding energy. If the energy furnished exceeds the binding energy, the bonds break out completely, producing a gas of free atoms and/or molecules.

The *plasma* is the next state: the fourth, reached by furnishing enough energy to break the atoms themselves, or rather to kick off at least the outer atomic electron, producing a mixture of electrons and ions. For doing so, one has to heat or to compress, to bombard with energetic radiation or particles, or to subject the medium to high electric fields, as we shall see in more detail in Section 2.4. One (or several) of these ionisation agents acts in most regions of the Universe.

But (generally) such is not the case in the thin atmospheric layer of the small planet Earth, where human beings live. This medium is not ionised because it is a very special place: it is relatively cold; it is protected from the solar ionising radiation by an atmosphere; and when an atom happens nevertheless

Figure 2.1 Solids, liquids and gases abound on the Earth, but most of the Universe is made of plasma: the fourth state of matter. (Production of vapour; drawing by Jean Effel, *La Création du Monde*, 1971, copyright Adagp, Paris, 2007.)

to be ionised, the particle concentration is so high that ions and electrons meet frequently enough to recombine into neutral atoms. This is why our everyday experience of plasmas is so limited: we can see them occasionally in lightning, in some flames, inside fluorescent tubes or neon signs, but most of the visible plasmas lie farther away and are seen in sky displays, in auroras, comets and stars.

This chapter introduces briefly some tools of plasma physics that are essential for understanding the solar wind and its interaction with objects. In addition to introducing classical concepts, we give some hints on two subjects that lie at the frontier of traditional plasma physics: non-Maxwellian distributions, which are ubiquitous in the heliosphere – fooling our intuition and raising questions still unanswered – and ionisation processes. The aim is to furnish a tool kit for dealing with the major processes at work in the heliosphere, with the necessary limitations – in space and scope – of such a kit. We have privileged insight, at the expense of rigor and completeness. More may be found in several excellent texts, for example [17], [6], [10], [4], [18], [16] and [7].

2.1 What is a plasma?

In any gas there are always a few atoms or molecules that manage to lose one electron, producing some small degree of ionisation. Being ionised is therefore not sufficient to qualify as a plasma. A useful definition may be instead that a plasma is:

- a gas[1] containing charged particles (together with neutral ones), which
- is quasi-neutral
- and exhibits collective behaviour.

We will explain these three properties below. For simplicity, we consider a plasma made of electrons (charge $-e$, mass m_e) and one species of singly charged ions (charge $+e$, mass m_i) of equal concentrations n. We do not consider complex plasmas containing a large quantity of heavily charged ions or of dust particles, which are rare in the heliosphere. We also assume the particles to be non-relativistic and non-degenerate; relativistic and degenerate plasmas will be discussed briefly later.

The concentration n is the average number of electrons (or ions) per unit volume. This assertion assumes implicitly that there are many particles in any volume considered, i.e. we shall consider spatial scales L greater than the average distance between particles, whose order of magnitude is

$$\langle r \rangle \sim n^{-1/3} \quad \text{(average distance between particles)}. \tag{2.1}$$

The temperature T characterises the agitation of the particles. In thermal equilibrium, the particles' velocities along each space co-ordinate (x, y, z) are Gaussian distributed around zero (in the frame where the bulk of them is at rest), with mean square values $\langle v_x^2 \rangle = \langle v_y^2 \rangle = \langle v_z^2 \rangle = k_B T/m$ for a particle species of mass m; in this case, the average kinetic energy per particle is

$$m \langle v^2 \rangle / 2 = 3 k_B T / 2. \tag{2.2}$$

However, an important property of space plasmas is their frequent lack of thermal equilibrium, even locally. Not only may electrons and ions have different bulk velocities and temperatures, but the particles' velocities may not be Gaussian distributed. In that case, one may still formally define a *kinetic temperature* for each particle species from (2.2), even though it is not a thermal equilibrium temperature. We shall return later to this point, which has basic applications in the solar corona (Section 4.6) and the solar wind (Section 5.5 and Problem 5.7.6). Meanwhile, we will assume that, even in the absence of thermal equilibrium, the particles have a typical random speed of the order of magnitude of $\sqrt{k_B T/m_{e,i}}$ (for the electrons and ions respectively).[2]

---

[1] This restrictive definition is adequate for space plasmas. We do not consider plasma crystals [15].

[2] This assumption is not as trivial as it might seem. Consider for example a power law velocity distribution, so that the probability for the speed to lie in the range $[v, v + dv]$ varies as $v^{-\alpha}$ (with $\alpha > 0$) for $v_1 < v < v_2$, with $v_1 \ll v_2$. The most probable speed is v_1, whereas you can show as an exercise that the root mean square speed – from which the kinetic temperature is defined – is of the order of magnitude of v_1 or v_2 depending on whether α is greater or smaller than 3, and the median speed is still very different. This example is extreme, since power law distributions are par excellence scale-free, but it is not academic since many processes produce similar distributions, as we shall see later in this book.

For each species of (non-relativistic and non-interacting) particles of number density n and mass m, the pressure is determined by the average random kinetic energy as

$$P = nm\langle v^2 \rangle /3 = 2w_{th}/3 \quad \text{pressure} \quad (v \ll c) \tag{2.3}$$

where w_{th} is the energy density of the particles; this is just the average flux of momentum along one space direction. This is equivalent to

$$P = nk_B T \tag{2.4}$$

with the kinetic temperature defined in (2.2). This definition of the pressure does not require the particles to be necessarily in thermal equilibrium. Note that in the simple plasma defined above (in which electrons and ions have the same number density n), the total particle pressure is the sum of the pressure of electrons and ions, that is $P = 2nk_B T$, where T is their kinetic temperature (or the average of them if they are not equal).

2.1.1 Gaseous plasma

For an assembly of charged particles to qualify as a gas, the particles must move freely, which means that random motions should largely overrun mutual interactions. The latter involve the Coulomb force; for two particles of charge $\pm e$ distant by r, the energy of interaction is of modulus $e^2/4\pi\epsilon_0 r$. The plasma thus behaves as a gas if the energy of interaction of two particles distant by the average interparticle distance $\langle r \rangle$ is much smaller than the average kinetic energy per particle, i.e.

$$e^2/4\pi\epsilon_0 \langle r \rangle \ll k_B T.$$

Substituting $\langle r \rangle \sim n^{-1/3}$, and introducing the coupling parameter Γ defined as the ratio of the average energy of interaction to $k_B T$, we deduce the condition

$$\Gamma \equiv \frac{n^{1/3} e^2}{4\pi\epsilon_0 k_B T} \ll 1 \quad \text{(gaseous plasma)}. \tag{2.5}$$

In the solar wind, Γ is of the order of magnitude of $10^{-8} - 10^{-7}$ at 1 AU, and varies weakly with heliocentric distance.

2.1.2 Quasi-neutrality

Debye shielding

Since charges of opposite signs attract each other, whereas charges of like signs repel each other, the Coulomb force tends to establish electric neutrality. The random agitation, however, mixes the particles, destroying this neutrality. The competition between both effects produces small regions that are non-neutral. The hotter the plasma, the greater the agitation and therefore the larger the maximum size of the non-neutral regions. On the other hand, the denser the medium, the greater the Coulomb force that keeps the plasma neutral, and therefore the smaller the size of the non-neutral regions.

To estimate this size, consider a region of size L in which the electrons are strongly depleted, so that it contains a total electric charge of order of magnitude $Q \sim ne \times L^3$. This produces an electric potential at the boundary of the region, of order of magnitude

$$\phi \sim Q/(\epsilon_0 L) \sim neL^2/\epsilon_0.$$

For random agitation to produce spontaneously such a structure, the corresponding energy per particle $\sim k_B T$ must be at least equal to the potential energy per particle $e\phi$, i.e. $k_B T \geq ne^2 L^2/\epsilon_0$. We deduce (in order of magnitude) the maximum size of non-neutral regions

$$L_D = \left(\frac{\epsilon_0 k_B T}{ne^2}\right)^{1/2}, \tag{2.6}$$

the so-called *Debye length*.

Detailed calculations show that indeed when a charge is put in an equilibrium plasma, it attracts ambient charges of opposite signs and repels charges of like signs, so that it is surrounded by a region of size L_D where the attracted particles are concentrated and the repelled ones are depleted, producing a charge distribution that shields the electrostatic field of the original charge. More precisely, the electrostatic potential at distance r of a charge q in an equilibrium plasma is

$$\Phi(r) = \frac{q}{4\pi\epsilon_0 r} e^{-r/L_{D*}} \tag{2.7}$$

where $L_{D*} = L_D/\sqrt{2}$ (because electrons and ions both contribute to the shielding). At distances $r \ll L_D$, the electric potential around the charge q is nearly the Coulomb one, whereas at $r \gg L_D$, the charge is completely shielded by the charges of the ambient plasma, and the potential vanishes. Thus the plasma is quasi-neutral at scales greater than L_D.

This holds also for the charges of the plasma itself, and we have here a first hint as to a fundamental plasma property: its collective behaviour. Any charge in the plasma is 'dressed' by the other ones – a dressing of far-reaching consequences.

Numerically, $L_D \simeq 69\sqrt{T/n}$ in SI units, which comes to about 10 m in the solar wind at 1 AU from the Sun ($n \sim 5 \times 10^6$ m^{-3}, $T \sim 10^5$ K). Therefore, we have not to worry about the quasi-neutrality of the solar wind, except when dealing with scales smaller than tens of metres – a problem that occurs in the environment of space probes (Section 7.2).

It is worth noting that the Debye shielding requires several conditions to be met:

- a region of size L_D must contain many particles, i.e. $nL_D^3 \gg 1$; with the definition (2.5) of Γ and the expression (2.6) of L_D, this condition reads: $(4\pi\Gamma)^{3/2} \ll 1$;

- the electric disturbance produced by the charge q on the ambient particles must not be greater than their average kinetic energy, otherwise they are not capable of shielding it;

- the charge q is at rest;

- the plasma is in thermal equilibrium.

Non-equilibrium plasma

The latter condition is in practice rarely met in space plasmas. Consider first the case of partial equilibrium, namely when the different particle species are each in equilibrium but at different temperatures. Since the quasi-neutrality is ensured by the Coulomb force and destroyed by the random agitation, Debye shielding is mainly produced by the less agitated particles, so that L_D is determined by the colder species.

In complete absence of equilibrium, the shielding is mainly provided by the slower particles of each species. More precisely (Problem 2.5.1), if the charge produces a sufficiently small perturbation, then the shielding length L_{D*} is determined by the average of $1/v^2$ for each species, as

$$1/L_{D*}^2 = 1/L_{D_e}^2 + 1/L_{D_i}^2 \qquad (2.8)$$

with

$$1/L_{D_{e,i}}^2 = ne^2 \langle v^{-2} \rangle_{e,i} / (\epsilon_0 m_{e,i}) \qquad (2.9)$$

where the subscripts e and i stand for electrons and ions respectively. At equilibrium at temperature T, $\langle v^{-2} \rangle = m/(k_B T)$, so that the shielding length reduces to $L_{D*} = L_D/\sqrt{2}$ with L_D given in (2.6).[3]

In essence, Debye shielding is not determined by the random kinetic energy of the particles, but by the average of the inverse of that kinetic energy.

Non-linear shielding

What happens when, in addition to the plasma not being in equilibrium, the electric disturbance produced by the charge is large, i.e. the Coulomb potential energy is not small compared to the kinetic energy? In that case, we shall see later that the particles produce a different contribution to the shielding, depending on whether they are attracted or repelled.[4] The resulting distribution of the attracted particles then depends on the geometry of the problem (see for

---

[3] For each species of mass m and temperature T, $\langle v^{-2} \rangle = \int_0^\infty dv\, e^{-mv^2/2k_B T} / \int_0^\infty dv\, v^2 e^{-mv^2/2k_B T}$.

[4] This is so because in order to shield the charge q, repelled particles have just to decrease their number density, which can be achieved by a mere deviation of their trajectories; attracted particles, on the other hand, have to increase their number density in order to shield the charge q, which requires some of them to change their incoming trajectories into closed orbits around q – a performance that requires collisions and cannot be achieved in the absence of equilibrium.

What is a plasma? 47

example [11] and [8]). We shall return to this point in Section 2.3, and shall see examples of application in Section 7.2, when calculating the electric charge of objects immersed in the solar wind.

Shielding of a moving charge

What happens if the charge q is moving? The answer depends on the value of its speed v compared to the most probable speeds of the plasma particles, whose order of magnitude is $(k_B T/m_{e,i})^{1/2}$ for respectively electrons and ions. Because electrons are much lighter than ions, these speeds satisfy the inequality $v_{thi} \ll v_{the}$. If the speed $v \ll v_{thi}$, then the plasma electrons and ions are fast enough to keep up with the charge motion, so that the shielding is not affected. On the other hand, if $v_{thi} \ll v \ll v_{the}$, then the plasma electrons are still fast enough to keep up with the charge motion, but the ions move too slowly to do so. In that case, the shielding is provided by the electrons only. Finally, if the charge q moves faster than the electrons (and the ions), then the bulk of the plasma particles cannot catch up with it, and therefore cannot shield it. Instead, the charge motion produces plasma waves, a novel kind of dressing to which we shall return in Section 2.3.

Timescale for shielding

This disappearance of Debye shielding (or rather its transformation into a new kind of dressing) occurs when the charge moves fast from the point of view of the electrons. A related problem is what happens when a charge q is suddenly put in a plasma initially at equilibrium. The plasma particles will take some time to distribute themselves in order to provide shielding. How long? Electrons, moving faster than ions, are the first to shield the charge. For doing so they must travel a distance of the order of L_D. At the most probable speed v_{the}, this takes the time $\tau \sim L_D/v_{the}$. With the expression (2.6) of L_D and $v_{the} \sim (k_B T/m_e)^{1/2}$, we find $\tau \sim (\epsilon_0 m_e/ne^2)^{1/2} \equiv 1/\omega_p$, where ω_p is the so-called (angular) *plasma frequency*, a basic plasma parameter to which we shall return later.

We get here a second hint as to the collective behaviour of plasmas. Not only are the charges dressed, but this dressing is highly dynamic, with a timescale of the order of magnitude of $1/\omega_p$.

This has an important implication. Consider an electromagnetic wave incident on a plasma. The variable electric field of the wave tends to destroy the plasma quasi-neutrality. But if the wave frequency is smaller than the plasma frequency, the disturbance has a timescale large enough that the plasma particles are capable of catching up with it and of restoring the quasi-neutrality. If they succeed, the electric field is cancelled and the wave does not propagate in the plasma. We shall see in Section 2.3 that, indeed, electromagnetic waves propagate in a plasma only at frequencies greater than the plasma frequency.

2.1.3 Collisions of charged particles

We now come to a further plasma property, which concerns collisions between particles.

Collisions serve to achieve equilibrium. They determine not only the time required to restore thermal equilibrium after a perturbation, but also the transport coefficients which control the response of the medium to various gradients in macroscopic properties:

- the diffusion coefficient, which determines the transport of particles in response to a gradient of concentration;

- the viscosity, which determines the transport of momentum in response to a gradient of velocity;

- the thermal conductivity, which determines the transport of heat in response to a gradient of temperature;

- the electric conductivity, which determines the transport of electric charge in response to an electric field.

The collisions thus play an important role, and a major difference between plasmas and neutral gases is the cross-section for particles' collisions. This has profound implications for plasma behaviour, which contradict the intuition acquired with neutral gases.

A reminder on collisions in neutral gases

Collisions between neutral particles have much in common with those of billiard balls. Macroscopic neutral particles collide when they come into contact, namely when they come closer than about their physical size. More precisely, two spheres of radius r collide when their centres come closer than $2r$, so that their cross-section for collision is the area of a circle of radius $2r$, i.e. $4\pi r^2$. The 'size' of an atom or a molecule relevant for collisions is not so clear-cut as the one of a billiard ball since the interaction involves induced dipoles in the distribution of electrons, a distribution determined by quantum mechanics. So, the cross-section for collisions between neutral atoms or molecules is somewhat greater than the 'billiard ball' value (taking for r a typical atomic size – see Section 2.4.1), but not by more than one order of magnitude. This yields the crude estimate

$$\sigma_{col} \sim 10^{-19} \text{ m}^2. \tag{2.10}$$

As in the case of billiard balls, most collisions between neutral atoms and molecules result in a large variation in momentum and energy (Fig. 2.2, left).

The mean collisional free path of particles is the average distance they have to travel in order to undergo one collision. A particle of cross-section σ_{col} travelling a distance l encounters all the particles contained in a cylinder of section σ_{col}

What is a plasma? 49

NEUTRAL PARTICLES   CHARGED PARTICLES

Collision   Collision

Trajectory   Trajectory

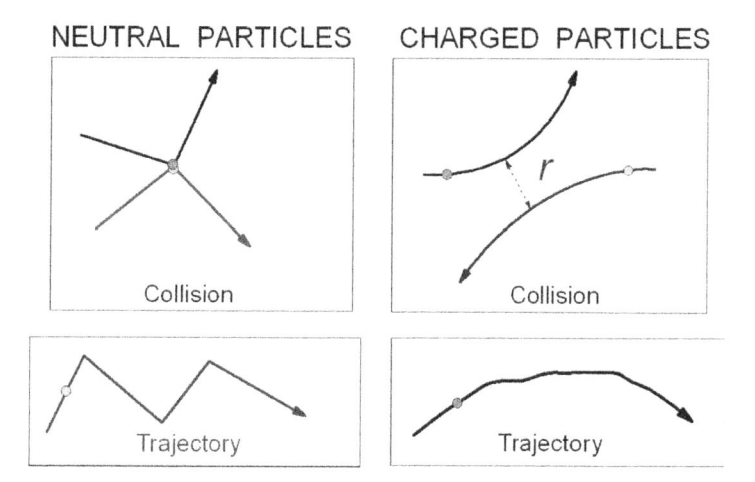

Figure 2.2 Collisions between neutral particles (left panels) and between charged particles (right panels). Collisions between neutrals occur, crudely, when they come into contact, and generally produce a large change in trajectory (top left panel). In contrast, charged particles interact even at large (but smaller than L_D) distances, via the Coulomb force (top right panel). The corresponding trajectories are sketched in the bottom panels.

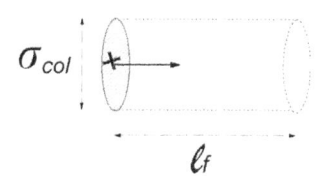

$$\sigma_{col} \qquad \ell_f$$

Figure 2.3 The collisional free path is the average distance travelled by a particle to undergo one collision. A particle of cross-section σ_{col} for collisions with particles of concentration n has the free path $l_f = 1/(n\sigma_{col})$.

and length l (Fig. 2.3), i.e. $n \times \sigma_{col} \times l$ particles of number density n. The collisional free path l_f corresponds to one collision, i.e.

$$l_f = (n\sigma_{col})^{-1} \qquad \text{collisional free path.} \qquad (2.11)$$

Near the surface of the Earth, the typical distance between particles is about 3×10^{-9} m, so that with the cross-section (2.10), the mean free path for collisions is of the order of magnitude 1 µm.

The *collision frequency* is the inverse of the average time between two collisions, that is the time for travelling the distance l_f. With a relative speed v, this yields

$$\nu_{col} = v/l_f = nv\sigma_{col}. \qquad (2.12)$$

Collisions between charged particles and neutrals

The cross-section for collisions between electrons and neutrals is given in order of magnitude by the above value (2.10). Because of their small mass, electrons move much faster than neutrals, so that the relative velocity is about their most probable speed $v_{the} \simeq (2k_BT/m_e)^{1/2}$. Hence the frequency of collision of electrons with neutrals of concentration n_n is given by substituting $v \sim v_{the}$ and the cross-section (2.10) in (2.12), which yields

$$\nu_{en} \sim 5 \times 10^{-16} n_n \sqrt{T}, \qquad (2.13)$$

a result we shall use in the context of planetary ionospheres (Section 7.1).

For collisions of ions with neutrals, the induced dipoles play a more important role, and the cross-section depends somewhat on the particle energy. At a small enough temperature (as for example near comets), the cross-section is proportional to the inverse of the relative speed, and a useful approximation for the frequency of collisions of ions with neutrals of concentration n_n is

$$\nu_{in} \sim 3 \times 10^{-15} n_n \sqrt{m_p/m_i}. \qquad (2.14)$$

Coulomb collisions

The mutual interaction of charged particles is basically very different. Consider two charges approaching each other (Fig. 2.2, right-hand panel). Since they interact via the Coulomb force, each 'encounter' generally deviates their trajectories, provided the particles come closer than the Debye length. (Farther away, the charges are shielded by the ambient plasma and no longer interact.) Each such encounter may thus be considered as a 'collision'.

What is the distance of closest approach required to produce a large perturbation in trajectory? Whatever the relative sign of the charges (in Fig. 2.2 the two charges are of like sign), the perturbation is large if the potential energy of interaction is at least equal to the average kinetic energy, i.e. $e^2/4\pi\epsilon_0 r \geq k_BT$, hence if the distance of closest approach is smaller than

$$r_L \equiv \frac{e^2}{4\pi\epsilon_0 k_B T} \qquad \text{(Landau radius)} \qquad (2.15)$$

in order of magnitude. Any encounter closer than this distance will result in a large perturbation in trajectory. We deduce that the effective cross-section for collisions producing a large perturbation is $\sigma_C \sim \pi r_L^2$, and the corresponding free path is

$$l_f \sim \left(n\pi r_L^2\right)^{-1} \qquad \text{(mean free path for large perturbations)}. \qquad (2.16)$$

What happens if the plasma is not in equilibrium? We may apply the same reasoning, but now k_BT has to be replaced by the kinetic energy of the particle,

What is a plasma?

$mv^2/2$, for a particle of (relative) speed v and (reduced) mass m. The effective distance for collisions producing a large perturbation is therefore in that case

$$r_{ef} = \frac{e^2}{4\pi\epsilon_0 \times mv^2/2} \propto v^{-2}.$$

An important result emerges: the faster the particle, the smaller the cross-section for collisions $\sim \pi r_{ef}^2 \propto v^{-4}$. Hence, fast particles undergo very few collisions. We shall return later to this point, which has basic consequences for plasma behaviour. Another interesting result is that if electrons and ions have similar temperatures, their collision cross-sections are similar, so that they have similar mean free paths. Since the collision frequency varies as v/l_f and electrons (being much lighter) move much faster, they have a much greater collision frequency.

How frequent are these close encounters producing large perturbations? Most particle encounters occur at distances of closest approach of the order of magnitude of the average distance between particles $\langle r \rangle \sim n^{-1/3}$. From the expression (2.15) of r_L and the definition (2.5) of the coupling parameter Γ, we have

$$r_L/\langle r \rangle = \Gamma. \tag{2.17}$$

Since $\Gamma \ll 1$, the distance r_L for producing a large perturbation is much smaller than the average interparticle distance, so that close encounters are very rare. Most encounters occur at much larger distances, resulting in small perturbations, so that the trajectory of charged particles is made of a succession of small deviations, rather than the zigzag path of neutrals (Fig. 2.2, bottom).

Figure 2.4 illustrates this property in a more realistic way. It shows the trajectory (projected on a plane) of a typical electron in a plasma with $\Gamma = 0.02$, from a numerical simulation [1] handling 2×10^6 particles in a box of size 10^2 times larger than the average distance between electrons.

Mean free path for collisions of charged particles

As a result of the numerous encounters at large distances, the cross-section for collisions of charged particles is greater than the value πr_L^2, which takes into account only the rare close encounters producing a large perturbation.

Consider the simple case of an electron that passes near a positive ion, with impact parameter p and velocity $\mathbf{v}_e$ and undergoes a small deviation (Fig. 2.5). Because of the large ion mass, we suppose it to be at rest. Most of the deviation of the electron takes place in the part of its trajectory where it is closest to the ion, i.e. at a distance of order of magnitude p from the ion, namely as it travels a distance of about p on each side of the ion, i.e. the distance $2p$ parallel to $\mathbf{v}_e$; this takes the time $\delta t = 2p/v_e$. In this part of the path, the Coulomb force on the electron is $F_\perp \simeq e^2/4\pi\epsilon_0 p^2$, roughly perpendicular to the original electron velocity $\mathbf{v}_e$. During the time δt, this force produces a change $\delta v_\perp$ in the

COULOMB ENCOUNTERS

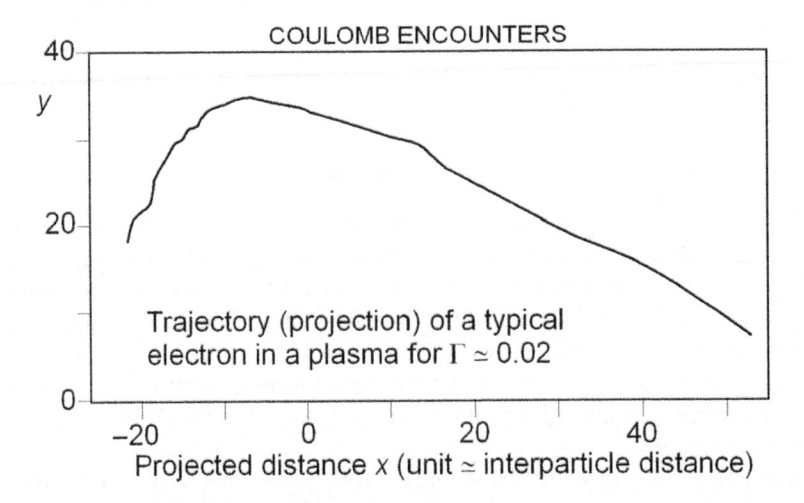

Figure 2.4 Typical trajectory (projected on a plane) of an electron in a plasma with $\Gamma = 0.02$, from a numerical simulation [1]: a box of size 110 arbitrary units contains 2×10^6 Maxwellian particles (electrons and ions); the trajectory shown is that of an electron having roughly the most probable speed. The mean free path (2.22) is nearly equal to the size of the box. (Courtesy A. Beck.)

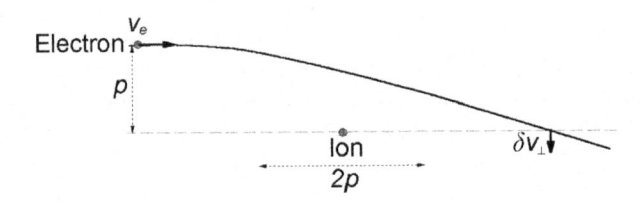

Figure 2.5 An electron of velocity $\mathbf{v}_e$ passing at distance p from an ion (of negligible velocity) and undergoing a small deviation.

electron velocity (perpendicular to $\mathbf{v}_e$) given by $m_e \delta v_\perp \simeq F_\perp \delta t$. Rearranging, this yields[5]

$$\delta v_\perp = v_e \times r_{Le}/p \quad \text{with} \quad r_{Le} = \frac{e^2}{4\pi\epsilon_0 \times m_e v_e^2/2}. \tag{2.18}$$

Statistically, the deviation may be in either sense with equal probability; hence the individual deviations do not add, but their squares do, as in a random walk. We thus calculate the mean total variation $\langle \Delta v_\perp^2 \rangle$ during a given time Δt, by integrating over encounters of various impact parameters p occurring during this time. From (2.18), each encounter of impact parameter p produces $\delta v_\perp^2 = (v_e r_{Le}/p)^2$. The number of encounters of impact parameter in the range

---
[5] An exact calculation turns out to yield the same result.

What is a plasma?

$[p, p+dp]$ (crossing the area $2\pi p dp$) during the time Δt is $dN = nv_e \times 2\pi p dp \times \Delta t$, so that

$$\langle \Delta v_\perp^2 \rangle = \int \delta v_\perp^2 \times dN = 2\pi \, nr_{Le}^2 \, v_e^3 \Delta t \int dp/p. \tag{2.19}$$

For impact parameters $p < r_{Le}$, the deviation is large, contrary to our assumption, whereas for $p > L_D$ the charges do not interact because of Debye shielding. Hence the integral (2.19) must be calculated in the range $r_{Le} < p < L_D$, which yields the factor $\ln(L_D/r_{Le})$.

The collision frequency is the inverse of the time Δt needed to produce a large deviation, i.e. to produce $\langle \Delta v_\perp^2 \rangle \simeq v_e^2$. Substituting this value into (2.19) yields the collision frequency $1/\Delta t$ between electrons and (singly charged) ions

$$\nu_{ei} \simeq nv_e \times 2\pi r_{Le}^2 \ln(L_D/r_{Le}) \tag{2.20}$$

whence the collisional free path

$$l_f \simeq \left[n \times 2\pi r_{Le}^2 \ln(L_D/r_{Le}) \right]^{-1}. \tag{2.21}$$

The mean value at equilibrium may be estimated by replacing $m_e v_e^2/2$ by the average kinetic energy $3k_B T/2$. From (2.15) and (2.18), we have $r_{Le} \simeq 2r_L/3$ and $L_D/r_{Le} \simeq 3/\left(4\sqrt{\pi}\Gamma^{3/2}\right)$, so that (2.21) yields the mean electron free path for collisions

$$l_f \simeq \left[n \times (4\pi/3) \, r_L^2 \ln(1/\Gamma) \right]^{-1} \tag{2.22}$$

where r_L is given by (2.15) and Γ by (2.5).[6] One can verify (Problem 2.5.2) in Fig. 2.4 that for a typical electron the velocity direction indeed changes significantly when the particle has travelled a distance given roughly by (2.22). This equation yields approximately (in SI units)

$$l_f \simeq \frac{10^9}{\ln(1/\Gamma)} \times \frac{T^2}{n}. \tag{2.23}$$

Comparing (2.22) with (2.16), we see that the cumulative effect of the numerous small deviations decreases the free path (and increases the collision frequency) by a factor of order of magnitude $\ln(1/\Gamma)$. For typical space plasmas that we shall encounter in this book, this factor lies approximately between 10 and 20.

In the solar wind at 1 AU from the Sun, we have $n \sim 5 \times 10^6$ m^{-3} and $T \sim 10^5$ K, so that the typical distance between particles is about 5 mm, whereas the mean free path is about 1 AU; collisions are thus very rare in the solar wind.

We considered for simplicity an electron encountering a singly charged ion. For an electron encountering an ion of charge Ze, the Coulomb force is greater by the factor Z, and therefore so is the radius r_L, producing an electron-free path smaller by the factor $1/Z^2$.

---

[6]We have approximated $\ln(0.6/\Gamma)$ by $\ln(1/\Gamma)$, which yields a very small error since $\Gamma \ll 1$.

Timescales for equilibrium

The scales $1/\nu_{ei}$ and l_f represent respectively the average time and distance for an average electron to change significantly the direction of its velocity due to the collisions with ions. Because of the large difference in mass between electrons and ions, this barely changes the particle energy.

Consider now collisions between electrons themselves. The calculation is slightly different since one can no longer assume one particle to be at rest, but we can make the calculation in the frame of the particles' centre of mass (using the reduced mass $m_e/2$), and the collision frequency is of the same order of magnitude. The major difference is that for particles of like mass, the collision now changes also the particle energy. Hence the values of ν_{ei} and l_f calculated above represent respectively (in order of magnitude) the collision frequency and the free path of electrons for change in *speed direction* (because of encounters with *ions and electrons*) and in *energy* (because of encounters with *electrons*).

Consider now the collisions between two ions. The result is the same as for collisions between two electrons, just replacing the electron properties by those of ions. Hence, if the temperatures are similar, the mutual collision frequency of ions is smaller than the above value by a factor equal to the ratio of their most probable speeds, that is about $(m_e/m_i)^{1/2}$, whereas the free path is the same as above.

Photon mean free path versus particle mean free path

It is interesting to compare the effective cross-section of electrons for collisions with charged particles σ_C, which is about one order of magnitude greater than πr_L^2 (because of the numerous large-distance encounters), with the effective cross-section of electrons for interaction with photons (the Thomson cross-section), given in (1.11). From (1.12) and (2.15), the ratio of both cross-sections is

$$\frac{\sigma_C}{\sigma_T} > \left(\frac{r_L}{r_e}\right)^2 \simeq \left(\frac{m_e c^2}{k_B T}\right)^2 \qquad (2.24)$$

where we have substituted the so-called classical electron radius $r_e = e^2/(4\pi\epsilon_0 m_e c^2)$. This ratio is much greater than unity for non-relativistic plasmas. Hence, plasmas interact more with plasmas than with radiation, and photon mean free paths in plasmas are generally much greater than charged particle free paths.

2.1.4 Plasma oscillations

Consider a volume of plasma that is initially quasi-neutral, and imagine that you displace all the electrons along the **x** axis by a distance x (Fig. 2.6). This produces a charge per unit volume equal to $\pm ne$ in two slabs of width x at the extremities; the electric field is equal to that between two capacitor plates of

What is a plasma?

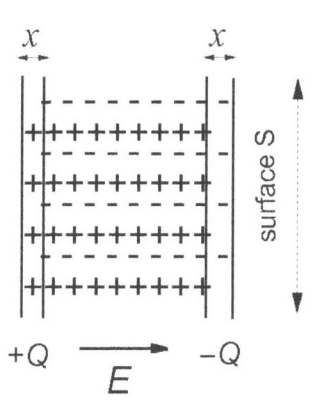

Figure 2.6 When electrons (in a plasma of density n) are displaced by x, the charge separation produces the electric field of a capacitor whose plates carry the charge of the electrons (or ions) contained in a plasma slab of width x, i.e. $\pm nex$ per unit area.

charge per unit area $\pm nex$, i.e

$$E = nex/\epsilon_0.$$

Each displaced electron is subject to a force $-eE$ along **x**, and moves according to

$$m_e \partial^2 x/\partial t^2 = -eE \quad \Rightarrow \quad \partial^2 x/\partial t^2 = -\omega_p^2 x$$

with

$$\omega_p = \left(\frac{ne^2}{\epsilon_0 m_e}\right)^{1/2} \quad \text{(plasma (angular) frequency)}. \quad (2.25)$$

This is the motion of a harmonic oscillator of (angular) frequency ω_p.

Charge separation in a plasma therefore makes electrons oscillate at the (angular) frequency ω_p. Ions, being much heavier, would oscillate more slowly (by the factor $(m_i/m_e)^{1/2}$), so that at the scale of the frequency ω_p, they barely move. Numerically, the *plasma frequency* is

$$f_p = \frac{1}{2\pi}\left(\frac{ne^2}{\epsilon_0 m_e}\right)^{1/2} \simeq 9\sqrt{n} \quad \text{plasma frequency} \quad (2.26)$$

in SI units, i.e. with f_p in Hz and n in m^{-3}. In the solar wind at 1 AU from the Sun, we have $f_p \sim 2 \times 10^4$ Hz; we shall see a direct illustration of the plasma frequency in Section 6.4. In the Earth's ionosphere (see Section 7.1), the plasma frequency is a few 10^6 Hz; electromagnetic waves of smaller frequency are reflected, a property which enabled early long-distance radio communications.

This oscillating behaviour holds under two conditions. First, the collisions between particles should not suppress the plasma oscillations. This requires the electron collision frequency to be much smaller than the plasma frequency. If the (gas) plasma is nearly completely ionised, this condition is always met since $\omega_p \sim v_{the}/L_D$, so that we have approximately

$$\frac{\nu_{ei}}{\omega_p} \sim \frac{L_D}{l_f} \simeq \Gamma^{3/2} \times \ln(1/\Gamma) \tag{2.27}$$

which is much smaller than unity in a gaseous plasma ($\Gamma \ll 1$). On the other hand, in a weakly ionised plasma, the collective behaviour requires that the collision frequency of electrons with neutrals be smaller than the plasma frequency, which requires the degree of ionisation to be large enough.

The second hypothesis made is that electrons move as a whole, i.e. that their random agitation is negligible. During a period of plasma oscillation $\sim 1/\omega_p$, the agitation displaces an electron by a distance equal to the most probable speed $\sim (k_B T/m_e)^{1/2}$ times $1/\omega_p$, i.e. the distance L_D. Hence, the plasma bulk oscillations occur only at scales much greater than the Debye length. It is only at such scales that a large number of particles can contribute cumulatively to produce a collective behaviour. We shall see in Section 2.3 that the electron random motion makes the plasma oscillations propagate as plasma waves, and also damps them if the scale becomes comparable with (or smaller than) the Debye length.

The origin of that collective behaviour is very different from the one in a neutral gas. In a neutral gas, the coupling between particles is due to their mutual collisions. In a plasma, the coupling is due to the mean electric field produced by particles. The collective behaviour therefore requires that the close encounters yielding large perturbations to this mean field be rare enough.

2.1.5 Non-classical plasmas

Quantum degeneracy

The above estimates are based on the assumption that the plasma behaves classically. If the concentration of particles is too high, however, their distance may involve scales so small that quantum effects act. Basically this is because, from Heisenberg's uncertainty relations, localising the particles in a small region Δx gives them a momentum $\Delta p \sim \hbar/\Delta x$. If the density is high, then Δx is small, and the corresponding Δp yields a high energy.

Let us estimate this effect. Pauli's exclusion principle tells us that two plasma particles (which are fermions) cannot be in the same quantum state; hence each particle must be localised within a region of size smaller than about half the average interparticle distance, i.e. $\Delta x \sim n^{-1/3}/2$. A compression at density n therefore produces the momentum $p \sim \hbar/\Delta x \sim 2\hbar n^{1/3}$ per particle. The corresponding energy is $p^2/(2m) \sim 2\hbar^2 n^{2/3}/m$ per (non-relativistic) particle of mass m. Because of the small electron mass, this energy is much greater for electrons than for ions.

Hence, compressing a plasma gives to each electron an energy of about $2\hbar^2 n^{2/3}/m_e$, where n is their number density. A detailed calculation confirms

What is a plasma?

this order of magnitude estimate; the exact values are $p_F = \hbar \left(3\pi^2 n\right)^{1/3}$ for the largest momentum, whence the so-called *Fermi energy* $W_F = p_F^2/(2m_e)$:

$$W_F = \hbar^2 (3\pi^2 n)^{2/3}/(2m_e). \tag{2.28}$$

The greater the particle density, the smaller the region that is available to an electron, and the higher the resulting Fermi energy. If the Fermi energy becomes greater than $k_B T$, then the total energy is determined by the Fermi energy instead of $k_B T$, and the electrons are said to be *degenerate*. This occurs if the temperature is smaller than $W_F/k_B \equiv T_F$, the so-called *Fermi temperature*. In that case, the Fermi temperature (instead of the kinetic temperature) determines the particle pressure, the coupling parameter Γ and the Debye length, so that these quantities become independent of the kinetic temperature, and only depend on the density.

We shall not encounter degenerate plasmas in this book, except when determining the limits of stellar (see Section 3.1) and planetary masses (see Section 7.1), which involve the Fermi energy.

Relativistic particles

Finally, to determine whether the particles are relativistic, we have to compare the average kinetic energy $\sim k_B T$ (or $k_B T_F$ if they are degenerate) with the rest mass energy mc^2.

For relativistic particles, the Fermi energy and temperature must be calculated with the relativistic energy–momentum relation. In particular if the particles are ultra-relativistic ($v \simeq c$), the energy of a particle of momentum p is $W \simeq pc$ (instead of $p^2/2m$), so that the electron Fermi energy is now $W_F \simeq p_F c$ (instead of $p_F^2/(2m_e)$). The Fermi energy of ultra-relativistic particles therefore varies as $n^{1/3}$ instead of $n^{2/3}$.

Likewise, the pressure of ultra-relativistic particles is 1/3 of their energy density (instead of the factor 2/3 relevant in the non-relativistic case). We shall not encounter relativistic plasmas in this book (but only individual relativistic particles), but we shall use the pressure of photons (which are par excellence relativistic particles) when studying the solar interior (Section 3.1) and the dynamics of heliospheric dust grains (Section 7.4).

2.1.6 Summary

Gaseous plasmas have $\Gamma \ll 1$, i.e. the (Coulomb) interaction energy of the particles is much smaller than the kinetic energy. The slow decrease with distance of the Coulomb force has two major consequences. First, any particle interacts simultaneously with a large number of particles and modifies the medium so that each particle may be regarded as being 'dressed' by the other particles. This dressing makes plasmas quasi-neutral on large spatial ($L > L_D$) and temporal ($t > \omega_p^{-1}$) scales, and produces a collective behaviour. Second, the particle collisional free path increases strongly with speed, so that fast particles tend to be nearly collisionless.

2.2 Dynamics of a charged particle

In this section, we consider briefly the dynamics of a charged particle in electric (**E**) and magnetic (**B**) fields which are given a priori, i.e. which are negligibly modified by the moving charge itself. Most of the results will therefore be applicable to a small minority of particles which do not affect the bulk of the plasma, for example cosmic rays. The coupling between the magnetic field and the bulk plasma will be considered later.

Magnetic fields are ubiquitous in the Universe, and we shall focus on them. But why is this so? Indeed, relativity theory tells us that electric and magnetic fields are symmetrical in that they transform into each other upon a change of reference frame.

2.2.1 The key role of the magnetic field

As we shall see below, the key role of the magnetic field stems from two facts:

- plasmas (made of electric charges) are ubiquitous in the Universe, whereas magnetic charges (the so-called *magnetic monopoles*) are absent,

- one generally considers non-relativistic ($V \ll c$) changes of reference frames.

The absence of magnetic monopoles[7] – whereas electric charges are ubiquitous – is at the origin of the asymmetry in Maxwell's equations:

$$\nabla \cdot \mathbf{E} = \rho_e/\epsilon_0 \quad \nabla \times \mathbf{E} = -\partial \mathbf{B}/\partial t \qquad (2.29)$$

$$\nabla \cdot \mathbf{B} = 0 \quad \nabla \times \mathbf{B} = \mu_0 \mathbf{J} + \left(1/c^2\right) \partial \mathbf{E}/\partial t \qquad (2.30)$$

which contain electric charges (ρ_e) and currents (**J**), but no magnetic charges and currents.

We have seen that plasmas are quasi-neutral on large-scales, so that the large-scale electric field nearly vanishes in the reference frame where the plasma is at rest. On the other hand, positive and negative electric charges moving differently yield electric currents, which produce magnetic fields.

Consider a plasma of (non-relativistic) bulk velocity **V** with respect to a 'laboratory' frame $\mathcal{R}$, where the electric and magnetic fields are respectively **E** and **B**. The fields in the plasma frame $\mathcal{R}'$ are given by the Lorentz transformations as

$$\mathbf{E}' = \mathbf{E} + \mathbf{V} \times \mathbf{B} \quad \mathbf{B}' = \mathbf{B} - \mathbf{V} \times \mathbf{E}/c^2 \qquad (2.31)$$

where we have neglected terms of order V^2/c^2. Since we have $\mathbf{E}' = 0$ in the

---

[7]You would get a magnetic charge if you could separate the two poles of a bar magnet. If these magnetic monopoles do exist, they have not yet been detected, which sets an upper limit on their concentration; see for example [13].

Dynamics of a charged particle

plasma frame $\mathcal{R}'$, the fields in the frame $\mathcal{R}$ are from (2.31)

$$\mathbf{E} = -\mathbf{V} \times \mathbf{B} \qquad \mathbf{B} = \mathbf{B}' - \mathbf{V} \times (\mathbf{V} \times \mathbf{B})/c^2 \simeq \mathbf{B}', \qquad (2.32)$$

again neglecting terms of order V^2/c^2.[8]

Hence the magnetic field plays a privileged role: it is independent of the reference frame; in contrast, the electric field depends on the reference frame and is nearly zero in the plasma frame.[9] The latter therefore appears as a natural reference frame, and the Faraday concept of magnetic field lines acquires a basic physical meaning since if two points are connected by a field line in this frame, they are so connected in another reference frame (for non-relativistic Lorentz transformations). Magnetic field lines – the pillars of magnetohydrodynamics – have still more interesting properties, which we shall study in the next section.

Another basic property of **B** is that it is a *pseudo-vector*, since its sense depends on the usual convention of right-handed co-ordinate systems. If one changes the co-ordinate system according to $\mathbf{x} \to \mathbf{x}' = -\mathbf{x}$ (a reflection about the origin, making the co-ordinate system left-handed), the components of true vectors (as a velocity or a force) transform as $v'_x = -v_x$, leaving the physical direction of the vectors unchanged. The Lorentz force $\mathbf{F} = q\mathbf{v} \times \mathbf{B}$ is also a true vector, so that the inversion of the co-ordinate changes the components of both **F** and **v**; therefore it does not change the components of **B**, whose physical direction is thus reversed. Formally, the magnetic field is analogous to a vortex. *Mirror asymmetry* plays a key role in magnetic field generation, and we shall encounter applications of this property in Sections 3.3 and 4.2.

2.2.2 Basic charge motion in constant and uniform fields

The basic equation of motion for a particle of charge q and velocity **v** subjected to the fields **E** and **B** is

$$\frac{d(m\mathbf{v})}{dt} = q(\mathbf{E} + \mathbf{v} \times \mathbf{B}) \qquad (2.33)$$

with the relativistic mass

$$m = \gamma m_0 \qquad \gamma = (1 - v^2/c^2)^{-1} \qquad (2.34)$$

where m_0 is the particle rest mass and γ the Lorentz factor.

Uniform magnetic field

If $\mathbf{E} = 0$ and **B** is constant, the Lorentz force reduces to $q\mathbf{v} \times \mathbf{B}$, perpendicular to the velocity; it produces a curvature of the particle path, but no change in

---

[8] We shall sometimes consider particles moving individually at relativistic velocities, but we shall not consider reference frames moving at relativistic velocities with respect to the bulk plasma.

[9] By 'nearly zero', we mean of amplitude small with respect to $|\mathbf{V} \times \mathbf{B}|$, and on a large scale.

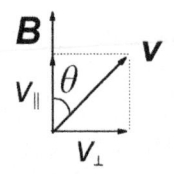

Figure 2.7 Components of the velocity parallel and perpendicular to the magnetic field; $v_\perp = v \sin\theta$, where θ is the so-called *pitch angle*.

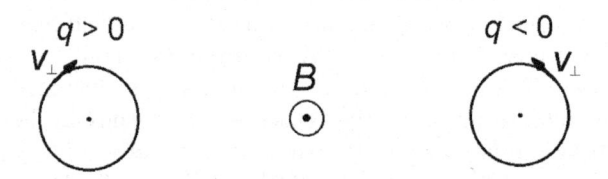

Figure 2.8 Gyration of a charge in a magnetic field (pointing out of the paper).

the speed v and thus in the relativistic mass m. Therefore in this case the motion of a relativistic particle is the same as that of a non-relativistic particle of (constant) mass $m = \gamma m_0$. Since the force vanishes along **B**, $v_\parallel$ is constant; since v is constant too, so is the angle θ between **v** and **B** (Fig. 2.7). On the other hand, in the plane $\perp$ **B**, the force produces a circular motion of radius r_g and (angular) frequency $\omega_g = v_\perp/r_g$, given by equating the acceleration $qv_\perp B/m$ to the centrifugal acceleration $v_\perp^2/r_g$, so that

$$r_g = \frac{mv_\perp}{|q|B} \quad \text{(Larmor radius)} \tag{2.35}$$

$$\omega_g = \frac{|q|B}{m} \quad \text{((angular) gyrofrequency)} \tag{2.36}$$

with particles of negative (positive) charge gyrating in the direct (opposite to direct) sense. Hence the magnetic field generated by the particle is opposite to the imposed field (Fig. 2.8): the plasma is diamagnetic.

The *Larmor radius* (or *radius of gyration*) and the *gyrofrequency* (or *cyclotron frequency*) set the scales below which the individual particle gyration plays an important role. Numerically, the cyclotron frequency is $f_g = \omega_g/2\pi \simeq 2.8 \times 10^{10} B$ in SI units for electrons (and smaller by the factor m_e/m_p for protons). In the Earth's environment, the Larmor radius is about a few centimetres for electrons and 1 m for protons; it is greater by more than five orders of magnitude in the solar wind.

The resulting path is a helix of constant pitch around a magnetic line of force. Since particles having the same value of mv/q and pitch angle have the same trajectory, high-energy particles (see Section 8.2) are often quantified by their so-called *rigidity* defined as $pc/|q|$ (with $p = mv$), expressed in volts since it has the dimension of energy per charge.

Dynamics of a charged particle

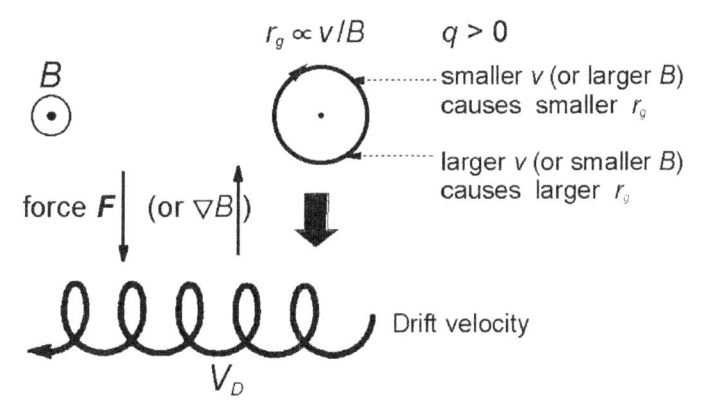

Figure 2.9 A force $\mathbf{F} \perp \mathbf{B}$ accelerates the particles along one half of the orbit and decelerates them along the other half. This makes the Larmor radius ($r_g \propto v/B$) greater near the bottom of the orbit than near the top (when $\mathbf{F}$ is downwards), which deforms the orbit, producing a drift. A gradient of B in the direction $\perp \mathbf{B}$ has a similar effect.

Electric field or applied force

How is this trajectory changed if the electric field does not vanish? Let $\mathbf{E}_\perp$ be the electric field in the direction $\perp \mathbf{B}$ in some frame $\mathcal{R}$. Consider now the reference frame $\mathcal{R}'$ moving at velocity

$$\mathbf{V}_D = \frac{\mathbf{E} \times \mathbf{B}}{B^2} \qquad (2.37)$$

with respect to $\mathcal{R}$. In $\mathcal{R}'$ the electric field $\perp \mathbf{B}$ is $\mathbf{E}'_\perp = \mathbf{E}_\perp + \mathbf{V}_D \times \mathbf{B} = 0$, since from (2.37) $\mathbf{V}_D \times \mathbf{B} = -\mathbf{E}_\perp$. Hence the motion in the plane $\perp \mathbf{B}$ reduces to the gyration found above. Going back to the frame $\mathcal{R}$, the motion in the plane $\perp \mathbf{B}$ is therefore the superposition of the gyration found above and a drift of velocity $\mathbf{V}_D$ given by (2.37). This velocity is the same for all charged particles, making the plasma move as a whole.

This drift velocity may be interpreted in either of two ways. The first way is that it produces a Lorentz force $q\mathbf{V}_D \times \mathbf{B}$ which balances the electric force $q\mathbf{E}_\perp$, so that for an observer moving at $\mathbf{V}_D$ the electric field has been transformed away. The other interpretation is sketched in Fig. 2.9. The force $q\mathbf{E}_\perp$ accelerates the particle during the part of the circular orbit where it moves in the same sense as the force, and decelerates it when it moves the other way. Hence the particle gyrates faster (thus with a greater Larmor radius) near the bottom of the orbit than near the top (when the force is downwards), producing a drift to the left when the gyration is clockwise; reversing either $\mathbf{B}$, the force or q reverses the drift.

This result can be applied to other forces by replacing in (2.37) $q\mathbf{E}$ by a general force $\mathbf{F}$, which therefore produces a drift

$$\mathbf{V}_D = (\mathbf{F}/q) \times \mathbf{B}/B^2. \qquad (2.38)$$

2.2.3 Non-uniform magnetic field

The magnetic field is generally non-uniform. If the non-uniformity is weak, namely if the field does not change much over a distance equal to the gyroradius (or during a time equal to the inverse of the gyrofrequency), the motion can be approximated by the gyration found above, around a point which is moving. This instantaneous centre of gyration is called the *guiding centre* of the particle. We calculate below its motion by considering separately the variation in magnetic field strength perpendicular and parallel to the magnetic field direction.

Drift produced by a variation of $B \perp \mathbf{B}$

If the magnetic field strength varies in the direction $\perp \mathbf{B}$, the magnetic field lines are curved, which forces the particles to follow curved paths along $\mathbf{B}$. If the radius of curvature of the field line is R_c, the centrifugal force on a particle of parallel velocity $v_\parallel$ is $F = mv_\parallel^2/R_c$, pointing opposite to the centre of curvature. This effective force produces a drift velocity given by (2.38). The particle gyration produces an additional drift because the gradient in B causes the Larmor radius ($r_g \propto 1/B$) to be larger during one half of the orbit than during the other half, which deforms the orbit. This has a similar effect as an applied force (Fig. 2.9) and produces an additional drift velocity. If the magnetic field is essentially produced by exterior currents, we have $\nabla \times \mathbf{B} = 0$, whence $1/R_c = |\nabla_\perp B|/B$, where the symbol $\nabla_\perp$ denotes the component of the gradient in the direction $\perp \mathbf{B}$, and $\nabla_\perp B$ points towards the centre of curvature. Finally one finds a total drift velocity equal to

$$\mathbf{V}_D = \left(\frac{mv_\perp^2}{2} + mv_\parallel^2\right) \frac{\mathbf{B} \times \nabla_\perp B}{qB^3}. \tag{2.39}$$

For a particle of energy W, we have in order of magnitude $V_D \sim W |\mathbf{B} \times \nabla_\perp B|/qB^3 \sim W/(qBR_c)$. Electrons and ions drift in opposite senses, producing an electric current.

Variation of $B \parallel \mathbf{B}$

Consider now a magnetic field oriented primarily along $\mathbf{z}$ with approximate cylindrical symmetry, and whose strength varies along $\mathbf{B}$. Let us assume for example $dB/dz > 0$ (Fig. 2.10). Magnetic flux tubes, whose surface is everywhere parallel to $\mathbf{B}$, have approximate cylindrical symmetry, and since the magnetic flux is a constant along a flux tube (because $\nabla \cdot \mathbf{B} = 0$), the field lines converge towards positive z, i.e. the radial component $B_r < 0$. Hence the Lorentz force has a component along $\mathbf{z}$

$$F_z = |qv_\perp| B_r \tag{2.40}$$

which has the same sign as B_r (here negative) whatever the sign of q (since the gyration speed $v_\perp$ changes of sense as q changes of sign).

Dynamics of a charged particle   63

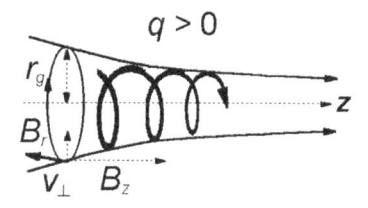

Figure 2.10 When the magnetic field strength varies along **B**, the Lorentz force on a gyrating particle has a component that decelerates (accelerates) it when it moves towards increasing (decreasing) B.

Hence, the motion is slowed down if the charge moves towards stronger B, and is accelerated if the charge moves towards smaller B. From $\nabla \cdot \mathbf{B} = 0$, we have $B_r = -(r/2)\, dB/dz$ at distance r,[10] which we substitute into (2.40) with $r = r_g = mv_\perp/qB$ to yield the force $\parallel \mathbf{B}$

$$F_\parallel = -\mu \nabla_\parallel B \qquad (2.41)$$

where the symbol $\nabla_\parallel$ denotes the component of the gradient in the direction of **B**, and

$$\mu = \frac{mv_\perp^2/2}{B} \quad \text{(magnetic moment)}. \qquad (2.42)$$

Magnetic moment

The force (2.41) is the usual force on a small diamagnetic magnet lying in a gradient of magnetic field strength. Similarly, the drift velocity produced by a gradient of B in the direction $\perp \mathbf{B}$ (ignoring the effect of curvature) corresponds to a force that may be written from (2.38)–(2.39) as $F_\perp = -\mu \nabla_\perp B$.

The quantity μ is called the *magnetic moment* of the particle. Indeed, the gyration of the charge averaged over one gyration corresponds to an electric current $I = |q|/(2\pi/\omega_g)$; since the loop area is $s = \pi r_g^2$ and $\omega_g r_g^2 = mv_\perp^2/qB$, the magnetic moment (2.42) is equal to $\mu = I \times s$; furthermore, since opposite charges gyrate in opposite senses, the sense of the current is independent of the sign of q. Hence, in average over one rotation, the particle gyration is equivalent to a current loop of magnetic moment μ given by (2.42) and pointing always opposite to **B** (Fig. 2.11). This illustrates the already mentioned plasma diamagnetism.

The potential energy of a magnetic dipole in a magnetic field **B** is $-\boldsymbol{\mu} \cdot \mathbf{B}$, and the corresponding force is formally $\nabla(\boldsymbol{\mu} \cdot \mathbf{B}) = -\nabla(\mu B)$ since $\boldsymbol{\mu}$ points opposite to **B**; hence the result found above that the force is $-\mu \nabla B$ in a (weak) gradient of magnetic strength suggests that $\mu = $ constant. In fact, one may prove

---

[10] To prove this, draw a cylinder of radius r and length dz along the z axis. The outward magnetic flux crossing its bounded surface is $(2\pi r dz) B_r + \pi r^2 [B_z(z+dz) - B_z(z)]$, which is equal to zero since $\nabla \cdot \mathbf{B} = 0$.

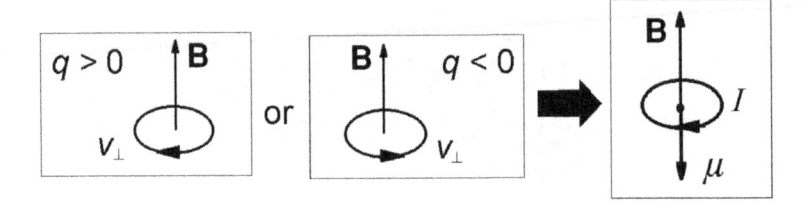

Figure 2.11 Magnetic moment produced by the gyration of a charge in a magnetic field. Whatever the sign of the charge, the current has the same direction and is equivalent (in average over one gyration) to a magnetic moment opposite to the imposed magnetic field.

a stronger result: the magnetic moment of a particle gyrating in a magnetic field remains nearly constant in both space and time when the magnetic field varies slowly at the scale of the gyration. Note that, for a relativistic particle, the conserved quantity is instead the magnetic flux Br_g^2 across the loop, so that the invariant is the quantity $\gamma\mu$ rather than μ. This invariance is an example of Lenz's law: electrical circuits change their currents in order to counteract externally caused changes of the enclosed magnetic fluxes.

Magnetic mirrors

This has an important consequence. When a particle is moving towards increasing magnetic field strength (converging magnetic field lines), the Lorentz force slows down the motion along **B**. The perpendicular energy $mv^2 \sin^2 \theta$ increases with B, keeping μ constant; since v remains constant because energy is conserved, θ increases, until $\theta = \pi/2$; at this point the particle is reflected back towards the weaker field. A region of increasing magnetic field thus acts as a mirror for charged particles.

Particles may therefore be trapped between two regions of strong magnetic field. This occurs close to magnetised planets having a dipolar magnetic field, where the increasing magnetic field strength in both hemispheres mirrors particles (see Appendix and Problem 2.5.3). Such particles may be viewed as small magnets (of magnetic moment pointing locally opposite to **B**), which are repelled by the large 'magnet' responsible of the planetary magnetic field. When approaching the planet's positive pole (with their own positive pole pointing ahead) they are repelled back towards the planet's negative pole. Since they approach it with their negative pole ahead, they are again repelled, and keep on oscillating between the poles.

So the particles not only gyrate around field lines, but also bounce between regions of high magnetic field. Furthermore, the transverse gradient and curvature of the field lines produces a drift velocity given by (2.39). With a dipolar magnetic field, $\mathbf{B} \times \nabla_\perp B$ is in the azimuthal direction, making the particles drift in longitude (Fig. 2.12).

Dynamics of a charged particle

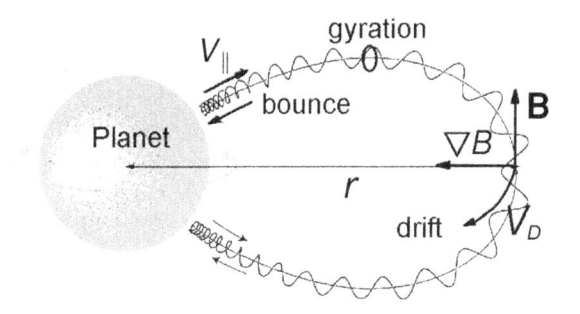

Figure 2.12 Charged particles can be trapped in the *magnetic bottle* formed by a dipolar planetary magnetic field. Their motion is the superposition of three components: the *gyration* around a field line, the *bounce* between magnetic mirrors in opposite hemispheres, and an azimuthal *drift* produced by the transverse magnetic field gradient.

2.2.4 Adiabatic invariants

The near invariance of μ is a particular case of adiabatic invariance, associated to periodic generalised co-ordinates of Hamiltonian systems [12]. When a parameter varies slowly at the scale of the period, the action integral varies much less than this parameter, and is called an *adiabatic invariant* – by analogy with thermodynamics where entropy is such an invariant in slow adiabatic processes. A classical example is the oscillating pendulum, whose adiabatic invariant is the energy divided by the frequency; indeed, if one changes slowly the length of a pendulum, the frequency varies in proportion to the energy. The same holds for a particle gyrating in a magnetic field. Since the particle motion has three degrees of freedom, there may be three adiabatic invariants if the system has several periodicities.

In this way, a particle trapped in a dipolar magnetic field has three adiabatic invariants associated to the three periodic motions:

- the gyration around magnetic field lines (speed $v_\perp$, period $T_1 \sim r_g/v_\perp$), whose adiabatic invariant is μ, given by (2.42),[11]

- the bounce between mirror points (speed $v_\parallel$, period $T_2 \sim r/v_\parallel \gg T_1$, where r is the distance to the planet), whose adiabatic invariant is the integral of the longitudinal momentum $mv_\parallel$ along the path between mirror points,

- the azimuthal drift produced by the gradient in magnetic field strength perpendicular to **B** (speed V_D given by (2.39), whence in order of magnitude $V_D \sim mv_\perp^2/(qBr) \sim v_\perp r_g/r$, period $T_3 \sim r/V_D \gg T_2$), whose adiabatic invariant is the magnetic flux across the area encircled by the drift path.

---
[11] Or rather $\gamma\mu$, if the particles are relativistic.

This adiabatic invariance has a number a consequences. It enables particles to remain trapped for a long time on the same magnetic shell in dipolar planetary magnetic fields; this is responsible for the long life of radiation belts around planets. We shall apply these concepts to the trapping and acceleration of particles in the contexts of magnetospheres (Section 7.3) and cosmic rays (Section 8.2).

2.2.5 Summary

Charged particles gyrate around magnetic field lines, keeping their magnetic moment invariant when the magnetic field varies weakly at the scale of the gyration. A weak longitudinal increase in magnetic field strength acts as a magnetic mirror. A force, or a weak transverse magnetic gradient, produces a small transverse drift.

2.3 Many particles: from kinetics to magnetohydrodynamics

A plasma is made of a large number of particles. In classical mechanics, the state is defined by the position **r** and velocity **v** of each particle at time t, and the evolution is determined by the equation of motion of each particle. To make the problem tractable, one has to decrease the number of variables. This is done by making averages, in two main ways:

- the *kinetic description* retains some microscopic properties by considering as the basic quantity the velocity distribution (for each particle species); basically, this amounts to replacing the equations of motion for each particle by a differential equation on the velocity distribution,

- the *fluid description* deals with a few macroscopic quantities as the mean density of particles (or of mass), the mean velocity, the pressure or the temperature, etc., which represent averages over the velocity distribution (for each particle species); basically, this amounts to replacing the velocity distribution – a function generally defined by an infinite number of parameters – by a few parameters, which is permissible only near thermodynamic equilibrium.

2.3.1 Elements of plasma kinetics

We define the particle velocity distribution[12] so that the number of particles in the volume element $[x, x + dx], [y, y + dy], [z, z + dz]$, and with velocities in the range $[v_x, v_x + dv_x], [v_y, v_y + dv_y], [v_z, v_z + dv_z]$ at time t is

$$d^6 N = f(\mathbf{r}, \mathbf{v}, t) \times d^3 r \times d^3 v \qquad (2.43)$$

---

[12] Beware that there are many subtleties in this definition, as discussed for example in [6].

where

$$d^3r = dxdydz \tag{2.44}$$
$$d^3v = dv_x dv_y dv_z \tag{2.45}$$

are the volumes in the space of positions and the space of velocities, respectively. A position $\mathbf{r}$ and velocity $\mathbf{v}$ thus correspond to a 'point' in a phase space of six dimensions $[x, y, z, v_x, v_y, v_z]$, which we denote $[\mathbf{r}, \mathbf{v}]$.[13]

This description is more complete than the fluid description which deals with averages of f as

$$n = \int d^3v\, f(\mathbf{v}) \qquad \text{(particle density)} \tag{2.46}$$

$$\mathbf{V} \equiv \langle \mathbf{v} \rangle = \int d^3v\, f(\mathbf{v}) \times \mathbf{v}/n \qquad \text{(velocity)} \tag{2.47}$$

and higher-order moments, for each species of particles; we have not written explicitly the dependence in $\mathbf{r}$ and t, to simplify the notations. These moments are macroscopic quantities defined in the ordinary space of three dimensions $[x, y, z]$.

A reminder: the Maxwellian distribution

In the special case of thermodynamic equilibrium at temperature T, statistical mechanics tells us that, as we already noted, the velocity is Gaussian distributed along each co-ordinate (in the frame where the mean velocity vanishes), as

$$\begin{aligned} f(\mathbf{v}) &= A e^{-mv_x^2/(2k_B T)} \times e^{-mv_y^2/(2k_B T)} \times e^{-mv_z^2/(2k_B T)} \\ &= A \exp[-mv^2/(2k_B T)] \end{aligned} \tag{2.48}$$

for particles of mass m, where $A = n[m/(2\pi k_B T)]^{3/2}$ to ensure the normalisation (2.46). This is the Maxwell–Boltzmann distribution.

Beware that the probability for the speed $v = (v_x^2 + v_y^2 + v_z^2)^{1/2}$ to lie in the range $[v, v+dv]$ is not $f(\mathbf{v})\,dv$ but

$$f(\mathbf{v}) \times 4\pi v^2 dv \tag{2.49}$$

since this range corresponds to a volume of velocity space equal to that of a spherical shell of radius v and width dv, that is $d^3v = 4\pi v^2 dv$. Throughout this book, the notation $f(\mathbf{v})$ (which reduces to $f(v)$ when the distribution is isotropic) denotes the distribution defined by (2.43).

With the Maxwellian distribution (2.48), the distribution in speeds (v) thus varies as $v^2 e^{-mv^2/(2k_B T)}$, so that the most probable speed (the one at which the derivative of $v^2 e^{-mv^2/(2k_B T)}$ vanishes) is

$$v_{th} = (2k_B T/m)^{1/2}. \tag{2.50}$$

---

[13]We consider below non-relativistic motions. In the relativistic case, one represents f in terms of $\mathbf{r}$ and $\mathbf{p} = m\mathbf{v}$ instead of $\mathbf{r}$ and $\mathbf{v}$.

On the other hand, the mean square speed[14] is $\langle v^2 \rangle = 3k_BT/m$, so that, as we already noted, the average kinetic energy per particle is

$$W = m\langle v^2 \rangle/2 = 3k_BT/2. \tag{2.51}$$

Evolution of f

From the definition (2.43), the velocity distribution f is a density in the (six-dimensional) phase space $[\mathbf{r}, \mathbf{v}]$, just as n (or ρ) is the particle (or mass) density in the ordinary (three-dimensional) space of positions. Let us study how f evolves.

As t varies, the distribution f varies, while the position $\mathbf{r}$ and the velocity $\mathbf{v}$ of a particle vary as

$$d\mathbf{r} = \mathbf{v}dt \tag{2.52}$$
$$d\mathbf{v} = \mathbf{a}dt \tag{2.53}$$

where the acceleration vector is from the equation of motion (2.33)

$$\mathbf{a} = d\mathbf{v}/dt = q\left(\mathbf{E} + \mathbf{v} \times \mathbf{B}\right)/m \tag{2.54}$$

for particles of charge q, mass m and velocity $\mathbf{v}$, in the fields $\mathbf{E}$ and $\mathbf{B}$, to which must be added in general a gravitational acceleration.

Therefore, the evolution of f may be seen from two different points of view: the variation with time at a fixed position and velocity (the so-called *Eulerian* point of view), and the variation following particles in their motion (the so-called *Lagrangian* point of view). In the latter viewpoint, the variation has two origins: the time variation proper (at fixed co-ordinates $[\mathbf{r}, \mathbf{v}]$), and the variation of the co-ordinates $[\mathbf{r}, \mathbf{v}]$ themselves.

The convective derivative

A similar distinction holds in fluid mechanics. Assume for example that you wish to analyse the composition of water in a river. You may do so in two ways. You may sit on the bank, and so observe the time evolution at a fixed position; by convention, we note the variation so observed as $\partial/\partial t$. A second method is to embark on a boat that drifts following the river motion; the observed variation is then noted d/dt. Both variations are related in one dimension (x) by the fact that a quantity n is a function of x and t: $n = n(x,t)$ with $x = x_0 + V_x t$ (V_x being the fluid velocity). Hence $dn/dt = \partial n/\partial t + \partial n/\partial x \times dx/dt = \partial n/\partial t + V_x \partial n/\partial x$. In three dimensions, the time variation as observed following a fluid moving at velocity $\mathbf{V}$ is therefore

$$\frac{d}{dt} = \frac{\partial}{\partial t} + (\mathbf{V} \cdot \nabla) \quad \text{(convective derivative)} \tag{2.55}$$

with the usual notation $\mathbf{V} \cdot \nabla = V_x \partial/\partial x + V_y \partial/\partial y + V_z \partial/\partial z$.

---

[14] Defined as $\langle v^2 \rangle = \int d^3v\, v^2\, f(\mathbf{v})/n$.

Vlasov equation

A basic result of statistical mechanics is *Liouville's theorem*, a consequence of which is that *in the absence of collisions, f is invariant following the motion in the six-dimensional phase space*. In other words, $df/dt = 0$, where the derivative must be understood as a convective derivative *in the six-dimensional phase space*. Just as the convective derivative in ordinary three-dimensional space [**r**] is given by (2.55), the convective derivative in the six-dimensional phase space [**r**, **v**] is given by

$$\frac{d}{dt} = \frac{\partial}{\partial t} + \mathbf{v} \cdot \frac{\partial}{\partial \mathbf{r}} + \mathbf{a} \cdot \frac{\partial}{\partial \mathbf{v}}. \tag{2.56}$$

Substituting (2.56), $df/dt = 0$ may be written

$$\frac{\partial f}{\partial t} + \mathbf{v} \cdot \frac{\partial f}{\partial \mathbf{r}} + \mathbf{a} \cdot \frac{\partial f}{\partial \mathbf{v}} = 0 \quad \text{(Vlasov equation)}. \tag{2.57}$$

Here the acceleration **a** is given in (2.54), where in the general case, the electric and magnetic field are the mean fields produced by all the plasma particles, and one must add a gravitational acceleration if it is not negligible.

This means that in the absence of collisions, the velocity distribution (the density in the six-dimensional phase space) behaves as an incompressible (six-dimensional) fluid. Note that since $\partial \mathbf{a}/\partial \mathbf{v} = 0$ (any component of **a** is independent on the velocity along the same direction because the Lorentz force is perpendicular to the velocity),[15] we have $\mathbf{a}\partial f/\partial \mathbf{v} = \partial/\partial \mathbf{v}\,(\mathbf{a}f)$, so that (2.57) is equivalent to a *continuity equation* (in the six-dimensional phase space): $\partial f/\partial t + \partial\,(\mathbf{v}f)\,/\partial \mathbf{r} + \partial\,(\mathbf{a}f)\,/\partial \mathbf{v} = 0$.

A reminder: the continuity equation

The most basic equation of fluid mechanics is the continuity equation, which merely states the conservation of the number of particles or of the mass. It may be derived as follows. In the absence of creation or destruction of particles, the time variation of the number of particles in a *fixed* volume v is the opposite of the outward flux of particles crossing the surface Σ bounding this volume

$$\frac{\partial}{\partial t}\int_v d^3r\, n = -\int_\Sigma d\mathbf{S}\cdot n\mathbf{V} = -\int_v d^3r\,\nabla\cdot(n\mathbf{V}) \tag{2.58}$$

where the second equality has been obtained by transforming the surface integral into a volume integral by Gauss's theorem. Since this is true for any arbitrary (fixed) volume v, we have

$$\frac{\partial n}{\partial t} + \nabla\cdot(n\mathbf{V}) = 0 \quad \text{(continuity equation)}. \tag{2.59}$$

In the particular case when the fluid is incompressible, we have $dn/dt = 0$, so that, using (2.55), the continuity equation is equivalent to $\nabla\cdot\mathbf{V} = 0$.

---

[15] By $\partial \mathbf{a}/\partial \mathbf{v} = 0$, we mean $\partial a_i/\partial v_i = 0$ for $i = x, y, z$. This property also holds with a gravitational force, but not with a friction force.

Other forms of the invariance of f

There are several other equivalent ways of expressing the conservation of f along particle trajectories. Since the number of particles in a volume of (six-dimensional) phase space is conserved, and so is f, so is this volume. The total volume in phase space therefore remains constant, whatever its change of shape as the system evolves.

Another consequence of the conservation of f along particle trajectories is that the motion of charges in given electric and magnetic fields (plus possibly a gravitational field) may be calculated by expressing f in terms of the constants of motion (energy, magnetic moment, ...); this is often called *Jeans' theorem*.

One must be careful to apply these results within their limits of application. In particular, they do not hold in the following cases:

- when the number of particles is not conserved (for example because of ionisation or recombination),

- when the acceleration varies with the velocity as $\partial \mathbf{a}/\partial \mathbf{v} \neq 0$,

- when collisions act,

- for values of $\mathbf{r}$ and $\mathbf{v}$ that are not accessible along particle trajectories, given the constants of motion.

When collisions are not negligible, $df/dt \neq 0$, which produces a non-zero term $(\partial f/\partial t)_c$ on the right-hand side of (2.57). In neutral gases, collisions involve two-particle encounters producing large perturbations, and this yields the *Boltzmann equation*. In plasmas, collisions act through the accumulation of small-angle Coulomb encounters, and this yields the *Fokker–Planck equation*.

Basic illustration: effect of a force on the velocity distribution

Most textbook applications of the Vlasov equation consider waves. We study here a stationary problem: the effect of a (conservative) force on the distribution of particles.

In its simplest form, the problem may be stated as follows. We know that in equilibrium at temperature T the density of particles subjected to a force deriving from a potential ψ is proportional to $e^{-\psi/k_B T}$ – the Boltzmann factor. How is this result changed in absence of equilibrium? This problem arises when measuring particles aboard a spacecraft, and on a larger scale (with subtle differences) when calculating the distribution of particles near a planet or a star, and the production of a wind.

Let us assume for simplicity that the particle velocity distribution is isotropic (i.e. depends only on the modulus v of the velocity) at some position. An isotropic velocity distribution may be expressed as a function of the particle energy only, which is a constant of motion. To further simplify the problem, let us assume that the potential depends on one co-ordinate only, for example the distance r from an object. For a given particle, the total (conserved) energy

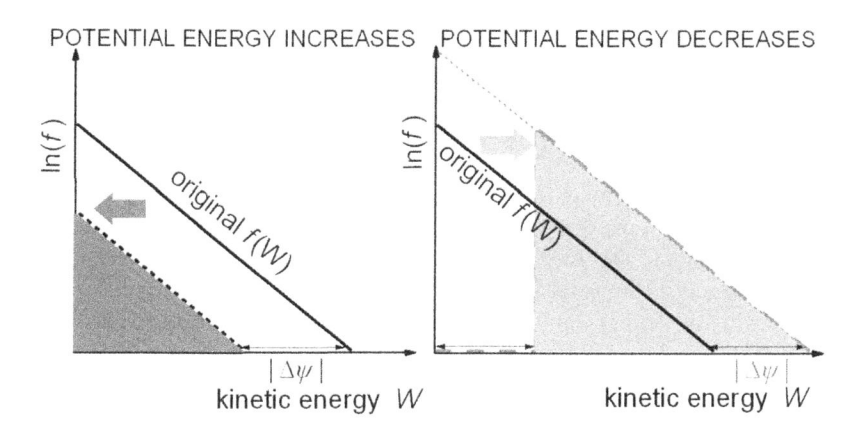

Figure 2.13 How the particle velocity distribution is modified when the potential energy increases (left) or decreases (right) by $|\Delta\psi|$, in the special case when the original distribution is a Maxwellian of temperature T. For $\Delta\psi > 0$ (particles coming against the force, left panel), the translation in energy decreases f by the Boltzmann factor $e^{-\Delta\psi/k_B T}$. For $\Delta\psi < 0$ (particles accelerated by the force, right panel) the translation in energy produces a hole in the distribution (thick dashed grey line); if collisions did act, they would fill the hole (thin dotted grey line).

is the sum of the kinetic energy W plus the potential energy $\psi(r)$. If ψ varies by $\Delta\psi$ between r_0 and r, then a particle of original kinetic energy W_0 at r_0 will have at r the kinetic energy $W_0 - \Delta\psi$, which is smaller or greater than W_0 depending on whether $\Delta\psi > 0$ (particles coming against the force) or $\Delta\psi < 0$ (particles coming in the direction of the force). From the conservation of f along particle trajectories, the velocity distribution at distance r (expressed as a function of the kinetic energy W) is thus related to the original distribution by the relation

$$f(r, W) = f(r_0, W + \Delta\psi(r)) \tag{2.60}$$

for values of $[r, W]$ accessible from r_0.

Therefore, the distribution $f(r, W)$ at some distance r is deduced from the original distribution at r_0 by a translation in energy of amplitude $\Delta\psi(r)$ – the variation in potential energy. This is sketched in Fig. 2.13 in the simple case when the original distribution is a Maxwellian of temperature T, i.e. $\propto e^{-W/k_B T}$, so that $\ln(f)$ as a function of W is originally a straight line.

The final distribution depends strongly on the sign of $\Delta\psi$. When the potential energy increases (particles coming against the force), the particle kinetic energy decreases, so that the kinetic energy distribution is translated to the left (Fig. 2.13, left). Therefore, with a Maxwellian original distribution, the particle density is reduced by the factor $e^{-\Delta\psi}$, the usual Boltzmann factor. Hence, in this case the *collisionless kinetic* description gives exactly the same result as if there were enough collisions to ensure equilibrium.

This is not so, however, when the potential energy decreases (particles moving in the direction of the force; Fig. 2.13, right). In that case, the translation in energy produces a hole in the velocity distribution, so that the density of these particles does not simply increase by the Boltzmann factor $e^{-\Delta\psi} \equiv e^{|\Delta\psi|}$ as in equilibrium. The origin of this hole is that particles of nearly zero kinetic energy at r_0 are accelerated to a kinetic energy equal to $|\Delta\psi(r)|$ at distance r, so that there is no particle of kinetic energy smaller than this value. In contrast, with collisions, the redistribution between degrees of freedom populates the hole, producing the usual Boltzmann factor.

This result illustrates the difference in Debye shielding for attracted and repelled particles in the absence of equilibrium, and has consequences on measurements of velocity distributions in space. We shall encounter a similar problem when calculating the plasma distribution in the solar corona (Section 4.6), and the solar wind acceleration (Section 5.5), with important differences: the large size of the system ensures electric quasi-neutrality; it enables collisions to populate some orbits, suppressing the hole; and the original distribution is generally not a Maxwellian.

This last point introduces a basic consequence of the rarity of collisions in space. Consider in more detail the left-hand panel of Fig. 2.13 ($\Delta\psi > 0$). The translation of the original Maxwellian distribution $f(W) \propto e^{-W/k_B T}$ yields a straight line (in log co-ordinates) having the same slope, i.e. a Maxwellian of the same temperature. But think what happens if the original distribution is not Maxwellian, a frequent situation in space. Then we no longer have a straight line, i.e. the slope (in log co-ordinates) does depend on the energy, so that the translation in energy does change the shape, thereby changing the effective temperature. Indeed, the potential filtrates the particles, letting only the fastest ones climb the potential barrier (Problem 2.5.4). This is a purely kinetic effect, completely outside the scope of the usual fluid description, and is of far-reaching consequences (Section 4.6.4).

2.3.2 First-aid kit for space plasma fluids

The infinite hierarchy of fluid equations

The simplest fluid picture describes each particle species by three macroscopic quantities:

- the particle density n defined by (2.46),
- the velocity $\mathbf{V} = \langle \mathbf{v} \rangle$ defined by (2.47),
- the pressure.

If the particle velocity distribution is isotropic in the frame where the mean velocity vanishes, the pressure is a scalar defined by

$$P = \frac{m}{3} \int d^3v \, f(\mathbf{v}) (\mathbf{v} - \langle \mathbf{v} \rangle)^2 \quad \text{(pressure)} \qquad (2.61)$$

for non-relativistic particles of mass m. The temperature is defined from $k_B T = P/n = Pm/\rho$, which, for a Maxwellian, coincides with the usual thermodynamic temperature. This generalises the definitions (2.2) and (2.3) to frames where the mean velocity does not vanish.

In this simplified picture, for a plasma containing n electrons and n (singly charged) ions per unit volume at the same temperature, the mass density is $\rho \simeq n m_i$ since the mass is essentially carried by the ions, but both species contribute equally to the pressure so that $P = 2n k_B T$. Hence $P = \rho k_B T/m$ where $m \simeq m_p/2$ is the average mass per particle.

With these three unknowns: ρ, $\mathbf{V}$ and P (or equivalently, T, since $P = \rho k_B T/m$), three equations are required to solve the problem. We have already written the first fluid equation – the continuity equation (2.59), for the particle number density n; an equivalent equation holds for the mass density ρ. The continuity equation may be obtained more formally by integrating over the velocities the equation of evolution of f; elastic collisions do not change the result since they do not change the number of particles.

In a stationary case, the continuity equation means that the mass entering a flow tube (a tube everywhere parallel to the fluid velocity) across a given section equals the mass leaving across another section. This yields

$$\rho V s = \text{constant} \qquad (2.62)$$

if s is the cross-section area. In the particular case when the medium is incompressible ($\rho =$ constant), this means that the flow lines diverge (converge) when the speed decreases (increases), or alternatively that the flow accelerates in a constricted tube – properties that are well known in hydrodynamics. In spherical symmetry, where the velocity is radial and the flow tubes vary as the square of the distance r, this yields $\rho V r^2 =$ constant.

Similarly, the fluid equation of motion may be derived in two ways. The most intuitive way is to note that for a fluid parcel of volume unity and mass ρ, the force $\rho d\mathbf{V}/dt$ (following the parcel's motion) is equal to the sum of the pressure force $-\nabla P$ and the gravity force $-\rho \nabla \Phi_G$, plus electric and magnetic forces if charges and currents are not negligible. Substituting the convective derivative (2.55), this yields the fluid equation of motion

$$\frac{\partial \mathbf{V}}{\partial t} + (\mathbf{V} \cdot \nabla) \mathbf{V} = -\frac{\nabla P}{\rho} - \nabla \Phi_G \quad \text{(fluid equation of motion)} \qquad (2.63)$$

(where we have omitted for the moment the electromagnetic force and also the viscosity force, which will be considered later). This equation of motion may be obtained more formally by multiplying the equation of evolution of f by $\mathbf{V}$ and integrating over the velocities. With a velocity distribution that is isotropic in the frame where the mean velocity vanishes, this yields (2.63), to which must be added electric and magnetic forces if there is a finite density of charge and current.

An important problem emerges, which is perhaps one of the most difficult problems of space plasma physics. With the continuity equation and the fluid equation of motion, we have only two equations for the three unknowns ρ, $\mathbf{V}$ and P (or equivalently, T, since $P = \rho k_B T/m$). One might think naively that this problem could be solved by going a step further in the averaging process. Indeed, the continuity equation stems from averaging the equation on f, the equation of motion stems from averaging the equation on f multiplied by $\mathbf{V}$, and similarly another equation (involving the energy) stems from averaging the equation on f multiplied by $\mathbf{V}^2$. Unfortunately this does not work, because just as the continuity equation determines ρ in terms of $\mathbf{V}$, the equation of motion (2.63) determines $\mathbf{V}$ in terms of ρ and P (or T), a third equation will involve a moment of higher order, in terms of which P (or T) will be expressed, and so on up to an infinite number of moments.

The fluid equations therefore constitute a ladder having an infinite number of steps. This is not surprising since one cannot replace an infinite number of unknowns (which a velocity distribution represents effectively in the general case) by a finite number of unknowns – for example ρ, $\mathbf{V}$ and P – without a miracle. The root of the problem is the transport of energy, which we shall examine below. Meanwhile, let us examine some cases when the miracle comes true.

Bernoulli's theorem

Let us make two assumptions:

- the problem is stationary, so that $\partial/\partial t = 0$,

- P is a function of ρ only (for example P and ρ obey a relation $P \propto \rho^\gamma$), or the fluid is incompressible.

In this case, we can define the enthalpy

$$H = \int dP/\rho \quad \text{(enthalpy)}. \tag{2.64}$$

We deduce $\nabla P/\rho = \nabla H$, so that multiplying the equation of motion (2.63) by $\mathbf{V}$, we obtain

$$\mathbf{V} \cdot \nabla \left[V^2/2 + H + \Phi_G\right] = 0. \tag{2.65}$$

This means that we have along flow lines

$$V^2/2 + H + \Phi_G = \text{constant} \quad \text{(Bernoulli's theorem)}. \tag{2.66}$$

Bernoulli's theorem is a pillar of fluid dynamics. It may be used to solve a host of problems, from domestic plumbing to astrophysics, including how wings of aeroplanes, insects and birds produce lifts (or crashes), how termites and prairie dogs design the ventilation of their homes,[16] or why your shower curtain engulfs you every morning.[17]

---

[16] See for example Vogel, S. 1998, *Cats' Paws and Catapults*, London, Penguin Books.
[17] The latter explanation is still under debate.

Let us calculate the enthalpy. Consider first an incompressible medium; from $P = \rho k_B T/m$ with $\rho =$ constant, we have $H = P/\rho = k_B T/m$.

Then consider the polytrope case $P \propto \rho^\gamma$. The case $\gamma = 1$ must be considered separately; we then have $T =$ constant, so that from $P = \rho k_B T/m$

$$H = \int dP/\rho = (k_B T/m) \ln \rho \qquad \text{(isothermal)}. \qquad (2.67)$$

On the other hand, if $P \propto \rho^\gamma$ with $\gamma \neq 1$, we have $P \propto T^{\gamma/(\gamma-1)}$, so that

$$H = \int dP/\rho = \frac{\gamma}{\gamma - 1} \int \frac{P}{\rho} \frac{dT}{T} = \frac{\gamma}{\gamma - 1} \frac{k_B T}{m}. \qquad (2.68)$$

With these values of H, the Bernoulli theorem (2.66) yields along flow lines, for a polytrope $P \propto \rho^\gamma$:

$$\frac{V^2}{2} + \frac{k_B T}{m} \ln \rho + \Phi_G = \text{constant} \qquad \gamma = 1 \qquad (2.69)$$

$$\frac{V^2}{2} + \frac{\gamma}{\gamma - 1} \frac{k_B T}{m} + \Phi_G = \text{constant} \qquad \gamma \neq 1. \qquad (2.70)$$

Equation (2.69) holds when transformations are so slow that *isothermal* equilibrium has enough time to establish everywhere.

On the other hand, (2.70) holds in the opposite case when transformations are so fast that heat has no time to flow: such processes are called *adiabatic*; in that case, $\gamma = c_p/c_v = 1 + 2/N$ – the ratio of specific heats for (non-relativistic) particles having N space degrees of freedom, so that $\gamma = 5/3$ for $N = 3$. Bernoulli's theorem then represents the conservation of the fluid energy per unit mass, which is the sum of the bulk kinetic energy $V^2/2$, plus the thermal energy $3k_B T/2$, plus the work $k_B T$ expended on compression, plus the gravitational energy Φ_G. Beware that (2.70) holds in the stationary case. In a time-dependent case, the adiabatic fluid energy equation simply reads $d(P\rho^{-\gamma})/dt = 0$ with $P = \rho k_B T/m$ and d/dt the convective derivative. With the help of the continuity equation and the equation of motion, it may be put under various (more or less complicated) forms. These three equations are known as the *Euler equations*.

The same result may be obtained formally by assuming the particle velocity distribution to be a Maxwellian of temperature T centred on a mean velocity $\mathbf{V}$, and making averages of the equation of evolution of f multiplied by $\mathbf{V}^2$. Since such a Maxwellian distribution is characterised by only three parameters – the density, the mean velocity and the temperature (or the pressure) – no miracle is needed to reduce its evolution to three equations involving these three unknowns. The establishment of a Maxwellian distribution, however, generally requires some process – such as collisions between particles – to ensure equilibrium, and is not so easily achieved in plasmas as in neutral gases, due to the nature of collisions; we shall return to this point later. Note that the relation $P \propto \rho^\gamma$ further reduces the number of independent unknowns to two, so that the equations of motion and of energy then become equivalent.

Sound waves and their plasma counterparts

Assume that the only force is the pressure force, and consider small perturbations around the simple solution having the velocity $\mathbf{V}_0 = 0$ and uniform mass density ρ_0 and temperature T_0 (and pressure $P_0 = \rho_0 k_B T_0/m$). Assume the perturbations to be fast enough for the behaviour to be adiabatic, i.e. $P \propto \rho^\gamma$, so that we have $\nabla P = (dP/d\rho) \nabla \rho$ with

$$dP/d\rho = \gamma P/\rho = \gamma k_B T/m. \tag{2.71}$$

Let us now write the continuity equation and the fluid equation of motion with the perturbed quantities $\mathbf{V} = \mathbf{V}_1$, $\rho = \rho_0 + \rho_1$, $P = P_0 + P_1$, where the subscript 1 denotes small perturbations of the initial solution. To first order in the perturbation, this yields

$$\frac{\partial \rho_1}{\partial t} + \rho_0 \nabla \cdot \mathbf{V}_1 = 0 \tag{2.72}$$

$$\rho_0 \frac{\partial \mathbf{V}_1}{\partial t} = -\nabla P_1 = -\frac{dP}{d\rho} \nabla \rho_1. \tag{2.73}$$

Taking the time derivative of (2.72) and substituting (2.73), we obtain

$$\frac{\partial^2 \rho_1}{\partial t^2} = V_S^2 \nabla^2 \rho_1 \tag{2.74}$$

$$V_S = \left(\frac{dP}{d\rho}\right)^{1/2} \tag{2.75}$$

and equations similar to (2.74) for the perturbations P_1 and $\mathbf{V}_1$. This has a plane wave solution varying as $e^{i(\mathbf{k}\cdot\mathbf{r}-\omega t)}$, which propagates at the speed $\omega/k = V_S$. With $P \propto \rho^\gamma$, we have from (2.71)

$$V_S = (\gamma P/\rho)^{1/2}. \tag{2.76}$$

This wave produces small perturbations in the fluid mass density and velocity; it is a *sound wave*, propagating at the *sound speed* $V_S = (\gamma k_B T/m)^{1/2}$ in a gas of particles of mass m at temperature T. Note that from (2.73), $\mathbf{V}_1$ is parallel to $\nabla \rho_1$. In the Fourier space $[\omega, \mathbf{k}]$ (see [3]), $\partial/\partial t$ transforms into $-i\omega$ and ∇ into $i\mathbf{k}$, so that (2.73) yields $-i\omega \rho_0 \mathbf{V}_1 = -V_S^2 i \mathbf{k} \rho_1$, whence $\mathbf{V}_1 \parallel \mathbf{k}$. This shows that in a sound wave, the velocity perturbation is parallel to the wave vector: this is called a *longitudinal wave*.

In a plasma, two major differences arise. First, the pressure is provided by electrons and ions in proportion of their temperatures, whereas the mass is essentially provided by the ions. The corresponding wave – called *ion–acoustic wave* – behaves differently depending on whether or not the electrons and ions move together, i.e. whether or not the plasma remains neutral.

If the plasma remains neutral, which holds at scales greater than L_D (i.e. wave numbers $k \ll 1/L_D$), the wave is a simple generalisation of the sound wave in a neutral gas, so that the phase speed is

$$V_S = \left(\frac{\gamma_e P_e + \gamma_i P_i}{\rho}\right)^{1/2} = \left(\frac{\gamma_e k_B T_e + \gamma_i k_B T_i}{m_i}\right)^{1/2} \tag{2.77}$$

where the subscripts e and i refer to electrons and ions respectively. If $\gamma_e \sim \gamma_i \sim \gamma$ and $T_e \sim T_i \sim T$, the phase speed reduces to $(\gamma k_B T/m)^{1/2}$ (in order of magnitude) with $m \simeq m_i/2$ (the average particle mass), as for a neutral gas. In this case, however, since the wave phase speed is of the same order of magnitude as the most probable speed of ions, the bulk of the ions move together with the wave so that they are subjected to a nearly constant force along their motion; this accelerates them efficiently at the expense of the wave, which is thus damped; this process is called *Landau damping*, and we shall return to it in Section 2.3.4. For the wave not to be damped, the electrons must be much hotter than the ions, to produce a wave speed much greater than the ion most probable speed so that the wave is no longer damped.

When the wave number $k \geq 1/L_D$, the plasma does not remain neutral, i.e. the electrons and ions do not move together; in the large k limit, the ions then perform plasma oscillations at their characteristic frequency

$$\omega_{pi} = \left(ne^2/\epsilon_0 m_i\right)^{1/2}. \tag{2.78}$$

The picture is then like ordinary plasma oscillations (Section 2.1.4), but with the role of electrons and ions reversed.

The second major modification that plasmas introduce in the sound wave arises when a magnetic field is present. We shall consider this point later.

Shocks

Sound waves (and their plasma generalisations) enable fluids to adapt gently to compressions. Hence, for motions at a speed smaller than the sound speed, the fluid behaves as if it were roughly incompressible. On the other hand, for larger speeds, nasty things may happen. To understand this, suppose you agitate your hand. In doing so, you compress the surrounding air, and your hand is able to move because the gas ahead goes out of the way. It can do so because the compression is transmitted farther away by the sound waves emitted by your moving hand, thereby transmitting to the gas ahead the information that your hand is approaching.

But suppose you try to move it faster than the sound speed. In that case, sound waves do not propagate fast enough to transmit the information that your hand is moving. Hence the gas far ahead, being not aware of the motion, is not perturbed. In contrast, close to your hand, the gas is compressed by the motion and moves at the same speed; the (adiabatic) compression also increases the local temperature (and sound speed), so that the information propagates just ahead of your moving hand.

Therefore the gas separates into two regions. Far away, it remains undisturbed; in the frame of the moving object that is approaching at the speed $-V_1$,

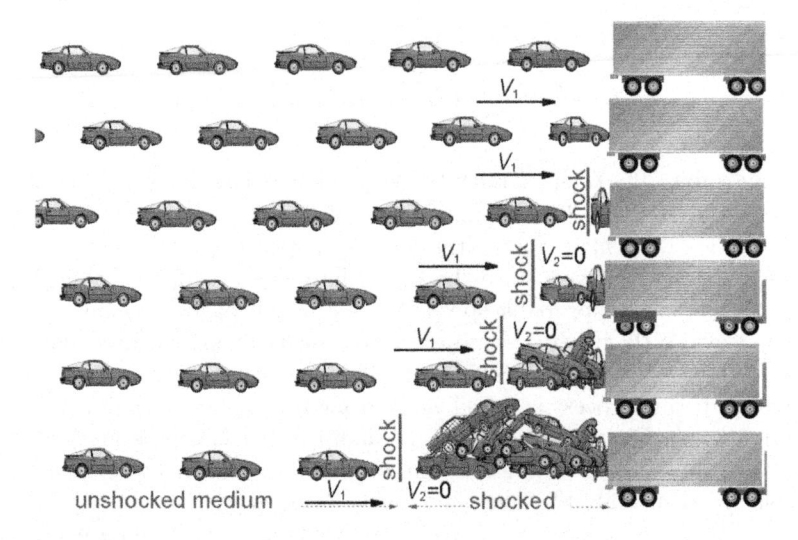

Figure 2.14 A simple shock: moving cars, whose drivers are asleep, are impacting a stopped truck. The rows show the state of the system at several consecutive times. A shock forms, separating two different states: upstream, the unperturbed fluid (the row of cars) moving at uniform velocity; downstream, the fluid that has stopped and undergone an irreversible transition. The shock moves towards the left, propagating information on the presence of the obstacle.

the gas moves at V_1, faster than the local sound speed. On the other hand, just before the object, the flow adapts gently, moving at a speed that is locally subsonic, and stopping at the object. In between lies a transition, at some distance ahead of the object, where the gas velocity changes from supersonic to subsonic. It is this transition that transmits the information on the presence of the obstacle ahead, and it does so at a supersonic speed – a performance that the small amplitude sound waves cannot achieve.

Contrary to the sound waves, this transition – called a *shock* – is not a reversible process, and it can transmit information faster than the sound speed. The irreversibility involves some dissipation, which, in the usual case of neutral gases, can be achieved via collisions between particles; in this case, the width of the transition is thus the scale at which the ideal fluid equations no longer hold, that is the mean free path of particles for collisions.

We have defined a shock as a large amplitude irreversible perturbation enabling propagation of information faster than the small amplitude compressible waves. An extreme case arises when no such waves do exist. Figure 2.14 illustrates such an example, that I have borrowed from [2], where an insightful introduction to shocks in space may be found. Imagine a stream of equally spaced vehicles on a straight freeway. Now assume that the drivers have fallen asleep, with the speeds of the cars somehow blocked at their original speed V_1, whereas a large truck suddenly stops in the middle of the lane. The system evolves as

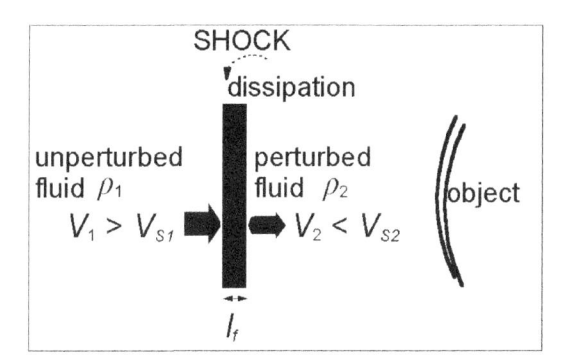

Figure 2.15 A plane shock before a moving object, in the reference frame where the shock (and the object) is at rest. The unperturbed fluid upstream (left) of uniform mass density ρ_1 moves at the supersonic speed $V_1 > V_{S1}$ (the upstream sound speed). At the shock, these properties change abruptly to ρ_2, $V_2 < V_{S2}$. Between the shock and the object, the fluid slows down smoothly, stopping at the object, and being diverted sideways (not shown). The width of the transition is about the scale l_f at which the ideal fluid equations no longer hold.

shown on the successive rows, from top to bottom. The accumulation of crashes quickly produces two states separated by an abrupt transition: a shock. Upstream, the medium (the regular row of cars) is unperturbed, being unaware of the obstacle ahead. Downstream accumulates a hump of crashed cars. The transition – a shock – moves to the left as more vehicles stop and crash.

This example is a limiting case when there is no small amplitude (reversible) wave propagating information, but it has several basic properties of shocks. The unperturbed medium is supersonic in the sense that it moves faster than the velocity at which information propagates (which is zero here), and its motion becomes abruptly subsonic (zero here) at the shock, where it undergoes an irreversible dissipative transition. The shock is the only way that information can propagate, and it does so at a speed that is effectively supersonic (in the above sense), and is determined by the initial speed of the vehicles, the separation between them and their compression upon crashing.

A more general case is sketched in Fig. 2.15, which is drawn in the frame where the shock (and the object ahead of which it lies) is at rest. Upstream (left), the unperturbed medium has the supersonic speed $V_1 > V_{S1}$ (the sound speed). At the shock the medium undergoes an abrupt dissipative transition, becoming subsonic, of speed $V_2 < V_{S2}$ (the sound speed just downstream of the shock). On this downstream side, the subsonic velocity enables the information on the presence of the obstacle to propagate, so that the speed can decrease smoothly up to the surface of the object where it vanishes. Furthermore, the stream lines are diverted sideways; this is not shown in Fig. 2.15, which is one-dimensional.[18]

---

[18] Namely, the radius of curvature of the shock (and of the object) is assumed to be much larger than the scales shown.

The fluid properties on both sides of the shock are related by the three ideal fluid equations. First, mass conservation tells us that the mass flux ρV is the same on both sides of the transition. Second, the equation of motion tells us that the variation in the flux of momentum ρV^2 between both sides balances the variation in pressure P. Finally, (adiabatic) energy conservation tells us that the sum of the density of kinetic energy $V^2/2$, plus the enthalpy $5P/(2\rho)$ (with $\gamma = 5/3$) is the same on both sides. This yields

$$\rho_1 V_1 = \rho_2 V_2 \tag{2.79}$$
$$\rho_1 V_1^2 + P_1 = \rho_2 V_2^2 + P_2 \tag{2.80}$$
$$\frac{V_1^2}{2} + \frac{5P_1}{2\rho_1} = \frac{V_2^2}{2} + \frac{5P_2}{2\rho_2} \tag{2.81}$$

where the subscripts 1 and 2 refer respectively to the values upstream and just downstream of the shock. These are called the *Rankine–Hugoniot relations*. The decrease in speed from upstream to downstream is accompanied by an increase in density, pressure and temperature. The Mach number $M = V/V_S$, with the sound speed $V_S = (\gamma P/\rho)^{1/2} = (\gamma k_B T/m)^{1/2}$, changes at the shock from $M_1 > 1$ to $M_2 < 1$, with from (2.79)–(2.81):

$$\frac{V_2}{V_1} = \frac{\rho_1}{\rho_2} = \frac{1 + 3/M_1^2}{4}. \tag{2.82}$$

In the particular case when $V_1 \gg V_{S1}$, we have

$$V_2/V_1 = \rho_1/\rho_2 \simeq 1/4 \tag{2.83}$$
$$V_2/V_{S2} \simeq \sqrt{1/5} \tag{2.84}$$
$$k_B T_2 \simeq m V_1^2/5. \tag{2.85}$$

In that case, one sees that most of the upstream kinetic energy is converted into downstream enthalpy, so that the downstream temperature can be very large.

This holds for neutral gases. In space plasmas, some complications arise, requiring tools that we have not yet introduced. We shall consider these complications when studying shocks in the solar wind (in Section 6.3), and its interaction with solar system objects (Section 7.2) and with the interstellar medium (Section 8.2).

Transport of momentum and heat

We now go a step further into the theory and examine how the ideal fluid equations are modified when the fluid is neither isothermal nor adiabatic. Or, from a microscopic point of view, when the particle velocity distribution is not Maxwellian.

Being not Maxwellian has two main consequences.

The first one concerns the property of isotropy, and is relatively trivial. In the presence of a magnetic field **B**, the velocity distribution tends to acquire a cylindrical symmetry around **B**, so that the pressure is no longer a scalar, even to zero order, being different in the directions parallel and perpendicular to **B**. This introduces a complication, but does not present basic difficulties if the velocity distribution remains Maxwellian in the parallel and perpendicular directions; it is then called *a bi-Maxwellian*[19] and defined by two temperatures: $T_\parallel$ and $T_\perp$.

In this case, the (isotropic) adiabatic relation $P \propto \rho^\gamma$ may be generalised in the following way. Just as, for individual particle motions, the ratio of the particle perpendicular energy to the magnetic field $mv_\perp^2/2B$ is an adiabatic invariant, so the ratio of the fluid perpendicular temperature to the magnetic field $T_\perp/B = $ constant. Similarly, we know that the adiabatic invariant of the particle bounce motion is the product of the parallel velocity $v_\parallel$ by the length L of the bounce path. Let us apply this invariance to a magnetic tube of length L and section s; conservation of mass yields $\rho L s = $ constant, and conservation of magnetic flux yields $Bs = $ constant. Hence $\rho L/B = $ constant, so that the invariance of $v_\parallel L$ is equivalent to $v_\parallel B/\rho = $ constant. Finally, therefore, for an anisotropic Maxwellian distribution, the adiabatic isotropic relation $P \propto \rho^\gamma$ is replaced by the so-called *CGL relations*[20]

$$T_\perp \propto B \tag{2.86}$$

$$T_\parallel \propto (\rho/B)^2. \tag{2.87}$$

Unfortunately, this simple scheme does not hold when the velocity distribution is not close to an anisotropic Maxwellian. And still worse, the conditions for the distribution to be an anisotropic Maxwellian are difficult to realise: there must be enough collisions to produce Maxwellians in both the parallel and perpendicular directions, but not so many that the parallel and perpendicular temperatures become equal. Furthermore, even though the gyration around the magnetic field comes to the rescue of collisions for providing quasi-equilibrium in the direction $\perp$ **B**, this is not so in the direction $\parallel$ **B**. Therefore, the velocity distributions in space plasmas are generally not bi-Maxwellian, except if the free path for collisions is much smaller than the scale of variation, at least in the direction $\parallel$ **B**.

We now come to the second consequence of not being Maxwellian. In ordinary gases, small perturbations to the Maxwellian are studied by performing

---

[19]The bi-Maxwellian distribution has the form $f(\mathbf{v}) = A e^{-mv_\parallel^2/2k_B T_\parallel} \times e^{-mv_\perp^2/2k_B T_\perp}$, with $A = n(m/2\pi k_B)^{3/2} T_\parallel^{-1/2} T_\perp^{-1}$, in order to ensure the normalisation $n = \int_{-\infty}^{+\infty} dv_\parallel \int_0^\infty 2\pi v_\perp dv_\perp f(\mathbf{v})$; $v_\parallel$ and $v_\perp$ are the velocity components in the directions respectively $\parallel$ **B** and $\perp$ **B**.

[20]When B is constant, the CGL relation (2.87) yields $T_\parallel \propto \rho^2$, so that the parallel pressure $P_\parallel \propto \rho T_\parallel \propto \rho^\gamma$ with $\gamma = 3$. Since for particles having N degrees of freedom, the adiabatic index $\gamma = 1 + 2/N$, (2.87) then corresponds to an adiabatic compression with 1 degree of freedom. This is not surprising since with constant B, the magnetic flux tube must keep a constant section, and thus can only be compressed (or expanded) along **B**.

an expansion in a small parameter: the ratio of the mean free path l_f to the scale L of variation of the medium, which is related to the extent by which the velocity distribution differs from a Maxwellian. Such an expansion yields the ideal fluid equations, plus small corrections representing a viscosity force and a heat flux. The equation of motion and the Bernoulli theorem are then replaced by

$$\frac{\partial \mathbf{V}}{\partial t} + (\mathbf{V} \cdot \nabla) \mathbf{V} = -\frac{\nabla P}{\rho} - \nabla \Phi_G + \mathbf{F}_{vis} \quad (2.88)$$

$$\rho \mathbf{V} \cdot \nabla \left(\frac{V^2}{2} + \frac{\gamma}{\gamma - 1} \frac{k_B T}{m} + \Phi_G \right) = -\nabla \cdot \mathbf{Q}. \quad (2.89)$$

The viscosity force $\mathbf{F}_{vis}$ tends to reduce the gradients in velocity. It may be written approximately

$$\mathbf{F}_{vis} \simeq \nu \nabla^2 \mathbf{V} + \nu' \nabla (\nabla \cdot \mathbf{V}) \quad (2.90)$$

where ν is the *kinematic viscosity*, and the second term vanishes if the medium is incompressible. The heat flux tends to reduce the gradients in temperature, and may be written (in this nearly Maxwellian approximation)

$$\mathbf{Q} = -\kappa \nabla T \quad (2.91)$$

where κ is the thermal conductivity. These transport terms are produced by the particle agitation and collisions which enable them to share their momentum and energy. The transport terms therefore increase with the gradients, the random speeds and the free path for collisions. We make below a simplified estimate.

The motion of the particles may be viewed as a random walk at the speed v_{th}, with individual random steps of length equal to the collision free path l_f. The average distance travelled in this way is zero, but the mean square is not, being $\langle d^2 \rangle = p \times l_f^2$ for p random steps. Travelling a distance $L \gg l_f$ therefore requires a number of steps given by $L^2 = p \times l_f^2$ and therefore a time $\tau \sim p \times l_f / v_{th}$, i.e.

$$\tau \sim \frac{L^2}{v_{th} l_f} \qquad l_f \ll L. \quad (2.92)$$

This enables us to estimate the coefficient of viscosity and the thermal conductivity. Consider first the equation of motion (2.88), and assume that the main contributions come from the time variation and the shear viscosity, so that in order of magnitude $\partial \mathbf{V}/\partial t \sim \nu \nabla^2 \mathbf{V}$. For a velocity varying at the scale L, we have $|\nabla^2 \mathbf{V}| \sim V/L^2$, so that $\partial \mathbf{V}/\partial t \sim \nu V/L^2$. This means that the viscosity can suppress a velocity variation of scale L in a time $\tau \sim L^2/\nu$. Since this is achieved by the diffusion of particles, which diffuse over a distance L in a time given by (2.92), both times are equal, so that

$$\nu \sim v_{th} l_f \qquad l_f \ll L. \quad (2.93)$$

The importance of viscosity is quantified by the *Reynolds number*, which represents the ratio of the inertial term $|(\mathbf{V} \cdot \nabla) \mathbf{V}|$ in the fluid equation of motion (2.88) to the viscosity force. If the velocity varies at the scale L, we have $|(\mathbf{V} \cdot \nabla)\mathbf{V}| \sim V^2/L$, while the viscosity force (2.90) is roughly $F_{vis} \sim \nu V/L^2$. The ratio is therefore

$$\frac{\text{inertia}}{\text{viscosity}} \sim VL/\nu \equiv R \quad \text{(Reynolds number)}. \quad (2.94)$$

With the expression (2.93) of the viscosity, we have $R \sim (V/v_{th}) \times (L/l_f)$, so that if $V > v_{th}$, the Reynolds number must be much greater than unity to ensure $l_f \ll L$ so that the fluid equation of motion (2.88) holds. Therefore, viscous forces are generally much less important than inertial effects.[21] This is still more true in space and astronomy, because of the extremely large scales.

Consider now the energy equation (2.89). This equation holds in the simple case when there is no time variation, so let us consider another simple case: when the fluid velocity is much smaller than the sound speed so that the dynamical terms are negligible (and the medium behaves as nearly incompressible). In that case, the divergence of the heat flux simply balances the variation in the density of kinetic energy per unit time, so that the energy equation becomes

$$n \frac{\partial}{\partial t} \left(\frac{3}{2} k_B T \right) = \kappa \nabla^2 T \quad \text{for} \quad V \ll V_S. \quad (2.95)$$

With the order of magnitude estimate $\nabla^2 T \sim T/L^2$ where L is the scale of variation, (2.95) shows that the heat flux makes the temperature diffuse over a distance L in a time $\tau \sim 3nk_B L^2/(2\kappa)$. Since we have seen that particles diffuse over a distance L in a time given by (2.92), the thermal conductivity is

$$\kappa \sim \frac{3}{2} n k_B v_{th} l_f \quad l_f \ll L. \quad (2.96)$$

Despite the simplicity of our approach, the above estimates of the viscosity and of the thermal conductivity turn out to be accurate to a factor of order unity in a collisional medium.

Transport in plasmas

How do these results apply in space plasmas? Because of the large ion-to-electron mass ratio, whereas ions and electrons generally have similar temperatures, ions have a much greater kinetic momentum mv_{th} than electrons, but a much smaller kinetic speed v_{th}, whereas the free paths are similar. Hence:

- ions transport momentum, and determine the viscosity,

- electrons transport heat, and determine the thermal conductivity.

---

[21] Beware that the viscosity force, however small quantitatively, may have important qualitative consequences, as we shall see in Section 6.4. Furthermore, we must be careful not to rely too heavily on our intuition, which is based on a familiarity with high Reynolds numbers. This point is nicely addressed by Purcell, E. M. 1977, *Am. J. Phys.* **45** 3.

Because the Reynolds number is generally very large in space plasmas, viscosity is in general negligible.

This is not so, however, for the heat flux. Consider the energy equation (2.89), and compare the transport of heat by thermal conduction, $\nabla \cdot \mathbf{Q}$, to the transport of heat by the fluid bulk motion, $\propto \rho \mathbf{V} \cdot \nabla k_B T / m$. For variations at the scale L, we make the order of magnitude estimate $\nabla \sim 1/L$, which yields

$$\frac{\text{conduction}}{\text{advection}} \sim \frac{\kappa T / L^2}{\rho V k_B T / (mL)} \sim \frac{v_{th} l_f}{VL} \tag{2.97}$$

where we have substituted the expression (2.96) of κ and $n = \rho/m$. If the same particle species did produce the viscosity and the heat conductivity, this ratio would be roughly the inverse of the Reynolds number, and would thus be very small. However, heat conduction in plasmas is provided by the electrons, so that the speed in (2.97) is that of electrons, i.e. much greater than that of ions. Furthermore, even though space plasmas have often a bulk motion faster than the ions' most probable speed, the bulk motion is generally slower than the most probable speed of electrons. Hence, heat conductivity is often important in plasmas, even when $l_f \ll L$.

Let us estimate the thermal conductivity in a (collisional) plasma. Substituting the numerical values in (2.96), with $v_{th} = v_{the}$ and the mean free path (2.22), we have

$$\kappa \simeq \frac{10^{-10}}{\ln 1/\Gamma} \times T^{5/2} \ \text{W m}^{-1} \text{K}^{-1} \tag{2.98}$$

in SI units, where T is the electron temperature. In space plasmas, we have $\ln 1/\Gamma \sim 10\text{--}20$, so that in order of magnitude $\kappa \sim 10^{-11} \times T^{5/2}$ W m^{-1}K^{-1} (SI units). For example, the solar corona, with $T \sim 10^6$ K, has a heat conductivity $\kappa \simeq 10^4$ W m^{-1}K^{-1}, of the same order of magnitude as the heat conductivity of brass [5].

Beware of fluid equations in space plasmas

These results, however, must be applied with extreme caution in space plasmas, for two reasons. First, the magnetic field has been neglected. The magnetic field does not affect the thermal conductivity along its direction. However, since in general the particle gyroradii are much smaller than the free paths, the conductivity is strongly reduced in the direction $\perp \mathbf{B}$. Second, even in the direction $\parallel \mathbf{B}$, the thermal conductivity (2.96) only holds when the free path is much smaller than the scales of variation. How much smaller? This question is still not fully solved, and we shall return to it in Sections 4.6 (in the context of the solar corona) and 5.5 (in the context of the solar wind). In practice, values of l_f/L as small as about 10^{-3} are required. The basic reason is the fast increase of the particle free path with speed. The free path l_f given by (2.22) is the value for particles of speed equal to the most probable speed. Since for a particle of velocity v, the free path $\propto v^4$, particles moving, say, three times faster have a

Figure 2.16 The basic difference between the fluid (left) and the kinetic (right) picture, and why the former is often inappropriate in space plasmas. The fluid picture implies a process that enables the particles to transport heat in bulk. However, heat is mainly carried by the particles moving faster than average, which are nearly collisionless and require a kinetic description. (Drawing by F. Meyer.)

free path greater by a factor of $3^4 \sim 10^2$. And since heat is transported by the faster particles, these fast particles must have a small free path for (2.96) to hold.

There lies the difficulty of applying fluid equations in space plasmas. The particles that transport heat are those for which the classical theory of heat transport does not apply! Stated in more precise terms, the fast variation of the free path with speed prevents the expansion of the moment equations in terms of the small parameter l_f/L to converge. Even though the fluid continuity equation and equation of motion do hold (provided the particle pressure is roughly a scalar, and including the electromagnetic contribution if it is not negligible), their solution requires another (local) fluid equation: the energy equation, which generally does not hold because the heat flux is not a simple function of the local derivatives. Another way of understanding this point is to note that for the heat flux to depend on a local derivative, there must be a process – such as collisions – that effectively localises the particles (Fig. 2.16).[22]

It is often said that the fluid picture nevertheless applies because the particles are localised by the gyration around the magnetic field and by the various plasma instabilities. However, the gyration of the particles only localises them in the direction perpendicular to the magnetic field, and the instabilities drive them towards stable configurations, which do not necessarily correspond to local thermal equilibrium.

2.3.3 Elements of magnetohydrodynamics

So far, we have ignored the contribution of the electromagnetic force to the fluid motion. This is permissible if the electric charge and currents carried by

---
[22]Adapted from Meyer-Vernet, N. 2001, *Planet. Space Sci.* **49** 247.

the plasma particles are vanishing or negligible. We have seen that the large-scale electric charge is generally negligible, but this is not necessarily so for the current. And since currents produce magnetic fields, which themselves act on the currents, the particles and the magnetic field are closely coupled. This confers special properties to the medium, which are generally studied with a fluid description. This is the subject of *magnetohydrodynamics* (MHD).

We neglect the charge separation, which is permissible at timescales $T \gg 1/\omega_p$ and spatial scales $L \gg L_D$, and consider the plasma as a single fluid moving at the non-relativistic speed $V \ll c$ and carrying the electric current **J**. We also assume that the transport coefficients are similar in the directions parallel and perpendicular to **B**. In the presence of an electric current, a further transport coefficient acts: electric conductivity, which tends to reduce the gradients in electric potential.

Plasma electric conductivity

The origin of the electric current is the slight difference in bulk motion of ions and electrons in the presence of an electric field **E**. In the absence of a magnetic field, each electron is accelerated as $m_e d\mathbf{v}/dt = -e\mathbf{E}$. The ions, being more massive, are less easily accelerated, producing a slight difference $\Delta\mathbf{v}$ between the electron and ion velocities, and thus an electric current of density

$$\mathbf{J} = -ne\Delta\mathbf{v} \qquad (2.99)$$

for n electrons and n (singly charged) ions per unit volume. Because of electron–ion collisions, of frequency ν_{ei}, the electrons lose their velocity excess $\Delta\mathbf{v}$ in an average time $\Delta t \simeq 1/\nu_{ei}$. At equilibrium (which requires **E** weak enough, see Section 5.4.5), the momentum gained by an electron per second $-e\mathbf{E}$ is balanced by the momentum transferred per second to the ions, $m_e\Delta\mathbf{v}/\Delta t$, so that

$$-e\mathbf{E} = m_e\Delta\mathbf{v} \times \nu_{ei}.$$

Eliminating $\Delta\mathbf{v}$ by using (2.99), we deduce $\mathbf{J} = \left(ne^2/m_e\nu_{ei}\right)\mathbf{E}$. This may be written

$$\mathbf{J} = \sigma\mathbf{E} \qquad (2.100)$$

$$\sigma = \frac{ne^2}{m_e\nu_{ei}} \quad \text{(electric conductivity)}. \qquad (2.101)$$

Substituting $\nu_{ei} = v_{the}/l_f$ with the free path (2.23) and $\ln(1/\Gamma) \sim 10$–20, we find $\sigma \sim 3 \times 10^{-4} \times T^{3/2}$. A more exact calculation yields

$$\sigma \simeq 6 \times 10^{-4} \times T^{3/2} \ (\Omega\ \text{m})^{-1}. \qquad (2.102)$$

With a temperature $T \simeq 10^6$ K, this yields $\sigma \simeq 0.6 \times 10^6\ \Omega^{-1}\ \text{m}^{-1}$, nearly equal to the electric conductivity of mercury. Note that σ depends only on the temperature, being independent of the density. This is not surprising since with

Many particles: from kinetics to magnetohydrodynamics 87

more particles, the current increases, but the collisional losses increase in the same proportion.

An important remark is in order. The above estimate assumes the electrons to follow straight lines between two collisions, and thus neglects the effects of the magnetic field. This is permissible only in the direction $\parallel \mathbf{B}$, or if the mean free path is much smaller than the radius of gyration, so that the curvature of the trajectories may be neglected. Therefore, (2.101) represents the electric conductivity:

- in the direction $\parallel \mathbf{B}$,
- in the directions $\perp \mathbf{B}$, if $l_f \ll r_g$.

In general, the opposite inequality holds, so that the particle gyration reduces strongly the electric conductivity in the directions $\perp \mathbf{B}$.

Magnetic diffusion

Let us consider an important consequence of the electric conductivity.

For slow time variations and non-relativistic plasma bulk speeds, we may neglect the term $(1/c^2)\, \partial \mathbf{E}/\partial t$ compared to $\nabla \times \mathbf{B}$ in Maxwell's equation (2.30),[23] so that we have

$$\nabla \times \mathbf{E} = -\partial \mathbf{B}/\partial t \qquad (2.103)$$

$$\nabla \times \mathbf{B} = \mu_0 \mathbf{J} \qquad (2.104)$$

in a 'laboratory' frame $\mathcal{R}$. In the frame $\mathcal{R}'$ of a plasma moving at velocity $\mathbf{V}$ with respect to $\mathcal{R}$, the electric field is $\mathbf{E}' = \mathbf{E} + \mathbf{V} \times \mathbf{B}$, so that the electric current is $\mathbf{J} = \sigma \mathbf{E}'$, i.e.

$$\mathbf{J} = \sigma\,(\mathbf{E} + \mathbf{V} \times \mathbf{B}). \qquad (2.105)$$

This yields $\mathbf{E} = \mathbf{J}/\sigma - \mathbf{V} \times \mathbf{B}$, which we substitute into (2.104), to yield

$$\partial \mathbf{B}/\partial t = -\nabla \times (\mathbf{J}/\sigma - \mathbf{V} \times \mathbf{B}). \qquad (2.106)$$

Eliminating $\mathbf{J}$ with the help of (2.104) and using the vector identity $\nabla \times (\nabla \times \mathbf{B}) = -\nabla^2 \mathbf{B}$ (since $\nabla \cdot \mathbf{B} = 0$), we deduce

$$\frac{\partial \mathbf{B}}{\partial t} = \frac{\nabla^2 \mathbf{B}}{\mu_0 \sigma} + \nabla \times (\mathbf{V} \times \mathbf{B}). \qquad (2.107)$$

This equation contains two contributions to the magnetic field variation:

- a diffusion, produced by the conductive losses,
- a convection, produced by the plasma bulk motion.

---
[23] This requires that the timescale τ, length scale L and mean velocity V satisfy $E/\tau c^2 \ll B/L$, i.e. with $E \sim VB$, $\tau \gg LV/c^2$.

For magnetic variations at the scale L, we have $|\nabla^2 \mathbf{B}| \sim B/L^2$, so that the first term on the right-hand side of (2.107) is of order of magnitude $B/(\mu_0 \sigma L^2)$. Its contribution yields $\partial B/\partial t \sim B/(\mu_0 \sigma L^2)$, making the magnetic field vary on a timescale

$$\tau_\sigma \sim \mu_0 \sigma L^2 \qquad (2.108)$$

proportional to the square of the spatial scale of variation, as in usual diffusion processes. The second term on the right-hand side of (2.107) is produced by the bulk speed, and is of order of magnitude VB/L. Either of these two effects can be dominant, depending on the relevant time and length scales. The ratio between both terms is the non-dimensional number

$$\frac{\text{convection}}{\text{diffusion}} \sim \mu_0 \sigma L V \equiv R_m \qquad \text{(magnetic Reynolds number)}. \qquad (2.109)$$

Magnetic diffusion is therefore negligible if $R_m \gg 1$; R_m is called the *magnetic Reynolds number*, in analogy with the fluid Reynolds number whose value quantifies the importance of viscosity.

Basically, the magnetic field diffuses in a conductive medium because the electric currents produce a joule energy loss, which converts magnetic energy into heat. To understand this, consider the following order-of-magnitude estimate. The rate of energy dissipation per unit volume is $\mathbf{J} \cdot \mathbf{E} = J^2/\sigma$. From Maxwell's equation (2.104), the current corresponding to a magnetic field of scale L is of the order of magnitude: $J \sim B/\mu_0 L$, and dissipates energy at the rate $(B/\mu_0 L)^2/\sigma$ per unit volume. During the diffusion time $\tau_\sigma \sim \mu_0 \sigma L^2$, the energy dissipated per unit volume is thus $\sim B^2/\mu_0$, equal (in order of magnitude) to the initial density of magnetic energy.

There is a major difference between laboratory experiments – on which our intuition is based – and astrophysics. In the laboratory, we have $R_m < 1$ so that magnetic diffusion dominates, and diffusion acts so quickly that the electric currents are mainly determined by the electric conductivity. In astrophysics, the opposite inequality holds because of the large scales and velocities.

Frozen-in magnetic field

When the magnetic Reynolds number is so large that the electric conductivity may be considered as infinite, the induction equation (2.107) reduces to[24]

$$\frac{\partial \mathbf{B}}{\partial t} = \nabla \times (\mathbf{V} \times \mathbf{B}). \qquad (2.110)$$

Consider a closed contour drawn in the fluid, and the magnetic flux that it embraces. As the fluid moves, the contour is displaced and deformed, but one may prove from (2.110) that the flux embraced remains constant. This is known as *Alfvén's theorem*, and is picturesquely expressed by saying that the magnetic

---

[24] The vorticity field $\nabla \times \mathbf{V}$ satisfies the same equation as $\mathbf{B}$ in the limit of an infinitely large Reynolds number (no viscosity).

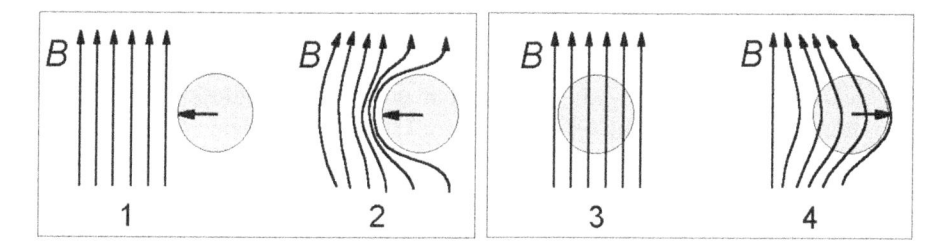

Figure 2.17 A consequence of flux freezing. If we try to introduce a piece of conductor into a magnetic field, the magnetic field lines bend away, avoiding the conductor (1, 2), until magnetic diffusion lets the magnetic field penetrate (3). Conversely, if we try to remove the bar from the magnetic field, the field lines follow the motion, remaining frozen into the conductor (4).

lines of force are 'frozen' in the fluid. The fluid can move freely along the magnetic field lines, but any motion of the fluid perpendicular to the field lines carries them with the fluid.[25]

Basically, this is because if the fluid moves across the magnetic field **B**, the motion induces an electric field of amplitude proportional to the component of the velocity $\perp$ **B**. If the conductivity is infinite, the electric field must vanish for the electric current to remain finite, and so does this velocity component.

An important consequence of magnetic flux freezing is that in a conducting fluid, one may increase the magnetic field by stretching the field lines. Indeed, consider a small magnetic flux tube of section s and length l, that is carried and deformed as the fluid moves. Alfvén's theorem tells us that the magnetic flux across the tube, which is constant along the tube at each time, remains constant too as the tube moves with the fluid, so that $Bs = $ constant. Conservation of mass yields $\rho s l = $ constant, so that $B \propto \rho l$. If the velocity is much smaller than the sound speed, the density ρ remains roughly constant, so that $B \propto l$, i.e. the magnetic field strength increases with the length of the tube. We shall see in Section 3.3 that this property has important applications in the production of the cosmic magnetic fields.

The freezing of the magnetic field in a conductor is a concept that has important consequences in astrophysics, where we have generally $R_m \gg 1$, but to which we are not accustomed in the laboratory where the opposite inequality generally holds. However, similar effects hold to a certain extent, albeit on different scales, with the conductors we encounter in ordinary life (Fig. 2.17). If we try to insert a copper bar, 0.1 m thick, say, in a magnetic field, the magnetic field lines do not penetrate it immediately, because the electric currents induced in the conductor produce a magnetic field opposing the external magnetic field.

---

[25] Saying that field lines are moving with the fluid is a way of identifying the motion of field lines rather than a statement of fact, since this motion cannot be defined unambiguously from electromagnetic theory alone; the concept of moving magnetic field lines might produce apparent paradoxes if it is applied without care.

It takes about 1 s for the currents to die away so that the external field enters the bar. Meanwhile, the field lines are deformed as shown (Fig. 2.17, left, and Problem 2.5.5). If you quickly pull the bar out of the magnetic field, electric currents are again induced, tending to trap the magnetic field within the bar. The magnetic field lines move with the bar, remaining inside for about 1 s, until the decay of the currents enables the magnetic field to disappear from the bar (Fig. 2.17, right).

This manifestation of magnetic field freezing in conductors occurs on much larger scales in astrophysics, so that the time of field decay is much larger than the timescale of motion. An important consequence is that plasmas in space, remaining tied to the magnetic field lines, do not mix easily across the magnetic field.

In the particular case when the magnetic field does not vary with time, Maxwell's equation (2.29) yields $\nabla \times \mathbf{E} = 0$, whence $\mathbf{E} = -\nabla \Phi_E$ where Φ_E is the electric potential, so that $\mathbf{E}$ is perpendicular to equipotential surfaces. With the approximation $\mathbf{E} \simeq -\mathbf{V} \times \mathbf{B}$, $\mathbf{E}$ is perpendicular to both $\mathbf{V}$ and $\mathbf{B}$, so that $\mathbf{V}$ and $\mathbf{B}$ lie on equipotential surfaces. Hence in this case, stream lines and magnetic field lines are equipotential.

Magnetic forces

Consider now the electromagnetic force that must be added in the fluid equation of motion (2.63), when the plasma electric currents are not negligible.

With a vanishing large-scale electric charge, the electromagnetic force per unit volume is $\mathbf{J} \times \mathbf{B}$. Eliminating the current density with the aid of Maxwell's equation (2.104), the force per unit volume is

$$\frac{1}{\mu_0}(\nabla \times \mathbf{B}) \times \mathbf{B} = \frac{1}{\mu_0}(\mathbf{B} \cdot \nabla)\mathbf{B} - \nabla\left(\frac{B^2}{2\mu_0}\right). \qquad (2.111)$$

This is the superposition of:

- a tension force along the field lines equal to B^2/μ_0 (per unit cross-section area normal to them),

- the gradient of a magnetic pressure equal to $B^2/2\mu_0$.

The magnetic force acting on a conducting medium may therefore be pictured in two equivalent ways. The force is the sum of the Lorentz forces $q\mathbf{v} \times \mathbf{B}$ acting on all the moving charges. Alternatively, since the charges are equivalent to a current, itself related to the magnetic field, the force can be described in terms of stresses in the magnetic field. Let us examine these stresses in more detail.

The magnetic tension arises from the curvature of the field lines, and vanishes when they are straight since in that case $(\mathbf{B} \cdot \nabla)\mathbf{B} = 0$. To understand its origin, consider the simple case of a magnetic field having a cylindrical symmetry, as the one produced by a current flowing along an axis. In this case, the magnetic field follows circles perpendicular to the axis, and depends only on the distance r from

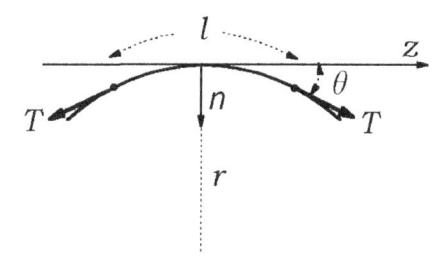

Many particles: from kinetics to magnetohydrodynamics

Figure 2.18 The magnetic tension along the field lines may be pictured as a tension force acting on an elastic wire.

this axis. The contribution of the tension to the volume force is $(\mathbf{B} \cdot \nabla)\mathbf{B}/\mu_0 = \mathbf{n}B^2/\mu_0 r$, where $\mathbf{n}$ is a unit vector pointing towards the axis (Fig. 2.18). This contribution is produced by the magnetic tension acting along the field lines, just as for a stretched string. Indeed, when a piece of string of small length l and cross-section area s is stretched with a tension force $\mathbf{T}s$, producing a radius of curvature r (Fig. 2.18), the (downward) vertical force at each extremity is $Ts \times \sin\theta \simeq Ts\theta$, whereas the net horizontal force vanishes. The net force (normal to the string) is $2Ts\theta \simeq Tsl/r$, since the length $l \simeq 2r\theta$, so that the force per unit volume is T/r. The magnetic tension $T = B^2/\mu_0$ can therefore be viewed as the tension force per unit cross-section normal to the field lines.

In the general case when the magnetic field strength varies along the field lines, the term $(\mathbf{B} \cdot \nabla)\mathbf{B}$ has also a component $\parallel \mathbf{B}$, which balances the component $\parallel \mathbf{B}$ of the gradient in magnetic pressure, so that the net magnetic force is $\perp \mathbf{B}$ (as it should be), and may be expressed as

$$\frac{B^2}{\mu_0}\frac{\mathbf{n}}{R_c} - \nabla_\perp\left(\frac{B^2}{2\mu_0}\right) \quad \text{(magnetic force per unit volume)} \quad (2.112)$$

where $\mathbf{n}$ is a unit vector pointing towards the centre of curvature, R_c is the radius of curvature and $\nabla_\perp$ denotes the component of the gradient in the plane $\perp \mathbf{B}$. The magnetic tension tends to oppose the curvature of the field lines and to shorten them, just as does the tension of an elastic string. The magnetic pressure tends to oppose the compression of the field lines and to expel the plasma from regions of high magnetic field, just as the ordinary gas pressure tends to expel matter from high pressure regions. Hence, the plasma and the magnetic field conspire to keep the plasma+magnetic pressure constant, by putting matter where the magnetic field is weak and vice versa.

Finally, therefore, the equilibrium and dynamics of a magnetised plasma are governed by three terms:

- inertia, corresponding to the density of bulk kinetic energy $\rho V^2/2$,

- thermal pressure, corresponding to the density of random kinetic energy $\sim \rho k_B T/m$,

- magnetic forces, corresponding to the density of magnetic energy $B^2/2\mu_0$.

If the first term dominates the third, the motion is not significantly affected by the magnetic field, and it controls the field lines. In contrast, if the third term dominates, the magnetic field controls the motion. Equating the first and third terms gives a speed

$$V = \frac{B}{(\mu_0 \rho)^{1/2}} \equiv V_A \quad \text{(Alfvén speed)}. \tag{2.113}$$

This is the typical speed to which the magnetic field can accelerate the plasma; we shall return to it later.

On the other hand, when the bulk velocity of the medium vanishes, the nature of the equilibrium is determined by the ratio of the second to the third terms:

$$\beta = \frac{n k_B T}{B^2 / 2\mu_0}. \tag{2.114}$$

In that case, the equilibrium is governed by the magnetic field if $\beta \ll 1$, and by the plasma if $\beta \gg 1$.

Magnetohydrodynamic waves

Just as a stretched string supports waves, in which transverse motions produced by the string's tension propagate along the string, so a magnetised plasma supports transverse waves, known as *Alfvén waves*, in which transverse motions of the field lines produced by the magnetic tension propagate along the field lines.

On a stretched string, the phase speed is $v = \sqrt{T/\rho}$, where T is the tension force per unit cross-section area, and ρ is the mass density of the string. Similarly, the phase speed of an Alfvén wave is $\sqrt{T/\rho}$, where the tension $T = B^2/\mu_0$ is produced by the magnetic field, and the mass density ρ is provided by the plasma which moves with the field lines because of flux freezing. With this analogy, the phase speed is $\sqrt{B^2/\mu_0 \rho}$, the Alfvén speed defined in (2.113), and oriented along the field lines. This result may also be found as follows.

Consider a magnetic field oriented along $\mathbf{z}$, i.e. $\mathbf{B} = B_z$, and deform the field lines as shown in Fig. 2.18. If $x(z)$ is the amplitude of the displacement normal to $\mathbf{z}$, the magnetic field component along $\mathbf{x}$ is

$$B_x = B_z dx/dz \simeq B dx/dz. \tag{2.115}$$

The magnetic tension produces a restoring force per unit volume given by the first term of (2.112) as $F_x = B^2/(\mu_0 R_c)$ where the radius of curvature $R_c = (d^2x/dz^2)^{-1}$. Since the field lines move together with the plasma of mass density ρ, this yields the equation of motion

$$\rho \frac{d^2 x}{dt^2} = \frac{B^2}{\mu_0} \frac{d^2 x}{dz^2}. \tag{2.116}$$

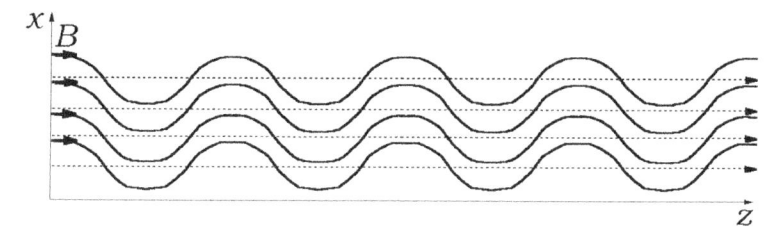

Figure 2.19 Sketch of the magnetic field lines in an Alfvén wave; the undisturbed magnetic field lines are shown as dotted lines.

There are two solutions of the form[26]

$$x \propto e^{-i(\omega t - kz)} \qquad \omega/k = \pm V_A \qquad V_A \equiv B/\sqrt{\mu_0 \rho} \qquad (2.117)$$

which are plane waves propagating along **z** at the phase speed V_A, perturbations of **B** that are $\perp \mathbf{B}$.

The material moves with the field line at right angles to the direction of propagation at the speed $v = dx/dt = -i\omega x$ (in Fourier space), so that the kinetic energy per unit volume is $\rho \mid v^2 \mid /2 = \rho \omega^2 \mid x^2 \mid /2$. From (2.115) and (2.117), the magnetic field variation in the wave is $B_x = Bikx$, so that the wave magnetic energy per unit volume is $\mid B_x^2 \mid /2\mu_0 = k^2 B^2 \mid x^2 \mid /2\mu_0$. Since $\omega/k = V_A$, both energies are equal.

The flux of energy carried by the wave is thus equally shared by the mechanic and magnetic energies, and equal to twice the flux of mechanical energy $V_A \times \rho \mid v^2 \mid$.[27]

Since the wave vector **k** is normal to the fluid velocity **v**, we have in Fourier space $\nabla \cdot \mathbf{v} = i\mathbf{k} \cdot \mathbf{v} = 0$, so that the continuity equation yields $\rho =$ constant. Hence, the mass density is not perturbed in these waves, as might be expected from Fig. 2.19, which shows that the deformation of the field lines does not change the volume of the flux tubes.

Finally, even though our calculation assumes small perturbations, these waves can exist with a large amplitude, provided the medium is incompressible and adiabatic. In this case, Alfvén waves may propagate at constant speed in a homogeneous medium without any distortion or attenuation. These waves are of great importance in astronomy, as they transport perturbations along the magnetic field over long distances.

The above calculation assumes a frozen-in magnetic field and non-relativistic speeds, which require in particular $k \ll 1/r_g$, $\omega \ll \omega_g$ and $V_A \ll c$, and consider a special geometry: perturbations of speed and magnetic field that are $\perp \mathbf{B}$ and propagate along **B**. For other directions, one finds three MHD modes: a generalisation of the Alfvén wave propagating at an angle with the magnetic

---

[26] In Fourier space, with the usual convention that the physical quantity corresponding to, say, the displacement noted x in Fourier space is equal to the real part of x.

[27] We have neglected the terms involving the electric field in Maxwell's equations, so that the electric energy is absent. This approximation is acceptable if $V_A \ll c$.

field, and two modes which, contrary to the Alfvén mode, involve some plasma compression, so that the restoring forces are the magnetic stresses plus the gas pressure gradient.

The generalised Alfvén wave has a wavefront that is not necessarily $\perp \mathbf{B}$, and propagates at the Alfvén speed V_A in the direction $\parallel \mathbf{B}$ whatever the angle θ between $\mathbf{k}$ and $\mathbf{B}$, so that $\omega/k = V_A \cos\theta$; the perturbations of velocity and magnetic field are perpendicular to both $\mathbf{k}$ and $\mathbf{B}$, so that a given magnetic field line still looks like a plucked string.

Of the two compressive modes, one propagates faster than the other, so that they are called the fast and slow *magnetosonic waves*. In the so-called *fast wave*, the particle pressure and the magnetic forces act roughly in phase, and the propagation depends strongly on the angle θ between $\mathbf{k}$ and $\mathbf{B}$:

- When $\theta \to 0$ the phase speed $\omega/k \to \max(V_A, V_S)$.

- When $\theta \to \pi/2$ the velocity perturbation is $\parallel \mathbf{k}$ just like a sound wave, with the effect of the gas pressure supplemented with that of the magnetic pressure. This is thus a longitudinal wave propagating $\perp \mathbf{B}$, in which the field lines move parallel to themselves, with alternating compressions and rarefactions of the gas and field, so that the restoring force is produced by the sum of the gas and the magnetic pressure. The phase speed may be calculated by generalising our calculation of the sound waves (2.76), as $(\gamma P/\rho + \gamma_M P_M/\rho)^{1/2}$ where $P_M = B^2/2\mu_0$ is the magnetic pressure and γ_M is the corresponding adiabatic index ($\gamma_M = 2$ for this two-dimensional compression normal to $\mathbf{B}$). Therefore, $\omega/k \to \left(V_A^2 + V_S^2\right)^{1/2}$.

The so-called *slow wave* has the restoring forces acting roughly out of phase:

- When $\theta \to 0$ the phase speed $\omega/k \to \min(V_A, V_S)$.

- When $\theta \to \pi/2$ the phase speed vanishes.

Non-ideal magnetohydrodynamics

The concept of a frozen-in magnetic field has been derived by assuming that the electric field vanishes in the plasma frame. This is not exactly so, for several reasons.

First of all, even if the conductivity is very large, it is never infinite. This point is not a mere semantic distinction because the electric resistivity, however small, makes the magnetic field diffuse on a timescale that is proportional to the square of the length scale. Diffusion thus acts very quickly if small scales arise, even with a very large conductivity. Let L be the typical scale of variation, so that $R_m \gg 1$ and the diffusion term $\nabla^2 \mathbf{B}/\mu_0 \sigma$ in (2.107) is negligible. Now suppose that some effect produces a variation at a smaller spatial scale $l \ll L$ so that $\mu_0 \sigma l V < 1$. In that case, the diffusion term $\nabla^2 \mathbf{B}/\mu_0 \sigma \sim B/\left(\mu_0 \sigma l^2\right)$

becomes larger than the motional term $\nabla \times (\mathbf{V} \times \mathbf{B}) \sim VB/l$, so that the electric conductivity is no longer negligible.[28]

With a finite electric conductivity, there is an electric field in the frame of the plasma equal to $\mathbf{J}/\sigma$. This is not, however, the only component of the electric field, because in deriving (2.100)–(2.101), we have neglected:

- the gradient of the electron pressure P_e, which produces a contribution to the electric field in the plasma frame equal to $-\nabla P_e/ne$, which balances the electron pressure force, to maintain the plasma neutral,

- the effect of the magnetic field on the electron trajectories, which produces an additional contribution to the electric field equal to $\mathbf{J} \times \mathbf{B}/ne$, in order to produce a force on the electrons that balances the Lorentz force,[29]

- the effects of the electron inertia on the current, which produces an additional contribution to the electric field of order of magnitude $VBc/(\omega_p L)$, which is therefore negligible for scales $L > c/\omega_p$.[30]

We shall see that the electron pressure, and the corresponding large-scale electric field, play an important role in the solar wind.

Another important consequence of the finite electric field is that the breakdown of the frozen magnetic field concept may produce important changes in the topology of the magnetic field. This is called magnetic *reconnection*, and is sketched in Fig. 2.20.[31] When field lines of different directions are pushed together, producing large gradients in a magnetic field, the magnetic field may disappear quickly in a small region, the magnetic energy being converted into other forms of energy, and the lines reconnect to form a new topology, so that the connectivity of plasma parcels by field lines changes. This enables the magnetic field to pass to a state of lower energy, releasing energy in producing plasma jets and high-energy particles, in addition to heating. This change in topology of the field lines enables different plasmas to mix.

Normal magnetic dissipation acts at the timescale $\tau_\sigma \sim \mu_0 \sigma L^2$. This may be compared to the collision time, which may be written, with the aid of (2.101), as

$$\frac{1}{\nu} \sim \tau_\sigma \times \left(\frac{c/\omega_p}{L}\right)^2$$

---

[28] A similar effect acts in hydrodynamics with the viscosity. In a non-viscous fluid, objects move without friction. Nevertheless, the friction force on an object of cross-section S moving at speed V in a fluid of density ρ is of order of magnitude $\rho V^2 S$ for a very large range of fluid viscosities including extremely small ones. We will return to this apparent paradox in Section 6.4.

[29] This term is called the *Hall electric field*. When it is not negligible, but the other terms are, the electric field in the laboratory frame is $\mathbf{E} = -(\mathbf{V} - \mathbf{J}/ne) \times \mathbf{B}$. If $\mathbf{V}_i$ and $\mathbf{V}_e$ are respectively the bulk speeds of ions and electrons, the current is $\mathbf{J} = ne(\mathbf{V}_i - \mathbf{V}_e)$, so that since $\mathbf{V} \simeq \mathbf{V}_i$ because the ions carry the mass, $\mathbf{E} = -\mathbf{V}_e \times \mathbf{B}$ and the frozen-in approximation holds for the electron gas, instead of holding for the plasma as a whole.

[30] The so-called *electron inertial length*.

[31] Beware that Fig. 2.20 is a simplistic view that not only ignores the three-dimensional nature of the phenomenon, but uses an MHD concept (moving field lines to which the plasma is attached) under conditions when it does not apply.

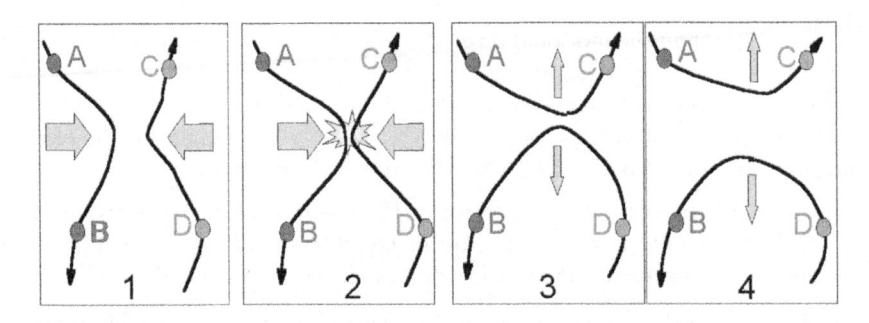

Figure 2.20 A naive view of magnetic reconnection. Two field lines of opposite directions are pushed together; the fluid parcels A and B lie on one field line, C and D lie on the other one (1). After reconnection (2), the fluid parcels A and C lie on a field line, the fluid parcels B and D lie on another one (3), which are separating from each other (4).

so that for scales greater than the electron inertial length c/ω_p (at which the electron inertia may be neglected), the collision time is smaller than the time τ_σ for magnetic resistive dissipation.

In nearly collisionless plasmas, small scales arise, which may be smaller than all the plasma characteristic scales: the gyroradii and inertial lengths for all particle species, and even the Debye length. In this case, all the fluid and MHD approximations break down, and one must describe the plasma in a kinetic way. Reconnection then acts faster than the scales τ_σ and $1/\nu$, acting instead on timescales that are typically a fraction of the *Alfvén time* (the time of displacement at the Alfvén speed)

$$\tau_A = L/V_A = (\tau_\sigma/R_M) \times (V/V_A). \tag{2.118}$$

2.3.4 Waves and instabilities

We have seen that perturbations in the magnetic field and/or the plasma pressure may drive several kinds of waves. In deriving the properties of these waves, however, we have neglected the electron inertia, and we have pictured the plasma as a fluid. Even if the unperturbed medium has Maxwellian velocity distributions, these approximations are not acceptable at frequencies equal to or greater than the plasma characteristic frequencies (the gyrofrequencies and the plasma frequency), or wavelengths smaller than the plasma characteristic scales (the Debye length, the gyroradii and inertial lengths).

Electromagnetic waves

At frequencies near the plasma frequency or greater, the inertia of the electrons is no longer negligible, but, because the ions have much greater mass, we may neglect their motion. We also neglect the electron random motion (which is acceptable if the wave phase speed is much greater than the electron most probable speed), and the ambient magnetic field (which is acceptable at frequencies much greater than the electron cyclotron frequency), and assume that the wave weakly perturbs the medium. In this case, we may assume all the electrons to acquire the same velocity **v** in the wave electromagnetic field, so that they obey the fluid equation of motion

$$m_e \partial \mathbf{v}/\partial t = -e\mathbf{E} \tag{2.119}$$

where **E** is the wave electric field; we have also neglected the Lorentz force produced by the wave magnetic field and the term $(\mathbf{v} \cdot \nabla)\mathbf{v}$ since they are of second order in the (small) perturbation. This electron motion produces a current density

$$\mathbf{J} = -ne\mathbf{v} \quad \Rightarrow \quad \frac{\partial \mathbf{J}}{\partial t} = \omega_p^2 \epsilon_0 \mathbf{E} \tag{2.120}$$

where n is the unperturbed electron density, ω_p the corresponding plasma frequency, and we have substituted (2.119).

Consider a wave satisfying $\nabla \cdot \mathbf{E} = 0$, so that from Maxwell's equation (2.29), the density of electric charge vanishes. In this case, we have $\nabla \times (\nabla \times \mathbf{E}) = -\nabla^2 \mathbf{E}$, so that Maxwell's equations (2.29) and (2.30) yield

$$\nabla \times \frac{\partial \mathbf{B}}{\partial t} = \nabla^2 \mathbf{E} = \mu_0 \frac{\partial \mathbf{J}}{\partial t} + \frac{1}{c^2}\frac{\partial^2 \mathbf{E}}{\partial t^2}.$$

Substituting the current (2.120), this yields

$$\frac{\partial^2 \mathbf{E}}{\partial t^2} - c^2 \nabla^2 \mathbf{E} = -\omega_p^2 \mathbf{E} \tag{2.121}$$

which has a plane wave solution $\propto e^{-i(\omega t - \mathbf{k}\cdot \mathbf{r})}$ with

$$\omega^2 = \omega_p^2 + k^2 c^2. \tag{2.122}$$

From Maxwell's equations (2.29), and since we have assumed $\nabla \cdot \mathbf{E} = 0$, this is a transverse wave ($\mathbf{k} \perp \mathbf{E} \perp \mathbf{B}$) as is light propagating in vacuum, but we see from (2.122) that the wave has a phase speed $\omega/k = c/\sqrt{1 - \omega_p^2/\omega^2}$ that depends on the frequency and wave number, and propagates only at frequencies greater than the plasma frequency; for $\omega < \omega_p$, k is imaginary, so that the wave decays in space, as does light reflected from a mirror. Since the phase speed $\omega/k > c$, it is generally much greater than the speeds of individual particles, which can thus be safely neglected. This is why the wave is neither affected by the pressure of the particles nor damped.[32]

If the ambient magnetic field is not negligible, the electrons gyrate in this field. This affects the wave for frequencies of the order of magnitude of the electron gyrofrequency or smaller. In particular, for propagation along the ambient

---
[32] In the absence of collisions.

magnetic field, the wave is split into two waves in which the electrons and the wave electric (and magnetic) field gyrate around the ambient magnetic field at the wave frequency.[33] The mode rotating in the same sense as do the electrons in the ambient magnetic field (the direct sense) is more easily emitted and absorbed by them. In the frequency range $\omega < \omega_{ge} < \omega_p$, one finds that the delay of propagation increases towards low frequencies, giving rise to a characteristic whistle, so that this mode is called a *whistler*.

Langmuir waves

The electromagnetic waves studied above have $\nabla \cdot \mathbf{E} = 0$, so that there is no variation in the density of electric charge. We have seen in Section 2.1 that variations in the density of electric charge make electrons oscillate in bulk when the random agitation is negligible. The random agitation has two major consequences. The first consequence is that it produces a pressure force that makes these oscillations propagate. This may be understood by picturing the electrons as a fluid of number density n and pressure $P = nk_BT$, so that their equation of motion in the wave field is

$$m_e \partial \mathbf{v}/\partial t = -e\mathbf{E} - \nabla P/n \qquad (2.123)$$

in the absence of ambient magnetic field. Because of the small timescale, we assume the electrons to behave as an adiabatic fluid (with 1 degree of freedom – the wave direction of propagation), so that $P \propto n^\gamma$ with $\gamma = 3$. This yields $\nabla P = 3k_BT \nabla n$. Because of the pressure term, we now have instead of (2.120)

$$\frac{\partial \mathbf{J}}{\partial t} = \omega_p^2 \epsilon_0 \mathbf{E} + \frac{3}{2} v_{the}^2 e \nabla n. \qquad (2.124)$$

Because of the small timescale, the ions do not move, so that the variation in electric charge density is $\partial \rho_e/\partial t = -e\partial n/\partial t$; hence the continuity equation applied to the electric charge and current yields $\nabla \cdot \mathbf{J} = e\partial n/\partial t$. Taking the divergence of (2.124) and expressing $\mathbf{J}$, $\mathbf{E}$ and ∇n in terms of the charge density ρ_e with the help of Maxwell's equation $\nabla \cdot \mathbf{E} = \rho_e/\epsilon_0$, we find

$$\frac{\partial^2 \rho_e}{\partial t^2} - \frac{3}{2} v_{the}^2 \nabla^2 \rho_e = -\omega_p^2 \rho_e. \qquad (2.125)$$

This equation has a plane wave solution $\propto e^{-i(\omega t - \mathbf{k}\cdot\mathbf{r})}$ with

$$\omega^2 \simeq \omega_p^2 + 3k^2 v_{the}^2/2 \simeq \omega_p^2 \left(1 + 3k^2 L_D^2\right). \qquad (2.126)$$

---

[33] These waves have respectively a *right-hand* and a *left-hand* circular polarisation, with respect to the direction of the ambient magnetic field. Beware that a number of different conventions are in use to label these waves, so that the same wave is given a different handedness depending on the context. For physicists, right-hand and left-hand generally refer to the direction of the wave vector $\mathbf{k}$; for radio astronomers, the same words refer to the direction from which the wave is coming, i.e. $-\mathbf{k}$, so that plasma physicists and radioastronomers use the same label only if $\mathbf{B}$ and $\mathbf{k}$ have opposite directions; furthermore, old textbooks call these waves respectively *ordinary* and *extraordinary*, whereas the latter names now denote electromagnetic waves of linear polarisation that propagate at an angle to the magnetic field.

The wave electric field (and also the particle velocity perturbation due to the wave) is parallel to **k**; it is due to the charge separation as the electrons oscillate whereas the massive ions are barely set in motion at these high frequencies. Whereas the electromagnetic wave found previously is a simple generalisation of the electromagnetic wave in vacuum, with the plasma acting as a dielectric medium (of refractive index $\sqrt{1 - \omega_p^2/\omega^2}$ for $\omega \gg \omega_{ge}$ and a birefringent medium otherwise), this longitudinal wave is entirely new. It is called a *Langmuir wave*, and is simply the Langmuir oscillation that propagates because of the electron thermal motion. One sees from (2.126) that the wavelength is greater than L_D for $\omega \sim \omega_p$, and tends to infinity as $\omega \to \omega_p$ where the wave reduces to a plasma oscillation. We shall see an illustration of these properties in Section 6.4.

Landau damping

The second consequence of the thermal agitation is that the Langmuir waves are damped. This is called *Landau damping*. This damping does not appear in (2.126), which gives a real value of k for any frequency $\omega > \omega_p$, because our derivation pictured the electrons as a fluid, while Landau damping comes from the individual motions of the particles and therefore requires a kinetic picture. This damping process is subtle since it produces losses without introducing any explicit damping term in the equation of motion, and appears in a wide range of problems outside plasma physics, from Saturn's rings to fireflies.[34]

A simple way of understanding this process is to picture it as a resonance between the wave and the electrons whose velocity equals the wave phase speed. These electrons see a constant electric field, and are therefore in resonance with the wave. The electrons moving slightly slower than the wave are accelerated, whereas those moving slightly faster are decelerated. With a Maxwellian velocity distribution (and more generally with a distribution whose derivative is negative for a velocity equal to the phase speed), there are more slower particles than faster ones, so that the net effect is to damp the wave. The damping is greater when there are more electrons having a speed close to that of the wave, which comes true when the phase speed is smaller than or close to the electron thermal speed. One sees from (2.126) that this happens when $k \gtrsim 1/L_D$. Hence, the Langmuir wave propagates at frequencies above but close to the plasma frequency, and is heavily damped at larger frequencies.

Conversely, if the velocity distribution has a positive derivative for a velocity equal to the wave phase speed, the wave grows. This happens for example when a beam of particles of velocity **v** propagates in a plasma faster than the electron thermal speed; the beam excites Langmuir waves of phase speed $\omega/k \simeq v$ directed along **v**, converting the energy of the beam into wave energy. This is an example of one of the numerous instabilities that arise in non-equilibrium plasmas, and we shall see an illustration of it in Section 6.4.

---

[34] See for example Meyer-Vernet, N. and B. Sicardy 1987, *Icarus* **69** 157, and Sagan, D. 1994, *Am. J. Phys.* **62** 450.

A similar resonance occurs for other kinds of plasma waves when the phase speed coincides with that of plasma particles. An example is cyclotron damping. For particles moving at velocity $v_\parallel$ along the ambient magnetic field **B**, the wave frequency is Doppler-shifted to the frequency $\omega - k_\parallel v_\parallel$ where $k_\parallel$ is the component of the wave vector along **B**, so that at some velocity it may coincide with the cyclotron frequency (or a harmonic), i.e. $\omega - k_\parallel v_\parallel = n\omega_g$. The wave is then damped when it has an electric field component perpendicular to **B**, so that particles experience a perturbing force which oscillates at the cyclotron frequency (or a harmonic).

2.3.5 Summary

The fast increase with speed of the collisional free path makes fast particles virtually collisionless, and therefore easily driven out of equilibrium. Two effects come to the rescue of collisions for tending to restore equilibrium: the first effect is the gyration around the magnetic field, but it acts only across the magnetic field; the second one is due to plasma instabilities which, however, only prevent the velocity distributions from becoming too crazy, but do not oblige them to be Maxwellian. This is why dilute plasmas have often velocity distributions which are non-Maxwellian but not too crazy. Since fluid descriptions – including MHD – assume velocity distributions to be nearly Maxwellian (or bi-Maxwellian), dilute plasmas often require a kinetic description.

Both fluid and kinetic descriptions involve the conservation of particles, momentum and energy. Since momentum and energy are carried respectively by plasma ions and electrons, of which the latter move much faster, thermal conductivity generally plays a more important role than viscosity, and is rarely negligible. Therefore, the major difficulty of fluid descriptions is to model correctly the transport of energy. Whereas in kinetic descriptions the heat flux is calculated self-consistently, fluid descriptions use various approximations of it. Basically, three kinds of fluid approximations are made, depending on the conditions: for very slow or very fast changes, the plasma is assumed to be isothermal or adiabatic respectively, corresponding to a heat flux that is respectively infinite or zero; for intermediate cases, the heat flux is approximated by the collisional transport. These approximations are valid only if the particle velocity distributions are close to Maxwellian.

When the electric field in the plasma frame is small enough, the plasma and magnetic field lines may be pictured as being tied together, so that the plasma can only move along the magnetic field but not across it. As a consequence, space plasmas tend to be organised by the magnetic field lines and do not mix easily across them. The magnetic forces on the plasma may be described as the superposition of a pressure acting across the field lines, which tend to expel the plasma from regions of strong magnetic field, plus a tension acting along the field lines, which tends to shorten and unbend them.

2.4 Basic tools for ionisation

What is the origin of ionisation in the solar interior, the solar wind and the planetary environments? A full answer to this question requires highly polished calculational techniques using the tools of quantum mechanics, and is outside the scope of this book; we shall give instead order of magnitude estimates based on elementary considerations [14], [20]. We will do so with the naive point of view of merely supplementing classical physics by the Heisenberg uncertainty relations.

2.4.1 Energy of ionisation and the size of the hydrogen atom

Let us estimate the size and energy of the hydrogen atom in its most stable state: the fundamental one. The H atom is made of a proton and an electron bound together by an attractive Coulomb force. The potential energy of the electron at distance r from the proton is $W_E = -e^2/(4\pi\epsilon_0 r)$. Because of the small size of the system, the kinetic energy of the electron is determined by Heisenberg's uncertainty relation, which says that an electron confined in a small region of size Δr has the momentum $\Delta p \sim \hbar/\Delta r$, whence the kinetic energy $W_{th} = (\Delta p)^2/2m_e$. Since in the H atom the electron is confined in a region of size $\Delta r \sim r$, we have $W_{th} \sim \hbar^2/(2m_e r^2)$, so that the total energy of the electron is

$$W \sim \frac{-e^2}{4\pi\epsilon_0 r} + \frac{\hbar^2}{2m_e r^2}. \tag{2.127}$$

The most stable state is the one of minimum energy, which arises for a distance r so that $dW/dr = 0$, i.e.

$$r \sim \frac{4\pi\epsilon_0 \hbar^2}{e^2 m_e} \equiv r_{Bohr} \quad \text{(Bohr's radius)}. \tag{2.128}$$

Substituting the numerical constants, we have

$$r_{Bohr} = \hbar/(\alpha m_e c) \simeq 0.53 \times 10^{-10} \text{ m} \tag{2.129}$$

where

$$\alpha = e^2/(4\pi\epsilon_0 \hbar c) \simeq 1/137 \quad \text{(fine structure constant)}. \tag{2.130}$$

For $r = r_{Bohr}$, we have $W_{th} = -W_E/2$ in accord with the Virial theorem (Section 3.1.1), and the total energy (2.127) is equal to minus

$$\frac{e^4 m_e}{8\epsilon_0^2 \hbar^2} \equiv W_{Bohr} \quad \text{(Bohr's energy)}. \tag{2.131}$$

Substituting the numerical constants, we have $W_{Bohr} \simeq 2.2 \times 10^{-18}$ J, which comes to about 13.6 eV.

This is the energy required to ionise a hydrogen atom from its fundamental state. By mere luck, this order-of-magnitude estimate gives the exact result. For atoms made of a nucleus of charge Ze surrounded by Z electrons, an electron of the outer shell sees the nucleus charge shielded by the charge of the $Z - 1$ other electrons, so that the energy required to strip an outer electron is of the same order of magnitude as for ionising hydrogen.

This is no longer true, however, as more electrons are stripped, so that producing highly charged ions requires a large energy. Very crudely, stripping an element of charge Ze of its last electron (of potential energy $-Ze^2/(4\pi\epsilon_0 r)$ at distance r), to produce a bare nucleus, requires an energy that is greater than the Bohr energy by the factor Z^2.

2.4.2 Ionisation by compressing or heating

These results furnish hints as to how a medium may be ionised.

One way is to compress, so that the average distance between ions becomes smaller than the sum of the radii of two atoms; this somehow crashes the atoms. For hydrogen, this happens when the ion number density n satisfies $n^{-1/3} < 2r_{Bohr}$, i.e. when the mass density satisfies $\rho > m_p/(2r_{Bohr})^3 \simeq 1.5 \times 10^3$ kg m^{-3}. We shall see in the next chapter that the density in the central parts of the Sun largely exceeds this value, producing ionisation.

Less dense media may be ionised by furnishing to atoms the ionisation energy W_{Bohr}. This may be done in several ways. One way is to heat. One might think naively that significant ionisation requires heating at a temperature so that the thermal energy $k_B T > E_{Bohr}$. This is, however, a classical point of view, and in fact a smaller energy is required because ionisation increases considerably the phase space accessible to an electron and therefore its number of possible states. This may be understood as follows.

At equilibrium, the ratio of the number of free electrons in some volume of phase space to the number of electrons bound in an H atom is proportional to $e^{-\Delta W/k_B T}$ (where ΔW is the difference in total energy between both states) times the ratio of the number of possible states for respectively a free electron and a bound one. Let $n_i = n_e$ be the ion (or electron) number density and n_n the number density of neutrals. In the volume V, a recombining electron may do so with either of the $n_i V$ ions and have one of two spin states, so that its number of possible states is $2n_i V$. On the other hand, the number of possible states of a free electron at temperature T and thermal speed $v_e \sim (k_B T/m_e)^{1/2}$ is roughly twice the ratio of the available volume V to the 'private' volume of a free electron (that is roughly the cube of its wavelength $h/m_e v_e$). With the approximation $\Delta W \sim W_{Bohr}$, this yields the degree of ionisation at thermal equilibrium $n_e/n_n \sim n_i^{-1}(h/m_e v_e)^{-3} e^{-W_{Bohr}/k_B T}$. A more exact calculation[35] yields nearly the same result:

$$\frac{n_i n_e}{n_n} \simeq \left(\frac{2\pi m_e k_B T}{h^2}\right)^{3/2} e^{-W_{Bohr}/k_B T} \qquad (2.132)$$

---

[35] Integrating over the electron velocity distribution.

Basic tools for ionisation

which is a simplified version of the so-called *Saha formula*. The term before the exponential is of the order of magnitude of $(T/T_F)^{3/2}$ where T_F is the Fermi temperature (see Section 2.1), which is much smaller than T in non-degenerate media. Hence, in general, ionisation is already significant at a temperature much smaller than $W_{Bohr}/k_B \simeq 1.6 \times 10^5$ K.[36]

Let us apply (2.132) to the solar corona. With $n_e \sim 10^{14}$ m^{-3} and $T \sim 10^6$ K, we find $n_e/n_n \sim 2 \times 10^{16}$, which shows that the corona is virtually completely ionised. One must be careful, however, in applying this formula since the corona is not in thermal equilibrium.

Planetary atmospheres are in general too cold for being thermally ionised.

2.4.3 Radiative ionisation and recombination

Ionisation and recombination

Another way of ionising particles is to subject them to photons of energy greater than the energy of ionisation.

The ionisation rate per atom is proportional to the flux of ionising photons F, and may be written

$$\Lambda_{ion} = F\sigma_{ion} \qquad (2.133)$$

in s^{-1}. Since the flux of photons F is expressed in m^{-2} s^{-1}, σ_{ion} – the *cross-section for ionisation* – has the dimension of an area. With a concentration n_n of neutrals, the ionisation rate per unit volume is therefore $dn_i/dt = n_n\Lambda_{ion} = n_n F\sigma_{ion}$.

Once ionised, the ions may recombine with electrons. For hydrogen, the radiative ionisation and recombination processes may be written

$$\text{H} + \text{ph} \rightleftharpoons \text{H}^+ + e^-.$$

For a given ion, electrons recombine at a rate proportional to their flux $\sim n_e v_{the}$ so that the rate of recombination per ion is

$$\Lambda_{rec} = n_e v_{the} \sigma_{rec} = n_e \beta \qquad (2.134)$$

in s^{-1}, where σ_{rec} has the dimension of an area, and may be expressed through the parameter $\beta = v_{the}\sigma_{rec}$, the *recombination coefficient*. At equilibrium, in the absence of bulk motion and of other ways of producing or suppressing particles, the rate of ionisation per unit volume $n_n\Lambda_{ion}$ balances the rate of recombination per unit volume $n_i\Lambda_{rec}$, so that since $n_e = n_i$ for singly ionised ions

$$n_n F\sigma_{ion} = n_e^2 \beta \quad \Rightarrow \quad n_e = \left(\frac{n_n F\sigma_{ion}}{\beta}\right)^{1/2}. \qquad (2.135)$$

---

[36] By a factor that is easily shown to be approximately $1/\ln(T/T_F)^{3/2} \ll 1$.

Cross-sections

To estimate the order of magnitude of the cross-sections, we consider incident photons of energy $W_{ph} \geq W_{Bohr}$, but still of the order of magnitude of W_{Bohr}, because, if the energy of the incident photon is much greater than W_{Bohr}, conservation of both energy and impulsion makes the probability of ionisation very small. Hence the electron liberated has the energy $W_{ph} - W_{Bohr} \leq W_{Bohr}$. Likewise, we consider recombining electrons of energy $\leq W_{Bohr}$. Now,

- in order for a photon to produce ionisation, it must (1) pass 'close enough' to an atom, and (2) be absorbed in liberating an electron,

- in order for a free electron to recombine radiatively, it must (1) pass 'close enough' to an ion, and (2) become bound in producing a photon.

In the frame of quantum mechanics, 'close enough' means closer than the quantum uncertainty on the position, that is $\hbar/p$ for a particle of momentum p. Therefore,

- for the ionising photon of energy $\sim W_{Bohr}$ and momentum $\sim W_{Bohr}/c$, this distance is $\hbar c/W_{Bohr}$,

- for the recombining electron of momentum $m_e v_{the}$, this distance is $\hbar/m_e v_{the}$.

We deduce the cross-sections for ionisation and recombination

$$\sigma_{ion} \sim \pi \left(\hbar c/W_{Bohr}\right)^2 \times P \qquad (2.136)$$

$$\sigma_{rec} \sim \pi \left(\hbar/m_e v_{the}\right)^2 \times P \qquad (2.137)$$

where P is the probability of absorption or emission of a photon by the electron during the time uncertainty corresponding to the energy involved, i.e. $\Delta t \sim \hbar/W_{Bohr}$.

We estimate P in a semi-classical way, regarding the bound electron as a harmonic oscillator of angular frequency $\omega = W_{Bohr}/\hbar$ and momentum $\hbar/r_{Bohr}$ so that the velocity is $v \sim \hbar/(m_e r_{Bohr}) \equiv \alpha c$, where α is the fine structure constant (2.130). For a harmonic oscillator, the speed varies by $\Delta v \sim v$ in a quarter of period, that is the time $\Delta t \sim 1/\omega$, and energy is radiated at a rate given by Larmor's formula as

$$\frac{dW}{dt} = \frac{e^2}{6\pi\epsilon_0 c^3} \left(\frac{dv}{dt}\right)^2. \qquad (2.138)$$

Writing $dv/dt \sim \Delta v/\Delta t$ with $\Delta v \sim v \sim \alpha c$ and $\Delta t \sim 1/\omega$, we find that the oscillator radiates during Δt the energy

$$\Delta W \sim \frac{dW}{dt} \times \Delta t \sim \frac{e^2 \alpha^2 \omega}{6\pi\epsilon_0 c}. \qquad (2.139)$$

Quantum mechanics tells us that it does so in discrete steps by emitting photons of energy $\hbar\omega$ with the probability $P = \Delta W/(\hbar\omega)$. From (2.139), this yields

$$P \sim \alpha^3. \qquad (2.140)$$

Basic tools for ionisation

Correct calculations give cross-sections that are roughly three times greater than the values (2.136)–(2.137) with our estimate $P \sim \alpha^3$, i.e.

$$\sigma_{ion} \simeq 10 \left(\hbar c/W_{Bohr}\right)^2 \times \alpha^3 \simeq 10^{-21} \text{ m}^2 \quad \text{(ionisation)} \quad (2.141)$$

$$\sigma_{rec} \simeq 10 \left(\hbar/m_e v_{the}\right)^2 \times \alpha^3 \quad \text{(recombination)}. \quad (2.142)$$

The cross-section (2.141) holds for ionisation of hydrogen by photons of energy $\sim W_{Bohr}$; for heavier atoms, the cross-section is of the same order of magnitude for the liberation of an outer electron. The recombination cross-section (2.142) holds for radiative recombination on the fundamental level of electrons of energy $\leq W_{Bohr}$. The corresponding recombination coefficient is thus given by

$$\beta_{rec} \simeq v_{the}\sigma_{rec} \simeq 10^{-17}/\sqrt{T} \text{ m}^3 \text{s}^{-1} \quad (2.143)$$

where T is the electron temperature. We shall use this recombination coefficient to understand why the solar wind is ionised (Section 2.5.7), and the radiative ionisation cross-section to estimate the basic properties of planetary ionospheres (Section 7.1), for deriving comet's properties (Section 7.5) and when studying the interaction of the solar wind with the interstellar medium (Section 8.1).

2.4.4 Non-radiative ionisation and recombination

Ionisation by particle impact

Another way to produce ionisation is by bombarding with particles. For the impact of a particle to ionise an atom, the kinetic energy of the relative motion of the particle must exceed the ionisation energy $\sim W_{Bohr}$. For an electron of mass m_e this requires $m_e v^2/2 > W_{Bohr}$, i.e. $v > \alpha c$. In this case, $\hbar/m_e v < r_{Bohr}$, so that the effective interaction distance between the electron and the atom (of approximate size r_{Bohr}) is no longer $\hbar/m_e v$ but rather r_{Bohr}. Hence we expect that the cross-section be of order of magnitude πr_{Bohr}^2. This holds a fortiori for a particle heavier than an electron.

However, the condition $v > \alpha c$ is not sufficient for producing ionisation, because, for the probability of interaction to be significant, the time of interaction $\Delta t \sim r_{Bohr}/v$ must ensure that the energy $\hbar/\Delta t$ be roughly equal to W_{Bohr}, i.e. $v \sim r_{Bohr} W_{Bohr}/\hbar \sim \alpha c$.

For an electron (of mass m_e), this requires that the electron kinetic energy be $m_e v^2/2 \sim W_{Bohr}$. However, for an ion or an atom, the mass is $m \gg m_e$, so that the kinetic energy required to ensure $v \sim \alpha c$ is greater by the factor $m/m_e \gg 1$.

Hence in space, impact ionisation is generally produced by electrons of energy of the order of magnitude of W_{Bohr}, with an effective cross-section

$$\sigma_{ion} \sim \pi r_{Bohr}^2$$

or by very energetic ions or atoms. With electrons of number density n and temperature $T \sim W_{Bohr}/k_B$, the electron flux is about $nv_{the} \sim n\alpha c$, and the rate of ionisation per atom is $\Lambda_{ion} \sim n v_{the} \times \pi r_{Bohr}^2$. Substituting the electron flux and the Bohr radius (2.129), we find the rate of impact ionisation per atom per electron (in order of magnitude)

$$\Lambda_{ion}/n \sim \frac{\pi \hbar^2}{\alpha m_e^2 c} \sim 2 \times 10^{-14} \text{ m}^3 \text{ s}^{-1}. \tag{2.144}$$

Dissociative recombination

We have seen that the cross-section for radiative recombination is extremely small, smaller by many orders of magnitude than the square of the typical atomic size, because of the small probability of photon emission. However, recombination occurs much more easily when the ions can dissociate. Instead of producing a photon, recombination then produces several atoms, by a reaction of the form

$$AB^+ + e^- \longmapsto A + B.$$

The dissociation into several components replaces the emission of a photon to conserve simultaneously the energy and the impulsion. Since no emission of photon is required, the cross-section (in order of magnitude) is given by the value (2.137) with $P = 1$, i.e.

$$\sigma_{rec} \sim \pi \left(\hbar / m_e v_{the} \right)^2 \tag{2.145}$$

if $\hbar/m_e v_{the} > r_{Bohr}$ i.e. if $m_e v^2/2 < W_{Bohr}$. This yields the coefficient of dissociative recombination

$$\beta = v_{the} \sigma_{rec} \sim 10^{-11}/\sqrt{T} \quad \text{(dissociative recombination)} \tag{2.146}$$

where T is the electron temperature (assumed smaller than $W_{Bohr}/k_B \sim 1.6 \times 10^5$ K).

This process is more effective than radiative recombination, by a factor of about six orders of magnitude. Hence it is the dominant process when molecular ions are present. In particular, it is the dominant recombination process in the ionospheres of inner planets (Section 7.1.4) and of comets (Section 7.5.2), whose atmospheres are made of complex molecules.

Charge exchange

Ionisation may also be produced by exchange of an electron between a neutral and an ion, as

$$A^+ + B \longmapsto B^+ + A$$

in the simple case when the ion is singly ionised. The atom B gives an electron to the ion A^+, and becomes ionised. This does not change the number of ions and free electrons, but changes the chemical nature of the ions and their speeds since the final ion has the properties of the original neutral and vice versa.

Problems

If F_A is the incident flux of ions A, the rate of ionisation per atom is $F_A \sigma_{ex}$ where σ_{ex} is the cross-section for charge exchange. As for the collision frequency for exchange of momentum between ions and neutrals (Section 2.1), the cross-section is heavily influenced by the electric dipoles induced in the atom and in the ion. For protons and hydrogen atoms with a relative speed of the order of magnitude of the typical solar wind speed, this produces a cross-section greater by two orders of magnitude than the Bohr radius squared, i.e.

$$\sigma_{ex} \simeq 2 \times 10^{-19} \text{ m}^2. \qquad (2.147)$$

This process plays an important role in the heliosphere when the solar wind encounters a large flux of neutrals; we shall see that this holds with neutrals of planetary and cometary origin (Section 7.5.6), as well as with interstellar neutrals (Section 8.1.2), with the interesting further consequence that when the original ion is highly charged, the final ion is left in a highly excited state whose de-excitation produces ultraviolet and X-ray emission.

2.5 Problems

2.5.1 Linear Debye shielding in a non-equilibrium plasma

In this problem, we generalise the Debye shielding to a non-equilibrium plasma, assuming small perturbations, and prove (2.8)–(2.9).

Consider an electron arriving from infinity (where its velocity is v) towards a point charge at the origin that produces the electrostatic potential $\Phi_E(r)$ (with $\Phi_E \to 0$ for $r \to \infty$). Show that if the point charge perturbs weakly the electron (i.e. if $e\Phi_E \ll m_e v^2$), then the electron velocity at distance r is changed by δv given by $\delta v/v = e\Phi_E(r)/(m_e v^2)$.

This velocity change is associated with a perturbation in electron number density around the point charge. For example, if $\Phi_E > 0$ the electrons are attracted and their trajectories are bent toward the charge, which increases their density; since, however, their velocity increases, they spend less time within a given region, which reduces this effect. Show that the net change in electron density is given by $\delta n_e/n = \delta v/v$.

Apply this result to the ions (changing the mass and charge), and deduce that the perturbations in electron and ion densities are given by

$$\delta n_e/n = e\Phi_E(r) \langle v^{-2} \rangle_e / m_e \qquad \delta n_i/n = -e\Phi_E(r) \langle v^{-2} \rangle_i / m_i \qquad (2.148)$$

where the brackets denote averages over the velocities at infinity, for electrons and ions respectively.

Deduce the shielding length from Poisson's equation.

Think about the limitations of this calculation.[37] The relation between δn and δv depends on the symmetry of the problem. How are the results changed with a different geometry? Show that if the point charge is replaced by a

---
[37] See Meyer-Vernet, N. 1993, *Am. J. Phys.* **61** 249, and references therein.

long wire, then the shielding disappears, whereas with a plane, there is an anti shielding. What happens when one takes into account perturbations that are not small? What kinds of particles are ignored in the above calculation?

Hints

We deduce $\delta v/v$ from the conservation of total (electrostatic + kinetic) particle energy between infinity and distance r.

To prove that the perturbations in density and speed are related by $\delta n/n = \delta v/v$ (in spherical symmetry), imagine a fictitious sphere of radius r collecting particles arriving on its surface. The particles arriving at grazing incidence with speed $v(r)$ have an impact parameter $p = r \times v(r)/v$, from conservation of angular momentum between infinity (where the speed is v) and distance r. The number of particles collected per second is $nv\pi p^2$. This number is also equal to $n(r) v(r) \times \pi r^2$ if the particles have density $n(r)$ and speed $v(r)$ at distance r (because for each surface element of the collector, half the particles are incident from one side, and their average perpendicular velocity is $v(r)/2$). Whence $n(r)/n = v(r)/v$.

2.5.2 Mean free path in a plasma

Verify from Fig. 2.4, with the plasma properties indicated in the caption, that for the velocity direction to change appreciably, the particle must travel a distance given roughly by (2.22).

2.5.3 Particles trapped in a planetary magnetic field

Consider a particle moving along a magnetic line of force near a planet of radius R_P, having a dipolar magnetic field (cf. Appendix and Fig. 2.12). Let θ_0 be the pitch angle in the magnetic equatorial plane at distance $r = LR_P$ from the planet, and B_P the magnetic field strength close to the planet in the polar regions. When the particle comes close enough to the planet, it is absorbed because of collisions with the atmosphere. When $L \gg 1$, the line of force crosses the planet surface in the polar regions.

Show that particles can be reflected between the north and south polar regions if

$$\theta_0 \geq \arcsin(B_0/B_P)^{1/2}. \tag{2.149}$$

Calculate this limit angle for $L = 6$.

Show that the bounce motion between the north and south regions follows the equation

$$mv_\parallel^2 + \mu B = \text{constant} \tag{2.150}$$

of a one-dimensional oscillator of potential energy μB.

Give an order of magnitude of the expression of the three periods of motion of a trapped particle (respectively T_1, T_2, T_3 for gyration, bounce, drift), and

Problems 109

show that $T_1 \ll T_2 \ll T_3$. Deduce that, in practice, the adiabatic invariant associated to the gyration (the particle magnetic moment μ) is more invariant than the one associated to the bounce motion, itself more invariant than the one associated to the drift.

With the numerical parameters relevant for the Earth (Appendix), estimate the drift velocity produced by the planet's gravitational field, and show that its ratio to the drift produced by the magnetic gradient is of the order of magnitude of the particle gravitational potential energy to the thermal energy. Deduce that it is in practice completely negligible.

2.5.4 Filtration of particles in the absence of equilibrium

In this problem, you will prove a very general result. Let a particle velocity distribution in some region be made of a superposition of Maxwellians of different temperatures. In absence of collisions, the particle velocity distribution in another region where the potential energy of particles is greater has a greater effective temperature. This result holds, for example:

- for the velocity distribution measured on a spacecraft, of particles that are repelled by the spacecraft electrostatic potential (Section 7.2),

- for the environment of a planet or a star, subjected to the body's gravitational potential (and electrostatic field).

Consider a velocity distribution that is a sum of Maxwellians of densities $n_{\alpha 0}$ and temperatures T_α, and give a formal expression of its effective temperature T_0 (defined from (2.51)).

Show that in the absence of collisions the distribution at a position where the potential energy of the particles has increased by $\Delta \psi > 0$ is again a sum of Maxwellians having the same temperatures T_α, but with densities $n_\alpha = n_{\alpha 0} e^{-\Delta \psi / k_B T_\alpha}$. Deduce that the effective temperature is greater than T_0.[38]

Prove this result graphically, by redrawing Fig. 2.13 with an initial distribution having more fast particles than a Maxwellian, i.e. whose slope flattens as energy increases.

Hints

The effective temperature of a distribution made of a sum of Maxwellians of densities n_α and temperatures T_α is

$$T = \frac{\sum_\alpha n_\alpha T_\alpha}{\sum_\alpha n_\alpha}. \qquad (2.151)$$

A particular application is studied in detail in Section 4.6.

---

[38] A general analytical proof may be found in Meyer-Vernet, N. 1995, *Icarus* **116** 202.

2.5.5 Freezing of magnetic field lines

Consider a bar of copper, of diameter L, and imagine you try to put it in a region of strong magnetic field (Fig. 2.17, 1, 2, 3). How long will it take for the magnetic field to penetrate into the bar? Conversely, once the magnetic field has penetrated into the bar, if you try to remove the bar, how long will it take for the magnetic field to disappear from the bar? At what speed should you move the bar for the effects shown on Fig. 2.17 to take place? Figure out the corresponding length and timescales for a cosmic object.

Hint

The electric conductivity of copper is about 0.6×10^8 mho.

2.5.6 Alfvén wave

Consider a small-amplitude Alfvén wave propagating along the ambient magnetic field in a uniform plasma at a speed much smaller than the velocity of light. Show that the force produced by the gradient in magnetic pressure is negligible. Show that the electric energy is negligible. Calculate the drift velocity of the particles, and comment.

2.5.7 Why is the solar wind ionised?

We have seen that the corona is made essentially of hydrogen, and is so hot that it is virtually completely ionised, and that the solar wind is produced by the expansion of the corona. Use the radiative recombination coefficient to understand why the solar wind remains ionised throughout the heliosphere.

Hints

The solar wind density is about $n \sim 5 \times 10^6 \, (d_\oplus/d)^2$ m^{-3} at distance d from the Sun, where $d_\oplus \simeq 1.5 \times 10^{11}$ m is the Sun–Earth distance (1 AU); the electron temperature is about $T \sim 10^5$ K; the size of the heliosphere is of the order of magnitude of 10^2 AU.

References

[1] Beck, A. and F. Pantellini 2007, N-Body plasma simulation, in preparation.

[2] Burgess, D. 1995, in *Introduction to Space Physics*, ed. Kivelson M. et al., Cambridge University Press, p. 129.

[3] Bracewell, R. N. 1978, *The Fourier Transform and its Applications*, New York, McGraw-Hill.

[4] Chen, F. F. 1984, *Introduction to Plasma Physics and Controlled Fusion*, New York, Plenum Press.

References

[5] Childs, W. H. J. 1972, *Physical Constants*, New York, Chapman & Hall.

[6] Clemmow, P. C. and J. P. Dougherty 1969, *Electrodynamics of Particles and Plasmas*, New York, Addison-Wesley.

[7] Davidson, P. A. 2001, *An Introduction to Magnetohydrodynamics*, Cambridge University Press.

[8] Garrett, A. J. M. 1988, Screening of point charges by an ideal plasma in two and three dimensions, *Phys. Rev.* **A37** 4354.

[9] Gordon, J. E. 1976, *The New Science of Strong Materials*, New York, Wiley.

[10] Ichimaru, S. 1973, *Basic Principles of Plasma Physics*, Reading MA, Benjamin.

[11] Laframboise, J. G. and L. W. Parker 1973, Probe design for orbit-limited current collection, *Phys. Fluids* **16** 629.

[12] Landau, L. D. and E. M. Lifchitz 1960, *Mechanics*, New York, Pergamon.

[13] MACRO Collaboration 2002, A combined analysis technique for the search of fast magnetic monopoles with the MACRO detector, *Astropart. Phys.* **18** 27.

[14] Migdal, A. B. 1977, *Qualitative Methods in Quantum Theory*, Reading MA, Benjamin.

[15] Morfill, G. E. and H. Thomas 1996, Plasma crystal, *J. Vac. Sci. Technol.* **A14** 490.

[16] Siscoe, G. L. 1983, in *Solar–Terrestrial Physics*, ed. R. L. Carovillano and J. M. Forbes, Dordrecht, The Netherlands, D. Reidel, p. 11.

[17] Spitzer, L. Jr 1967, *Physics of Fully Ionized Gases*, New York, Wiley.

[18] Sturrock, P. A. 1994, *Plasma Physics*, Cambridge University Press.

[19] Tabor, D. 1996, *Gases, Liquids and Solids*, Cambridge University Press.

[20] Thirring, W. 1981, *A Course in Mathematical Physics*, vol. 3, *Quantum Mechanics of Atoms and Molecules*, New York, Springer, Ch. 1.

3

Anatomy of the Sun

Ἥλιος γὰρ οὐχ ὑπερβήσεται μέτρα. εἰ δὲ μή, Ἐρινύες μιν Δίκης ἐπίκουροι ἐξευρήσουσιν.

Heraclitus of Ephesus[1]

What makes the Sun behave as it does? How does it shine? What causes its magnetic activity? And finally, how is its atmosphere produced? We will address these questions in a simplified way, using fundamental physics and the tools of plasma physics introduced in Chapter 2. There are large gaps in our understanding, however. Whereas we understand reasonably well how the Sun shines, its magnetic activity is still not correctly understood. In the spirit of this book, I have put aside the non-essential points and tried to highlight the main physics; far more complete accounts at various levels may be found in [37], [20], [38], [47], [8], [27], [53] and [11].

3.1 An (almost) ordinary star

The mass of the Sun is 2×10^{30} kg – a typical value for a normal star, that is an object which shines steadily for a fairly long time (by astronomical standards) by burning hydrogen; in the astronomer's jargon this is called a *main-sequence star* [15]. The largest mass observed for a normal star is of the order of $10^2 M_\odot$, while the lower limit is slightly below $10^{-1} M_\odot$.

Less massive objects may shine at the beginning of their life, but they rapidly cool into obscurity. Those lighter objects include *brown dwarfs* and giant planets [7]; brown dwarfs are now known to be very common objects, straddling the realms of planets and stars; in some sense they are aborted stars, while their physics and chemistry [6] are very similar to those of giant planets [18].[2]

---

[1] The Sun must not break the Law; if it does, the Erinyes, the handmaidens of Justice, will punish him. (Trans. L. M. Celnikier.)

[2] The largest giant planet in the solar system (Jupiter) has a mass of about $10^{-3} \times M_\odot$; a large number of objects of mass similar or larger have been discovered around other stars [18].

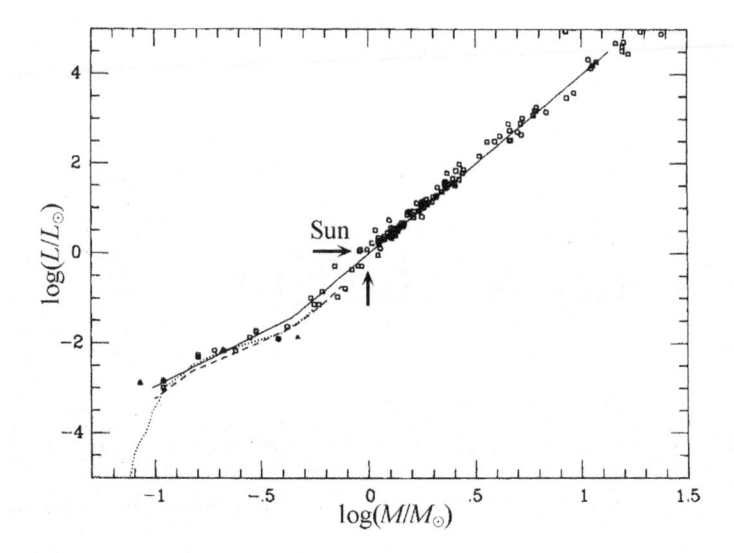

Figure 3.1 Mass–luminosity diagram for main-sequence stars, normalised to solar values. The lines are theoretical fits to the data. (Adapted from [32].)

Why do the masses of normal stars lie within such narrow limits? Furthermore, normal stars follow well-defined empirical laws; in particular their luminosity increases with mass as $L \propto M^\alpha$, where α is found to vary from 3 to 4 (Fig. 3.1). And normal stars are fairly stable structures: we know from palaeontological records that the Sun has been shining fairly steadily for more than 4×10^9 years – an appreciable fraction of the age of the Universe. Can we understand these properties from basic principles?

3.1.1 Hydrostatic equilibrium of a large ball of plasma

Everyday observation tells us that gases tend to fill all the available volume. This is no longer true at cosmic scales, because of gravitation. Large cosmic gas clouds tend to contract under their own weight and thus to heat up, eventually forming stars when the temperature at the centre becomes sufficiently high to ignite the nuclear fusion of hydrogen. In essence, the Sun is a self-gravitating fusion reactor – a huge ball of gas, whose hot central part generates most of the energy output.

Virial theorem

The stability of the Sun and other ordinary stars suggests that they are in hydrostatic equilibrium, namely that pressure forces balance gravitational ones. Consider a spherically symmetric object of mass M and radius R. Let ρ be the mass density and P the pressure at distance r from the centre. A unit volume is subjected to the net outward pressure force $-dP/dr$ and to the inward gravitational force $\rho M_r G/r^2$, where M_r is the mass within the radius r. Hydrostatic

An (almost) ordinary star

equilibrium yields

$$dP/dr = -\rho M_r G/r^2. \tag{3.1}$$

Multiplying both sides by $4\pi r^3$ and integrating by parts the left-hand side between $r=0$ and $r=R$ (where the pressure is negligible), one obtains

$$3\int_0^R P \times 4\pi r^2 dr = \int_0^R \rho(M_r G/r) \times 4\pi r^2 dr. \tag{3.2}$$

The left-hand side is just three times the product of the average pressure $\langle P \rangle$ in the object by its volume V. The right-hand side is the opposite of the object's gravitational energy W_g, since the energy released when a unit mass comes from infinity to the distance r is $M_r G/r$. Hence

$$3\langle P \rangle V = -W_g. \tag{3.3}$$

This is a general form of the so-called *Virial theorem*. We have seen in Section 2.1 that for an ideal non-relativistic gas, the pressure is related to the density of particle thermal energy w_{th} by

$$P = 2w_{th}/3 \tag{3.4}$$

whose average over the object is

$$\langle P \rangle = \frac{2}{3}\langle w_{th} \rangle = \frac{2}{3}\frac{W_{th}}{V},$$

W_{th} being the total thermal energy of the object; from (3.3), this yields

$$2W_{th} = -W_g \tag{3.5}$$

so that the total energy is

$$W_{th} + W_g = W_g/2 = -W_{th}.$$

It is negative, which means that the object is bound.

Temperature within the star

Because of the huge gravity within the star, the pressure is so great that it produces ionisation, as we saw in Section 2.4. For an order of magnitude estimate, we assume the object of volume V to be made of N protons and N (free) electrons at uniform temperature T, so that the total pressure P is

$$P \sim 2Nk_B T/V. \tag{3.6}$$

To estimate the gravitational energy, given by the opposite of the right-hand side of (3.2), we substitute $\rho = m_p N/V$ with $V = 4\pi R^3/3$, $M_r = \rho 4\pi r^3/3$, and integrate, which yields

$$W_g \sim -3N^2 m_p^2 G/5R. \tag{3.7}$$

Substituting (3.6) and (3.7) into (3.3) and rearranging, we deduce the temperature of the object as a function of its radius R and total number N of protons

$$T \sim \frac{m_p^2 G}{10 k_B} \frac{N}{R}. \qquad (3.8)$$

Since the object has been assumed uniform, T has the meaning of an average temperature. Let us apply (3.8) to the Sun, whose mass and radius are given in Table 1.1, so that the total number of protons is

$$N_\odot = M_\odot / m_p \simeq 1.2 \times 10^{57}. \qquad (3.9)$$

Substituting this number and the solar radius in (3.8) yields the Sun's average temperature $T \sim 2.3 \times 10^6$ K.

This value is several orders of magnitude greater than the effective temperature of 5800 K at the surface. The central temperature T_c is still greater. How much greater? To get a rough idea, let us approximate the temperature variation with r by a linear law, that is $T(r) \simeq T_c(1 - r/R)$; this yields for the average temperature $T \simeq T_c \int_0^R (1 - r/R) 4\pi r^2 dr / V$, whence $T \simeq T_c/4$. With the value of T found above, we find a central temperature $T_c \sim 10^7$ K. Note that elaborate numerical calculations yield instead $T_c \simeq 1.6 \times 10^7$ K (Fig. 3.2), and an average temperature in the solar interior about twice greater than our rough estimate.

By approximating the Sun as a ball of free electrons and protons in hydrostatic equilibrium, we have estimated its central temperature from its mass and radius; we shall see below that this temperature is sufficient to sustain thermonuclear fusion of hydrogen. Let us now try to understand the origin of the solar luminosity.

3.1.2 Luminosity

Radiative energy

A hot body in thermal equilibrium at temperature T contains photons of energy density

$$w_{ph} = aT^4 \qquad (3.10)$$

where $a = 4\sigma_S/c$ is the radiation constant (σ_S is Stefan's constant). The total radiative energy of an object of volume V is thus

$$W_{ph} = aT^4 V. \qquad (3.11)$$

Because of the outward temperature decrease within the star, the photons moving outwards carry on average a slightly greater energy than those moving inwards, so that there is a net outward radiation flux. This flux is the main agent of energy transport in most of the solar interior. In some sense, a star can be thought of as a 'leaky box' containing photons. At a temperature slightly above

10^7 K, a typical photon transports radiation in the X-ray range, which diffuses to the surface where it escapes as radiation at the much lower temperature T_{eff} – in the visible range. This diffusion of photons takes some time, t_{ph}, so that the radiative luminosity is approximately

$$L \sim \frac{\text{total radiative energy } W_{ph}}{\text{mean time } t_{ph} \text{ to reach the surface}}. \tag{3.12}$$

Mean free path of photons

If the photons were free to escape, they would take a time of only about R/c to reach the surface. The solar material is, however, very opaque, so that photons travel only a short distance before interacting with other particles. When a beam of photons traverses a layer of thickness dx, whose opacity is κ and mass density is ρ, scattering removes part of the radiation from the original direction so that a fraction $\kappa \rho dx$ is lost from the beam. This means that over a distance x sufficiently small for $\kappa \rho$ to remain constant, the number of photons decreases by the factor $e^{-\kappa \rho x}$. The mean distance travelled by a photon before it is strongly scattered – its mean free path – is

$$l_{ph} \equiv 1/(\kappa \rho). \tag{3.13}$$

If $l_{ph} \ll R$, photons execute random walks inside the star, only crossing the surface by chance; this random walk (see Section 2.3) increases their escape time by the factor R/l_{ph}, so that the mean time to reach the surface is

$$t_{ph} \sim (R/c) \times (R/l_{ph}) = R^2/(l_{ph} c). \tag{3.14}$$

Substituting the free path (3.13) with the average density $\rho = Nm_p/V$, we find the order of magnitude of the photon escape time

$$t_{ph} \sim NR^2 \kappa m_p/(Vc) \tag{3.15}$$

where κ is the mean opacity.

Luminosity versus mass

We deduce an order of magnitude of the luminosity $L \sim W_{ph}/t_{ph}$ by using (3.11), the mean temperature (3.8), the escape time (3.15), the volume $V = 4\pi R^3/3$, and rearranging as

$$L \sim (Nm_p)^3 \left(\frac{4\pi}{300}\right)^2 \frac{ac}{\kappa} \left(\frac{m_p G}{k_B}\right)^4. \tag{3.16}$$

From the assumption of radiative equilibrium, we have thus derived a simple luminosity–mass law $L \propto M^3$ (recalling that the mass of the star is $M = Nm_p$), when the opacity is independent on mass.

To proceed further, we need to know the opacity of the stellar material, i.e. how photons interact with stellar matter; this is one of the essential factors in stellar structure. The opacity is caused by a multitude of atomic processes; for relatively massive stars (whose matter is fully ionised), the main opacity agent is photon scattering by free electrons. Approximating the scattering cross-section by the Thomson value σ_T, we have

$$l_{ph} = 1/(n\sigma_T) \tag{3.17}$$

where n is the proton (or electron) number density. Using (3.13) with $\rho = n m_p$, this yields the opacity

$$\kappa \sim \sigma_T/m_p. \tag{3.18}$$

Substituting this opacity in (3.16) with σ_T given by (1.11) and (1.12), we deduce the luminosity as a function of the total number of protons N in the star. Normalising to the solar value $N_\odot$ given by (3.9) and substituting the numerical constants, we obtain finally

$$L \sim 3.5 \times 10^{26} \left(\frac{N}{N_\odot}\right)^3 \text{ W}. \tag{3.19}$$

For a solar mass object, this yields a luminosity of 3.5×10^{26} W, very close to the actual solar value given in Table 1.1.

Our simple order-of-magnitude estimate, which is little more than an improved dimensional analysis, turns out to be very good. It should not be so good, because in most of the solar interior (as in other stars of similar or smaller mass), photon interaction with heavy ions produces an opacity greater than the Thomson scattering value by about an order of magnitude, in spite of the low abundances of these ions;[3] in that case (3.18) is a lower limit for the opacity, rather than an actual value.

From the assumption of hydrostatic and thermal equilibrium, we have estimated the luminosity of a normal star as a function of its mass and opacity, and applied the result to the simple case when the opacity is mainly due to (Thomson) scattering by free electrons. But we have not yet studied the basic question: where does the energy lost by radiation come from?

3.1.3 Energy source and timescales

Nuclear fusion

At the core temperature of more than 10^7 degrees, the large thermal speeds of nuclei enable them to come sufficiently close together in spite of their electrostatic repulsion, and they undergo thermonuclear fusion; this process is the

---

[3] Several effects contribute to stellar opacity: (Thomson) scattering by free electrons, photoionisation (free–bound transitions), free–free transitions of electrons as they pass close to ions and bound–bound transitions. Thomson scattering dominates in the solar core, which is virtually completely ionised. However, in most of the interior, photoionisation is dominant because, even where hydrogen and helium are significantly ionised, heavy elements have still some strongly bound inner electrons that can be stripped out by X-ray photons.

most important energy source for normal stars. This might appear surprising, since the thermal energy of nuclei does not enable them to approach each other much closer than the Landau radius (2.15), which at this temperature is of the order of magnitude of 10^{-12} m – larger by three orders of magnitude than a nuclear radius. Even though the Landau radius decreases as the inverse of the particle energy, very few particles will have enough energy to come within the range of nuclear forces. This reasoning, however, ignores quantum effects. Quantum uncertainties enable bridging of the gap, in that tunnelling effects increase considerably the probability of crossing the Coulomb barrier.

The nuclear reaction proceeds in several steps, whose net result is the conversion of four protons into a helium nucleus – made of two neutrons and two protons. This yields a net mass change of

$$\Delta m = (4m_p - m_{He})/4 \simeq 7 \times 10^{-3} m_p \text{ per proton} \qquad (3.20)$$

releases the energy

$$\Delta W_{nu} = \Delta m \times c^2 \text{ per proton} \qquad (3.21)$$

and also produces neutrinos which escape, carrying away part of the energy; the remaining energy, that is most of the energy released, is available for the star's energy balance.

Timescales

How long can the Sun radiate its luminosity $L_\odot$ by using this source of energy? About 10% of the solar mass is available for hydrogen fusion, providing an energy reservoir of $\Delta W_{nu} \times N_\odot/10$. The solar nuclear lifetime is thus

$$\tau_{nu} \sim \frac{\Delta W_{nu} \times N_\odot/10}{L_\odot} \qquad (3.22)$$

$$\sim \frac{7 \times 10^{-4} \times M_\odot c^2}{L_\odot} \sim 10^{10} \text{ years.}$$

Hence the present Sun is roughly halfway through its hydrogen burning phase.

In the absence of nuclear reactions, the Sun might still radiate, but this would be at the expense of its gravitational energy so that it would shrink. We have seen in Section 3.1 that the gravitational energy varies as $W_g \propto -1/R$, the mean temperature as $T \propto 1/R$, while the total energy equals $W_g/2$. As the body contracts, therefore, the total energy decreases (escaping as radiation), whereas the temperature increases. The loss of energy makes the star hotter! The solar material behaves as if its heat capacity were negative; this is because only half of the released gravitational energy is radiated, while the other half heats the body. We shall return to this property later.

How long could the Sun radiate at its present luminosity by consuming its store of gravitational energy? This time, often called the *Kelvin–Helmholtz* time, is equal to the available gravitational energy divided by the luminosity:

$$\tau_{KH} = \frac{|W_g|}{L} \sim \frac{M^2 G}{RL} \qquad (3.23)$$

$$\sim 3 \times 10^7 \text{ years for the Sun.}$$

Since the thermal and gravitational energies are of the same orders of magnitude (because of the Virial theorem), this time is also the time needed by the star to settle to equilibrium after a thermal perturbation. Since energy is transported by radiation (except in the outer region), this is also the timescale for radiative transport, which is much greater than the photon random walk time t_{ph} because photons are in equilibrium with particles, whose energy is much greater. (Problem 3.4.2).

We have assumed the Sun to be in hydrostatic equilibrium. What would happen otherwise? Suppose that gravitation dominates pressure. A small mass element at a distance r would fall towards the centre according to $d^2 r/dt^2 \sim -M_r G/r^2$. The star would then collapse in a timescale τ_{hydr} whose order of magnitude is given by $R/\tau_{hydr}^2 \sim MG/R^2$, whence

$$\tau_{hydr} \sim \left(\frac{R^3}{MG}\right)^{1/2} \qquad (3.24)$$

$$\sim 2 \times 10^3 \text{ s for the Sun.}$$

This is the typical time for the star to react to a small perturbation of the hydrostatic equilibrium; this is also the typical pulsation period of the star. Note that since the order of magnitude of the sound speed is $V_S \sim (k_B T/m_p)^{1/2}$ and $k_B T/m_p \sim MG/R$, this time is of the order of R/V_S – the time for an acoustic wave to travel across the star; indeed, small departures from hydrostatic equilibrium are restored at the speed of sound.

Finally, therefore, we have for normal stars

$$\tau_{hydr} \ll \tau_{KH} \sim \tau_{thermal} \sim \tau_{radiation} \ll \tau_{nu}.$$

This justifies a posteriori our assumption of hydrostatic and thermal equilibrium, and also shows that in response to a perturbation, hydrostatic equilibrium is restored very quickly. This has an important consequence on stellar stability.

Stability of the Sun

We have seen that the solar luminosity is sustained by hydrogen fusion. Why then does the Sun not explode like an H-bomb? The key is the negative effective heat capacity. Imagine a perturbation causing the Sun to produce more nuclear energy than it can radiate; this increases the total energy, equal to $W_g/2$ in hydrostatic equilibrium; since $W_g \propto -1/R$ and $T \propto 1/R$, this makes the Sun expand and cool; and in turn this cooling decreases the production of nuclear energy and cures the original problem. Conversely, a decrease in nuclear energy production would cause the Sun to contract and heat up, thereby increasing the nuclear energy production and curing the original problem.

Finally, therefore, the fact that the stellar material behaves as an ideal gas confined by self-gravity provides a safety valve against runaway nuclear reactions and enables stars to shine quietly. In an apparent paradox, nuclear fusion does not heat the star; rather it keeps it cool. By replacing the energy lost by radiation, it prevents the total energy from decreasing which would make the star contract and heat up. Because of hydrostatic and thermal equilibrium, the solar radius has just the right value to produce a core temperature adequate for nuclear energy production to balance the radiated energy. The luminosity is not determined by the nuclear production rate but by the rate at which the star can radiate.

3.1.4 The mass of a normal star

> The Baroness weighed about 350 pounds and in consequence was deeply respected.
>
> Voltaire, *Candide*[4]

What makes the Sun as big as it is? Why do stellar masses only vary from about ten times less than the Sun to a hundred times more? In essence, an object that is too light has not enough self-gravity to compress its central region to the high temperature required for igniting nuclear fusion. On the other hand, an object that is too massive is compressed to a temperature so high that radiation pressure dominates gas pressure, making the structure unstable. Let us put these arguments on a quantitative basis.

The simplified calculations we made above neglect two contributions to the pressure:

- the Fermi (electron degeneracy) pressure,
- the radiation pressure.

Fermi pressure is relatively small for normal stars, but it becomes important for small objects. On the other hand, radiation pressure is negligible for the Sun (Problem 3.4.2), but it may become important for large objects.

Minimum mass

Let us first estimate the minimum mass of a normal star. In the spirit of the order of magnitude estimates made in this section, we again picture the star as a homogeneous ball of radius R containing N protons (and the same number of electrons). A ball of gas (sufficiently cold for radiation pressure to be negligible) will remain in equilibrium if its self-gravity is balanced by the sum of the thermal and Fermi pressures. In this case, the gravitational energy per particle must be comparable to the thermal energy plus the Fermi energy. The

---

[4]Trans. L. M. Celnikier.

thermal energy is about $k_B T$, the gravitational one W_g/N is given from (3.7), and we saw in Section 2.1 that the Fermi energy per electron of number density n is $(3\pi^2 n)^{2/3}\hbar^2/(2m_e)$. Substituting $n = N/V$ and rearranging, we get

$$k_B T \sim 3N m_p^2 G/5R - (9\pi N/4)^{2/3}\hbar^2/(2m_e R^2). \tag{3.25}$$

When R is large, the thermal energy exceeds the Fermi energy, so that $T \propto 1/R$ as we found previously. As the radius shrinks, however, the contribution of the Fermi term rises, and the temperature will reach a maximum when the derivative with respect to R of the right-hand side of (3.25) is zero; this occurs at $R \simeq 6\hbar^2/(m_e m_p^2 G N^{1/3})$, for which the temperature is

$$k_B T_{max} \sim 0.05 \times N^{4/3} m_e (m_p^2 G/\hbar)^2. \tag{3.26}$$

In essence, contraction increases the Fermi energy at the expense of the gravitational energy, so that the temperature no longer increases.

For the object to become a normal star, the temperature at the centre T_c must reach a value at least equal to the minimum temperature required for igniting hydrogen fusion, which is of the order of 10^7 K. In the spirit of our simple estimate, T_{max} given by (3.26) has the meaning of the average temperature T in the star's interior. Using our previous estimate $T_c \simeq 4 \times T$, the ignition condition $T_c > 10^7$ K can be rewritten as $T_{max} > 10^7/4$ K. Substituting this value in (3.26) gives the minimum number of protons in a normal star:

$$N_{min} \sim \left(\frac{k_B \times 5 \times 10^7}{m_e c^2}\right)^{3/4} \times \left(\frac{\hbar c}{m_p^2 G}\right)^{3/2} \sim 0.6 \times 10^{56}. \tag{3.27}$$

A more accurate calculation gives $N_{min} \simeq 0.8 \times 10^{56}$ protons, that is $0.7 \times 10^{-1} M_\odot$.

Maximum mass

At the other end of the mass range, consider now what happens when the radiation pressure is not negligible. To enlighten the effect of radiation pressure on stability, let us apply the Virial theorem under the general form (3.3) to a gas of photons. Photons are par excellence relativistic particles, and we saw in Section 2.1 that for such particles the pressure is related to the energy density by

$$P_{ph} = w_{ph}/3. \tag{3.28}$$

Substituting this pressure in (3.3), we find

$$W_{ph} = -W_g. \tag{3.29}$$

Hence if the star contained only photons, the total energy would be $W_{ph} + W_g = 0$. This is in sharp contrast with non-relativistic particles, for which the Virial theorem has the form (3.5), producing a negative total energy. This means that if the energy of photons is dominant over that of non-relativistic particles, the total energy will be small compared to both W_g and W_{ph}, so that a small perturbation will be sufficient to make the total energy positive and thus to destabilise the system.

When does this happen? In order of magnitude, the ratio of radiation and thermal particle energy densities is

$$\frac{w_{ph}}{w_{th}} \sim \frac{aT^4}{nk_BT} \sim \frac{aT^3 \times 4\pi R^3/3}{Nk_B}$$

$$\sim N^2 \frac{4\pi a}{3k_B} \left(\frac{m_p^2 G}{10 k_B}\right)^3 \quad (3.30)$$

where we have used the photon energy density (3.10), and substituted the mean temperature (3.8). Very roughly, one expects the star to be stable when radiation is not dominant, i.e. when $w_{ph}/w_{th} < 1$. Using this condition and substituting the expression of the radiation constant a in (3.30), we deduce the maximum number of protons in a normal star

$$N_{max} \sim 20 \times \left(\frac{\hbar c}{m_p^2 G}\right)^{3/2} \sim 4 \times 10^{58}. \quad (3.31)$$

This gives a maximum mass of about 40 solar masses for a normal star. A more accurate calculation gives a value about twice as great.

Finally, therefore, we can understand from simple quantitative arguments why the masses of normal stars lie in a narrow range corresponding to a number of protons around $(\hbar c/m_p^2 G)^{3/2} \simeq 2.2 \times 10^{57}$ [53], [27]. With many fewer protons, the Fermi energy prevents the temperature reaching the value required for igniting hydrogen fusion, whereas with many more protons, the radiative energy makes the star unstable. And the major difference between normal stars and brown dwarfs emerges from basic physics: as the infant star contracts, the contraction is stopped either by electron degeneracy or by hydrogen burning, whichever comes first; the first case produces a brown dwarf, the second case produces a star like our Sun [6].

3.2 Structure and dynamics

We have derived order-of-magnitude estimates of the basic solar properties from simple principles, making drastic approximations to enlighten the basic physics: we have pictured the Sun as a homogeneous ball of protons and electrons, assumed the equation of state to be that of an ideal gas, and the energy to be transported by radiation only. This is not perfectly true. In particular, the Sun contains other elements than hydrogen, it shows some internal structure and small deviations from the ideal gas law. Furthermore, we shall see below that in the outer part of the Sun, the energy transport is convective rather than radiative – a point that has important consequences on the solar atmosphere and ultimately on the solar wind.

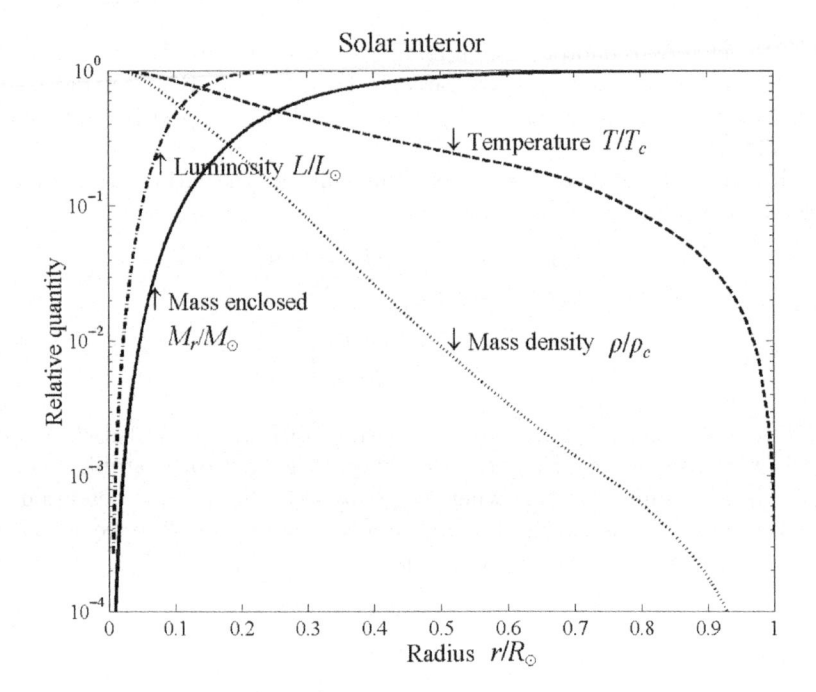

Figure 3.2 Variation of solar properties with distance from the centre, from [4] inward of 0.95 solar radii, and from [9] elsewhere. The temperature and mass density are normalised to the values at the centre: $T_c = 1.57 \times 10^7$ K and $\rho_c = 1.52 \times 10^5$ kg m^{-3}. The mass M_r enclosed within r and the luminosity are normalised to the values at distance $R_\odot$.

3.2.1 Modelling the solar interior

Figure 3.2 shows how the temperature, mass density, luminosity and mass enclosed within radial distance r vary with r. It is based on a standard solar model [4], which we have completed from 0.95 $R_\odot$ outwards with the temperature given by [9]. The hydrogen mass fraction (not shown) is about $X \simeq 0.73$ except in the core where nuclear reactions take place.[5] The mean particle mass (not shown) is $\mu \simeq 0.7 m_p$ in most of the solar interior; it increases in the outer layers as ionisation of H and He decreases, approaching 1.25 close to the surface where H and He are no longer significantly ionised.

Standard solar models are based on a set of assumptions on the main physics governing the interior, that include spherical symmetry, hydrostatic and thermal equilibrium (apart from the slow evolution of the star), composition, equation of state, production of nuclear energy, and energy transport. A model is then computed by starting with a chemically homogeneous Sun with some initial

---

[5] Due to hydrogen fusion into helium, the hydrogen mass fraction decreases in the core as the Sun ages; it is presently only 0.34 near the centre [4]; and we have already said that there is also a small concentration of heavy elements.

composition, evolving to the present solar age of 4.6×10^9 years, and adjusted to agree with the radius, luminosity and composition observed at the solar surface.

Because of the huge opacity, the solar interior is invisible to us, but it can be probed in two ways. The first one is solar seismology, that is measuring solar oscillations. Helioseismology reveals the solar interior from its oscillation modes, somewhat as the sound of a glass can reveal properties of the liquid it contains. Solar seismology has now reached unprecedented precision and shows a stunning agreement with current solar models, on which it places strong constraints [3].

The second way (in principle) of probing the solar interior is by measuring the neutrinos that accompany nuclear fusion and escape from the Sun. Solar neutrinos have posed a major problem which has haunted solar and particle physicists for three decades, when the observed fluxes were substantially below model predictions. This question has now been settled in favour of the standard solar models, the culprit being inadequate neutrino physics. Rather ironically, solar neutrino observations, which were initially intended to use neutrinos (whose properties were thought to be well known) to test solar models, have finally revealed novel neutrino physics, and the Sun is now sufficiently well understood to be exploited as a neutrino factory for studying fundamental physics [2].

What does Fig. 3.2 tell us? One sees that more than 90% of the luminosity is generated in the inner fifth of the solar radius, which contains one-third of the mass and whose temperature exceeds 10^7 K. The temperature decreases outwards, roughly as the mean density to the power $1/3$, up to the outer 30% of the radius, where the temperature gradient begins to steepen.

In that region, more precisely outwards of $0.71 R_\odot$, an important change occurs in the solar properties. Recall that we have assumed strict spherical symmetry, i.e. that all properties are constant on concentric spheres. Although this may be true in average, small fluctuations are bound to arise. These perturbations may be ignored if they do not grow; if they grow, however, they give rise to macroscopic local motions that, even if spherically symmetrical on average, modify the structure in mixing the material and transporting energy. This is what happens in that region of the Sun where convection develops (Fig. 3.3).

Convection is common in fluids heated from below, where cold matter overlies hot. Since colder fluids tend to be heavier and thus to sink, these structures may be unstable to overturning motions, producing convection. Even though the convection region contains only a few per cent of the solar mass, convection plays a major role in the Sun, by structuring the upper layers, producing a magnetic field, and ultimately heating the atmosphere.

3.2.2 Convective instability

When do these local perturbations grow? Let us represent a local perturbation by a small bubble having different properties from its surroundings. Consider a

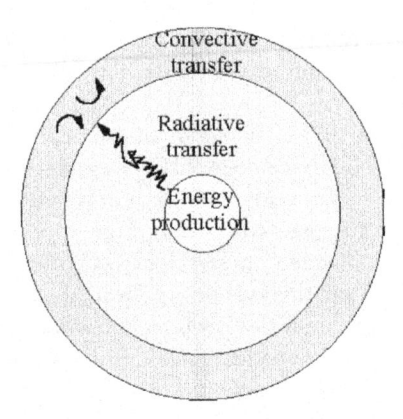

Figure 3.3 The Sun sketched as an onion. Most of the luminosity is produced in the central core. The energy transfer is radiative inward of $0.71R_\odot$, and convective outwards.

bubble slightly hotter; since its internal pressure P must match the exterior one,[6] and $P \propto \rho T$, it is lighter than its surroundings and tends to rise; conversely, a bubble colder than its surroundings tends to sink. To test the stability, we must determine whether the displaced bubble will return to its original position or continue its motion.

The key point is that pressure and temperature decrease upwards, while gravity acts inwards. Consider a rising hot bubble. In rising it encounters a smaller external pressure which makes it expand, so that its temperature decreases. If this temperature decrease is less than the external one, the bubble remains hotter than its surroundings and continues to rise; if on the other hand its temperature decrease is greater than the external one, then at some point the bubble becomes colder than outside, which makes it sink down and return to its original position. In the former case, the slightest non-uniformity will produce bulk motions and the structure is unstable. Since hot bubbles rise whereas cold ones sink, this yields a net upward heat transport without net mass transport.

Criterion for instability

Let us assume that the expansion of a rising bubble is sufficiently fast to be adiabatic, and that changes in composition are negligible. We then have within the bubble $P \propto \rho T \propto \rho^\gamma$, whence $T \propto P^{(\gamma-1)/\gamma}$ where $\gamma = c_p/c_v$ is the ratio of specific heats. Thus, along the radial distance r

$$\left(\frac{dT}{dr}\right)_{adiabatic} = \frac{\gamma - 1}{\gamma} \frac{T}{P} \frac{dP}{dr}. \qquad (3.32)$$

The bubble will remain buoyant and continue to rise if it remains hotter than its surroundings, i.e. if its rate of temperature decrease given by (3.32) is less

---

[6] If not, it will quickly expand or contract to restore pressure equilibrium, since the time to restore pressure equilibrium is much shorter than the time taken to exchange heat.

Structure and dynamics

than dT/dr outside, i.e., since the derivatives are negative,

$$\left|\frac{dT}{dr}\right| > \left|\left(\frac{dT}{dr}\right)_{adiabatic}\right|. \qquad (3.33)$$

If this so-called *Schwarzschild criterion* (named after Karl Schwarzschild[7]) is true, then any random motion will precipitate convection.

Substituting in (3.32) the pressure gradient at hydrostatic equilibrium, $dP/dr = -\rho g$ with $P = \rho k_B T/\mu$, we find the adiabatic temperature gradient

$$\left(\frac{dT}{dr}\right)_{adiabatic} = -\frac{\gamma - 1}{\gamma}\frac{T}{H} = -\frac{g}{c_p} \qquad (3.34)$$

where

$$g = M_r G/r^2 \qquad (3.35)$$

$$H = \frac{k_B T}{\mu g} \qquad (3.36)$$

$$c_p = \frac{\gamma}{\gamma - 1}\frac{k_B}{\mu} \qquad (3.37)$$

are respectively the gravitational acceleration at distance r, the local pressure scale height and the specific heat at constant pressure; μ is the mean particle mass; for pure atomic hydrogen, $\mu = m_p$, whereas for fully ionised hydrogen $\mu = m_p/2$; for an ideal monoatomic gas, $\gamma = 5/3$, which yields $(\gamma - 1)/\gamma = 0.4$.

Applying the criterion

In the absence of convection, the actual temperature variation dT/dr is determined by radiative transfer. To determine the radiative energy flux F_{ph} at distance r outside the energy production region, we note that the energy scattered per unit time per unit surface in a shell of width dr: $\kappa \rho F_{ph} dr$ is the difference between the energy fluxes at r and at $r+dr$. Since the photon energy is the product of the impulsion by c, the above difference is just the variation in radiation pressure dP_{ph} times c. With a small radiative anisotropy, the pressure of photons is given by (3.28) with the energy density (3.10), i.e. $P_{ph} = aT^4/3$. This yields $-\kappa \rho F_{ph} = c \times d\left(aT^4/3\right)/dr$, whence the radiative flux at distance r

$$F_{ph} = -\frac{4ac}{3\kappa \rho}T^3 \frac{dT}{dr}. \qquad (3.38)$$

In the absence of convection, this radiative flux is related to the luminosity at r by

$$F_{ph} = \frac{L}{4\pi r^2} \qquad (3.39)$$

---
[7]Schwarzschild, K. 1906, *Nachr. Kgl. Ges. Wiss. Göttingen, Math.-Phys. Klasse* **1** 41.

so that the temperature gradient is finally

$$\frac{dT}{dr} = -\frac{3\kappa\rho L}{16\pi a c r^2 T^3}. \tag{3.40}$$

Comparing this gradient with the adiabatic value (3.34), we see that convective instability will be favoured by either a low temperature and a high density and opacity (which tend to steepen the radiative temperature gradient), or a large specific heat c_p, which tends to make the adiabatic gradient milder. This happens in the outer 30% of the solar radius, where the opacity becomes far greater than in the inner region, while ρ/T^3 changes much less. Furthermore in the upper layers, hydrogen and helium become only partially ionised, which increases not only the opacity but also the specific heat because the degrees of ionisation depend on temperature.

In essence, at some distance the huge opacity increase hampers the escape of radiation, making the outward temperature decrease steepen, up to the value where the medium becomes convectively unstable; convection then develops and takes over as the primary mechanism of energy transport.

3.2.3 Convective energy transfer

How effective is convection for heat transport?

Solar convection is a very complex problem, and each of the three fronts of attack – theory, experiment and numerical simulation – encounters great difficulties. First of all, the Reynolds number (Section 2.3) is very large.[8] The non-linear inertial term in the fluid equation of motion is far greater than the one due to viscosity, which is thus unable to damp out small perturbations, so that the motion becomes turbulent.

And turbulence is not understood, even in cases far simpler than the Sun. In order not to discourage the reader, we shall sidestep it in this section by using an old trick, postponing its discussion until Section 6.4. To increase further the theoretical difficulties, the pressure scale height is significantly smaller than the size of the system, which is thus strongly stratified. The experimental front encounters difficulties, too, because the properties (including the non-dimensional numbers which determine the physics) are hugely different from those encountered in our terrestrial laboratories. And finally, the range of relevant scales is so wide that numerical simulations cannot handle the full problem, even with the most powerful computers presently available.

The mixing-length approach

The old trick we shall use is the so-called *mixing-length* picture. This approach hides the physics that is not understood under global parameters and enables

---

[8] At the base of the convection zone, we have from Fig. 3.2: $\rho \simeq 200$ kg m^{-3} and $T \simeq 2 \times 10^6$ K, so that the coefficient of viscosity $\nu_{vis} \sim (k_B T/m_p)^{1/2} l_f \sim 10^{-4}$ in order of magnitude; with a scale $\sim 0.1 R_\odot$, this yields a huge Reynolds number.

one to estimate global properties. It is derived from the traditional view of turbulence as many swirling 'eddies', to which we shall return in Section 6.4.

In this picture, the fluid is viewed as made of blobs that move up or down and eventually lose their identity by merging with their environment. The vertical distance l a blob travels before dissolving is called the mixing length. When a blob dissolves, it shares with its surroundings its heat excess (if it is rising) or deficiency (if it is sinking), so that energy flows upwards.

The energy delivered by a blob when it dissolves is determined by its temperature difference ΔT_{conv} with its surroundings. Since the blob has started at ambient temperature, ΔT_{conv} is the difference between:

- the change within the blob, which is adiabatic, of temperature gradient $(dT/dr)_{adiabatic}$ given by (3.34)

- and the change in the ambient medium, of temperature gradient dT/dr,

whence over the distance l

$$\Delta T_{conv} = \left[\left(\frac{dT}{dr}\right)_{adiabatic} - \left(\frac{dT}{dr}\right)\right] \times l. \tag{3.41}$$

Since the blob remains in pressure equilibrium with its surroundings, the energy transferred is the difference in enthalpy, namely $\rho c_p \Delta T_{conv}$ per unit volume, which yields the net energy flux

$$F_{conv} = \rho c_p \Delta T_{conv} \times v_{conv} \tag{3.42}$$

where v_{conv} is the average speed of the blobs.

The key parameter in this picture is the mixing length l. One generally assumes that l is roughly equal to the local pressure scale height H, to a factor of order unity.[9] A heuristic justification is to note that when matter has moved upwards by a pressure scale height, its pressure has decreased by a factor of about three, so that its volume has increased by the same order of magnitude. Hence blobs whose radial cross-section occupied initially a significant fraction of the surface of a domain at some altitude will cover entirely this domain after having risen by one or two scale heights, and will thus mix together (Fig. 3.4).

Let us now estimate the convection speed v_{conv}. If the motion of the blobs is driven by the buoyancy force produced by their density deficit $\Delta \rho_{conv}$ with respect to the ambient medium, the kinetic energy density $\sim \rho v_{conv}^2$ equals the work of this force along the vertical distance l, i.e. $\rho v_{conv}^2 \sim |\Delta \rho_{conv}| gl$. Since $\rho T \propto P$ for both the blob and the exterior and their pressures are equal, we have $\Delta \rho_{conv}/\rho = -\Delta T_{conv}/T$, whence, substituting $l \sim H$ and rearranging

$$v_{conv} \sim \left(\frac{k_B \Delta T_{conv}}{\mu}\right)^{1/2}. \tag{3.43}$$

---

[9]In most models, the value of l/H required to reproduce the solar properties lies between 1 and 2.

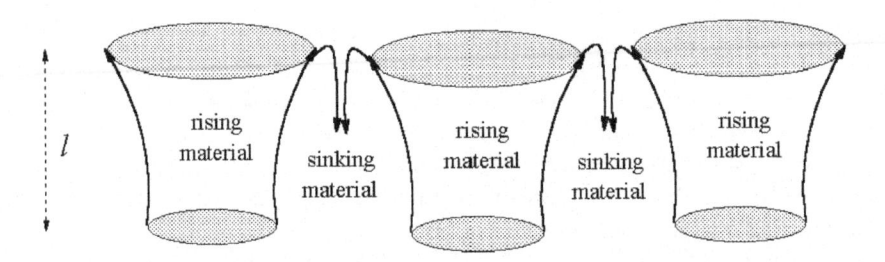

Figure 3.4 Heuristic sketch suggesting why the mixing length l should be of the order of the pressure scale height.

If $\Delta T_{conv} \ll T$, this value is much smaller than the sound speed $V_S = (\gamma k_B T/\mu)^{1/2}$, so that convection does not perturb the hydrostatic equilibrium.

Finally, with the convection speed (3.43), the convective energy flux (3.42) is

$$F_{conv} \sim \rho \frac{\gamma}{\gamma - 1} \left(\frac{k_B \Delta T_{conv}}{\mu} \right)^{3/2}. \qquad (3.44)$$

It is smaller by a factor of about $(\Delta T_{conv}/T)^{3/2}$ than the maximum flux $F_0 \sim \rho c_p T V_S$ that could be carried by fluid motion if matter were to move at the speed of sound.

Application to convection within the Sun

Let us tentatively apply these results to the Sun. First, let us estimate the temperature excess of a rising blob. An upper limit to the convective energy flux is the total flux determined by the luminosity, i.e. at distance r

$$F_{conv} \leq \frac{L}{4\pi r^2}.$$

Substituting this value into (3.44), we find an order-of-magnitude estimate of the temperature difference of a convective blob

$$\Delta T_{conv} \leq \left(\frac{L}{4\pi r^2 \rho \gamma/(\gamma - 1)} \right)^{2/3} \frac{\mu}{k_B}. \qquad (3.45)$$

Near the base of the solar convection zone ($r \simeq 0.7 R_\odot$), we have from Fig. 3.2: $L \simeq L_\odot$, $\rho \simeq 200 \text{ kg m}^{-3}$, $M_r \simeq M_\odot$ and $T \simeq 2 \times 10^6$ K, whence the scale height $H \simeq 0.1 R_\odot$, so that (3.45) yields $\Delta T_{conv} \lesssim 0.5$ K (in order of magnitude), which is very small compared to the temperature itself.

Let us compare:

- the excess of the actual temperature gradient over the adiabatic value, equal to $|\, dT/dr - (dT/dr)_{adiabatic}\,| \equiv \Delta T_{conv}/l$, from (3.41)

- with the temperature gradient itself $|\, dT/dr\, | \sim \langle T \rangle / R_\odot$.

We see that the ratio of both quantities is about $(\Delta T_{conv}/\langle T \rangle) \times R_\odot/l$, which is a very small number since $(\Delta T_{conv}/\langle T \rangle) \sim 10^{-7}$ and $R_\odot/l \sim 10$.

Structure and dynamics 131

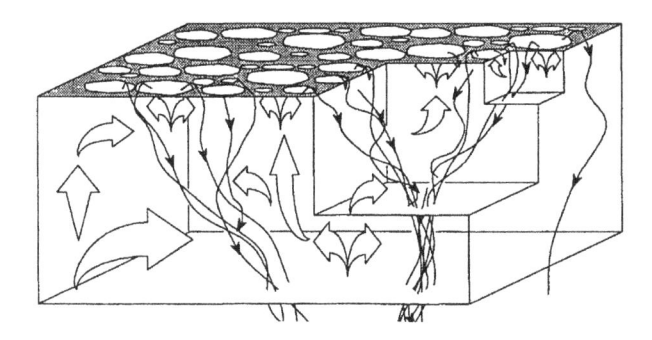

Figure 3.5 (Speculative) sketch of thermal convection in a stratified medium. The white arrows illustrate the widening of upflows, whereas the thin black arrows show the narrowing and successive merging of downflows; the boxes illustrate how the same process might occur on different scales. (Reprinted with permission from [44] © 1990 by Annual Reviews.)

We conclude, therefore, that when convection takes place, the actual temperature gradient remains very close to the adiabatic value. This means that convection is so effective at transporting heat that a very small super adiabaticity is enough to carry out the whole energy flux. In short, the temperature gradient has to be steeper than the adiabatic value for convection to begin, but once begun, convection keeps the actual gradient close to the adiabatic one. Hence, the adiabatic gradient (3.34) is not only a limiting value, but a roughly actual value, convection acting in some sense as a valve.

This has a fortunate consequence: our lack of detailed understanding of convection has no dramatic effect on stellar models since in most of the convective zone, convection keeps the temperature gradient close to adiabatic. We shall see below that this is no longer true in the close vicinity of the solar surface, where the density and opacity become so small that a large superadiabadicity is required to transport the energy flux.

With the above value of ΔT_{conv}, the convective speed near the base of the convection zone is from (3.43): $v_{conv} \sim 70$ m s^{-1}. The lifetime of a blob is thus $\tau_{conv} \sim l/v_{conv} \sim 10^6$ s; this is very short compared to the timescale of stellar evolution, so that convection is very effective at mixing stellar material.

Beyond the mixing-length picture

One should not get the impression that the mixing-length picture is the key for understanding solar convection. It is only a convenient way of deriving scaling properties and introducing convection into stellar models. Detailed numerical simulations, which are beginning to acquire a high degree of sophistication, draw a somewhat different picture. In particular the simulation of the subsurface layers show a large asymmetry between upward and downward motions, with concentrated downflows extending over the whole convective domain (Fig. 3.5).

This asymmetry between ascending and descending matter is mainly due to the strong density stratification of the medium, which causes the cross-section of a flow tube to decrease with depth in order to conserve mass. As a result, a tube of rising material diverges, whereas a tube of sinking material converges. Matter whose radial cross-section occupied initially some fraction of a surface will cover it entirely after having risen by a few scale heights (Fig. 3.5), so that upflows cannot extend very far in height. In contrast, sinking matter contracts to a small fraction of its initial cross-section, which tends to produce extended filamentary downflows; they may merge into fewer flows, on successively larger scales at successively larger depths, forming a tree-like structure [44].

Conservation of angular momentum introduces an effect akin to what is observed in a bath tub. Small cyclonic motions around the centres of the downdrafts have their circular velocity amplified as sinking matter contracts. This implies that downflows should be highly turbulent, whereas, by the same argument, upflows may be smooth since expansion smoothes out the irregularities.

Finally, we found a lifetime for a convective cell of the order of $\tau_{conv} \sim 10^6$ s at the base of the convective zone. This is of the same order of magnitude as the solar rotation period of about 25 days. Hence solar rotation and convective motions should significantly influence each other. We shall examine below the motions induced by solar rotation.

3.2.4 The quiet photosphere

We said that the solar convective zone occupies the outer 30% of the solar radius. But first of all, how is the solar radius, that we took for granted up to now, defined from a basic point of view? And can we understand some observed aspects of the photosphere surveyed in Chapter 1 from the above estimates?

The solar surface

We have seen that the density of the solar material decreases outwards, without undergoing any abrupt transition. And yet we see a well-defined solar surface, the *photosphere*, where the solar interior 'ends' and the atmosphere 'begins'. As we noted in Section 1.2, the sharpness of this transition is due to the smallness of the pressure scale height H: with a mean molecular mass $\mu \simeq 1.25 m_p$ near the surface, (3.36) yields $H \simeq 150$ km $\simeq 2 \times 10^{-4} R_\odot$. Virtually all the light we receive from the Sun comes from there, because the material below is essentially opaque, whereas the material above is essentially transparent.

Let us see how this property determines the position of this layer. Understanding this region is important, because most of what we know about the Sun is based on analysis of radiation coming from there. The task is difficult, however, because most of the simplifications made to model the solar interior fail there (see [22]). A detailed treatment involving radiation transfer calculations [23] is outside the scope of this book, and we shall only make elementary estimates.

Structure and dynamics

The temperature is small there – smaller than the mean temperature in the solar interior by about three orders of magnitude (see Fig. 3.2); this is why the scale height is so small, causing the density to fall sharply with altitude so that the medium becomes nearly transparent to radiation. Going outwards, photons encounter fewer and fewer particles – their free path increases – and at some level, they have a fair probability of escaping directly, so that the light we see comes essentially from that level and those above. This means that between this radius r_s and infinite distance, the mean number of particles encountered by a photon is of order of magnitude unity, i.e.

$$\int_{r_s}^{\infty} dr/l_{ph}(r) \sim 1 \qquad (3.46)$$

where $l_{ph}(r)$ is the average free path of photons at distance r. Substituting from (3.13): $l_{ph} = 1/\kappa\rho$ with $\kappa \sim \kappa(r_s)$ and $\rho \propto e^{-r/H}$, we find

$$1 \sim \kappa(r_s)\rho(r_s) \int_{r_s}^{\infty} dr\, e^{-(r-r_s)/H}$$
$$\sim \kappa(r_s)\rho(r_s) H(r_s) \sim H(r_s)/l_{ph}(r_s).$$

Hence at the photosphere, the mean free path of photons is of the order of magnitude of the pressure scale height; equivalently, this is the level from which the radiation we receive is attenuated by e^{-1} by the overlying material.

We have based our estimate on the sharp outward density decrease. In fact, κ drops off with altitude even more rapidly than ρ because the opacity in the outer solar layers is produced by the negative ion H^-, that is atomic hydrogen with a second electron loosely attached to it.[10] This opacity falls dramatically at the photosphere, because ionisation becomes negligible (the electron-to-neutral density ratio is only about 10^{-4}), so that free electrons are no longer available for H^- formation.

With this opacity, the photosphere level is found at a mass density of $\rho \simeq 3 \times 10^{-4}$ kg m^{-3}.

Convection near the surface

What happens to convection near the solar surface? The sharp decrease in density and opacity is expected to change its nature.

Let us try to estimate the average temperature excess of a convective blob at the photosphere level. If convection were to transfer a significant part of the solar flux, i.e. if $F_{conv} \sim L_\odot/4\pi R_\odot^2$, then with $\rho \simeq 3 \times 10^{-4}$ kg m^{-3}, (3.45) would yield $\Delta T_{conv} \sim 2 \times 10^3$ K – a value that is not very small compared to the temperature itself; this suggests that convection becomes ineffective as a heat-transport process near the solar surface. As the medium becomes transparent,

---

[10] H^- ions play an important role there because hydrogen is mostly in its neutral atomic form, so that many H atoms are available, and heavy elements are partly ionised, making free electrons available for H^- formation.

the temperature excess of the blobs disappears quickly because of radiation, making convection less and less efficient, so that both radiation and convection compete to carry the energy; ultimately, very close to the surface, the medium becomes convectively stable.

With a convectively stable photosphere, why do we observe some convection at the surface in the form of granulation, as described in Section 1.2? This is because the convective motions have a finite velocity at the boundary of the convective layer, making them intrude into the stable layer above, so that the boundary of the convective zone is not sharp.

Let us try to interpret the observed properties of granulation as a surface extrapolation of convection in the upper convective zone, keeping in mind that the traditional mixing-length picture is expected to yield only scaling properties, and that the phenomenon is inherently non-local, being driven essentially by radiative cooling at the surface. To begin with, let us estimate the convection speed. With $\Delta T_{conv} \sim 2 \times 10^3$ K, (3.43) yields $v_{conv} \sim 5 \times 10^3$ m s^{-1}, which agrees in order of magnitude with the observed vertical velocity in granulation of about 2 km s^{-1}.

Granules

Can we understand the observed pattern of bright granules of rising material surrounded by dark narrow lanes of sinking matter described in Section 1.2? As we noted previously, conservation of mass implies that a tube of rising matter diverges. Because of the small scale height near the surface, this divergence must be strong, with the horizontal velocity increasing rapidly with distance from the axis of a flow tube. Since there is no force in the horizontal direction (because gravity acts vertically) this velocity increase must be accompanied by a pressure decrease (according to Bernoulli's theorem, cf. Section 2.3).

Hence one expects a pressure excess near the axis of the rising material, which deflects the flow into a nearly horizontal expansion (Fig. 3.4). The same is true in the intergranular regions to reduce the horizontal speed and deflect matter downwards. Bernoulli's theorem tells us that this pressure excess is $\Delta P \sim \rho v_x^2$ in order of magnitude, v_x being the maximum horizontal speed. Since the order of magnitude of the pressure is $P \sim \rho V_S^2$ where V_S is the local sound speed, we have $\Delta P/P \sim v_x^2/V_S^2$. Now, since the density $\rho \propto P/T$, we have for small perturbations

$$\Delta \rho / \rho = \Delta P / P - \Delta T / T. \tag{3.47}$$

Recall that the fluid rise is driven by the buoyancy force which requires $\Delta \rho < 0$. Therefore if v_x is not small compared to V_S, the (positive) term $\Delta P/P$ counteracts the temperature term, thereby braking the buoyancy. Hence, upflows are braked, which increases their widening because of mass conservation. The same effect accelerates the downflows since the positive ΔP increases the density excess; this acceleration increases their narrowing. The greater speed in downflows than in upflows must be compensated for by a smaller cross-section, since upflows and downflows must carry the same mass on average, and their density difference is not very large. These arguments, however naive, explain why we see wide bright upflows (granules) separated by dark narrow downflows (intergranules).

Since the opacity decreases strongly as the temperature decreases, we can see deeper where the temperature is low. Hence the dark lanes are seen deeper than the bright granules; this effect reduces the contrast between bright and dark structures, since the latter are seen at deeper levels, whose average temperature is greater.

The above elementary considerations on dynamics might also explain the maximum size observed for the granules. Consider a flow tube of rising material with cylindrical symmetry. We have seen that the tube widens; the rising mass entering the tube at altitude z where the tube's radius is x equals that at $z+dz$ where the radius is $x+dx$ (if $\partial/\partial t = 0$). Neglecting the variation in vertical velocity v_z, we thus have $d\left(\rho\pi x^2\right)/dz = 0$, whence

$$x^2 d\rho/dz + 2\rho x dx/dz = 0.$$

Substituting $d\rho/dz \sim -\rho/H$ and noting that the tube shape follows flow lines, so that the horizontal and vertical velocities satisfy $dx/dz = v_x/v_z$, we deduce

$$v_x/v_z \sim x/2H. \tag{3.48}$$

Hence the horizontal velocity grows with the size of the granule. But we have seen that when this velocity becomes of the order of the sound speed, the excess pressure brakes the flow. Thus $v_x < V_S \simeq 10 \text{ km s}^{-1}$, so that with the observed vertical velocity $v_z \sim 2 \text{ km s}^{-1}$, (3.48) gives the maximum granule size

$$x \leq 2H \times V_S/v_z \sim 10H.$$

Substituting the scale height found above, we obtain 1.5×10^3 km, reasonably close to observation.

Sophisticated numerical simulations are now able to reproduce impressively most of the observed properties of granulation [46]. The patterns observed at larger scales, however, are more difficult to explain; in particular, the most conspicuous large-scale pattern – the *supergranulation*, of about $(2-3) \times 10^4$ km in size and one day in duration – to which we shall return later.

3.2.5 Solar rotation

Before addressing the magnetic field, we must discuss the solar rotation, which plays a major role in its generation.

How the Sun rotates

As we said in Section 1.2, the Sun is rotating, with the equatorial regions rotating faster than the polar ones. Even though the solar rotation has been studied for centuries, how the solar interior rotates has remained a mystery until recently,

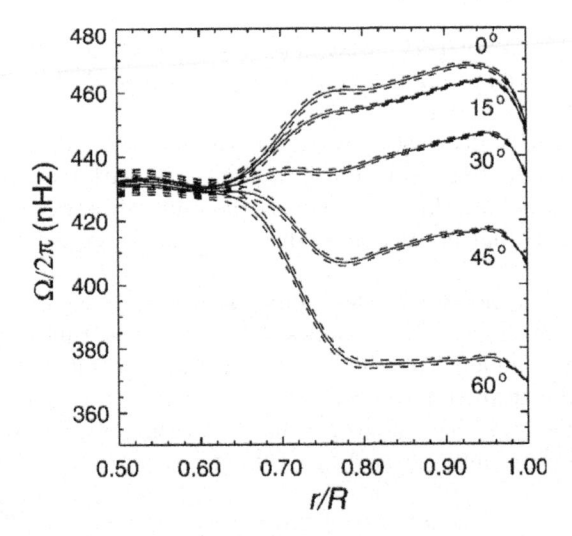

Figure 3.6 Solar rotation frequency (time-averaged) plotted against radius at different latitudes, deduced from helioseismic data (dashed lines represent 1 σ error bounds; reprinted with permission from [17]). The inner part of the Sun (not shown) is inferred to rotate approximately as a solid body. Copyright 2000 AAAS.

when observation of global solar oscillations opened a new window to the solar interior.

These solar oscillations result from standing waves, produced as waves – mainly acoustic waves – propagate within the Sun and are refracted and reflected due to the variations of properties with depth. This yields oscillations at discrete frequencies, somewhat as in musical instruments, and these eigenfrequencies reveal properties of the material where the waves are propagating. Rotation causes some modes to split into multiplets, whose analysis reveals how the rotation varies within the Sun [40].

Recent observations are shown in Fig. 3.6. One sees that the angular frequency observed near the solar surface, where rotation is faster at the equator than near the poles, extends through much of the convection zone (the outer 30% by radius), with little radial dependence except at the boundaries. Most of the radial variation takes place in a thin layer at the base of the convection zone, where the angular velocity adjusts to that of the deeper interior, which appears to rotate roughly as a solid body, at a rate somewhere between the equatorial and polar surface rates.

What produces the latitude dependence of rotation in the convective zone, and how is operated the transition between this region and the deeper interior which appears to be rotating quasi-uniformly? Conservation of angular momentum would suggest that the angular velocity should vary as a function of the lever arm, $r \cos \lambda$ (at distance r and latitude λ), implying a cylindrically symmetric rotation pattern (Problem 3.4.3). Figure 3.6 shows that this argument is inadequate. Solar differential rotation is far from being understood, even though numerical simulations yield promising results [12].

Some consequences of solar rotation

Solar rotation has many consequences. First of all, it affects the shape of the Sun, as the centrifugal force acts against gravity near the equator, and tends to transform it into an oblate spheroid. A hint as to the importance of this effect may be obtained by comparing the centrifugal force $\Omega^2 R_\odot$ to the gravitational force $M_\odot G/R_\odot^2$ (per unit mass)

$$\chi = \frac{\text{centrifugal force}}{\text{gravitational force}} = \frac{\Omega^2 R_\odot^3}{M_\odot G} \sim (\Omega \tau_{hydr})^2 \qquad (3.49)$$
$$\sim 2 \times 10^{-5}$$

where Ω is the angular rotation frequency and τ_{hydr} is the hydrostatic timescale (3.24). This is expected to produce a small oblateness, and calculations yield a relative difference of about 10^{-5} between the equatorial and polar solar radii, which comes to only 14 km – less than can be resolved with the best solar telescopes, so that measurements require sophisticated techniques [14].

Another basic effect of solar rotation is that it induces large-scale motions in the solar interior, because the centrifugal force modifies the hydrostatic equilibrium; the pressure gradient must change to compensate, and in turn this modifies the temperature gradient and the energy flux. The only way for the Sun to maintain a steady state despite this difference between equatorial and polar regions is by circulating matter and energy between them.

And finally, solar rotation is a key factor to solar magnetism.

3.3 Some guesses on solar magnetism

The title of this section is inspired by Richard Feynman (speaking of the origin of the Earth's magnetic field): 'Nobody really knows – there have only been some good guesses.'[11]

The origin of cosmic magnetic fields is a long-standing problem, and the solar system is no exception. Significant advances in our understanding of the Earth's magnetic field (and of planetary magnetic fields in general) have been made over four decades; this is due in a large part to major improvements in computational tools [19], and the basic mechanism is perhaps understood. This is not so, however, for the solar magnetism. A large part of the difficulty lies in the huge magnetic Reynolds number (Section 2.3) in the Sun, which makes the range of scales involved so wide that it cannot be handled by numerical simulations, even with the best computers available at the beginning of the twenty-first century.

---

[11] Feynman, R. P., R. B. Leighton and M. Sands 1964, *The Feynman Lectures on Physics*, New York, Addison-Wesley.

Many problems have to be solved. The first one is the origin of the solar cycle: what produces the large-scale magnetic field, which reverses every 11 years? A related issue is the phenomenology observed along the cycle, for example the pattern of appearance of sunspots – a point that we shall not address in detail (see [21]). A fundamental question is the origin of the intermittent structure of the magnetic field, which appears and disappears in permanence at the solar surface, and exhibits small regions of huge magnetic field.

To begin with, let us examine how a magnetic field might be produced and sustained in the conducting solar interior. The Sun is not special in this respect; this problem is generic for cosmic bodies that are both fluid and rotating [24], [30], [55]. This question is full of subtleties [34], and the treatment below captures only a minute part of it.

3.3.1 Elements of dynamo theory

Might the solar magnetic field be the remnant of some primordial field, dating back to the birth of the Sun? Let us apply naively the results of Section 2.3. A magnetic field of spatial scale L in a medium of electric conductivity σ decays at the timescale

$$\tau_\sigma = \frac{L^2}{\eta} \quad \text{with} \quad \eta = \frac{1}{\mu_0 \sigma} \tag{3.50}$$

where $\eta = (\mu_0 \sigma)^{-1}$ plays the role of a magnetic diffusivity. Estimating the conductivity from (2.102) with the temperature at the base of the convection zone, we find $\sigma \sim 10^6 \, \Omega^{-1} \, \mathrm{m}^{-1}$, whence the ohmic diffusivity $\eta \sim 1 \, \mathrm{m}^2 \, \mathrm{s}^{-1}$. With $L \sim 7 \times 10^8$ m (the solar radius), the decay time is $\tau_\sigma \sim 10^{18}$ s, that is 3×10^{10} years. We know, however, that the solar magnetic field reverses every 11 years, and that smaller structures change much faster. Even with L of the order of the granule's scale ($\sim 10^6$ m), the ohmic decay time is still vastly greater than the observed timescales. Clearly, we need a mechanism to destroy the field much faster and also to recreate it, for producing a cycle.

A current view is that some kind of dynamo process is at work, converting mechanical energy into magnetic energy. Indeed, ample kinetic energy is available in the solar rotation and convective motions. One may imagine that, starting from some small seed magnetic field, fluid motions induce electromotive forces that create just those currents required to generate the initial magnetic field. This would not be a perpetual motion machine since the energy lost via joule dissipation is furnished by the fluid motion.

How could such a dynamo work? From our school days, a dynamo evokes memories of a rotating frame placed between magnetic poles to produce an electric current. More practically, it evokes the bicycle dynamo, in which some of the work done by pedalling is used to spin a conducting coil in the presence of a magnetic field, producing a current in the lamp filament. The solar problem is basically different, for two reasons. First, the magnetic Reynolds number is huge:

$$R_m = \mu_0 \sigma v L \gg 1. \tag{3.51}$$

Some guesses on solar magnetism 139

Indeed, taking for the length L and the speed v the characteristic values we derived at the base of the convection zone, we find $R_m \sim 10^{10}$. We saw in Section 2.3 that in this case the magnetic field is frozen in the plasma; the opposite is true in our home dynamos, whose vastly smaller size generally yields $R_m < 1$, letting the magnetic field lines slip through the conductors. The second difference is that the Sun holds no magnets, discs or wires, but a continuous fluid, so that the electric current path cannot be prescribed a priori.

Basically, a self-excited solar dynamo should involve two items. First, fluid motions should be capable of increasing an initial magnetic field and of producing a cycle, with a much shorter timescale than the absurdly large diffusion time τ_σ estimated above. Second, these motions have to be maintained against the Lorentz force.

Hints on kinematic dynamos

Let us forget for the moment the back reaction of the Lorentz force on the motion, even though this is the most difficult part of the problem. Let us concentrate on the first point: finding fluid motions capable of producing an adequate magnetic field. This problem is formalised in the induction equation (Section 2.3)

$$\frac{\partial \mathbf{B}}{\partial t} = \frac{\nabla^2 \mathbf{B}}{\mu_0 \sigma} + \nabla \times (\mathbf{v} \times \mathbf{B}) \qquad (3.52)$$

which enables one to calculate the magnetic field $\mathbf{B}(\mathbf{r}, t)$ from the fluid velocity field $\mathbf{v}(\mathbf{r}, t)$. Since this equation is linear in $\mathbf{B}$, a magnetic field cannot be created 'from nothing' but can only be amplified (or reduced). For the same reason, this equation alone cannot determine how large the magnetic field can grow, and the solution (if any) will only be applicable if the magnetic field is sufficiently small to perturb negligibly the motion.

Let us recall the physical significance of the induction equation. Diffusive magnetic decay is embodied in the first term of the right-hand side, and acts at the timescale $\tau_\sigma = \mu_0 \sigma L^2$ for magnetic variations at scale L. If the velocity does not vary at smaller scales, the second term changes the field in a time L/v, which is shorter than τ_σ if $\mu_0 \sigma v L > 1$. A large value of R_m thus means that ohmic diffusion acts more slowly than fluid motion, so that the field may (not necessarily will) be amplified by fluid motions.

How can fluid motions amplify the magnetic field? We saw in Section 2.3 that this can be done by stretching the field lines. Imagine a tube of magnetic flux in a conducting fluid and assume that the fluid motion perpendicular to the field is not uniform (Fig. 3.7). For $R_m \gg 1$, the field lines are frozen in the fluid and thus dragged along by the flow. If the stretching does not change the mass density (which holds true if the speed is small compared to the sound speed), flux freezing implies that the magnetic field increases in proportion to the length of the tube. The increase in magnetic energy stems from the work done in stretching the tube against the tension of the field lines.

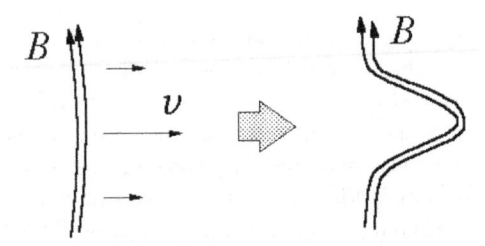

Figure 3.7 Distortion of a magnetic field by a non-uniform perpendicular velocity.

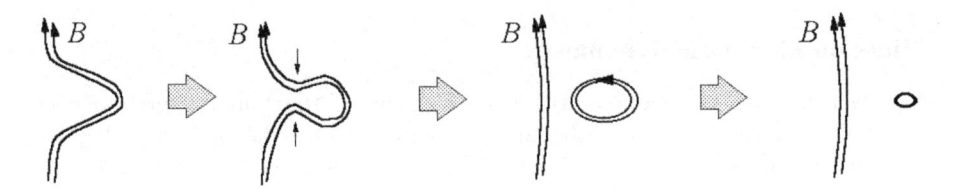

Figure 3.8 Increasing magnetic diffusion by producing magnetic variations at small scales; the field lines are pushed together locally, making them reconnect, followed by decay of the detached loop.

This is not the whole story, because the tube cannot be stretched indefinitely; to produce a dynamo, we should return (more or less) to the initial tube geometry, albeit with a greater magnetic field. Achieving this, however, would mean that the magnetic flux through a surface moving with the fluid could increase, in contradiction with flux freezing. There is a way out, however. Since the first term in (3.52) – which embodies magnetic diffusion – involves a second derivative, magnetic diffusion might proceed much faster if the transverse magnetic field gradients are much greater than those of the flow (Fig. 3.8).

We thus see that two basic ingredients enter in a kinematic fluid dynamo. First, amplification of the large-scale magnetic field by stretching the field lines. Second, magnetic diffusion and reconnection via the production of large magnetic gradients. Whether the dynamo works depends on the relative effectiveness of these two competing processes; in other words, an efficient dynamo requires a little magnetic field diffusion, but not too much.

Unfortunately, these two ingredients are not sufficient; in particular it has been proved that dynamos having a high degree of symmetry do not work; in particular neither a stationary axisymmetric dynamo nor a centrally symmetric one works; the same is true for a dynamo whose velocity field is restricted to a plane. Hence a successful dynamo has to be complicated ... and so is its analysis.

Some guesses on solar magnetism 141

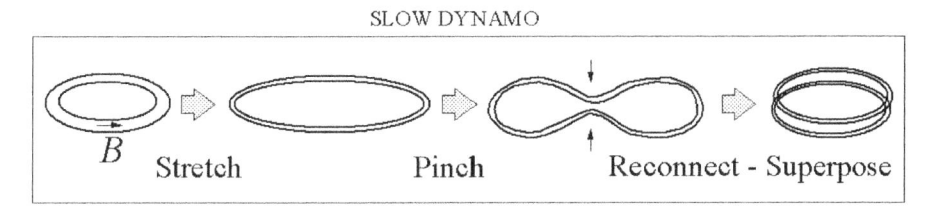

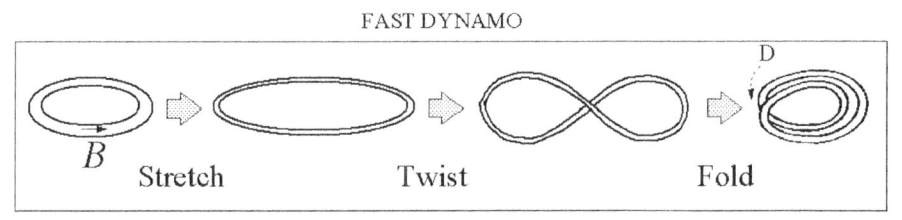

SLOW DYNAMO

B Stretch — Pinch — Reconnect - Superpose

FAST DYNAMO

B Stretch — Twist — Fold

Figure 3.9 Two thought experiments based on a loop of magnetic field frozen in a conductor. Top panel: one increases the magnetic field by stretching the loop; the loop is then pinched to promote diffusion and reconnection, producing two loops which are superimposed and packed together. Bottom panel: the stretched loop is twisted, folded over itself and repacked.

Two prototype dynamos

Figure 3.9 (top) shows a dynamo imagined by Alfvén [1]. Consider a loop of magnetic field immersed in a conducting fluid and stretch it to, say, twice its length; this halves the cross-section s and doubles the value of B. One then pinches the loop to produce large magnetic gradients, which accelerates magnetic diffusion. Reconnection of the field lines produces two separate loops, each resembling the initial one. Superposing these loops and packing them together reproduces the original loop, but with B amplified by a factor of two. The magnetic energy has increased, at the expense of the work done in driving the motion. This process can be repeated, but it is very slow since it cannot work faster than ohmic decay because of the central role played by magnetic diffusion.

The bottom panel of Fig. 3.9 shows an ingenious device which has been suggested to produce a faster dynamo, because its timescale appears to depend less critically on diffusion [49]. Instead of pinching the loop and having to wait for diffusion to cut the loop in two, we twist it to form a figure of eight and then fold it back to merge with itself. As in the above example, the total cross-section s is restored to its original value and contains twice the original magnetic flux. The process can be repeated, each step doubling the magnetic flux, so that after n steps the flux is amplified by a factor of $2^n = e^{n \ln 2}$. Since the number n of steps plays the role of time, the magnetic flux grows in an exponential way. If the typical length and speed are respectively L and v, the doubling time is L/v, so that this dynamo may proceed much faster than the previous one.

But does it really? The apparent simplicity masks a number of subtleties which embody many basic problems of dynamo action [25].

First of all, the topology of the field lines changes at each step, and a little diffusion around the point D (indicated by the arrow) is necessary to eliminate small-scale variations and reconnect the loops in order to recover the original topology.

A paper tape can enable one to visualise what happens: the twist out of the plane of the loop induces a twist of the tape around its own centre line; hence when the loop is broken and reconnected at D, the two loops thus created have the form of Möbius strips, each one having a twist of π. This intrinsic twist means that the magnetic field has acquired a component perpendicular to the plane of the loop, and therefore an helicity. Helicity is a basic feature of solar magnetic fields, to which we shall return in Section 4.2. In addition, the stretch in the plane of the loop flattens its cross-section into an ellipse whose short axis is perpendicular to the loop's plane, and decreases at each step, until the large gradients thus produced promote magnetic diffusion. Finally, consider two elements initially close to each other on the original loop. The stretch–twist–fold operations quickly separate them in a way that mimics turbulence.

This prototype dynamo may nevertheless be fast (i.e. act at the timescale L/v instead of τ_σ) since one can repeat each step without having to wait for diffusion to take place. Its major interest is that despite its apparent simplicity it captures several features that are thought to be essential in a dynamo process:

- magnetic amplification through the stretching of field lines produced by differential motions,
- twist and linkage of the field lines,
- appearance of fine structures that accelerate field diffusion,
- non-stationary motion (albeit stationary upon averaging),
- lack of symmetry and complex fluid trajectories at small scales.

Could a similar dynamo work in the Sun?

3.3.2 Solar kinematic dynamos

The alpha and the omega

The first step of our prototype dynamo can be achieved rather easily. Since the Sun does not rotate as a solid body, the solar rotation stretches the field lines, generating a large magnetic field. Figure 3.10 sketches how an initially vertical magnetic field line is distorted as it is drawn by the solar differential rotation in the simple case where the rotation varies only with latitude. The field lines are stretched along the rotation, and the net result is the production of an azimuthal magnetic field from a magnetic field roughly parallel to the rotation axis. In

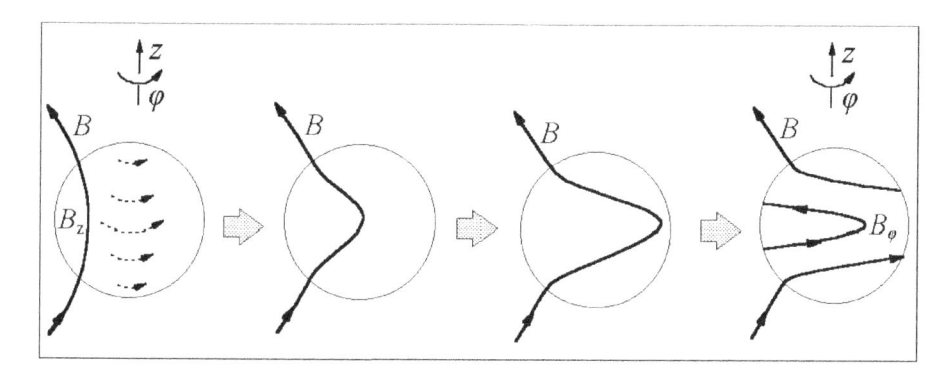

Figure 3.10 Producing a large azimuthal magnetic field by stretching a poloidal field on a rotating sphere that rotates faster at the equator than at the poles.

the Sun, the stretching is believed to be mainly realised near the layer where the rotation varies strongly with depth (Section 3.2.5), at the base of the convection zone.

Once a large-scale azimuthal magnetic field is produced from a poloidal one (i.e. oriented in meridian planes) through differential rotation (this step is known as the *omega* part of the dynamo process), we need a mechanism to reproduce a poloidal field from an azimuthal one, in order to complete the cycle.

This might be done as follows. Consider a large-scale azimuthal field line in a plane parallel to the equator (Fig. 3.11, left), and assume that there are small-scale motions – for example turbulence – lacking reflectional symmetry; the mean fluid helicity is in this case generally non-zero, i.e. the velocity satisfies $\langle \mathbf{v} \cdot (\nabla \times \mathbf{v}) \rangle \neq 0$. In other words, we have a set of small screw-like vortices – somewhat like cyclones in the Earth's atmosphere – with unequal numbers of right-handed and left-handed ones.[12] These motions produce small loops that are twisted as the one shown on Fig. 3.11 (middle). Diffusion acts faster than elsewhere at the base of the twisted loop (indicated by D in Fig. 3.11), because the gradients are large, finally making the loop detach from the parent line. Each detached loop corresponds, by Ampère's law, to an azimuthal current j_ϕ, and many of these taken together around the large-scale field B_ϕ are equivalent to a large-scale azimuthal current J_ϕ (right), which in turn produces a large-scale poloidal magnetic field.[13]

Finally, therefore, from a large-scale azimuthal magnetic field, small helical motions may produce a large-scale poloidal magnetic field, completing the cycle.

---

[12]This might be a consequence of solar rotation and convection. An old argument says that a rising blob expands laterally, producing a small loop of magnetic field. The Coriolis force (or conservation of angular momentum) makes the loop rotate more slowly, so that it acquires a spin opposite to that of the Sun. Sinking blobs acquire an opposite twist, but since their velocity is also opposite, the helicity $\langle \mathbf{v} \cdot (\nabla \times \mathbf{v}) \rangle$ has the same sign.

[13]In Fig. 3.11 the helicity is negative in a right-handed frame of reference and the twist angle is smaller than π (which may be true if the disturbances are short-lived); in this case J_ϕ is along B_ϕ, so that the component B_z is opposite to the one of Fig. 3.10.

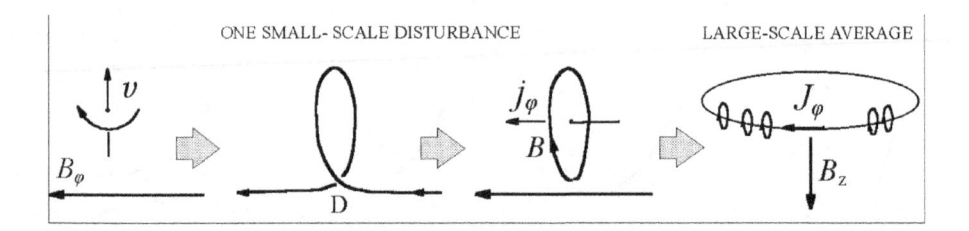

Figure 3.11 Producing a large-scale poloidal magnetic field from an azimuthal one through small-scale helical motions. A localised velocity disturbance having a helicity – for example a rising spinning blob – distorts a large-scale field B_ϕ (left), producing a small twisted loop (middle), which detaches and is equivalent to an electric current j_ϕ. Many such small loops located along a large circumference yield a large-scale current J_ϕ, producing a poloidal field (right).

This concept, originally due to Parker [29] and known as the *alpha* effect in the jargon of dynamo experts, is formalised with various refinements in the so-called *mean-field electrodynamics*. In this scheme, the velocity and magnetic field are split into a large-scale part which remains after averaging over the small-scales, and a fluctuating part having a much smaller scale and a zero average, which may come from turbulence. If the small-scale flow has a non-zero helicity, a large-scale poloidal component is produced, as found above in a heuristic way. This yields a periodic solution which some think might explain the solar cycle.[14]

Turbulent magnetic diffusion

As we have said, an efficient dynamo requires significant magnetic diffusion, which may occur if small enough scales are produced. Turbulence might solve this problem, by entangling the field lines, making the mean field spread rapidly over a large volume, while bringing together fields of opposite polarity (Fig. 3.12).

Turbulence may even do more. When the flow is chaotic, the trajectories of two neighbouring parcels separate exponentially, thereby stretching the field lines and increasing the magnetic field amplitude [13]; this entanglement of the field lines is accompanied by a decrease in spatial scales which promotes field diffusion. The small-scale flow has thus the major features of a dynamo device and may be a dynamo in its own right [31]; we shall return to this point later.

Let us consider the large-scale mean field. How fast can turbulence make it spread and decay? We may get some hints by noting that at large R_m, the magnetic field behaves in some way as a vector connecting two neighbouring fluid parcels. Consider a scale l_T and a speed v_T associated with turbulence, and assume that the fluid parcels separate as in a random walk of step length l_T and speed v_T. The time τ_T for diffusing over a distance $L \gg l_T$ is of the order of $\tau_T \sim L^2/(v_T l_T)$ (from the same reasoning as in Section 2.3). Since

---

[14] Although the method encounters many difficulties when the Reynolds number calculated with a scale and speed corresponding to turbulence is large, as is the case for the Sun.

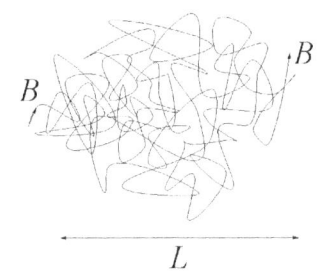

Figure 3.12 A naive sketch showing how turbulence entangles the magnetic field lines and enhances magnetic diffusion.

the entanglement of field lines mixes together lines of opposite polarity until the length scale is so small that ordinary diffusion operates and makes them annihilate, a magnetic field of spatial scale L will decay in the time

$$\tau_T = \frac{L^2}{\eta_T} \text{ with } \eta_T \sim v_T l_T. \tag{3.53}$$

This rough picture suggests that turbulence strongly increases the magnetic diffusivity, reducing the magnetic decay time (3.50) by a factor of $\mu_0 \sigma v_T l_T$, which is just the magnetic Reynolds number calculated with the scale and speed associated to turbulence. Taking for v_T and l_T the scales we determined at the bottom of the convection zone, we find $\eta_T \sim 5 \times 10^9$ m^2 s^{-1}. With L equal to the solar radius, (3.53) yields $\tau_T \sim 10^8$ s, about 3 years – a significant fraction of the solar cycle. It is thus no coincidence that this enhanced diffusion plays an important role in the mean-field dynamo theory.

This is not the whole story, however, because turbulent diffusion cannot be applied without caution to the magnetic field as if it were a scalar, and we have not yet addressed the more difficult question: how large can the magnetic field grow before the growth is halted by the effects of the Lorentz force? For doing so, let us first examine some subtle consequences of the induction equation (3.52) at small scales when $R_m \gg 1$.

3.3.3 Concentrating and expelling the magnetic field

Consider a very simple situation: a uniform magnetic field $\mathbf{B}_0$ in a conducting fluid where a cylindrical region $r \leq r_0$ is rotating as a solid body around its axis perpendicular to $\mathbf{B}_0$ (Fig. 3.13). How does this local rotation distort the magnetic field?

Let ω be the angular speed; the speed at distance r_0 from the axis is ωr_0, and the corresponding magnetic Reynolds number is $R_m = \mu_0 \sigma \omega r_0^2$. When $R_m \leq 1$, the diffusion term in the induction equation (3.52) dominates, so that the magnetic distortion cannot build up before diffusion counteracts it. As R_m increases, the magnetic distortion increases; Fig. 3.13 shows the shape of the field lines in the steady state, for values of R_m still sufficiently moderate to enable matter to stream significantly across the field [24].

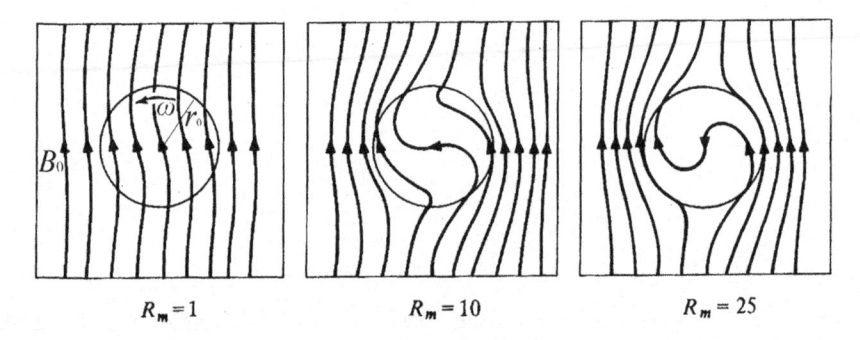

Figure 3.13 Distortion and partial expulsion of a magnetic field from a rotating cylinder. The rotation axis is perpendicular to the figure. The field distortion is small when $R_m \leq 1$; it increases with R_m, and for large R_m the field tends to be excluded from the rotating region. (Adapted from [24].)

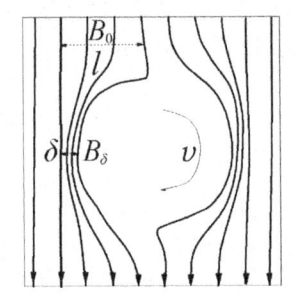

Figure 3.14 Magnetic flux amplification at the boundary of a rotating region.

When R_m increases more, the dragging of the field lines wraps them more and more until the gradients become so large that diffusion and field line reconnection tend to eradicate the field from the rotating region (Fig. 3.14). This field expulsion is related to the skin effect in electrical engineering. To understand this, consider a frame of reference rotating at the angular velocity ω; in this frame we have a magnetic field rotating at $-\omega$ outside a cylindrical conductor. Such a rotating field may be decomposed into two perpendicular components oscillating out of phase at the (angular) frequency ω. It is well known that at high frequencies the field is excluded from the conductor, penetrating only a small 'skin depth' $\delta = 2/(\mu_0 \sigma \omega)^{1/2}$.

At high R_m therefore, the magnetic field penetrates only a small distance δ into the rotating fluid. Expressing the above value of the skin depth in terms of the magnetic Reynolds number $R_m = \mu_0 \sigma \omega r_0^2$, we have

$$\delta \sim l/R_m^{1/2} \tag{3.54}$$

where l is the spatial size of the rotating region (r_0 in this case). The small width δ of the region where the magnetic field accumulates results in the balance between magnetic diffusion – acting with the timescale $\mu_0 \sigma \delta^2$, and advection –

acting with the timescale $\sim 1/\omega$; the value (3.54) of δ ensures that both timescales are comparable.

This result is a general property of fluid differential motion at large magnetic Reynolds numbers [51]. The magnetic flux concentrates in a thin layer of thickness given by (3.54), with

$$R_m = \mu_0 \sigma v l \qquad (3.55)$$

in terms of the typical dimension l and speed v. The thickness δ adjusts to be small enough for the diffusion time $\mu_0 \sigma \delta^2$ to match the timescale l/v for magnetic line motion over the structure (its turnover time), so that the two corresponding terms in the induction equation (3.52) balance each other.

This accumulation of field lines at the boundaries of the region corresponds to an increase in magnetic field amplitude. Let us estimate this increase with a (very) rough argument. Consider the field lines at some time t (Fig. 3.14). Let B_0 be the undisturbed magnetic field strength far from the eddy and B_δ the value in the thin region of thickness δ where the magnetic flux accumulates. Roughly speaking, two magnetic lines separated by about the eddy size l at a large distance from the eddy, where the magnetic field strength is B_0, become separated by only about δ in the concentration region at the edge of the eddy. In the two-dimensional problem studied here, the cross-section of a flux tube is proportional to the distance between the field lines. Hence flux conservation along a flux bundle yields $B_0 l \sim B_\delta \delta$ with $\delta \sim l R_m^{-1/2}$, whence

$$B_\delta \sim B_0 R_m^{1/2}. \qquad (3.56)$$

In this two-dimensional problem, the flux is concentrated over one dimension only – producing a sheet. In a three-dimensional problem, if the velocity field concentrates the magnetic flux along two dimensions, it produces a small tube, whose cross-section varies as the square of the width δ, so that the magnetic field is amplified by a factor of R_m instead of $R_m^{1/2}$.

In conclusion, for $R_m \gg 1$ a rotating eddy excludes the magnetic field from its interior. The magnetic flux accumulates at the boundaries, forming thin sheets or tubes of width smaller than the eddy's by a large factor, where the magnetic field strength is amplified by the same factor; from the above elementary argument, this factor is of the order of magnitude of $R_m^{1/2}$ if the contraction acts over one dimension, producing sheets, and even greater with other geometries. This magnetic field amplification near the boundaries takes place very rapidly: in about a turnover time.

This result may give a hint as to why the magnetic flux at the solar surface is concentrated into small regions of large magnetic field, at the boundaries of convective structures, where convection sweeps the magnetic field. The length δ is a major parameter in the dynamo process; it is the scale at which the field is no longer frozen and topological changes may occur, letting the lines of force sever and coalesce, so that the field may reassemble itself to produce the large-scale cycle.

How large can the magnetic field grow in this way? When the magnetic field increases, so does the Lorentz force; this may stop the motions that concentrate the magnetic flux. How does this affect dynamo action?

3.3.4 Lorentz force restriction on dynamo action

The Lorentz force is expected to modify the dynamics when the magnetic energy density at small scales $B_\delta^2/2\mu_0$ becomes comparable to the kinetic one $\rho v^2/2$; in this case, convective motions will no longer be capable of compressing the magnetic flux. Equating both values gives an upper limit to the small-scale field:

$$B_\delta \leq (\mu_0 \rho)^{1/2} v \equiv B_{eq}. \tag{3.57}$$

Once this limit is reached, the Lorentz force produced by the small-scale field should prevent any motion across itself.[15]

Using the relation (3.56) between the field at small and large scales, we deduce from (3.57) the limit on the large-scale field

$$B_0 \leq \frac{(\mu_0 \rho)^{1/2} v}{R_m^{1/2}} \equiv \frac{B_{eq}}{R_m^{1/2}}. \tag{3.58}$$

This figure is smaller by a factor of $R_m^{1/2}$ than the limit B_{eq} obtained if the large-scale field itself were allowed to reach equipartition; note that if magnetic compression acts over two dimensions, the limit is even smaller.

What happens if the large-scale field B_0 becomes greater than the limit (3.58)? The stretching properties of the small-scale flow are expected to be strongly affected, leading to a decrease in the turbulent diffusivity and in the effectiveness of the turbulent dynamo. Given the huge values of R_m in the Sun, (3.58) yields $B_0 \ll B_{eq}$, so that the Lorentz force is expected to shut down the turbulent dynamo process when the field produced at large scales is well below the equipartition value.

And yet the solar magnetic fields manage to be much above the equipartition values (Table 3.1). Even though there are possible solutions to this problem (we shall see one below) and the above reasoning is oversimplified, this raises some doubts on the capabilities of a turbulent dynamo to generate the solar cycle [50]. There is indeed considerable controversy over this question [31], to such an extent that other explanations of the solar cycle have been proposed. Magnetic turbulence is indeed far from being understood, and much of our present beliefs may turn out to be erroneous.

The essential conclusion of all this is that in the present state of the art, an efficient solar dynamo requires three ingredients: ROTATION, CONVECTION and ... OPTIMISM. Much remains to be done, even at the basic level, since even the very nature of the dynamo process itself is not known with certainty.

---

[15] This is expected to occur after a time of the order of the turnover time l/v; hence smaller eddies are affected first since, as we shall see in Section 6.4.2, smaller turbulent eddies have a smaller turnover time.

Some guesses on solar magnetism   149

Table 3.1 *Dynamic properties at the bottom of the convective zone and at the photosphere, and corresponding equipartition fields compared to the actual values*

Location	$0.7\,R_\odot$	$1.0\,R_\odot$
ρ	$200\ \mathrm{kg\,m^{-3}}$	$3\times10^{-4}\ \mathrm{kg\,m^{-3}}$
v	$70\ \mathrm{m\,s^{-1}}$	$10^3\ \mathrm{m\,s^{-1}}$
$B_{eq}=(\mu_0\rho)^{1/2}v$	1 T	0.02 T
B_{actual}	10 T (presumed)	0.1–0.3 T (measured)

3.3.5 Elementary physics of magnetic flux tubes

How can the observed solar fields (Table 3.1) be much above the equipartition values that should stop convection? Furthermore, the magnetic activity observed on the Sun strongly suggests continual emergence of magnetic field lines through the solar surface. How does this come about?

Solar flux tubes

The solar magnetic field often appears in small regions of great magnetic field, so that it can often be viewed as made of thin tubes of large magnetic flux, isolated from each other in a background of a much smaller field.

If the tube thickness is small compared to the other scales, the properties can be assumed to be uniform across it, and at equilibrium the total pressure inside (particle plus magnetic) matches the one outside. Furthermore, if the magnetic field is greater than the equipartition value B_{eq}, the tube cannot be deformed passively by turbulence and may be pictured as having an existence of its own. Elementary MHD at large magnetic Reynolds numbers (Section 2.3) tells us that an isolated flux tube has a number of interesting properties:

- its large magnetic field produces a large magnetic pressure, which tends to make it expand and become lighter (and colder) than its surroundings,
- magnetic tension tends to make it shorter, so that it can be pictured as a stretched rubber band,
- the matter within can flow along it but not across it,
- when twisted or knotted, it cannot easily get rid of its helicity (see Section 4.2), much like a garden hose.

These properties make flux tubes amenable to quite subtle gymnastics (Fig. 3.15) [30]; a detailed review may be found in [33]. The birth of regions of opposite magnetic polarities on the solar surface – which vary in size from packets of sunspots to much smaller field concentrations – may be interpreted as the emergence of flux tubes through the Sun's surface (Fig. 3.16), magnetic

Figure 3.15 A favourite pet of solar physicists: the isolated magnetic flux tube. (Drawing by F. Meyer.)

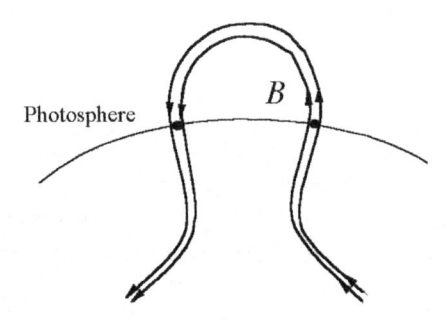

Figure 3.16 Naive picture of the emergence of a bipolar magnetic structure through the solar surface.

structures being pictured as the footprint areas where the tubes pierce the surface. Sunspots and related structures may form as azimuthal flux tubes emerge from deep within the Sun, which may explain why sunspots appear in pairs having opposite vertical magnetic fields, and (from Fig. 3.10) why they are concentrated near the equator, with pairs oriented roughly along the azimuthal direction; and finally why these pairs have opposite polarities in the two hemispheres, which reverse at each cycle. The observed pattern of emergence can be explained to some extent by detailed analysis, but a full comprehension would require understanding the solar dynamo itself, and the whole three-dimensional structure of solar convective flows.[16]

---

[16]In particular, it has been suggested that a meridional flow deep in the Sun might set the solar cycle period (see [16]).

Some guesses on solar magnetism 151

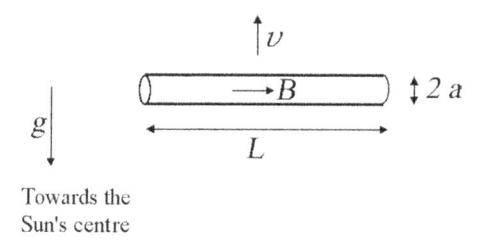

Figure 3.17 A tube of magnetic flux oriented in the horizontal direction.

Magnetic buoyancy

What makes flux tubes emerge from the solar interior? Just as for an emerging submarine, the starting physics is simple – being based on buoyancy – but the plumbing is complex.

Consider a thin flux tube immersed in the convection zone and oriented locally perpendicular to the solar gravity (Fig. 3.17), for example in the azimuthal direction. If its magnetic field amplitude B is much greater than the one outside, lateral pressure equilibrium between the interior (whose matter pressure is P) and the exterior (at pressure P_{ext}) yields

$$P + B^2/2\mu_0 = P_{ext}. \qquad (3.59)$$

The term $B^2/2\mu_0$ which compensates for the difference in internal and external matter pressures is the Lorentz force per unit surface on the thin current sheet bounding the tube (see Section 2.3.) If the temperature T in the tube is not too different from that outside, we have $P/\rho \simeq P_{ext}/\rho_{ext} \simeq k_B T/\mu$. Substituting this equality in (3.59), we find that the tube has a mass density smaller than outside by

$$\rho_{ext} - \rho \simeq \frac{B^2/2\mu_0}{k_B T/\mu}. \qquad (3.60)$$

The tube of volume V is thus subjected to an upward buoyancy force

$$F_B = (\rho_{ext} - \rho) V g \sim \frac{(B^2/2\mu_0) a^2 L}{H} \qquad (3.61)$$

where H is the local pressure scale height (3.36), and we have approximated the tube volume by $a^2 L$.

This buoyancy force, equal to the ratio of the tube's magnetic energy to the local pressure scale height, pushes it upwards. A portion of buoyant tube rising at vertical speed v is subjected to an aerodynamic drag force from the surrounding fluid that is of the order of

$$F_D \sim \frac{1}{2} \rho_{ext} v^2 a L \qquad (3.62)$$

where we have approximated the tube's longitudinal cross-section by aL, and assumed a drag coefficient of roughly unity. When this force balances the buoyancy force (3.61), the rising speed is therefore

$$v \sim v_A \left(a/H\right)^{1/2} \left(\rho/\rho_{ext}\right)^{1/2} \tag{3.63}$$

where $v_A = \left(B^2/\mu_0 \rho\right)^{1/2}$ is the Alfvén speed in the tube.

Actual tubes, if they do exist, are not expected to have the convenient linear shape of Fig. 3.17; a curvature will produce additional magnetic forces.

Furthermore the tubes are subjected to a number of instabilities, which make them break in parts or move as a whole. First, an untwisted magnetic tube, unlike a submarine or a gas bubble in water, has nothing to make it maintain its shape, and should break up rapidly. Hence, stable tubes are expected to be twisted, i.e. to have an azimuthal magnetic field component, which produces an inward Lorentz force helping the tube to maintain its shape (Problem 3.4.4) as does surface tension for bubbles in water. Second, the tubes are subjected to the convective instability we studied in Section 3.2.2, since in the convective zone a small upward displacement of a tube as a whole leads to an increase in its buoyancy. Since this kind of instability is driven by the buoyancy force, it is expected to arise with the timescale of normal convection. We found in Section 3.2.3 that this time is of the order of a month at the bottom of the convection zone; this is far smaller than the period of the solar cycle. Hence these tubes are not expected to remain there for a significant fraction of the solar cycle.

There is, however, a place where the tubes have enough time for being stretched before rising, so that the magnetic field can be amplified: the bottom of the convection zone, where the medium is expected to be convectively stable (although there are overshooting motions from the convective zone just above), and the strong variation of solar rotation with depth (Fig. 3.6) can stretch the flux tubes (see [10]).

This is not, however, the whole story. We can understand how the flux tubes may be stored, and how they may erupt, but we have assumed in the first place that the tubes have an existence of their own. For this to be so, their magnetic field must be significantly above the equipartition value, for the magnetic forces to dominate the other ones. This may be understood by comparing the drag force produced on a tube by convective motions to other forces, for example to the magnetic buoyancy force (3.61). The convective drag force is given by (3.62), where v now stands for the convection speed; expressing v as a function of $B_{eq} = \left(\mu_0 \rho_{ext}\right)^{1/2} v$, the drag force produced by convection can be written as $F_D \sim B_{eq}^2 aL/(2\mu_0)$. In order for the tube not to be dragged passively by convection, we must have $F_D < F_B$, whence in order of magnitude[17]

$$B > B_{eq} \left(\frac{H}{a}\right)^{1/2}. \tag{3.64}$$

---

[17]This condition is equivalent to requiring that the convection speed be smaller than the buoyancy speed (3.63).

Table 3.2 *Convective, thermal and magnetic energy densities at the bottom of the convection zone and at the photosphere (in $\mathrm{J\,m^{-3}}$)*

Location	$0.7\,R_\odot$	$1.0\,R_\odot$
$\rho v^2/2$	5×10^5	1.5×10^2
$\rho k_B T/\mu$	7×10^{12}	1.5×10^4
$B^2/2\mu_0$	4×10^7 (presumed)	$(0.4-4) \times 10^4$ (measured)

Since the flux tube has (by hypothesis) a radius much smaller than the scale height, we see that its magnetic field must be greater than the equipartition value by at least an order of magnitude for the convective drag force to be negligible. From the value of B_{eq} in Table 3.1 we thus expect an actual field magnitude of at least 10 T in the generating region. Other arguments based on flux tube dynamics give a value of that same order of magnitude [39].

Convective collapse

We have seen that convective motions can concentrate the magnetic flux and increase locally the magnetic energy, but no more than the value corresponding to the kinetic energy density. And yet both the values observed at the surface and those presumed deep in the Sun are greater by an order of magnitude than the equipartition values (Table 3.1). One key to the solution, which we hinted at in Chapter 1, lies in Table 3.2, which shows that the energy available in thermal motions (and in the potential energy of gravitation, since we saw that both are comparable) is of the same order or larger than the actual magnetic energy. How can the magnetic field be amplified to such a level?

Consider a thin vertical tube of flux below the solar surface, where the medium is convectively unstable. We have seen that the temperature stratification is strongly superadiabatic there, so that a small adiabatic downward displacement of matter within the tube makes it less hot than the (superadiabatic) surroundings. Because of its small width, the tube is in pressure balance with the outside at each altitude. Suppose that the magnetic energy in the tube is much smaller than the thermal energy. This means that the magnetic contribution to the total pressure in the tube is negligible. Just as we found in Section 3.2.2, lateral (total) pressure balance implies that the (cooler) displaced fluid is denser than its surroundings, which accelerates its downward motion, making the motion unstable. As matter moves downward in the tube and cools, the scale height is reduced, and the top of the tube is evacuated. This decrease in matter pressure at the top of the tube must be accompanied by an increase of the same amount in magnetic pressure to conserve (total) horizontal pressure balance with the exterior. This makes the magnetic field in the tube increase.

When the magnetic pressure in the tube becomes comparable to matter pressure, the instability is suppressed because the magnetic field can compensate for the changes in matter pressure. As a result, in the final state the magnetic pressure in the upper part of the tube should be comparable to matter pressure, i.e. the magnetic and thermal energy densities are comparable. Note that an upward displacement of matter within the tube is also unstable, but it makes the field decrease, so that in that case the instability tends to behave as ordinary convection and persists, finally dispersing the field [43].

Finally, therefore, this instability tends to produce either thin tubes with a large field strength – of energy roughly equal to the thermal energy – or a vanishing field.

3.3.6 Surface magnetic field

Small-scale magnetic field

The convective collapse of flux tubes might thus take over the job of magnetic energy concentration begun by convective motions (which is limited by the kinetic energy of convection), and produce structures of much greater magnetic energy density – comparable to the thermal energy density (Table 3.2). Note that other mechanisms may produce large magnetic fields. In particular, in the cold regions separating granules, the small scale height and the mechanical balance across vertical field lines tend to produce magnetic field structures of energy density equal to the ambient thermal energy density [35]. Furthermore, elaborate theoretical studies and simulations show a more complicated and subtler interaction between convection and magnetic fields, which – as we have already noted – may even produce a dynamo in its own right [52], [31].

These effects help us to understand the ubiquitous network of intense magnetic fields observed at the solar surface, which are concentrated at the edges of the convective structures. Magnetic concentration is especially effective at the edges of the supergranulation cells, which are larger than the granulation cells by about one order of magnitude in size and two orders of magnitude in duration, and are more deeply seated (Fig. 3.18, left). These fields appear to emerge at the solar surface, to be transported by bulk motions and to accumulate at the stagnation points in the flow, and finally to disperse as the convection cells evolve and disappear, whereas fields of opposite directions can merge and annihilate when they come close together (see [41]). Magnetic concentrations also accumulate at the boundaries of the granules, but because of the short lifetime of these structures, the resulting magnetic concentrations are not so conspicuous (see [36]).

How do these small magnetic structures appear on white-light images of the solar surface? Do they appear brighter or darker than the surrounding photosphere? Because of their large magnetic field ($B > B_{eq}$), the convective heat transport which brings heat from the solar interior is reduced in their vicinity since matter can move along the magnetic field but not across it, and thus cannot mix freely. This should make these tubes slightly colder than their surroundings, making them appear dark. A little reflection, however, shows that this is not so for two reasons. First, we have seen that these magnetic tubes are less dense than their surroundings; they are therefore more transparent, so

Some guesses on solar magnetism 155

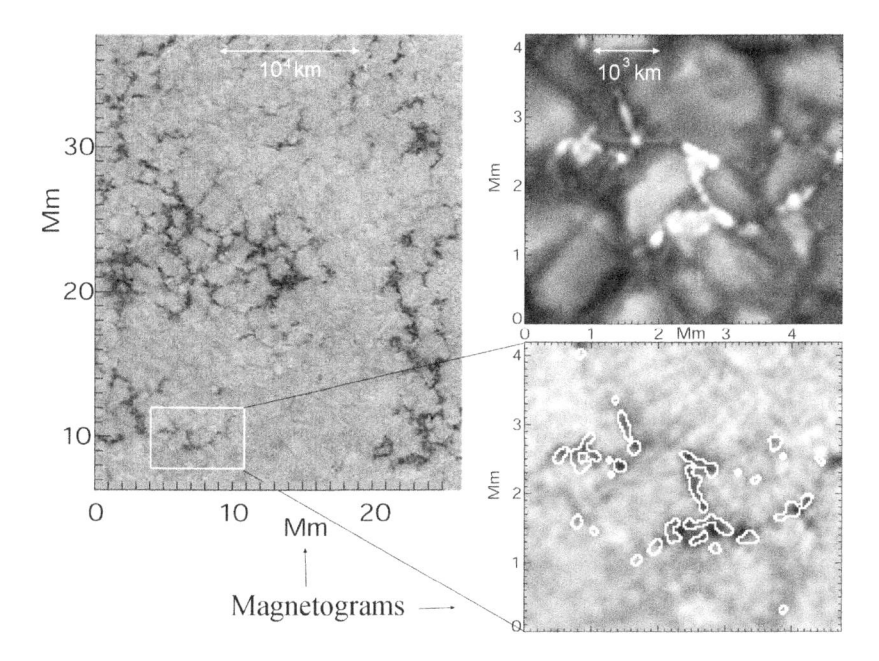

Figure 3.18 Left panel: magnetogram (dark corresponds to strong magnetic field) of a part of the solar surface, showing magnetic concentration at the edges of supergranulation structures. Right panel: high-resolution images, taken with the Swedish 1-metre solar telescope on La Palma, showing bright structures in the dark lanes bounding granules (top), and a magnetogram on which are superimposed contours delineating the brightest structures of the upper panel (bottom). (Adapted from [5].)

that when we look at them we can see much deeper into the Sun than when looking at the normal photosphere; we thus see a greater temperature and a greater brightness. Second, thin tubes can be heated effectively by radiation, which increases their temperature.

And indeed, these small magnetic structures do appear bright, as can be seen in Fig. 3.18 (right). It shows high-resolution ($\sim$70-km) images of the region delineated by the white box on the left-hand magnetogram. The upper image shows intensity in the so-called G-band[18] and the lower image is a magnetogram (darkest regions indicate $\sim$0.1-T magnetic field), showing that the bright optical structures (delineated by the white contours on the magnetogram) generally coincide with large magnetic flux densities. High-resolution magnetic measurements show a wide range of magnetic flux densities and spatial sizes [5].

---

[18] The so-called G-band (around 0.43 μm) is dominated by lines of the CH molecule and shows brightenings that coincide with high magnetic fields with an especially large contrast (Sanchez-Almeida, J. *et al.* 2001, *Ap. J.* **555** 978).

Sunspots

> The Sun is often pictured as a kind of human figure, but this has not any foundation. Other pictures show it covered with volcanoes or bubbling foam; but in actual fact we only see a plain yellow surface, on which appear from time to time several dark blotches that are called solar spots; they may be smoke or cinders from this huge furnace ...
>
> J. de Lalande, *Ladies' Astronomy*[19]

From our twenty-first-century vantage point, we can find one truth in this citation. Sunspots owe ultimately their existence to convection, even though their darkness stems from lack of convection, and convection is driven by the solar furnace. But the road to explaining them is tortuous and still not fully cleared.

Sunspots are the most spectacular of the relatively large and intense magnetic structures associated with the solar cycle and known as *active regions*. Sunspots are very large, of a size greater than the scale height by one or two orders of magnitude. Their magnetic field is great, with a magnetic pressure even larger than the particle thermal energy in the surrounding photosphere. They are colder by nearly 2×10^3 K than the surrounding photosphere. And finally they are (relatively) stable – often lasting for more than a month – so that they can be thought of as being in close to magnetohydrostatic equilibrium.

Sunspots are now measured in such detail that empirical knowledge on them far exceeds theoretical understanding, and has led to a whole taxonomy. One also measures oscillations and waves, and can even infer their internal structure using solar seismology. Sunspots may be growing by agglomeration of thin intense flux tubes. But how the tubes coalesce is not fully understood, nor is the apparent stability of the whole structure.

However, we may understand some of their basic properties with the help of the concepts introduced in this chapter. Let us first try to understand their most basic property: the huge magnetic field, of energy density greater than the thermal value in the normal photosphere. How can this be so?

Consider a simplified picture: a (large) vertical flux tube at the Sun's surface, with a magnetic field of much greater amplitude than outside. Horizontal pressure balance between the interior and the exterior yields

$$P_{int}(z) + B(z)^2/2\mu_0 = P_{ext}(z). \tag{3.65}$$

The great magnetic field amplitude requires that $P_{int} \ll P_{ext}$. Hence the decrease of P_{ext} with altitude must be matched by a decrease in B, making the tube fan out to conserve magnetic flux. Now, vertical pressure balance yields

$$\frac{dP_{int}}{dz} = -\rho_{int} g$$

$$\frac{dP_{ext}}{dz} = -\rho_{ext} g.$$

---

[19] Jérôme Lefrançois de Lalande, Bidault, Paris (1785). I am indebted to Peter Hingley, at the Royal Astronomical Society, for the copies dating back to this epoch he kindly provided.

Taking the derivative of (3.65) with respect to z, and substituting the above equations, we see that $dB/dz < 0$ requires $\rho_{int} < \rho_{ext}$. This means that the structure is not only colder than its surroundings, but also less dense. It is therefore less opaque, so that we observe deeper layers in sunspots than when we look outside. Indeed, it has been known for a long time that sunspots appear as depressions in the Sun's surface.[20] (They do not appear bright as do thin flux tubes because their coolness is too important to be compensated by the fact that we see deeper.)

The lower geometric altitude of sunspots has an important consequence. The large magnetic field observed in them refers to a deeper altitude than the normal photosphere; hence in the equation of horizontal balance (3.65) we must put for P_{ext} the pressure at a lower altitude than the normal photosphere, and thus a greater P_{ext}, since pressure increases with depth. This may explain why the magnetic pressure in sunspots can be greater than matter pressure at the normal photosphere level.[21]

Finally, what makes them so cold? As we have already noted, their large magnetic field impedes convection at their bottom, so that less heat arrives to them from the solar interior; furthermore, the fanning out of the tube with height spreads the heat flux over a greater area. With less heat arriving per unit surface, sunspots must be cooler than their surroundings.

The near blocking of the heat flux below sunspots raises an interesting question. Due to their low temperature, sunspots radiate only about one-fifth of the energy radiated by a similarly sized area of the normal photosphere. Do sunspots make the Sun radiate less, or does the blocked energy emerge elsewhere on the solar surface? In the former case, since the radiation of a star determines its internal structure (Section 3.1), does the emergence of a sunspot change the solar structure?

This question is addressed in Problem 3.4.5, where we can see that the solar convective zone conducts and stores heat very effectively. The great heat conductivity ensures that the energy diverted by sunspots is dispersed relatively rapidly in the convective zone; the large heat capacity ensures that the global Sun is not affected because the timescale for a global temperature variation in the convection zone is extremely large.

A kitchen analogy may be useful [45]. Imagine an electric heater plate made of a very thick piece of copper. Now, put on it a small piece of heat insulator material, for example a small ceramic tile, and wait until the heat lost from the plate balances the incident power. What happens? The blocking of heat transport makes the top of the insulating tile colder than the rest of the plate, but no hot ring forms around it because the heat blocked by the tile spreads quickly throughout the whole copper plate, making it slightly hotter than before. Contrary to the Sun, our heater plate has not a large enough heat capacity for its temperature to remain roughly constant; we may however simulate the solar case by removing the ceramic tile before the plate has had enough time to reach equilibrium.

---

[20]This is known as *Wilson depression*.
[21]The picture is complicated by the curvature of the field lines and by the variation in sunspot properties in the horizontal direction.

3.4 Problems

3.4.1 Conductive heat transfer in the solar interior

Show that the energy transfer by thermal conduction is negligible in the solar interior.

3.4.2 Timescale for radiative transport

Why is the timescale for radiative transport in the Sun given by the thermal timescale τ_{KH} instead of the time t_{ph} photons need to random walk to the surface?

Hints

Show that in order of magnitude

$$\frac{t_{ph}}{\tau_{KH}} \sim \frac{W_{ph}}{|W_g|} \tag{3.66}$$

and that W_{ph} is smaller than W_g by roughly three orders of magnitude. Deduce that in the solar interior, particles contain much more energy than photons. Since photons are in thermodynamic equilibrium with particles – a heat reservoir containing much more energy than themselves – they cannot random-walk independently of the particles [48]. Therefore, radiation ultimately diffuses at the timescale determined by the particle thermal energy τ_{KH}, which is much greater than the photon random-walk timescale τ_{ph}.

3.4.3 Solar differential rotation

Show that the fluid momentum equation in a frame rotating with angular velocity Ω_0 is

$$\frac{d\mathbf{v}}{dt} = -\frac{\nabla P'}{\rho} - \nabla \Phi - 2\mathbf{\Omega}_0 \times \mathbf{v} \tag{3.67}$$

where Φ is the gravitational potential, viscosity and magnetic forces are neglected and $P' = P - \rho(\mathbf{\Omega}_0 \times \mathbf{r})^2/2$ is an effective pressure incorporating the centrifugal force. Show that in the absence of meridian velocities (i.e. with $\mathbf{v}$ in the azimuthal direction) the left-hand-side term would vanish for a stationary flow. Assume that the left-hand-side term can be neglected and take the curl of the resulting equation; show that with an adiabatic structure in the convective zone, the azimuthal component of the velocity, v_ϕ is independent of the coordinate z parallel to $\mathbf{\Omega}_0$. Since this is contradicted by Fig. 3.6, one or several of the above hypotheses should be wrong [40].

3.4.4 Twisted magnetic flux tube

Consider a flux tube where the field is: $\mathbf{B} = ar\phi + \mathbf{z}$ for $r < R$ and zero elsewhere (in cylindrical coordinates, where ϕ and $\mathbf{z}$ are unit vectors). Show that the azimuthal magnetic field component is responsible for a magnetic tension equal to $-a^2\mathbf{r}/\mu_0$ and a magnetic pressure force having the same value, pointing radially inward. Interesting numerical simulations of the behaviour of such flux tubes in the Sun can be found in [54] and [26].

3.4.5 The heat flux blocked by sunspots

What is the fate of the energy flux blocked at the bottom of sunspots? The energy diverted does not seem to emerge as bright rings around them; on the contrary, dips in solar luminosity have been observed at the birth of large sunspots groups. Where does the blocked energy go?

To begin with, use Sections 3.1 and 3.2 to estimate how long a thermal disturbance takes to diffuse through the whole convective zone.

The convective zone contains so much matter that even a large amount of heat would produce a very small global temperature change. In other words, its huge thermal energy content W_{thC}, that is proportional to the product of its temperature by its mass, makes its thermal timescale $\tau_{thC} \sim W_{thC}/L_\odot$ very large. Calculate this timescale.

Hints

Timescale for heat to spread through the convective zone

Convective heat transport can be thought of as a random walk of convective eddies travelling at a speed equal to the convection speed v_{conv}, with a free path equal to the mixing length l. We know that the time for travelling a distance D is in this case $T_{conv} \sim (D/v_{conv}) \times (D/l)$. Putting the values we found at the base of the convection zone: $v_{conv} \sim 70$ m s^{-1} and $l \sim 0.1 R_\odot$, with $D \sim 0.3 R_\odot$ (the width of the convection zone), we find $T_{conv} \sim 3 \times 10^7$ s.

Timescale for global temperature change in the convective zone

We have seen that (because of hydrostatic equilibrium) the thermal timescale for the whole Sun is roughly equal to the Kelvin–Helmholtz time τ_{KH} (3.23) (originally introduced in terms of the gravitational energy). For the convective zone, this time is about two orders of magnitude smaller than for the whole Sun because it contains only a few per cent of the solar mass, at a temperature somewhat smaller (Fig. 3.2), i.e. $\tau_{thC} \sim 10^{-2} \times \tau_{KH}$. This yields a few 10^5 years – much larger than both the time T_{conv} for energy to disperse in the convective zone, and the lifetime of sunspots [45].

References

[1] Alfvén, H. 1950, Discussion of the origin of the terrestrial and solar magnetic fields, *Tellus* **2** 74.

[2] Bahcall, J. N. 2002, Solar models: an historical overview, *Nucl. Phys. B (Proc. Suppl.)* **118** 77.

[3] Bahcall, J. N. *et al.* 1997, Are standard solar models reliable? *Phys. Rev. Lett.* **78** 171.

[4] Bahcall, J. N. *et al.* 2001, Solar models: current epoch and time dependences, neutrinos, and helioseismological properties, *Ap. J.* **555** 990.

[5] Berger, T. E. *et al.* 2004, Solar magnetic elements at 0."1 resolution: general appearance and magnetic structure, *Astron. Astrophys.* **428** 613.

[6] Burrows, A. and J. Liebert 1993, The science of brown dwarfs, *Rev. Mod. Phys.* **65** 301.

[7] Burrows, A. *et al.* 2001, The theory of brown dwarfs and extrasolar giant planets, *Rev. Mod. Phys.* **73** 719.

[8] Celnikier, L. M. 1989, *Basics of Cosmic Structures*, Paris, Editions Frontières.

[9] Christensen-Dalsgaard, J., C. R. Proffitt and M. J. Thompson 1993, Effects of diffusion on solar models and their oscillation frequencies, *Ap. J.* **403** L75.

[10] Cline, K. S., N. H. Brummell and F. Cattaneo 2003, On the formation of magnetic structures by the combined action of velocity shear and magnetic buoyancy, *Ap. J.* **588** 630.

[11] Cox, A. N. *et al.* eds. 1991, *Solar Interior and Atmosphere*, Tucson AZ, University of Arizona Press.

[12] DeRosa, M. L., P. A. Gilman and J. Toomre 2002, Solar multiscale convection and rotation gradients studied in shallow spherical shells, *Ap. J.* **581** 1356.

[13] Galloway, D. J. and M. R. E. Proctor 1992, Numerical simulations of fast dynamos in smooth velocity fields with realistic diffusion, *Nature* **356** 691.

[14] Godier, S. and J. P. Rozelot 2000, The solar oblateness and its relationship with the structure of the tachocline and of the sun's subsurface, *Astron. Astrophys.* **355** 365.

[15] Harwit, M. 1998, *Astrophysical Concepts*, New York, Springer.

[16] Hathaway, D. H. *et al.* 2003, Evidence that a deep meridional flow sets the sunspot cycle period, *Ap. J.* **589** 665.

References

[17] Howe, R. *et al.* 2000, Dynamic variations at the base of the solar convection zone, *Science* **287** 2456.

[18] Hubbard, W. B., A. Burrows and J. I. Lunine 2002, Theory of giant planets, *Annu. Rev. Astron. Astrophys.* **40** 103.

[19] Jones, C. A. 2000, Convection-driven geodynamo models, *Phil. Trans. Roy. Soc. London A* **358** 873.

[20] Kippenhahn, R. and A. Weigert 1994, *Stellar Structure and Evolution*, New York, Springer.

[21] Knobloch, E. 1994, in *Lectures on Solar and Planetary Dynamos*, ed. M. R. E. Proctor and A. D. Gilbert, Cambridge University Press p. 331.

[22] Kurucz, R. L. 2002, A few things we do not know about the sun and F-G stars, *Baltic Astron.* **11** 101.

[23] Mihalas, D. 1978, *Stellar Atmospheres*, San Francisco CA, W. H. Freeman.

[24] Moffatt, H. K. 1978, *Magnetic Field Generation in Electrically Conducting Fluids*, Cambridge University Press.

[25] Moffatt, H. K and M. R. E. Proctor 1985, Topological constraints associated with fast dynamo action, *J. Fluid Mech.* **154** 493.

[26] Moreno-Insertis, F. and T. Emonet 1996, The rise of twisted magnetic tubes in a stratified medium, *Ap. J.* **472** L53.

[27] Nauenberg, M. and V. Weisskopf 1978, Why does the sun shine? *Am. J. Phys.* **46** 23.

[28] Pap, J. M. 1997, in *Past and Present Variability of the Solar-Terrestrial System: Measurement, Data Analysis and Theoretical Models*, Bologna, Italy, IOS Press, p. 1.

[29] Parker, E. N. 1955, Hydromagnetic dynamo models, *Ap. J.* **122** 293.

[30] Parker, E. N. 1979, *Cosmical Magnetic Fields: Their Origin and Activity*, Oxford UK, Clarendon Press.

[31] Proctor, M. 2003, in *Stellar Astrophysical Fluids Dynamics*, ed. M. J. Thompson and J. Christensen-Dalsgaard, Cambridge University Press, p. 143.

[32] Reid, N. 1987, The stellar mass function at low luminosities, *Mont. Not. Roy. Astron. Soc.* **225** 873.

[33] Roberts, B. 1990, in *Physics of Magnetic Flux Ropes, Geophys. Monogr. Ser.* **58**, ed. C. T. Russell, E. R. Priest and L. C. Lee, Washington DC, American Geophysical Union p. 113.

[34] Roberts, P. H. 1994, in *Lectures on Solar and Planetary Dynamos*, ed. M. R. E. Proctor and A. D. Gilbert, Cambridge University Press, p. 1.

[35] Sanchez-Almeida, J. 2001, Thermal relaxation of very small solar magnetic structures in intergranules: a process that produces kilogauss magnetic strengths, *Ap. J.* **556** 928.

[36] Sanchez-Almeida, J. et al. 2004, Bright points in the internetwork quiet sun, *Ap. J.* **609** L94.

[37] Schwarzschild, M. 1958, *Structure and Evolution of the Stars*, Princeton University Press.

[38] Schrijver, C. J. and C. Zwaan 2000, *Solar and Stellar Magnetic Activity*, Cambridge University Press.

[39] Schüssler, M. 1993, in *The Cosmic Dynamo, IAU Symp.* **157**, ed. F. Krause et al., New York, Kluwer, p. 27.

[40] Sekii, T. 2003, in *Stellar Astrophysical Fluids Dynamics*, ed. M. J. Thompson and J. Christensen-Dalsgaard, Cambridge University Press, p. 263.

[41] Simon, G. W., A. M. Title and N. O. Weiss 2001, Sustaining the sun's magnetic network with emerging bipoles, *Ap. J.* **561** 427.

[42] Solanki, S. K. 2002, in *From Solar Min to Max: Half a Solar Cycle with SoHO*, ed. A. Wilson, ESA SP-508, p. 173.

[43] Spruit, H. C. and E. G. Zweibel 1979, Convective instability of thin flux tubes, *Solar Phys.* **62** 15.

[44] Spruit, H. C. et al. 1990, Solar convection, *Annu. Rev. Astron. Astrophys.* **28** 263.

[45] Spruit, H. C. 1992, in *Sunspots: Theory and Observations*, ed. J. H. Thomas and N. O. Weiss, New York, Kluwer, p. 163.

[46] Stein, R. F. and Å. Nordlund 1998, Simulations of solar granulation, *Ap. J.* **499** 914.

[47] Stix, M. 2002, *The Sun*, New York, Springer.

[48] Stix, M. 2003, On the time scale of energy transport in the sun, *Solar Phys.* **212** 3.

[49] Vainshtein, S. I. and Ya. B. Zeldovich 1972, Origin of magnetic fields in astrophysics, *Sov. Phys. Usp.* **15** 159.

[50] Vainshtein, S. I. and F. Cattaneo 1992, Nonlinear restrictions on dynamo action, *Sov. Phys. Usp.* **393** 165.

References

[51] Weiss, N. O. 1966, The expulsion of magnetic flux by eddies, *Proc. Roy. Soc. London A* **293** 310.

[52] Weiss, N. O. 2003, in *Stellar Astrophysical Fluids Dynamics*, ed. M. J. Thompson and J. Christensen-Dalsgaard, Cambridge University Press, p. 329.

[53] Weisskopf, V. 1975, Of atoms, mountains, and stars: a study in qualitative physics, *Science* **187** 605.

[54] Wissink, J. G. *et al.* 2000, Numerical simulations of buoyant magnetic flux tubes, *Ap. J.* **536** 982.

[55] Zeldovich, Ya. B., A. A. Ruzmaikin and D. D. Sokoloff 1983, *Magnetic Fields in Astrophysics*, New York, Gordon & Breach.

4

The outer solar atmosphere

> To study the corona and chromosphere requires a firm optimism.
> H. C. Van de Hulst, 1950[1]

> DANGER!! ENTER AT YOUR OWN RISK!!
> J. Newmark, 2003[2]

Not only does the Sun try to hide its interior, but it seems to take perverse pleasure in disguising its atmosphere. Van de Hulst's remark, which refers to solar eclipses (that invariably occurred by cloudy weather or after observers have contracted nasty diseases) remains true half a century later, albeit for quite different reasons. And the latter citation reminds us that observation of the outer solar atmosphere is still a difficult art; in opening new windows, progress in techniques has revealed a hierarchy of intricate patterns, reminiscent of a rainforest in which there is a profusion of insects, birds, flowers and trees that we can see but not touch, and whose understanding requires much subtlety.

We took a quick tour of the solar atmosphere in Section 1.2, and we shall now enter into more detail. We will have to complement our plasma physics tool box with some additives, in order to find intellectual coherence in the baroque architecture revealed by observation. By necessity, however, I shall draw an outrageously simplified picture, keeping only a few basic aspects. Far more complete accounts are given for example in [22], [48], [4] and references therein; a lucid survey can be found in [33] for solar activity – of which a nice historical point of view is given in [1].

---

[1]Kuiper, G. P. ed. 1953, *The Sun*, Chicago IL, University of Chicago Press, p. 207.

[2]Heading of the web page on plasma diagnostics in the user's guide of the Extreme Ultraviolet Imaging Telescope on the spacecraft SoHO in 2003 (http://umbra.nascom.nasa.gov/eit/eit_guide/).

4.1 From the photosphere to the corona

4.1.1 The atmosphere in one dimension

Figure 4.1 shows how the average density and temperature vary with height above the photosphere, as inferred from observation.

At low heights the temperature, which we have seen to be significantly smaller than 10^4 K, is too small for producing significant ionisation (Section 2.4). I have plotted the number densities of neutral hydrogen and of electrons, together with the gas temperature, from a semi-empirical model of the average (undisturbed) solar chromosphere and transition region.[3]

At high heights, the large temperature ensures virtually complete ionisation, and I have plotted the electron density[4] and temperature.[5] I have chosen to plot the values observed in polar regions, near solar activity minimum, because the diagnostics there is relatively free from perturbations (and these regions furnish the major contribution to the fast wind); other regions are generally denser and hotter.

To fill the gap between the low-height and high-height models, I have devised an outrageously crude interpolation.[6]

A quick look at Fig. 4.1 reveals a roughly 2000-km thick (weakly ionised) *chromosphere* at a temperature of 4–7 $\times 10^3$ K, ending with a sharp jump into the million-degree *corona* (virtually fully ionised).

This apparent simplicity, however, masks a number of difficulties, so much so that Fig. 4.1 should be viewed as a starting point for reflection rather than a realistic description. There are two major reasons for that. For one thing, the densities and temperatures are not measured directly, but rather adapted by modellers in order to reproduce the diagnostics available. This is an uncertain game, because the diagnostics is based on radiation involving a large number of chemical species in various ionisation states, whereas the medium is not in local thermodynamic equilibrium. The problem is especially severe in the corona, where we shall see that the free path of particles is greater than the scale height, so that different species have different temperatures; in particular, some (still controversial) measurements suggest that protons may be hotter than electrons, and that heavier ions are still hotter; we shall return to this point later. Furthermore, the medium is not static, and in particular the plasma has a significant

---

[3] Model C in [20]; this model includes heights up to 2219 km, which corresponds to a normalised radial distance from the Sun's centre $r/R_\odot \simeq 1.003$.

[4] I have plotted $n_e = a_1 e^{a_2 R_\odot / r} \left[(R_\odot/r)^2 + a_3 (R_\odot/r)^3 + a_4 (R_\odot/r)^4 + a_5 (R_\odot/r)^5 \right]$ for $r/R_\odot > 1.02$, where n_e is in m^{-3}, r is the radial distance to the Sun's centre, $a_1 = 1.3 \times 10^{11}$, $a_2 = 4.8$, $a_3 = 0.3$, $a_4 = -7.2$, $a_5 = 12.3$, based on [53]; I have rounded up the parameters, multiplied the published value of a_1 by 10^8 (E. Sittler, personal communication), and transformed in SI units.

[5] I have plotted $T_e = 1.65 \times 10^6 / \left[(r/R_\odot)^{0.7} + (r/R_\odot)^{-4} \right]$ for $0.02 < r/R_\odot - 1 < 1.1$, and $T_e = 1.3 \times 10^6 (r/R_\odot)^{-0.4}$ for $r/R_\odot - 1 > 1.1$.

[6] In the range $0.0033 < \zeta < 0.02$, where $\zeta = r/R_\odot - 1$, I have plotted $T_e = 0.85 \times 10^6 - 10^2/(\zeta - 0.003) - 1.5 \times 10^9 (\zeta - 0.02)^2$, and $n_e = 8 \times 10^{13} \exp\left[-1.4 \times 10^7 (\zeta - 0.02)/T_e \right]$; the constant in the exponential is the numerical value of $\mu M_\odot G / k_B R_\odot$.

From the photosphere to the corona 167

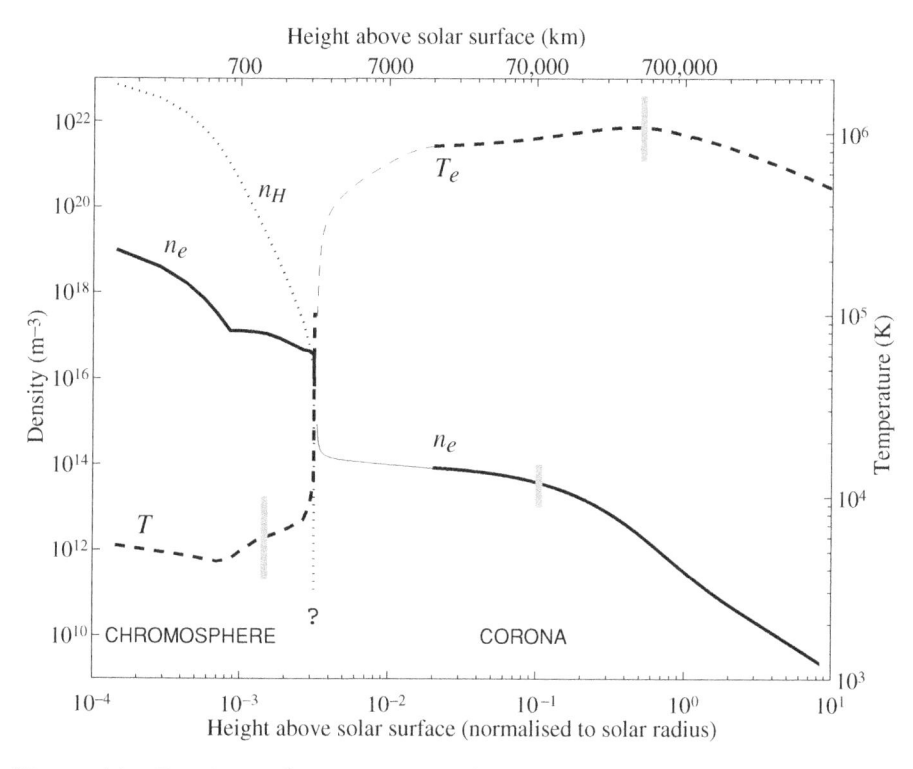

Figure 4.1 Density and temperature above the solar surface outside active regions versus the normalised height $r/R_\odot - 1$, from empirical models[3,4,5] (electron density n_e, solid lines; hydrogen density n_H, dotted lines; temperature T and electron temperature T_e, dashed lines). The coronal parameters apply to polar coronal holes. The grey bars show the range of values from alternative models. The thin lines represent a crude interpolation filling the gap between the inner and outer models,[6] and the question mark reminds us of the inadequacy of a one-dimensional static model, especially in the transition between the chromosphere and the corona.

mean outflow velocity at distances of a few solar radii. Finally, there is no agreement even on the basic mechanisms that determine the energy balance; we shall also return to this later. It is not an exaggeration to say that the outer solar atmosphere is neither adequately measured, nor correctly understood.

The second major reason to have doubts on the validity of Fig. 4.1 is that it is one-dimensional, and thus represents an atmosphere spherically symmetrical and time-stationary. This is not true (especially at low heights) as we shall see in detail below, and an average is of little significance, since the medium is full of structures whose scales are smaller than the resolution of observation.

For both of these reasons there is no agreed-upon model, and I have sketched with grey bars on the figure the scatter of measured or inferred values (see for example [2], [18], [58] and references therein).

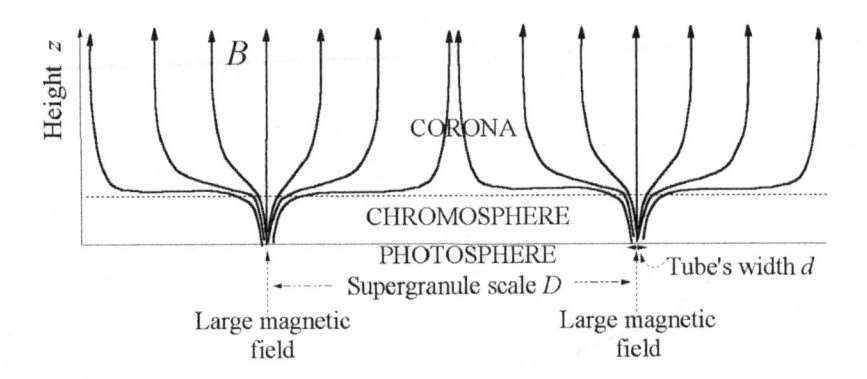

Figure 4.2 A two-dimensional sketch of magnetic field lines above the solar surface. The thin vertical bundles of large magnetic field located at the boundaries of the supergranular cells in the photosphere diverge rapidly with height (as the pressure of the surrounding gas drops), until they fill all the space available and confine each other.

4.1.2 One more dimension

Let us therefore drop the one-dimensional picture and consider one more dimension. We must remember that the magnetic field at the solar surface is far from uniform; we have seen in Section 3.3 that in the normal photosphere it tends to be concentrated at the boundaries of the granular cells, and especially in lanes at the boundaries of the supergranular cells. Figure 4.2 sketches a vertical cross-section above the solar surface, perpendicular to such a lane; regions of great magnetic field are roughly a distance D apart (the supergranular scale) along the horizontal direction shown. In this simplified picture, the other horizontal direction (normal to the figure) is along a lane so that the structure is approximately two-dimensional. It is also roughly plane-parallel because we consider heights much smaller than the solar radius.

As thin vertical flux bundles of large magnetic field emerge at the photosphere, the pressure in the surrounding gas, which decreases upwards, becomes insufficient to maintain their concentration, so that the bundles fan out with height. To make an order-of-magnitude estimate, we note that horizontal pressure balance requires the bundles' magnetic pressure $B^2/2\mu_0$ to vary roughly as the gas pressure $P \propto e^{-z/H}$, so that $B \propto e^{-z/2H}$. Since the magnetic flux does not change along the bundle (because $\nabla \cdot \mathbf{B} = 0$), the section varies as $1/B$; since in this two-dimensional picture the width of the bundle varies in proportion to the section, it increases as $e^{z/2H}$. The bundle encounters its closest neighbours at the height z where its width matches the distance D. This takes place at $z \sim 2H \ln(D/d)$ if the initial bundle's width is d. Taking $D \sim 30\,000$ km (the supergranular scale), $d \sim 100$ km and $H \sim 150$ km (from Section 3.2), we find that the bundles merge at a height of about 2000 km. Referring to Fig. 4.1, this corresponds to the upper chromosphere.

At this altitude the magnetic field, which varies inversely as the bundles' section, has decreased by a factor of about d/D from its value at the base of the tubes, which we saw in Section 3.3 to be of the order of magnitude of 0.1 T. This yields a magnetic field $\sim 0.1 \times d/D \sim 3 \times 10^{-4}$ T.[7]

Above this height, the funnels fill all the space available and confine each other, so that the configuration is expected to be more uniform. Since the gas pressure continues to drop (though with a greater scale height since the temperature increases), while the magnetic field should no longer decrease much up to $z \sim D$, the magnetic energy is expected to dominate that of the particles from the upper chromosphere to at least several $10^{-2} R_\odot$.

The canopy of nearly horizontal field lines shown in Fig. 4.2 above the regions of the photosphere that are less magnetised has an important consequence. We saw in Section 2.3 that heat conduction takes place predominantly along magnetic field lines. Hence heat conduction can easily take place above the photospheric flux bundles of vertical magnetic field, but it is hampered in the regions where the field is roughly horizontal. This means that conduction takes place mainly above the lanes of concentrated magnetic flux at the edges of supergranular cells. Therefore, one-dimensional models making an average over horizontal co-ordinates cannot handle correctly heat transport in the upper chromosphere [21].

This picture makes an important point, but it is grossly oversimplified. In particular we have assumed the flux bundles to be of like polarity (magnetic field pointing outwards in Fig. 4.2). However, when a magnetic flux tube emerges at the photosphere (Fig. 3.17), the magnetic field points outwards at one end and inwards at the other one, so that there is a salt-and-pepper mix of regions having different polarities on the Sun. Moreover, the lanes where the photospheric magnetic flux accumulates do have a finite width, and the magnetic field does not vanish outside them. To deal with these effects, we must consider one more space dimension.

4.1.3 Three dimensions in space

Figure 4.3 is an attempt to sketch the magnetic field in three dimensions above the solar surface. Regions of opposite magnetic polarity that are close together must be connected by magnetic field lines. Hence, in addition to funnels of open field lines, which remain open at least up to coronal altitudes, there are loops or arches of magnetic field closing on the Sun.[8] Indeed, loops are observed to be ubiquitous in the solar atmosphere (Figs. 1.13, 1.14 and 4.8); we shall return to them later.

4.1.4 ...and one dimension in time

The picture drawn in Fig. 4.3 still misses an important point. We have seen in Section 3.4 that the photosphere is a very dynamic place. In particular, the

---

[7]This holds outside active regions; we have seen that in active regions, the size of the magnetic field concentrations at the photosphere is much greater; this greater value of d yields a smaller merging altitude, and thus a greater magnetic field there.

[8]We shall see in the next section that this picture is still grossly oversimplified: the field lines are not so neatly aligned but are braided and twisted, which has further consequences.

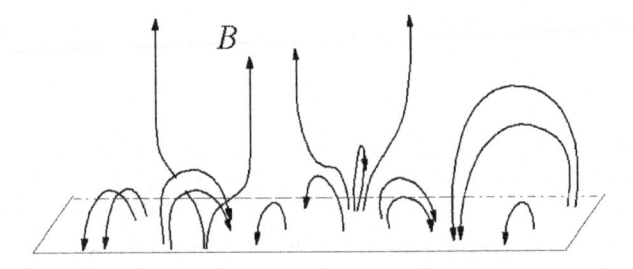

Figure 4.3 Sketch of field lines above the solar surface, with loops and arches connecting regions of opposite magnetic polarity, and open magnetic field funnels.

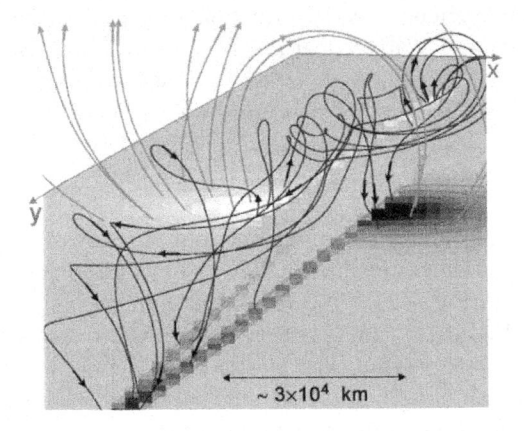

Figure 4.4 Magnetohydrodynamic simulation of the large-scale magnetic structure resulting from shearing motion at the photosphere (velocity parallel to **x**, varying along **y**). (Courtesy of G. Aulanier, based on [7].)

supergranular cells at the edges of which the magnetic flux accumulates are not permanent structures; they change over a few hours – having a lifetime of about a day – as the fluid motions make new magnetic field lines emerge, and draw and shuffle them in permanence. Shear motions distort the field lines, producing at higher altitudes helical structures as shown in Fig. 4.4 [7]. This has subtle consequences that we shall discuss in Section 4.2.

When the motions draw close together field lines of like polarity, this merely increases locally the magnetic field. When the lines have opposite directions, however, this produces geometries as sketched in Fig. 4.5 (continuous lines) in simple cases. The field then tends to relax to a new equilibrium having less energy (Fig. 4.5, dotted lines), by reconnection of field lines within a thin region (sketched in grey), as we saw in Section 2.3.

This has two major consequences. First, as we already saw in Section 3.3, this enables the magnetic field to slip through the plasma – through transient

From the photosphere to the corona 171

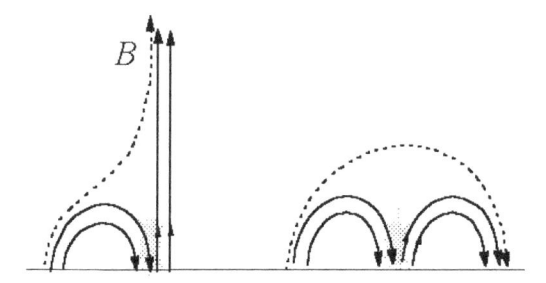

Figure 4.5 Schematic illustration of magnetic field merging in simple cases: two loops (right) and one loop and one unipolar field (left). The magnetic field lines before and after reconnection are shown respectively in continuous and dotted lines, and the reconnection site is sketched in grey.

formation of small scales – so that the topology of the field lines can change (for example, in Fig. 4.5 (right), two small loops give way to a large one). This takes place on a timescale whose order of magnitude may be as small as a few transit times L/V_A (L being a typical size and V_A the Alfvén velocity), thus much faster than ordinary ohmic diffusion.

Second, the magnetic energy so liberated goes into the particles; this not only heats the medium but drives bulk motions, ejecting the plasma in an explosive way. A large variety of explosive events are indeed observed at various spatial scales in the solar atmosphere; we shall return to them in Section 4.5.

The permanent dragging of the base of the field lines by photospheric fluid motions has another, gentler, consequence. Acting somewhat as a bow on the strings of a rudimentary musical instrument, it excites Alfvén waves which, as we have seen in Section 2.3, propagate along magnetic field lines in the form of motions at right angles to the lines, that bend and unbend them; we shall return to these waves in Section 4.6.3.

What should one conclude from all this? We have seen that above the chromosphere – in the low corona – the magnetic energy dominates, so that it is the main driver of the medium. The thermal pressure forces are thus not sufficient to balance any gradient in magnetic pressure, so that the magnetic field is left to its own devices: magnetic pressure forces must be balanced by magnetic tension; we shall examine this point more thoroughly below. As a result, regions of very different gas densities and temperatures may coexist without disturbing the mechanic equilibrium, because the magnetic field has only to vary by a very small amount to maintain the global balance of forces (the structures shown in Fig. 4.9 below are a good illustration of this property).[9]

The predominance of magnetic energy makes the solar atmosphere highly structured in space and variable in time; it does not share the nice (approximate) spherical symmetry characteristic of the solar interior, which is governed by gravitation. The chromosphere and its transition to the corona must therefore be

[9]This gentle adaptation works, however, up to a point, which we shall examine more thoroughly in Section 4.5, and of which Fig. 4.10 is an illustration.

Figure 4.6 'White tongue today, bad weather on the way.' (Drawing by Cham, from *Grand Album de caricatures par Cham, Cours d'Astronomie*, Plon, Paris.)

thought of not as well-defined physical layers but rather as temperature regimes, consisting of a number of varying structures, including small (unresolved) ones. Indeed, though most of the solar surface is called the *quiet Sun*, it is everything but quiet. In this respect, the main difference with the so-called *active regions* – which are located above sunspots and adjacent structures where the magnetic field is correlated on a large scale – is in the reduced size and energy (and helicity, too, as we shall see below) of the structures.

4.1.5 A (tentative) look at the solar jungle

> I looked below, and saw with my physical eye all that domestic individuality which I had hitherto merely inferred with the understanding. And how poor and shadowy was the inferred conjecture in comparison with the reality which I now beheld!
> Edwin A. Abbott, *Flatland – A Romance of Many Dimensions*, 1884

Unfortunately, contrary to the hero of Flatland transported to the Land of Three Dimensions, solar observers do not yet have three-dimensional pictures at their disposal. Since the medium is nearly transparent in most of the wavelength range, any image integrates the emission along the line of sight; hence three-dimensional structures can only be inferred, and it is crucial to resist the temptation of over-interpreting the data (Fig. 4.6).

These observations reveal a forest of structures having virtually every size and lifetime. A flavour of them may be appreciated in recent reviews [4] and [19] and by looking at Figs. 4.7, 4.8 and 4.9.

From the photosphere to the corona 173

Figure 4.7 Three pictures of size $144\,000 \times 280\,000$ km^2 on the Sun north polar cap (taken nearly simultaneously on 31 August 1996 with the ultraviolet spectrograph SUMER on SoHO), thought to reveal gas near temperatures respectively just below 3×10^4 K (lower panel), 2.4×10^5 K (middle) and 1.1×10^6 K (upper panel). (Adapted from [59].)

Figure 4.7 shows a very quiet region seen at three different wavelengths (revealing matter in three different temperature ranges) [59]. The pictures, taken at solar activity minimum, show a part of the polar region and encompass a large coronal hole.

The lower panel is thought to represent matter just below 3×10^4 K; it shows a ubiquitous pattern – known as the *chromospheric network* – that outlines the edges of supergranular cells where the photospheric magnetic flux accumulates; it may be made of open funnels and loops in the chromosphere, as sketched in Figs. 4.2 and 4.3.

The middle panel, thought to reveal matter around 2×10^5 K, has a completely different aspect. A number of luminous spikes, somewhat like blades of grass in a meadow, shoot up on the limb, whereas darker bush-like structures

Figure 4.8 Image taken by TRACE in November 1999 in a narrow spectral band around 171 Å, revealing plasma around 10^6 K. (Image by Stanford-Lockheed Institute for Space Research and NASA.)

outline the boundaries of the supergranular cells; these individual spikes, that continuously emerge and disappear, are generically known as *spicules*; we shall return to them in Section 4.4.1.[10]

A still different picture emerges from inspection of the upper panel, around 10^6 K, i.e. at coronal temperatures. The network has been washed out (recall Fig. 4.2), but a number of *bright points* appear, some of which seem to be at the base of luminous *plume*-like structures extending roughly radially outwards. These bright points lie above the boundaries of supergranular cells, have lifetimes in the range of a few hours (but vary on timescales of minutes) and are inferred to be clusters of miniature loops [23]. This time variability is part of a whole spectrum of phenomena to which we shall return in Section 4.5.

Figure 4.7 shows the quieter part of the Sun. Active regions exhibit larger structures, including huge *loops* (Fig. 4.8), magnetically levitated cold ribbons called *prominences* (Fig. 4.9), and time variability on larger scales, including giant eruptions in the form of *flares* and/or *coronal mass ejections* (Fig. 4.10).

In order to try to understand these pictures, let us assemble a few basic tools.

4.2 Force balance and magnetic structures

Table 4.1 summarises the physical parameters in the solar atmosphere outside active regions, estimated with the density and temperatures of Fig. 4.1 and magnetic field amplitudes inferred in Section 4.1.2. Active regions have a greater density, temperature and magnetic field, but the ratio β of gas pressure to magnetic forces is not too different in order of magnitude. The question marks accompanying the values of B are a reminder of the difficulty of measuring the small coronal magnetic fields (see [30], [43]) far from regions that are active or

---

[10]Figure 4.7 (middle) shows what specialists call *macrospicules*.

Force balance and magnetic structures 175

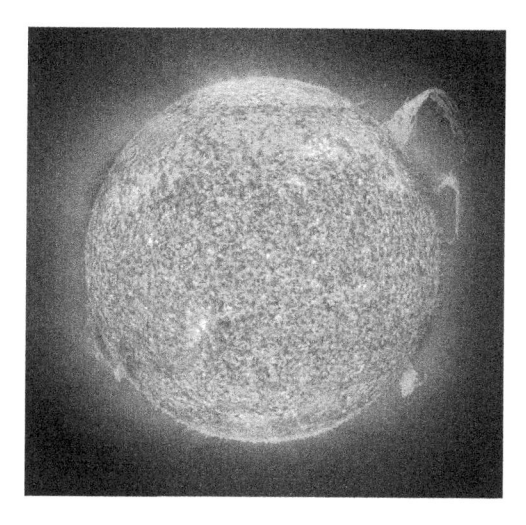

Figure 4.9 Image taken on the spacecraft SoHO with the EIT telescope in a narrow spectral band around 304 Å, revealing plasma around 7×10^4 K, on 11 January 1998 showing several prominences. (Image from SoHO/EIT consortium, ESA and NASA.)

special. Note that in both the photosphere and the low chromosphere, which are weakly ionised, the gas pressure, and thus also the parameter β, is dominated by neutrals (which are closely coupled to charged particles by collisions).[11]

Most of the observed structures have elongated shapes and appear to delineate magnetic field lines. This is because the particle gyroradii are much smaller than the major scales of variation, and the large electrical conductivity makes the magnetic Reynolds number huge for typical sizes and speeds. Hence the plasma tends to move along the field lines, but not across them. Furthermore, we have seen that heat flows essentially along magnetic field lines; this makes the gas more isothermal along field lines than across them, so that the temperature also tends to delineate them.

4.2.1 Forces

The tendency of both plasma and heat to flow along the field lines enables one to make one-dimensional estimates along the direction of the magnetic field **B**. Consider first the balance of forces and assume the plasma to be in local thermal equilibrium, i.e. all particle species have Maxwellian velocity distributions with the same temperature T. Since the magnetic forces are perpendicular to **B**, the balance along it is only determined by:

- the gradient in gas pressure dP/ds, where s is the length along **B**,

- the projected gravitational force $-\rho g \cos\theta$, where θ is the angle of **B** to the (outward) radial, and $g = M_\odot G/r^2$.

---

[11] The number density of neutrals is typically 10^2 and 10^4 times greater than the electron density respectively in the low chromosphere and the photosphere.

Figure 4.10 Coronal mass ejection observed in white light with the LASCO coronagraph on the spacecraft SoHO on 27 February 2000. The shaded disc is a mask making an artificial eclipse in the instrument; the white circle sketches the size of the Sun. (Image from SoHO/LASCO consortium, ESA and NASA.)

With the pressure $P = \rho k_B T/\mu$ varying at the scale L, the pressure gradient is of the order of $P/L \sim \rho V_S^2/L$, V_S being the sound speed. If the fluid motion has typical speed v, the inertial term in the equation of motion is of the order of $\rho v^2/L$; hence, the inertia can be neglected if the speed satisfies $v \ll V_S$. In this quasi-static case, the balance of forces along **B** may thus be written

$$-\frac{dP}{ds} = \frac{P \cos\theta}{H} \quad \text{with} \quad H = \frac{k_B T}{\mu g}. \tag{4.1}$$

This yields $P \propto e^{-s\cos\theta/H}$, i.e. the pressure decreases exponentially with height ($s \cos\theta$) on the scale H. In the presence of perturbations, this pressure equilibrium takes some time to be established – the time for sound waves to propagate. For a structure of size L, this time is (in SI units)

$$\tau_P \sim L/V_S \sim (L/150)\, T^{-1/2}.$$

From Table 4.1, we see that for either a 10^3 km ($\sim 10^{-3} R_\odot$) sized structure in the chromosphere or a ten times greater one in the corona, this time is around 1 min.

The situation is different across **B**, since magnetic tension and magnetic pressure provide forces of the order of $B^2/\mu_0 L$ per unit volume, L being the curvature radius of the line or the scale of variation across the field (see Section 2.3). Once again, the inertia term – of order $\rho v^2/L$ – may be neglected if $\rho v^2 \ll B^2/\mu_0$, which is equivalent to $v \ll V_A$ where V_A is the Alfvén speed. If the opposite inequality holds, the magnetic field has not enough time

Force balance and magnetic structures 177

Table 4.1 *Evolution of physical parameters in the solar atmosphere outside active regions*

	Photosphere	Upper chromosphere	Lower corona	Corona
Height ($R_\odot$)	0.0	$2\text{–}5 \times 10^{-3}$	$10^{-2}\text{–}10^{-1}$	$10^{-1}\text{–}1$
Temperature T (K)	6×10^3	10^4	10^6	10^6
Scale height $H = k_B T/(\mu g)$ (m)	1.5×10^5	5×10^5	5×10^7	10^8
Sound speed $V_S = (\gamma k_B T/\mu)^{1/2}$ (m/s)	0.8×10^4	1.2×10^4	1.5×10^5	1.5×10^5
Magnetic field amplitude B (T)	0.1 (strong B)	$(2\text{–}10) \times 10^{-4}$?	$(2\text{–}10) \times 10^{-4}$?	10^{-4} ?
Ratio of pressure forces to magnetic forces $\beta = 2V_S^2/\gamma V_A^2$	~ 1	~ 1	< 1	< 1

to adapt to a velocity perturbation, which thus acts as an antenna exciting MHD waves.

What is the structure of the magnetic field when the medium moves so slowly ($v \ll V_A$) that it has ample time to adjust to perturbations? This problem of magnetostatic equilibrium and of the corresponding instabilities [24] is classic in laboratory devices, and its mastering is vital for achieving controlled nuclear fusion. The solar case, however, is different and encounters three major difficulties.

First, due to the large spatial scales, both the Reynolds number and the magnetic Reynolds number are so large that dissipation is negligible except in small localised regions. Second, gravity is responsible for a decrease of more than seven orders of magnitude in density between the photosphere and the low corona (see Fig. 4.1), so that the coronal magnetic structures have their feet anchored in the photosphere where matter energy dominates (gravity also produces instabilities). Third, the system has no rigid walls; not only can it exchange matter and energy with the dense photosphere (and the layers below), but it is open to infinity.

4.2.2 Force-free magnetic field

In the low corona, where magnetic forces dominate, the magnetic force $\mathbf{J} \times \mathbf{B}$ cannot be balanced by other terms than itself, so that it must be close to zero at equilibrium. The magnetic field must arrange itself for the curvature force to approximately balance the magnetic pressure force. In order to have $\mathbf{J} \times \mathbf{B} = 0$ with $B \neq 0$, we must have either:

- $\mathbf{J} = 0$. In this case, as we saw in Section 1.3.3, the magnetic field derives from a scalar potential obeying the Laplace equation, so that one can apply the well-known techniques and results of potential theory; in particular, the field within a volume is uniquely defined by the values of its normal component B_n on the bounding surface; furthermore, if B_n is prescribed on that closed surface, this *potential field* – which has no local currents – is the one having the minimum possible energy.

- or: $\mathbf{J} \parallel \mathbf{B}$, i.e. the lines of current and magnetic field are everywhere parallel. This may be written

$$\nabla \times \mathbf{B} = \mu_0 \mathbf{J} \tag{4.2}$$
$$= \alpha \mathbf{B} \quad \text{(force-free magnetic field)} \tag{4.3}$$

where α is a scalar function of position and time. Taking the divergence of (4.3) and using $\nabla \cdot \nabla \times = 0$, we deduce $\nabla \cdot (\alpha \mathbf{B}) = 0$, or $\mathbf{B} \cdot \nabla \alpha = 0$ (since $\nabla \cdot \mathbf{B} = 0$), i.e. α is constant along a field line.

To summarise, when magnetic forces are dominant, either the local current is zero and $\mathbf{B}$ derives from a scalar potential obeying Laplace's equation, or the local current is along $\mathbf{B}$.

What is the physical meaning of α? Let us integrate (4.3) over a surface S and transform the left-hand side surface integral into a line integral along the circumference C of S, by using Stokes' theorem; this yields

$$\int_C \mathbf{B} \cdot \mathbf{ds} = \int_S \alpha \, \mathbf{B} \cdot \mathbf{dS} \tag{4.4}$$

where $\mathbf{ds}$ is a line element along the contour C, and $\mathbf{dS}$ is normal to the element dS of the surface S, oriented as a right-handed screw with respect to the contour C. Equation (4.4) means that α measures approximately the ratio of the magnetic field integrated along a given circumference C, to the flux through the surface bounded by C. It is thus related to the twist in the magnetic field, in both amplitude and sense; positive (negative) α means that the field lines follow a right-handed (left-handed) screw. Let us put this on a quantitative footing.

Twist

To quantify the relation between the twist and the force-free parameter α, let us consider a magnetic field having a cylindrical and longitudinal symmetry about an axis z, i.e. being a function of r (the distance to the axis) only, so that $\mathbf{B}$ has only longitudinal (B_z) and azimuthal (B_ϕ) components. The field lines are helices lying on cylindrical surfaces (Fig. 4.11, left).

The curvature of the field lines yields a radial force $B_\phi^2/\mu_0 r$ (per unit volume) at distance r, pointing inwards, which tends to uncurl the field lines; since this tension must be balanced by the magnetic pressure force $-d/dr\left(B^2/2\mu_0\right)$, the magnetic field must decrease away from the axis. Because of Maxwell's equation (4.2), B_ϕ is related to the component J_z of the current density, while B_z is related to J_ϕ, so that if $\mathbf{J}$ and $\mathbf{B}$ are to be everywhere parallel, they must have both azimuthal and longitudinal components, i.e. the field must be twisted.

Force balance and magnetic structures

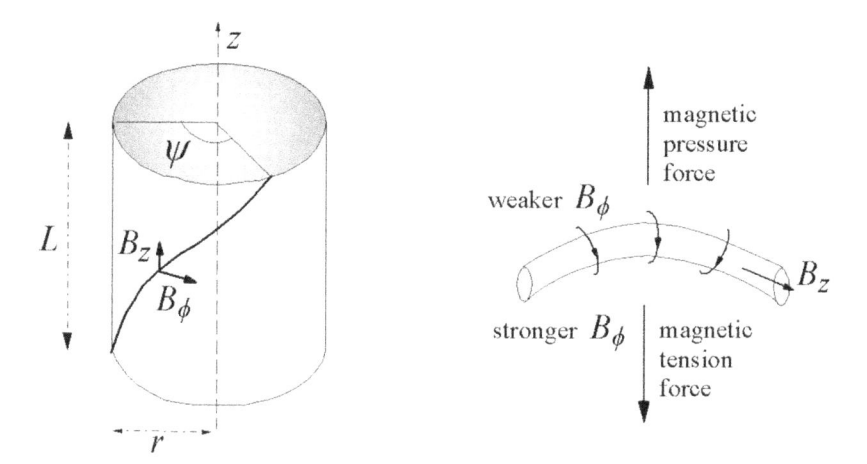

Figure 4.11 Left: a cylindrically symmetric flux tube, with a magnetic field line twisted by the angle ψ. Right: a sketch of the kink instability: the azimuthal field B_ϕ of a curved tube produces a magnetic pressure force which tends to increase the curvature; the axial field B_z is stabilising since its curvature produces a tension force which tends to uncurl the tube.

Let us calculate the angle ψ through which a line at radial distance r is twisted on going a distance L along z. We write the equation of field lines: $rd\phi/B_\phi = dz/B_z$, and integrate over ϕ and z; this gives $r\psi/B_\phi = L/B_z$, i.e.

$$\psi = (B_\phi/B_z)(L/r). \tag{4.5}$$

How is the twist angle ψ related to the force-free parameter α? Let us apply (4.4) to a cross-section of a tube of radius r, C being a circle of radius r and S the enclosed disc; since $\mathbf{B} \cdot \mathbf{dS} = B_z \times 2\pi r dr$, we get

$$B_\phi \times 2\pi r = \int_0^r \alpha B_z \times 2\pi r dr.$$

We may deduce an order-of-magnitude estimate by factoring α and B_z out of the integral (though they generally depend on r), to obtain $B_\phi r \sim \alpha B_z r^2/2$, whence

$$\alpha \sim B_\phi/rB_z \sim \psi/L \tag{4.6}$$

where we have substituted (4.5). Hence, with this geometry, α measures – in order of magnitude and in an average sense – the amount of twist ψ gained by a field line per unit length. Formal calculations of twisted structures may be found in [41].

Kink

Magnetic flux tubes cannot be twisted too much, because if the azimuthal component of the magnetic field becomes too large, an instability arises. Consider a

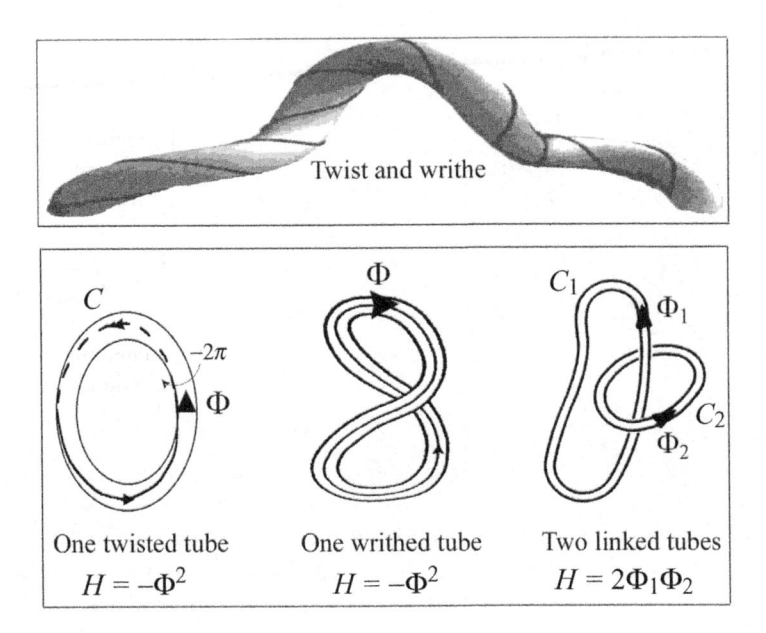

Figure 4.12 Top: how the extra twist of a flux tube appears as writhe. Bottom: three elementary examples of helicity of different signs, coming respectively from twist, writhe and linkage (from left to right); note that with a single tube (left and middle), changing the sense of **B** does not change H; with two tubes (right), changing the sense of **B** in both tubes does not change H. (The top sketch is adapted from Berger's paper in [10] and the bottom one from [38].)

magnetic field that is nearly azimuthal (B_ϕ), and suppose that the tube is bent slightly (Fig. 4.11, right). The field lines are pressed together on the concave side, and spaced out on the convex side. As a result, the magnetic field – and thus the magnetic pressure – increases on the concave side and decreases on the convex side. The greater magnetic pressure on the concave side produces a force normal to the tube axis which tends to increase the initial bending. This is known as the *kink instability*.

However, if the magnetic field has a component B_z along the axis, the bending produces an inward magnetic tension force which tends to counteract this effect. Whether or not the tube is unstable depends on the value of B_ϕ/B_z, on the geometry and on the boundary conditions. In practice, the tube will generally kink if the average twist angle ψ is greater than a few π radians, i.e. if it is twisted by significantly more than one full turn.

Hence, just as with an ordinary rope, twisting a magnetic tube tends to produce a screwed structure somewhat as in a phone cord. Such kinked helical structures are indeed observed. The kinking enables the tube to get rid of its unwanted extra twist, but the twist reappears in disguise: the axis of the tube coils into an helicoidal shape, i.e. the tube becomes writhed (Fig. 4.12, top). Suppose that the ends of the tube are fixed, being for example anchored at the photosphere. We can change both the twist of the tube around its axis and the coiling of the axis itself, but these deformations transform into each other. This introduces a fundamental notion: the magnetic helicity.

4.2.3 Magnetic helicity

Magnetic helicity derives from a topological concept whose applications are centuries old. The Incas had devised an ingenious method based on coloured knotted cords, called *quipus*, for recording information.[12] Applications of helicity in physics date back to Gauss, and to Kelvin's idea of describing atoms as knotted tubes of a fluid aether. If Kelvin had been correct, the topological invariance[13] and diversity of knots would have made them perfect tools to build a periodic table of elements. As often happens in science, this failed theory inspired rich developments in other fields, from particle physics to biology. Helicity plays an important role in solar plasmas for reasons we shall explain below. A clear topological discussion can be found in [38], and applications to space magnetic fields are nicely surveyed in the monograph [10].

Helicity – and more generally a lack of reflectional symmetry – is everywhere around us. We are made of left-handed proteins, and a DNA molecule can be viewed as a long curve that is intertwined millions of times, linked to other curves, and coiled in order to fit into a small space. We have already encountered helicity when discussing solar dynamos models: the density of fluid helicity is $\mathbf{v} \cdot \omega$, the scalar product of velocity and vorticity $\omega = \nabla \times \mathbf{v}$.

Definition

In MHD, the density of magnetic helicity is defined as $\mathbf{A} \cdot \mathbf{B}$, where $\mathbf{B}$ is the magnetic field and $\mathbf{A}$ is a vector potential which gives rise to it as $\mathbf{B} = \nabla \times \mathbf{A}$. We shall see that the integral of $\mathbf{A} \cdot \mathbf{B}$ in a volume that wholly contains the magnetic field measures the degree of linkage of the field lines, how they are braided, twisted and knotted. Its importance lies in the fact that in an ideal conductor, magnetic field lines are frozen in the medium, so that their topology cannot change; if two closed field lines are initially linked together p times, they must remain so linked for ever. Furthermore, we shall see that slight departures from perfect conductivity, that may change locally the linkage of flux tubes through reconnection events, do not change appreciably the global magnetic helicity, so that magnetic helicity is a very robust invariant in the solar interior and atmosphere.

Formally, the helicity in a volume V is defined as

$$H = \int_V d^3r \, \mathbf{A} \cdot \mathbf{B} \tag{4.7}$$

---

[12] The Inca recording system was based on combinations of colours, types and number of knots and their places on the cords. It was a base 10 positional system. For example, a cord with three cluster positions containing in order: 4 single knots, 5 single knots, and a long knot of 3 turns respectively might be interpreted in our notation as 453 ($4 \times 100 + 5 \times 10 + 3 \times 1$); Ascher, M. and Ascher, R. 1981, *Mathematics of the Incas: Code of the Quipu*, New York, Dover.

[13] Two structures are topologically invariant if they can be continuously deformed into each other.

where the volume V is simply connected (no holes) and completely contains **B**, whose normal component is assumed to vanish on the surface bounding V.[14] This reflects the non-local nature of helicity; clearly, the vector potential at a point cannot be determined by the field at that point alone, and in this sense it is non-local; in a more subtle way, the linkage of curves involves the behaviour of these curves at points that may be very far from each other. Without this constraint, one could choose an integration volume so small that it contained no linkage.

The strict invariance of H can be proved formally from the frozen field equation of MHD.

Twist, writhe and linkage

Helicity is easy to calculate in the simple case illustrated in Fig. 4.12 (bottom right, adapted from [24] cited in Chapter 3) of two thin flux tubes following the closed curves C_1 and C_2 and having respective fluxes Φ_1 and Φ_2. The helicity (4.7) is the sum of an integral over the volume of the tube number 1, plus an integral over the volume of the tube number 2. Since **B** follows the contours C_1 and C_2 and the magnetic flux (product of B by the tube's section) is infinitesimal and invariant along each of them, **B** d^3r may be replaced by $\Phi_1 \mathbf{dl}$ on C_1 and by $\Phi_2 \mathbf{dl}$ on C_2, so that (4.7) takes the form

$$H = \Phi_1 \int_{C_1} \mathbf{A} \cdot \mathbf{dl} + \Phi_2 \int_{C_2} \mathbf{A} \cdot \mathbf{dl}. \tag{4.8}$$

Now, by Stokes' theorem, the line integral of **A** along a contour C is equal to the flux of **B** across the surface delimited by C and oriented as a right-handed screw. Hence, for the geometry shown, where each tube encircles the other in the right-handed sense, we have $\int_{C_1} \mathbf{A} \cdot \mathbf{dl} = \Phi_2$ and vice versa, so that (4.8) yields $H = 2\Phi_1\Phi_2$. This can be generalised to tubes having a finite cross-section and winding several times around each other; the helicity is then proportional to the number of linkages of the lines, each having a sign determined by the relative orientation of the winding. The helicity is thus proportional to a topological invariant – the linkage coefficient – which does not change under continuous deformation.

This concept also holds for a single tube being twisted or writhed, as shown in Fig. 4.12 (left and middle) or having knots (Problem 4.8.2). To understand this, let us imagine a simple torus-like tube carrying flux Φ, whose field lines are unlinked circles parallel to the torus circular axis C. Such a tube has zero helicity. Now, let us cut it along a cross-section perpendicular to C, producing

---

[14]This is because the potential vector **A** is defined up to the gradient of an arbitrary function f, and adding ∇f to **A** changes H by $\int_V d^3r \, \nabla \cdot (f\mathbf{B})$ (since $\nabla \cdot \mathbf{B} = 0$), which, by Stokes' theorem, vanishes if the normal component of **B** vanishes on the surface bounding V. Otherwise, some subtle variants are required to define gauge invariant quantities (see the paper by M. A. Berger in [10]).

Force balance and magnetic structures 183

free ends; then let us twist these free ends through a relative angle 2π and reconnect them. Each field line in the new flux tube is a closed curve in the shape of an helix wrapped around the axis C. This is sketched in Fig. 4.12 (left), where the twist is left-handed, so that the helicity is $H = -\Phi^2$. More generally, twisting the free ends through a relative angle ψ (in the right-handed sense) increases H by $\Delta H = \Phi^2 \psi / 2\pi$.

Now, let us try a little experiment with a tape of paper (as we did in Section 3.3), or preferably with a rubber tape. Twist the free ends by a relative angle of 2π, and glue the ends as for the flux tube shown in Fig. 4.12 (left). Then try to remove the twist; you obtain a figure whose flattened shape is sketched in Fig. 4.12 (middle), namely you have transformed the twist of the tape about its centre line into a deformation of the centre line itself. This happens every day with our telephone cords. The interesting point is that the helicity has not changed. More generally, one can build an extremely complex knot having twist, torsion and linkage, and deform it. The helicity will be distributed differently, but the total value will not change [38]. Note that zero helicity does not mean an absence of twist, writhe or linkage, but that these different contributions cancel out in the expression of H.

Twists, writhes and linkages of magnetic field lines are associated with field-aligned currents. A simple case is illustrated by the magnetic flux tube of Fig. 4.11 (left). Since it is twisted by the angle ψ, its helicity is $H = (\psi/2\pi)\,\Phi^2$, and from (4.6) the parameter $\alpha = \mu_0 \mathbf{J}/\mathbf{B}$ is $\sim \psi/L$. Hence in this case $H \sim (L\alpha/2\pi)\,\Phi^2$ in order of magnitude. In more general cases, the force-free parameter α is related to H in a non-trivial way.

Magnetic helicity as a robust invariant

Given the huge magnetic Reynolds number in the solar environment, magnetic topology is conserved except during brief moments and in small regions where large magnetic gradients occur, that enable dissipation and reconnection of field lines. These events change not only the energy but also the magnetic topology, so that the magnetic helicity is no longer strictly conserved. However, and this is a very interesting point, it can be shown that H changes very little during such events (whose timescale is vastly smaller than the huge resistive dissipation timescale). This may be understood from dimensional arguments. In a volume V where the magnetic field is B, the magnetic energy scales as $W_B \propto B^2 V$, whereas the helicity scales as $H \propto \Phi^2 \propto B^2 S^2$ where S is a typical cross-section. Since $S^2 = VL$, we deduce that the magnetic helicity scales as

$$H \propto W_B L. \tag{4.9}$$

The medium contains structures spanning a large range of scales (see a discussion of turbulence in Section 6.4), and for large R_m, dissipation occurs at small scales. Since (4.9) suggests that in small-scale processes (L small), the helicity H should vary much less than the magnetic energy W_B, H is expected not to vary significantly during dissipation events.

The scaling (4.9) has another consequence: it should take much less energy to put helicity into large scales than into small scales; hence H should have a preference for being transferred to large scales in transformation processes. This suggests that helicity might play an important role in producing large-scale magnetic fields; we shall see later the importance of this property for building the large-scale structures involved in coronal mass ejections. (Note that magnetic helicity has a very different behaviour from the hydrodynamic helicity $\mathbf{v} \cdot \nabla \times \mathbf{v}$, which scales as v^2/L, i.e. as the kinetic energy W_k divided by L.)

The invariance of helicity, even in the presence of resistive effects, provides a constraint to find equilibrium states having the lowest possible energy. It can be inferred in this way that the state of minimum magnetic energy – in a volume V containing all the magnetic energy – is a force-free field for which α has the same value for all field lines [55]. This means that as field lines are deformed and even broken and reconnected, the helicity tends to spread out, with its distribution becoming more uniform.

Helicity transfers

This robust invariance of helicity raises a problem: if it is so difficult to remove – and to create – how can it be produced in the first place? To answer that question, two facts are crucial. First, the definition (4.7) of helicity concerns a volume at the boundary of which the normal component of $\mathbf{B}$ vanishes, so that it cannot be applied as such to the whole solar atmosphere because of the vertical magnetic field at the photosphere (some subtle variants are required). Second, since helicity can have both signs, one can easily create positive and negative helicities without changing the total value; for example, take a rope whose extremities are tied, and twist the middle; this is equivalent to transferring helicity from one half of the rope to the other.

Indeed, the solar atmosphere is far from being a closed system and helicity may be transferred from one region to the other and even locally from some scales to other ones. It can exchange helicity in three ways. First, flux tubes are emerging in permanence from the solar interior, thereby transferring helicity. Second, motions at the photosphere (where the magnetic field is not force-free) may twist and braid the flux tubes, thereby changing their helicity. Third, the corona is open to 'infinity', with which it may exchange helicity, for example by ejecting matter.

Helicity segregation in the Sun and its atmosphere

A simple case of helicity segregation arises through the solar differential rotation. Look at Fig. 3.10, which is an elementary picture of how differential rotation produces an azimuthal magnetic field B_ϕ from a poloidal one. Starting from an essentially dipolar magnetic field with positive B_z, differential rotation distorts the field lines into a left-handed screw – having a *negative* helicity – in the *northern* solar hemisphere, and to a right-handed screw – having a *positive* helicity – in the *southern* hemisphere. The amplitude of these helicities increases as the field winds up. One solar cycle later, the sense of the dipole has reversed, but the handedness of each screw has not changed.

Hence, while the magnetic flux through each hemisphere changes sign at each cycle as the dipolar magnetic field reverses, the solar differential rotation always produces negative magnetic helicity in the northern solar hemisphere and positive helicity in the southern one. Only the amplitude of these helicities changes during the solar cycle, not the sign.[15]

As magnetic loops emerge in permanence through the photosphere, they carry magnetic helicity, and since global helicity is conserved as they evolve (and even reconnect), coronal structures are expected to have a preferred negative helicity in the northern corona and a preferred positive helicity in the southern corona. Since the sign of the helicity should be still more strongly conserved than the helicity itself, it may survive the adventures endured by the structures during their life, and should therefore be observed in the corona and the solar wind. And this is indeed observed (see the review by D. Rust in [14] and references therein).

4.2.4 Inferences on magnetic structure in the low corona

What is the essential conclusion of all this? We have discussed (Section 3.4) the importance of intense magnetic flux tubes for the solar magnetism. As these tubes penetrate into the corona, they lose their great intensity and are left to their own devices, except at their feet, where various motions stress and shuffle them, so that they become twisted and braided. The associated field-aligned currents produce an excess of magnetic energy with respect to the potential field, so that both energy and helicity accumulate in the magnetic field.[16] Since helicity is virtually globally invariant in reconnection processes, it tends to build up as new tubes emerge, are stressed and interact. Furthermore, since helicity scales as magnetic energy times size, it tends to evolve towards larger helical structures as magnetic energy is dissipated.

In regions where the flux tubes have all the same sense (outwards or inwards), the field lines tend to open to 'infinity', and can therefore get rid relatively easily of their helicity (through wave emission), just like a bunch of ropes with free ends.

In other regions, the field lines form loops anchored at both ends, in which helicity builds up and tends to form large complicated flux ropes. What happens next? Before looking at this question, let us forget the magnetic field for a while, except for its providing a privileged direction along which heat can flow, and let us study the large-scale energy balance.

---

[15] Another source of helicity involves the action of the Coriolis force on rising loops within the Sun.

[16] In fact, the magnetic helicity does not fully characterise the topological complexity of the magnetic field, and other invariants can be defined.

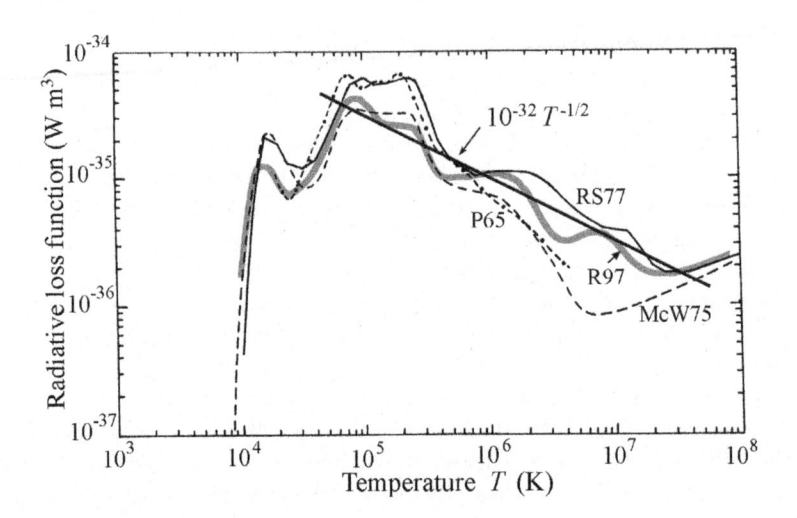

Figure 4.13 The radiative loss function of the solar atmosphere $F(T)$, from a number of studies cited in the text (labelled by first letters of author's name). The straight line is the analytical approximation used here.

4.3 Energy balance

In the absence of bulk speeds and of unspecified heating processes, energy balance in an (optically thin) atmosphere is determined by:

- conductive heat transport along the magnetic field, towards the cooler side (from (2.91) and (2.98))

$$Q_c = -q_0 T^{5/2} dT/ds \ \ \text{W m}^{-2} \quad \text{where} \quad q_0 \simeq 10^{-11} \quad (4.10)$$

- radiative losses

$$W_R = n^2 F(T) \ \ \text{W m}^{-3} \quad (4.11)$$

where $F(T)$ is shown in Fig. 4.13 and n is the electron or proton density.

Conductive heat transport acts either as an energy supply or as an energy loss, depending on the sign of dQ_c/ds, whereas radiative losses always cool the gas down. Let us study this in more detail.

4.3.1 Radiative losses

In a collisional plasma, radiative losses are proportional both to the density of the radiating ions and to that of the electrons that ionise and excite them, so that the power radiated by unit of volume, W_R, is proportional to the square of the plasma density n. When the plasma is close to thermal equilibrium, the so-called *radiative loss function* $F(T) = W_R/n^2$ can be calculated as a function

Energy balance

of temperature from the abundances of the radiating elements and from their ionisation fractions. Figure 4.13 shows values published in [45][17] and a more recent curve from J. Raymond (thick grey line).[18] One sees that the radiative losses jump sharply near 10^4 K, mainly due to hydrogen Lyα radiation, and peak around 10^5 K. At higher temperatures, the curve has a complex shape with several bumps due to the contribution of spectral lines of several heavy elements, and falls progressively as prominent atomic transitions disappear with increasing ionisation. Above 10^7 K, the radiation losses increase again due to bremsstrahlung radiation.

Substituting $n = P/(2k_BT)$ for a fully ionised gas of pressure P, the radiation loss rate may be written as

$$W_R = \left(\frac{P}{2k_BT}\right)^2 F(T). \tag{4.12}$$

We see in Fig. 4.13 that in the temperature range from 5×10^4 K to 5×10^7 K, the radiative loss function is fairly well approximated by

$$F(T) \simeq 10^{-32}/\sqrt{T} \text{ W m}^3. \tag{4.13}$$

We deduce an approximation of the radiation rate in this temperature range

$$W_R \simeq W_0 P^2 T^{-5/2} \text{ W m}^{-3} \quad \text{where } W_0 = 1.3 \times 10^{13} \tag{4.14}$$

in SI units.[19]

The decrease of radiative losses with increasing temperature above 5×10^4 K has an interesting consequence. Imagine that for some reason, heat conduction is suppressed and that some perturbation heats the gas, making the temperature increase above its equilibrium value; this temperature growth reduces the radiative losses, making the temperature grow further. Hence, heating the plasma above a few 10^4 K produces a runaway increase in temperature. Conversely, imagine that some perturbation makes the temperature fall off; this increases the radiative losses, so that the gas cools more, and finally collapses to a lower temperature. These instabilities, however, occur only when heat conduction may be neglected, because these variations in temperature strongly change heat conduction, which may stabilise the gas by enabling it to get rid of its heat excess or deficit.

4.3.2 Radiative and conductive timescales

In order to compare the relative importance of these processes, let us estimate how fast structures cool off by radiation and by heat conduction. Since the gas

---

[17] With values from Raymond, J. and Smith B. W. 1977, *Ap. J. Suppl.* **35** 419 (continuous line), [35] (dashed) and Pottasch, S. R. 1965, *B.A.N.* **18** 8 (dash-dotted).

[18] From Raymond and Smith 1977, with updated abundances.

[19] Recent calculations using abundances of elements relevant for closed coronal structures yield a similar approximation, but with a value of W_0 twice as great (Martens, P. C. H. et al. 2000, *Ap. J.* **537** 471) – a change of no important consequence on our order-of-magnitude estimates.

energy density is $3nk_BT$ (for electrons and protons), the timescale for temperature change due to radiation is given from (4.12) by (in SI units)

$$\tau_R = \frac{3nk_BT}{W_R} = \frac{6(k_BT)^2}{PF(T)} \sim \frac{T^{5/2}}{W_0 P}. \tag{4.15}$$

Using Table 4.1 and (4.14), we find in the low corona outside active regions: $\tau_R \sim 4 \times 10^4$ s, that is roughly 11 h. In active regions, which are typically hotter by a factor of two or three, and denser by a factor of ten or more, (4.15) yields a timescale of only 1–2 h.

On the other hand, the timescale for temperature change due to heat conduction is, from (4.10),

$$\tau_c = \frac{3nk_BTL}{q_0 T^{5/2} \times T/L} \sim \frac{PL^2}{q_0 T^{7/2}} \tag{4.16}$$

where L is the length of the structure and we have approximated dT/ds by T/L. Contrary to τ_R, τ_c depends on the scale of the structure; heat conduction tends to suppress small-scale temperature variation along the magnetic field. Consider a large coronal structure of height $L \sim 10^8$ m; with the values of Table 4.1 outside active regions (4.16) yields $\tau_c \sim 3 \times 10^3$ s – i.e. roughly 1 h, so that $\tau_c \ll \tau_R$. In active regions, the numerator and denominator in (4.16) are greater in similar proportions, so that the timescale τ_c is similar and is therefore of the order of magnitude of τ_R. Shorter structures cool even more rapidly.

4.3.3 Temperature structure

We conclude, therefore, that in a static corona heat conduction is the main cooling process outside active regions, whereas heat conduction and radiative losses are both important in active regions.

To get a feeling about the temperature structure from the chromosphere to the corona, let us consider the energy balance along a radial magnetic field line, where the geometry requires the flux tube cross-section to increase as the square of the radial distance r to the Sun. In the absence of macroscopic speeds and of any unspecified heating process, energy balance then yields, from the expression (4.10) of the heat flux,

$$\frac{q_0}{r^2} \frac{d}{dr}\left(r^2 T^{5/2} \frac{dT}{dr}\right) = W_R. \tag{4.17}$$

Consider this equation together with Fig. 4.1. Integrating it from some radius r_{in} located in the upper chromosphere (where $dT/dr > 0$) to some radius r_{out} in the outer corona (where $dT/dr < 0$), we get

$$r_{out}^2 T_{out}^{5/2} \left[\frac{dT}{dr}\right]_{r_{out}} - r_{in}^2 T_{in}^{5/2} \left[\frac{dT}{dr}\right]_{r_{in}} = \frac{1}{q_0} \int_{r_{in}}^{r_{out}} r^2 W_R dr.$$

Since dT/dr is negative at r_{out} and positive at r_{in}, the left-hand side of this equation is negative, whereas the right-hand side is positive. Clearly, something

Energy balance

is missing; we must add a negative term on the right-hand side to balance radiation and to produce the temperature gradients shown in Fig. 4.1. This means that something should heat the gas, producing an additional term in the energy balance (4.17), which should be replaced by

$$\frac{q_0}{r^2}\frac{d}{dr}\left(r^2 T^{5/2}\frac{dT}{dr}\right) = W_R - W_{heat}. \tag{4.18}$$

There is no agreement as to the identity of this heating term, nor as to how and where it operates. We shall return to this point in Section 4.6. Meanwhile, let us proceed with as few hypotheses as possible. Although the heating term cannot be dispensed with near the temperature maximum, we may try to neglect it at smaller and greater distances to the Sun.

Chromosphere–corona transition

Consider first the vicinity of the chromosphere–corona transition region, which takes place over a small altitude range, in which the distance to the Sun and the pressure do not change much. We may factorise r from the derivative in (4.17) and substitute the expression (4.14) of W_R, to obtain

$$\frac{d}{dr}\left(T^{5/2}\frac{dT}{dr}\right) = \frac{W_0 P_0^2}{q_0} T^{-5/2} \tag{4.19}$$

where P_0 is an average plasma pressure in this region. Rearranging, we get

$$R_\odot^2 \frac{d^2}{dz^2} T^{7/2} = A T^{-5/2} \quad \text{with} \quad A = \frac{7 W_0 P_0^2 R_\odot^2}{2 q_0} \simeq 10^{38} \tag{4.20}$$

where z is the height measured from some arbitrary base point, and we have substituted the numerical values of q_0 (4.10) and W_0 (4.14) with $P_0 \sim 10^{-2}$ Pa (from Fig. 4.1). The dimensional form of (4.20) suggests that the temperature should vary approximately as $T^6 \sim A z^2 / R_\odot^2$.

This has two interesting consequences. First, the huge value of A indicates that the temperature should increase rapidly with altitude, in agreement with the sharp transition region shown in Fig. 4.1, and with our small altitude range approximation. Second, since T varies as $A^{1/6}$, our results are weakly sensitive to the approximations made for the radiative losses and for the plasma pressure. To go a step further, let us substitute a trial solution $T = T_*(z/R_\odot)^{1/3}$ into (4.20) [35]; this yields $T_* = (36A/7)^{1/6}$, whence

$$T = 2.5 \times 10^6 \, (z/R_\odot)^{1/3} \text{ K} \quad \text{(at low heights)}. \tag{4.21}$$

This yields a temperature jump to about 2×10^5 K at a height of $10^{-3} R_\odot$ above the base point (that is located at the base of the transition region), to 5×10^5 K at $10^{-2} R_\odot$ and to 10^6 K at $10^{-1} R_\odot$. Compared to Fig. 4.1, this is not too bad a result for so simple a calculation. At greater heights, it is no longer permissible to neglect the heating term and the decrease in pressure, even to make an order of magnitude estimate. In particular, we have $dT/dz = 0$ at the temperature maximum, and the heating term is essential there to balance the radiative losses.

Outer corona

Consider now the outer corona (outside active regions), significantly above the maximum in temperature. We have seen that we may neglect radiation there, so that (4.17) reduces to

$$\frac{d}{dr}\left(r^2 T^{5/2} \frac{dT}{dr}\right) = 0 \tag{4.22}$$

which can be integrated to give $r^2 T^{5/2} dT/dr = $ constant, whence

$$\frac{d}{dr} T^{7/2} = \frac{\text{constant}}{r^2}. \tag{4.23}$$

Since from Fig. 4.1 the temperature at large distances is much smaller than in the low corona, we assume

$$T \to 0 \quad \text{for} \quad r \to \infty$$

so that (4.23) can be integrated to yield

$$T \propto r^{-2/7}. \quad \text{(at large distances)}. \tag{4.24}$$

Comparing again to Fig. 4.1, this is not too bad a result.

4.4 Some prominent species

Let us now examine more closely the population of the solar jungle, whose most prominent species are shown in Section 4.1.5.

4.4.1 Spicules

The tallest spike in the middle panel of Fig. 4.7 extends roughly 40 000 km ($0.06\,R_\odot$) above the surface; referring to Fig. 4.1, this means that it extends from the chromosphere to the corona, through a medium whose average parameters vary by many orders of magnitude; the other spikes, which are more typical, extend roughly four times less.

What are spicules? Detailed studies show that what appears as a burning 'prairie' when seen on the solar limb is actually made of clumps[20] clustered along the magnetic network at the boundaries of the supergranular cells. Individual spicules appear as sharp jet-like structures, with an average lifetime $\tau \sim 600$ s;

---

[20] They are then called *mottles*.

they are thought to be made of gas and plasma with $n_e \sim 10^{17}$ m^{-3} and (perhaps) $T_e \sim 10^4$ K; individual spicules are inferred to be (perhaps) $d \sim 150$ km across or more, to start at about 1500 km above the solar surface, and rise at an apparent speed of $v \sim 30$ km s^{-1} along magnetic field lines to an altitude of about 10^4 km, where the material dissolves and falls off. A number $N \sim 7 \times 10^4$ of them (with heights greater than 5000 km) have been estimated to be present on the Sun at any time.

Nobody knows what exactly spicules are. Might their energy be provided by the magnetic field in some way or another? With the order of magnitude $B \sim 10^{-3}$ T estimated above, the magnetic energy density is $B^2/2\mu_0 \sim 0.4$ J m^{-3}. The energy required to lift a spicule to a height $h \sim 10^7$ m is $n_e m_p g h \sim 0.4$ J m^{-3}. This suggests that the available magnetic energy is sufficient to lift spicules. The detailed mechanism, however, remains elusive, despite more than a century of observation and a large variety of theoretical models (see [54]). Growing them in our gardens is still a theoretician's dream.

4.4.2 Magnetic loops

Magnetic loops are observed – or rather inferred – virtually everywhere in the solar atmosphere, with a large range of temperatures and sizes, hotter loops generally extending to greater heights. They have a threaded structure; as resolution increases, each loop can be resolved into several finer ones (Fig. 4.8), down to the limit of resolution.

In some sense, the three regions empirically defined on the Sun may be distinguished by their loops. The so-called active regions have the most conspicuous and hottest loops, many of which seem fairly stable, lasting for days or weeks, even though their appearance may change over minutes; most of them tend to have a twisted structure, with a tendency to have a negative (positive) helicity in the northern (southern) hemisphere – in accordance with the discussion of Section 4.2.3. In contrast, the so-called coronal holes, which appear relatively dark on images, have a dominant direction of field lines, that extend far away without returning to the Sun; they have only small loops which in general do not reach coronal heights. The rest of the Sun – the so-called quiet Sun which occupies much of the surface – has a large variety of loops.

This ubiquity of loops in ultraviolet and X-ray images raises a question: why do we see them as discrete entities? In other words, since magnetic lines pervade the solar atmosphere, why do we see only some of them, and not a plain continuum? To answer this question, let us try to derive some basic properties of loops at equilibrium. Because transport processes take place along the magnetic field that delineates the loops, we may picture them as being thermally isolated – except at their foot points – and varying only along their length, which permits one-dimensional modelling.

Consider energy balance along a static loop of constant cross-section:

$$-dQ_c/ds = W_R - W_{heat} \tag{4.25}$$

where s is the arc length along the loop, and W_{heat} is the already mentioned unidentified heating rate. For a loop of constant cross-section, with the conductive heat flux (4.10) and the radiation rate (4.14), this yields

$$q_0 \frac{d}{ds}\left(T^{5/2}\frac{dT}{ds}\right) = W_0 P^2 T^{-5/2} - W_{heat}. \qquad (4.26)$$

This equation enables us to derive an order-of-magnitude scaling between the length L of the loop (from one foot to the apex) and its average temperature $\langle T \rangle$ and pressure P – assumed not to change much along the loop except near its foot points (a reasonable approximation if the height is not much greater than the scale height H corresponding to $\langle T \rangle$). For an order-of-magnitude estimate, we drop the heating term and write $dT/ds \sim \langle T \rangle /L$, to obtain

$$\langle T \rangle^{7/2}/L^2 \sim (7W_0/2q_0)\, P^2 T^{-5/2}$$

whence

$$\langle T \rangle \sim 1.4\,(W_0/q_0)^{1/6}\,(PL)^{1/3} \sim 1.4 \times 10^4\,(PL)^{1/3} \qquad (4.27)$$

where we have substituted the numerical values of q_0 (4.10) and W_0 (4.14). Recall that we are working in SI units; cgs afficionados should divide the constant by 10. Our simple order of magnitude estimate (4.27) is very close to the well-known scaling law derived by [45] with a particular model for the loop and the heating rate, and observed to describe rather well quiescent coronal loops in active regions (see [16] and references therein).

Let us consider a typical active region loop, with a density of about 10^{15} m^{-3} and length $L \sim 10^7$–10^8 m. The scaling relation (4.27) yields a temperature $\langle T \rangle \sim (0.9$–$2.8) \times 10^6$ K, which is in the ballpark of the observed values.

We can derive a further property from (4.26). Just as we saw previously for the general corona, the loops must be heated by some process if they are to exist at all. Indeed, consider the global energy balance along a loop, from one foot to the other. The loop loses energy by radiation from its entire volume, whereas heat conduction between the loop and the exterior acts only at its feet; since these feet are colder than the rest of the loop, normal heat conduction evacuates heat from the loop, not the reverse. Hence some heating process must compensate for these losses. In order of magnitude, (4.26) shows that the heating rate W_{heat} per unit volume should vary as

$$W_{heat} \sim W_0 P^2 \langle T \rangle^{-5/2} \sim 10^3 P^{7/6} L^{-5/6} \sim 3 \times 10^{-12} \langle T \rangle^{7/2} L^{-2} \qquad (4.28)$$

where we have substituted the numerical value of W_0, and the scaling (4.27).[21] In other words, the energy furnished to the loop by a – still unspecified – process is essentially radiated away. Equation (4.28) might help to explain why loops are observed as discrete entities. Increasing the heating rate of a loop of given length L increases its pressure and its density (because from (4.28) and (4.27), P varies with W_{heat} as a higher power law than does T), making it radiate more than its surroundings and thus emerge as a discrete loop.

---

[21] In cgs units the constant 10^3 becomes $10^{4.5}$ and 3×10^{-12} becomes 3×10^{-7}.

Note, finally, that loops may not be in hydrostatic equilibrium, and that they have been observed to oscillate, opening the way to a new discipline, coronal seismology; see the sobering discussion in [57] and references therein.

4.4.3 Prominences

Solar images taken in the H$_\alpha$ line of hydrogen (which is a traditional way of observing the chromosphere) reveal great elongated structures floating above the Sun's surface, that may last for weeks and even months, and quite often erupt. When seen above the limb, they appear as bright *prominences*. On the other hand, when seen on the disc, they show up as dark *filaments* (since they are between us and the bright disc). Prominences and filaments are the two names of ribbon-like structures made of dense and cool gas, that float approximately vertically at heights up to about 50 000 km in the corona (Fig. 4.9). Their typical inferred electron density is $n \sim 10^{16}$ m^{-3}, with a neutral density roughly 10 times greater so that the mass density $\rho \sim 10 n m_p \sim 2 \times 10^{-10}$ kg m^{-3}, and their temperature is around $T \sim 10^4$ K; they appear to be of thickness $\delta \sim 5000$ km, with other dimensions at least 10 times larger, so that their typical mass is $\rho\delta(10\delta)^2 \sim 10^{12}$ kg.

In short, prominences are bits of (dense and cold) chromosphere – having the mass of a small mountain – floating in the 100 times less dense and hotter corona. Furthermore, their vertical extent is far greater than the gravitational scale height corresponding to their temperature (Table 4.1). How can this be so?

As for most problems concerning the solar atmosphere, the key lies in the magnetic field. We have already seen that the magnetic energy is so large that the magnetic field can easily accommodate large variations in plasma parameters. And indeed, prominences are always observed where the magnetic field is roughly horizontal, connecting regions where the vertical component of the photospheric magnetic field changes sign. Furthermore, the field is observed to have some helicity.

Figure 4.14 (left) is a simple view of such a structure, where the magnetic field lines form a hammock where the prominence is resting. The curvature of the magnetic field lines produces a tension force that balances the weight of the prominence. In the region where the magnetic field lines make up a small hammock-shaped hollow, of curvature radius R, the balance of forces (per unit volume) yields

$$B^2/\mu_0 R = \rho g.$$

The radius of curvature of the lines supporting the prominence is thus about $R \sim B^2/\mu_0 \rho g$. Substituting the prominence mass density indicated above, $g \simeq 260$ m s^{-2} and $B \sim 10^{-3}$ T measured in these structures [29], we find $R \sim 2 \times 10^7$ m – greater than the typical width of prominences.

We conclude that the coronal magnetic field is amply able to support prominences, with a relatively small deformation. But how are the curved field lines kept in place? Probably not by the matter lying there, whose mass is grossly

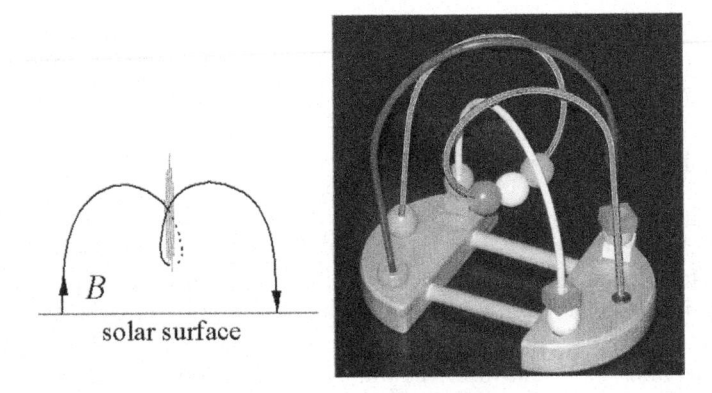

Figure 4.14 How a prominence may be supported against gravity in a cradle of twisted magnetic field lines connecting magnetic patches of opposite polarity on the Sun. Left: basic helical hammock, whose magnetic tension supports a weight. Right: helical hammock (light grey) supporting weights, magnetic arcade (dark grey) that prevents it from blowing away and the field line delineating the axis of the helix (white). (Courtesy G. Aulanier.)

insufficient. An answer is suggested in Fig. 4.14 (right), which symbolises (in black) the magnetic arcade that prevents the hammock from blowing away. And how do prominences form in the first place? How do they accommodate their radiation losses? These questions are still under debate (see [3], [33] and references therein.)

4.5 Time variability

4.5.1 Empirical facts

We have already mentioned the importance of time variability in the solar atmosphere, even outside the so-called active regions. Most of the observed variability involves the brightening, fading or deformation of loops, and/or transient matter ejection and energisation of particles, taking place in a wide range of space, temporal and energy scales.

These transient events have been given various names according to their size, appearance and wavelength of observation. The difficulties of observation, however, especially projection effects, often make similar phenomena appear different, and vice versa. Furthermore, there is some ambiguity in the definition of an 'event' since not all authors agree on what they mean by 'event' (see [11]).

Small scales

At small scales, one observes in permanence on the Sun, of total surface

$$S_\odot = 4\pi R_\odot^2 \simeq 6 \times 10^{18} \text{ m}^2 \tag{4.29}$$

about $N \sim 4 \times 10^4$ short-lived ($\tau \sim 60$ s) *explosive events*. They are inferred to be jets of average speed $V \sim 100$ km s^{-1}, temperature $T \sim 10^4 - 10^5$ K and mass $m \sim 10^6$ kg [17], i.e. an average kinetic energy $W \sim mV^2 \sim 10^{16}$ J per event. Since the kinetic energy of a proton moving at this speed satisfies $m_p V^2 \gg k_B T$, these events are highly non-thermal.

Recent measurements [60] have detected far more similar events, over a greater energy range, and the number[22] of events dN, arising per second per surface unit of the Sun in the energy range dW (and going outwards), as a function of their energy W follows the approximate power law

$$\frac{dN}{dW} \simeq \frac{2 \times 10^{17}}{W^3} \text{ s}^{-1} \text{ m}^{-2} \text{ J}^{-1} \quad \text{for} \quad 10^{15.7} < W < 10^{18.1} \text{ J}. \quad (4.30)$$

Since these jets have temperatures typical of the chromosphere, it may be interesting to see whether chromospheric spicules (Section 4.4.1) follow a similar distribution. If the energy per unit mass of a spicule is mainly the gravitational potential gh, the average energy is $W \sim n_e m_p V g h \sim 10^{17}$ J, where we have substituted the volume $V \sim d^2 \times h$, with the parameters and notations of Section 4.4.1. The observed number N of spicules, of duration τ, on the solar surface $S_\odot$, yields a point on the energy distribution at $N/(\tau S_\odot W) \sim 2 \times 10^{-34}$ s^{-1} m^{-2} J^{-1}, for $W \sim 10^{17}$ J (taking $\Delta W \sim W$). We have put it in the figure, and one can see that it is in close agreement with the distribution (4.30) – a result that may not be a coincidence.

Energy distribution

Jets and brightenings are observed at widely different energies, ranging from about 10^{16} to 10^{25} J or more, over the whole solar surface. The low-energy events are detected from Doppler shifts in ultraviolet lines, which are attributed to bulk flows of matter; on the other hand, the high-energy events are detected as X-ray brightenings, which are attributed to Bremsstrahlung radiation from energetic electrons. The energy distributions of all these events follow power laws of indices ranging from -1.5 to -3. This is illustrated in Fig. 4.15, on which we have plotted results from a number of studies, obtained in both quiet and active regions ([15], [51], [5], [6], [27] and [60]). Note that the event rate in active regions drops by about a factor of 20 near solar activity minimum [15] (when the surface of these regions also drops), but the power law index remains approximately the same.

Although these power laws tend to be systematically steeper at smaller energies, it is interesting that a single power law distribution

$$\frac{dN}{dW} \sim \frac{3}{W^2} \text{ s}^{-1} \text{ m}^{-2} \text{ J}^{-1} \quad \text{for} \quad 10^{16} < W < 10^{25} \text{ J} \quad (4.31)$$

approximates the superposition of all these results over nearly 10 orders of magnitude in energy, to better than a factor of 10. This distribution has several

---

[22] The value dN is the number of events of energy between W and $W + dW$ present per square metre of solar surface, divided by the duration of one event.

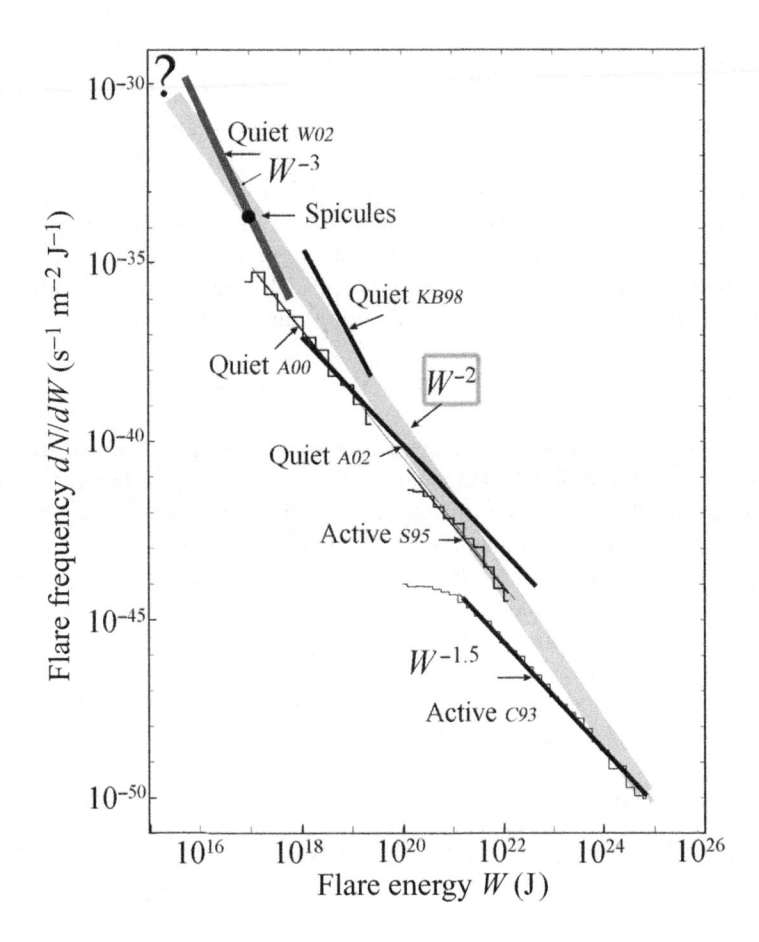

Figure 4.15 Frequency distribution of transient events in the solar upper atmosphere as a function of energy, in quiet and active regions, from a number of studies cited in the text (labelled by first letter of author's name). All events appear to follow power laws of index between 1 and 3 (dark grey lines); the superposition of them is close to a power law of index -2 (thick light grey line).

fundamental properties. It is scale-free (a fundamental property of power laws), and its first moment, i.e. the total energy that would be produced if the energy range extended from 0 to infinity: $\int_0^\infty dW\, W\, dN/dW \propto \int_0^\infty dW/W$, diverges logarithmically at both ends. In practice, this means that the total energy depends on both the smaller and larger energies, but that this dependence is extremely weak. Stated otherwise, with a smaller power law index, most of the energy would be concentrated in large events, whereas with a larger index, most of the energy would be concentrated in small events.

Integrating (4.31) over the range covered by the statistics, from $W_{min} = 10^{15.7}$ J to $W_{max} = 10^{25}$ J, we find for the total energy output the modest value of : $3 \times \ln(W_{max}/W_{min}) \simeq 60$ W m^{-2}. We shall return to this figure later.

Large scales

At the large end of the energy distribution lie the most powerful transient phenomena in the solar system; these events far outreach their solar birthplace, affecting the outskirts of the heliosphere, and we shall return to them in later chapters.

Observation shows essentially two classes of phenomena, that occur both separately and together. The first class is a *coronal mass ejection*, in which a gargantuan loop containing up to 10^{13} kg of matter rushes outwards from the Sun at several hundreds of kilometres per second; it may reach the Earth's environment in a few days if it travels in the right direction, sometimes with dramatic consequences. These events are responsible for the most damaging disturbances associated to what is commonly known as *space weather*. They are often associated with the eruption of a prominence and typically occur twice a day or more at solar activity maximum [14]. We shall return to them in Sections 6.3.2 and 7.3.3.

The second class is a large *flare*, which is detected as high-energy particles and radiation, and releases roughly the same energy: up to 10^{25} J (and sometimes more), in a time that may be as small as a few minutes (with variations at scales down to milliseconds) in a region of typically 20 000 km size, i.e. a volume $V \sim 10^{22}$ m^3. The plasma is heated and some particles are accelerated to high energies, emitting radiation in most of the electromagnetic spectrum, from (kilometric) radio wavelengths to gamma rays (of wavelength $\sim 10^{-13}$ m) – a range of wavelengths extending over more than 17 orders of magnitude. Whereas radio emission comes from moderately energetic electrons, gamma spectral lines reveal that nuclear reactions are taking place. Analysis of the emitted radiation strongly suggests that a very large part of the energy released in a large flare goes initially into accelerating electrons and ions [37].

An amount of 10^{25} J is what the Sun radiates normally in $10^{25}/L_\odot \sim 0.02$ s; hence these large flares do not change significantly the mean total solar output (even though the more energetic ones – as the one reported by Carrington – do increase the visible solar luminosity during an appreciable fraction of a second). This is not true in the X-ray and gamma-ray ranges, where flare emission is much greater than the normal value.

What is the maximum energy released by flares? The answer is unknown because the statistics is poor for giant flares and they tend to saturate the (X-ray) detectors; energies as great as 10^{27} J have been tentatively reported [25].

4.5.2 Hints from physics

These giant phenomena are powered by the magnetic field. This is strongly suggested by the unavailability of any other significant energy source, and by three observational clues [62]. First, large eruptions always occur at the boundary of large adjacent patches having opposite vertical magnetic fields; when such regions are pushed together, reconnection arises and liberates energy; if the region is large, so is the energy. Second, large eruptions occur in regions where the magnetic field has a complicated twisted and/or sheared large-scale structure; there is no way to get rid of the magnetic helicity, except by ejecting the whole structure. Third, these events are followed by a large-scale reorganisation of the structure, which presumably reveals a similar reorganisation of the magnetic field.

Let us make an order of magnitude estimate. In or near active regions, where these large events occur, we may assume a magnetic field of, say, 3×10^{-2} T in a volume $V \sim 10^{22}$ m^3; this produces a magnetic energy of $(B^2/2\mu_0)V \sim 4 \times 10^{24}$ J, which is close to the energy liberated in a large eruption.

In practice, however, not all the energy available can be released. In the absence of mass ejection, conservation of magnetic helicity puts a constraint on the final state, which tends to be a force-free field with a constant value of α, so that the energy available for dissipation is only the energy in excess of such a state. On the other hand, a coronal mass ejection ultimately transfers helicity in the interplanetary medium, so that the total energy available in that case is the one in excess of that of a potential field, and is thus larger.

The detailed process is far from being understood, and three major questions arise:

- how is the magnetic energy accumulated and stored?

- how is the eruption triggered?

- how is the released energy transferred to particles?

Just as in the case of a compressed spring, direct observation from outside does not tell us how energy is accumulated and stored in the stressed material, and finally released. Hence one can only make educated guesses as to the mechanisms at work [26], [32].

There is little doubt that the magnetic energy comes from solar rotation and convection, through emergence of magnetic flux tubes from the solar interior, and their stressing and tangling by motions at their feet in and below the photosphere. The basic state might be a *magnetic flux rope*, i.e. a twisted tube of flux (Fig. 4.16, top right), possibly formed from a sheared magnetic arcade (Fig. 4.16, top left).[23]

Energy is expected to build up slowly in the electric currents as the field lines are stressed. The energy cannot leak out because the feet of the lines

---

[23]The complexity of structures may enable magnetic energy and helicity to accumulate more rapidly. This is because with only one or two (unknotted) tubes, the twists and braids commute with each other; for example, when a tube is twisted by an angle ψ, twisting it by an opposite angle cancels this action (however much you have twisted it in the interval), just as in a linear random walk. This is not so with three tubes or more, that are braided or interwoven. In that case, removing the braiding requires performing all the operations in reverse without commuting them, which has a very small probability of occurring by chance. As a result, random motions of the feet of the tubes might increase the magnetic energy more rapidly than in a normal random walk process [9].

Time variability   199

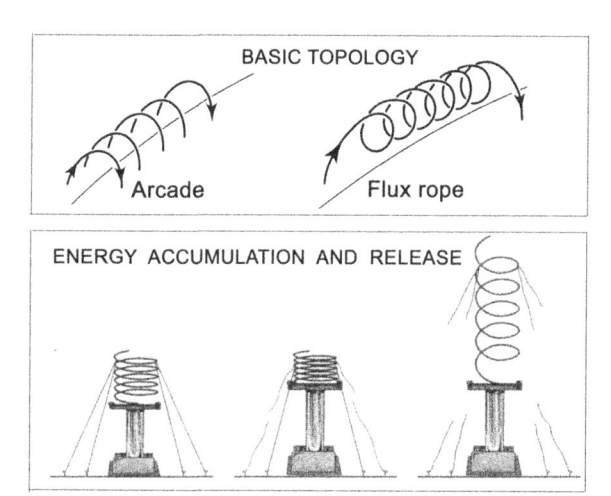

Figure 4.16 Basic topologies inferred in models of coronal mass ejections (top), and mechanical analogue of some models of energy accumulation and release (bottom). A spring is slowly compressed by being pushed from below while it is held by overlying rope tethers. The strain builds up until the tethers break, letting the spring uncoil explosively. (Adapted from [26].)

sink into dense regions whose characteristic scale of variation is much smaller than the wavelength of Alfvén waves. A flux rope such as the one in Fig. 4.16 (top right) would tend to grow in size and rise, but it is maintained by arcades of overlying magnetic field (see Fig. 4.14, right). At equilibrium, magnetic tension balances the magnetic pressure force. The structure finally breaks when the balance is tipped in favour of the outward pressure gradient, to produce a state that is energetically more favourable. This involves breaking and reconnection of the field lines [26]. Figure 4.16 (bottom) shows a mechanical analogue: a compressed spring, that uncoils when the ropes holding it break.

However, whereas in an impulsive flare, energy is rapidly released into heat and energetic particles, in a coronal mass ejection, magnetic energy is often progressively transferred to particles as matter progresses outwards. Indeed, the ejection of matter often starts at speeds much below the Sun's escape speed of $(2M_\odot G/R_\odot)^{1/2} \sim 500 \,\mathrm{km\,s^{-1}}$ and accelerates progressively, pushed by the Lorentz force, magnetic energy being used to lift matter against solar gravity and to give it bulk motion (and also heat) [56]. How can magnetic energy be progressively transferred to particles as the ejected matter travels and expands? Two hints can help us to answer this question: one observational, one theoretical. The observational hint is that these ejections show up as a giant expanding magnetic rope having a significant helicity travelling from the corona (Fig. 4.10) to interplanetary space [14]. The theoretical hint is the conservation of magnetic helicity. We have seen that for a twisted loop of size L, helicity and magnetic energy are related in order of magnitude by $H \propto W_B L$. Hence, as the loop expands, conservation of helicity requires that the magnetic energy decreases, making energy available to the particles. Note finally that since the helicity of the coronal ejections is ultimately transferred to interplanetary space, these events enable each hemisphere of the corona to shed some of the helicity continually provided to it [33].

Returning to the image of mass ejection shown in Fig. 4.10, one may interpret the bright core of the structure as prominence material that is trapped at the bottom of the helical field lines and dragged upward.

4.5.3 Further difficult questions

Let us return to the flare energy distribution (Fig. 4.15). All classes of events tend to be distributed as a power law of index between -1.5 and -3 over nearly 10 orders of magnitude in energy. Furthermore, despite the difficulties of observing small events, some hints indicate that small flares may be scaled down versions of larger ones. This suggests that similar mechanisms may be operating at different scales, and raises four basic questions:

- are the observed differences in power laws genuine or are they due to observational bias?

- what produces the power law distributions?

- what is the total flux of particles provided and is it relevant to aliment the solar wind?

- what is the total flux of energy provided and is it relevant for heating the corona and powering the solar wind?

Power law distribution

Are the reported differences in power law indices genuine? There are two obvious observational biases: first, the number of events detected depends on how an 'event' is defined, and second, one is never sure that all the energy produced has been detected. Indeed, the measured power law index has been found to vary with the observational scheme (see for example [27], [6] and [11]).

What produces the power law distribution(s)? This problem arises in a number of contexts, and its ubiquity has aroused much interest in widely different fields [39]. Prominent examples in the solar system are cosmic rays and earthquakes. Despite their widely different microscopic physics, their energy distributions are rather similar, exhibiting power law shapes of index about -2.5 and -2 respectively, over nearly 10 orders of magnitude in energies. The same is true of a number of other phenomena.

Several tentative explanations have been proposed. One of these is the so-called *Yule process*, which concerns quantities which increase in proportion to their number, as for example taxonomic species and cities. It is widely used in biology, but may be applied in a modified form to magnetic structures and plasma particles. A simple related version is the mechanism proposed by Fermi

Time variability

a long time ago to explain the power law distribution in energy of cosmic rays [31].[24] For solar flares [46], let us assume that energy tends to accumulate at a rate proportional to itself above some reference energy (introduced to avoid divergence), below which the rate tends to become constant, i.e. $dW/dt = \alpha(W + W_0)$. The energy accumulated at time t is

$$W = W_0 \left(e^{\alpha t} - 1\right) \tag{4.32}$$

where W_0 is the reference energy. From time to time, this energy is released, with the (Poisson) probability of flaring at time t

$$P(t) = dN/dt = \nu e^{-\nu t}. \tag{4.33}$$

The probability of flaring with energy W is thus

$$dN/dW = (dN/dt)/(dW/dt) = P\left[t\left(W\right)\right]/(dW/dt) \tag{4.34}$$

where $t(W)$ is obtained from (4.32).

Note that the rates of energy accumulation (α) and release (ν) act only through their ratio $\kappa = \nu/\alpha$, which measures how much energy accumulates during the average time $1/\nu$ separating two releases, since when $\kappa \gg 1$, the energy accumulated during the time $1/\nu$ is (from (4.32) with $e^{\alpha/\nu} \simeq 1 + \alpha/\nu$) about W_0/κ. The smaller the value of κ the greater the energy accumulated.

Calculating $t(W)$ from (4.32) we have $e^{-\nu t(W)} = (1 + W/W_0)^{-\kappa}$, which we substitute into (4.33)–(4.34) to obtain the flare energy distribution (normalised to unity)

$$\frac{dN}{dW} = W_m^{-1}\left(1 + \frac{W}{\kappa W_m}\right)^{-(\kappa+1)} \tag{4.35}$$

where $W_m = W_0/\kappa$.

The distribution (4.35) is called a *kappa* distribution. It has two interesting properties. First, as $\kappa \to \infty$ (infinitely small accumulation of energy between two average releases), it tends to a Boltzmann distribution $W_m^{-1} e^{-W/W_m}$, which is no surprise. Second, at high energies, it is proportional to the power law $W^{-(\kappa+1)}$. The mean energy is[25]

$$\langle W \rangle = \int_0^\infty dW\, W\, \frac{dN}{dW} = \frac{\kappa}{\kappa - 1} W_m. \tag{4.36}$$

Only if κ is large does the average energy $\langle W \rangle$ roughly equal the energy W_m accumulated during the average time separating releases. If κ is not much greater than unity, energy is not liberated sufficiently fast and the average energy $\langle W \rangle$ can be extremely large. It becomes infinite when $\kappa \to 1$, i.e. when the

---
[24] In this model, cosmic rays are accelerated at a constant rate in some confining volume, for example a galaxy, from which they escape by diffusion at a constant rate.
[25] Easily calculated by using the identity: $W = \kappa W_m \left[(1 + W/\kappa W_m) - 1\right]$.

release rate ν equals the rate α of energy accumulation. In practice, however, there is a limit on the time during which energy can build up according to (4.32), and to the value it may reach.

This kind of distribution is often encountered in space physics, and we shall encounter it again later in this book.

In answering one question, however, we raise three further ones. How can magnetic energy accumulate in the solar atmosphere at the same rate at all scales? And how can the rate of energy accumulation be proportional to the flare rate, over such a wide energy range? And finally, what physical process in the corona determines the energy W_m?

Another popular scheme avoids these difficulties by picturing eruptions as sand avalanches or other critical phenomena, for which no privileged scale exists, and where most of the changes take place through catastrophic events [34] (see [8] and references therein for applications in other contexts). Basically, this scheme assumes the magnetic field to be in a self-organised critical state analogous to a sand pile. As sand is added to a sand pile, the average slope increases until a state is reached where it remains approximately constant. When this critical state is achieved, addition of more sand triggers avalanches that readjust the local shape. As a grain of sand moves locally, it disturbs the equilibrium in its vicinity, so that the perturbation propagates in a chain reaction. As a result, small local perturbations give rise to avalanches of all sizes, whose probability distribution has no privileged scale – whence the power law shape.

Applying this scheme to the magnetic field of the solar atmosphere involves three ingredients. First, a forcing of the system by addition of energy – analogous to adding sand grains; this role might be played by random motions of the field lines at the photosphere and below, which transform solar convective energy into atmospheric magnetic energy. Second, a local instability arising when some local parameter – playing the role of the local slope in a sand pile – exceeds a critical value; this role might be played by magnetic reconnection, which arises when the local field gradient exceeds some threshold value (or by another kind of plasma instability). Third, a transport process which enables the instability to change the value of this parameter at nearby sites; this role might be played by the reorganisation of the magnetic field whose connectivity changes during reconnections.

But what value should the power law index have? And what determines the absolute amplitude of the distribution, i.e. how many events occur in a given energy range? Answering these questions should require a deeper understanding of the physics, of how the energy of flares scales with their size, and of their space filling factor – more precisely of their fractal dimensions.[26]

Even though a huge number of scenarios have been proposed, based on a still greater number of plasma instabilities and wave modes, none of them is basic enough to have a predictive value.

---

[26] Recent experiments with sand piles show that, despite early claims to the contrary, the power index does depend on the nature of the grains. For example, just as do risotto recipes, piles of rice grains behave differently depending on whether the rice is long-grain or short-grain (Frette, V. *et al.* 1996, Avalanche dynamics in a pile of rice, *Nature* **379** 49).

Mass flux

What is the total mass involved in flares of all sizes? Instead of introducing model-dependent scaling factors, let us make an estimate in a few cases where the mass can be estimated from observation. Consider first spicules. With the notations and parameters of Section 4.4.1, the mass that may be supplied to the corona via spicules is $Nn_e m_p h d^2/\tau \sim 4 \times 10^9$ kg s^{-1}. This is much greater than the mass loss through the solar wind. In other words, spicules supply to the corona much more material than escapes through the wind, so that most of it must return to the Sun. Next, consider the explosive events discussed in Section 4.5.1; the mass supplied is of the order of $Nm/\tau \sim 0.7 \times 10^9$ kg s^{-1}; this is roughly equal to the solar wind mass flux.

Both estimates indicate that only a small proportion of the mass available is required to sustain the wind. This suggests that the mass flux of the solar wind is not determined by the availability of solar material, but rather by other processes.

Energy flux

We have seen in Section 4.5.1 that with a flare energy distribution varying as W^{-2}, the total energy provided in the range covered by observation is only about 60 W m^{-2}, with a large uncertainty.

How is this figure affected by uncertainties on the power index and on the energy range? Let us assume the distribution (4.30) that varies as W^{-3} to continue down to some minimum energy W_{min}. In this case, the total energy would be $W_{tot} = \int_{W_{min}} dW\, W\, dN/dW \simeq [2 \times 10^{17}/W_{min}]$ W m^{-2}. If W_{min} were smaller by one order of magnitude than the lower end of the present observational range, the total energy would be multiplied by 10. We conclude that small flares might easily provide much energy to the corona if the power law index is close to -3. This is not so, however, if the index is ≥ -2. With the distribution (4.31) of index -2, producing the modest value of $W_{tot} \simeq 600$ W m^{-2} would require $W_{max}/W_{min} \simeq e^{200} \simeq 10^{87}$. Taking $W_{min} \sim 10^{-19}$ J (the thermal energy of a particle at $T \sim 10^4$ K), this would require $W_{max} \sim 10^{68}$ J – greater by more than 20 orders of magnitude than the total mass energy of the Sun!

4.6 Coronal heating: boojums at work?

We have seen that the solar corona is very hot, even outside the so-called active regions, and that this high temperature cannot be explained with the fluid energy equation without introducing an extra heating term, as yet unspecified.

There is still no agreement on this term, despite an unreasonable increase of the literature on the subject during the last decades, both in size and stodginess. The theories are so numerous that a new discipline has emerged: classification of theories; a recent record lists seven classes, each involving up to eight different processes [4]. Although this suggests irresistibly some 'anti-Ockham razor'[27]

---

[27] William of Ockham was a fourteenth-century philosopher, whose approach was – put simply – economy of hypotheses, i.e. to put aside complicated explanations when a simple one would do; this principle is known as *Ockham's razor*.

principle, some of these theories are perhaps right. Balanced reviews may be found in [63] and [57].

4.6.1 The energy budget and how to balance it

Let us first estimate the energy required, by assuming conservatively that the heat flux is given by (4.10).

Consider first the corona outside active regions. We have seen that the energy budget is dominated by the heat flux. From (4.10) and the temperature profile sketched in Fig. 4.1, we see that the major losses of the corona take place at its base, towards the chromosphere, since it is in this part of the profile that $T^{5/2}dT/ds$ is greatest; from Fig. 4.1, we have in order of magnitude $Q_c \sim 10^2$ W m^{-2}.

On the other hand, in active regions, we may evaluate the heating requirements from the scaling (4.28); the flux required for a loop of length L is $W_{heat} \times L$; eliminating L with the aid of (4.27), we get $W_{heat}L \sim 10 \times T^{1/2}P$. With the parameters given in Section 4.4.2 we find, in order of magnitude, a few 10^3 W m^{-2}; more accurate calculations yield values of the same order of magnitude [61]; heating the top of the chromosphere requires a still greater energy flux, because the density and the radiation losses are greater (recall Fig. 4.13). However, active regions represent a very small fraction of the total solar surface. Their greater heating is not surprising: the magnetic energy available is much greater, and, since they are essentially made of loops protruding from the solar vicinity, energy can accumulate without being steadily evacuated outwards as it does elsewhere along magnetic field lines open to the interplanetary medium.

Comparing these values to the total solar radiative flux $L_\odot/S_\odot \simeq 6.3\times 10^7$ W m^{-2}, we see that the heating of the solar atmosphere involves a very small fraction of the solar output.

Three main basic mechanisms have been proposed to explain the high temperature of the solar upper atmosphere:

- heating by dissipation of electric currents through small reconnection events,
- production of waves, and heating by dissipation of the associated energy,
- production of non-Maxwellian particle velocity distributions, and filtration of the particles by the solar gravitational potential.

The first two mechanisms rely ultimately on the transformation of magnetic and kinetic energy into heat through viscous and ohmic dissipation. Some variants invoke MHD turbulence, either for producing small scales that enable dissipation, or for producing MHD waves. In contrast, the third mechanism is non-dissipative: it relies on the non-equilibrium state of the atmosphere in the form of fast (non-thermal) particles, which are filtered by the solar gravitational potential.

While controversies are raging as to how the corona is heated, it is sobering to remark that, ultimately, all the proposed mechanisms may be based on the

Figure 4.17 Solving the coronal heating problem. (Drawing by F. Meyer.)

same phenomenon: accumulation of energy in the tortured magnetic field, and subsequent release, and indeed, it has been inferred that the heating increases in proportion to the magnetic energy density [16]. The release of magnetic energy is expected to produce heating, Alfvén waves and magnetic turbulence, and suprathermal particles. These effects are at the basis of the three mechanisms proposed, and might all be in action – although not necessarily in the same place and at the same time.

A similar sobering remark applies to the controversies about the proper way of describing the medium. Most schemes are based on fluid equations, whereas other ones are kinetic, and even sometimes collisionless; indeed, the most popular heating mechanism involving waves is kinetic, as is also the mechanism based on the filtration of particles. As we saw in Section 2.3, whether a fluid or a kinetic description is more adequate depends on the availability of thermalisation processes and on the role they play. Whereas the photosphere is highly collisional and can be adequately described by the usual fluid equations, this is not so in the outer atmosphere because of the strong density decrease and temperature increase with height. Each description of the medium may be right, if applied in the proper way and in the proper place.

4.6.2 Heating through reconnection events

This scheme, championed by E. Parker [40], is based on dissipation of magnetic energy through transient reconnection events. More precisely, if the low end of the flare energy distribution shown in Fig. 4.15 is extrapolated downwards, the individual events become so small and so numerous that the result is essentially continuous, and might hopefully heat the corona.

This speculation raises two questions. First, is there enough energy available? As we have seen in Section 4.5.3, the total energy provided depends strongly on the slope of the power law distribution at low energies. If the index is in the ballpark of -3, a minor extrapolation of the measurements downwards in energy may be sufficient. If, on the other hand, the index is ≥ -2, the prospect of heating by these events is poor. As we already noted, however, it is by no means certain that all the energy released can be detected. Whereas smaller flares are detected through jets of matter, larger flares are detected through the radiation produced by high-energy electrons. It would be surprising if lower-energy (but still non-thermal) particles would not also be produced, as also MHD waves, and, as we shall see below, both might be responsible for the large coronal temperature.

This raises a second question. If these small events heat the upper atmosphere, how do they manage to do so? What is the role played by bulk jets, which should produce shocks, with subsequent heating, acceleration of particles and wave emission? And what is the role played by waves and suprathermal particles?

Finally, the temperature profile in the solar atmosphere (Fig. 4.1) is reminiscent of the one in a granular fluid with inelastic collisions that is subjected to gravity and injection of energy at the base [42]. A similar effect might possibly occur in the low solar atmosphere with ionisation and recombination playing the role of inelastic collisions, and reconnection events being the heating agent.[28]

4.6.3 Heating by waves

Heating the corona through transient reconnection events assumes the atmosphere to respond to photospheric motions through quasi equilibrium states, separated by reconnection events arising when some threshold is reached. This requires the photospheric driving motions to act at timescales greater than the response time of the atmospheric magnetic field, i.e., roughly speaking, the driving speeds should satisfy the condition $v < V_A$ (V_A is the Alfvén speed). The values in Table 3.1 suggest that this might be so. If the opposite inequality holds, however, the driving motions excite Alfvén waves. Furthermore, reconnection events themselves are expected to emit Alfvén waves.

If these waves manage to propagate to great altitudes and dissipate there, they might heat the outer atmosphere. For this to be so, however, three conditions have to be met. First, enough wave energy must be produced. Second, this wave energy must reach the right places; in particular, heating has to be provided rather high in the atmosphere, where the temperature has a maximum. Third, enough energy must be transferred to the particles. Unfortunately, meeting all these constraints turns out to be difficult.

---

[28] Pantellini, F., personal communication, 2004.

Alfvén waves carry transverse perturbations in velocity and magnetic field, without density and pressure variations. They are not, however, the only waves a priori capable of heating the corona. Perturbations in pressure and longitudinal speed produce acoustic waves, which involve the compressibility of the medium, and carry variations in density and pressure.

Acoustic waves

In a non-magnetic medium, acoustic waves can propagate in any direction at the speed V_S. In the solar atmosphere, the magnetic field provides a privileged direction and additional forces, so that these waves are modified as outlined in Section 2.5. Furthermore, at low heights, strong magnetic flux tubes may guide the waves.

Suppose that such compressive waves manage to be emitted. What is the energy flux transported, and what happens to these waves as they propagate outwards?

Sound waves of speed amplitude δv carry an energy flux $(\rho \delta v^2 /2) V_S$. With the parameters of Section 3.3 and the sound speed from Table 4.1, we find that this flux might reach a few 10^5 W m^{-2}, much more than is needed for heating purposes. Because the viscosity is so small, there is no dissipation, so that this energy flux should remain constant. Hence, as these waves propagate in the low chromosphere, where the temperature changes very little whereas ρ drops sharply, the wave amplitude grows sharply as $\delta v \propto \rho^{-1/2}$. This makes the waves steepen rapidly into shocks, so that they quickly dissipate within a small region. In doing so, however, they are absorbed and do not reach the corona. Hence these waves might heat the low chromosphere, but there is little prospect for them to heat the corona.

Alfvén waves

Consider now Alfvén waves, which carry transverse velocity and magnetic field disturbances. The simplest way of generating them is to shake the field lines faster than the Alfvén speed; they may also be excited by reconnection events of geometry sketched in Fig. 4.5 (left) (see [47]), and by turbulence.

Alfvén waves are expected to traverse the solar atmosphere with little attenuation, because they are difficult to dissipate. This resistance to dissipation stems from their non-compressive nature, and from the small values of the viscosity and of the resistivity. In short, whereas compressible waves deliver easily their energy through formation of shocks, but (because of this property) cannot reach the remote place where they are needed, Alfvén waves, in contrast, easily reach the corona, but they stubbornly tend to keep their energy for themselves. Subtle scenarios must thus be invoked to enhance their damping. For example, transforming them (by non-linear coupling) into waves that damp more easily, and/or invoking subtle variants or adequate plasma instabilities.

This resistance of Alfvén waves to dissipation, however, is only true in the context of MHD. And indeed, a popular heating scheme involves resonant damping of kinetic Alfvén waves. A little digression is needed here, because such a collisionless dissipation mechanism appears at odds with ordinary MHD. Alfvén waves are driven by the elasticity of the field lines, whose inertia is provided by the particles attached to them by flux freezing – a pillar of MHD (Section 2.3). So, what kind of stuff are *kinetic* Alfvén waves? The answer lies in the realisation that flux freezing only holds at scales greater than the radius of gyration of the particles. Kinetic Alfvén waves are Alfvén waves whose wavelength is not greater than the radius of gyration of the particles, and which are thus outside the scope of MHD. Resonant damping of waves is also a kinetic effect (Section 2.5). A resonance occurs when a particle moves in such a way that it experiences nearly constant wave electric and magnetic fields in its own rest frame, so that energy can be exchanged very easily.

For heating ions, one must find a resonance with their own motion. Since ions gyrate around magnetic field lines, a cyclotron resonance occurs when the field of the wave rotates in the same sense and at the same frequency as do ions (as seen in their own rest frame). Since positively charged ions gyrate in the left-hand sense, the wave must be a (left-hand) circularly polarised Alfvén wave. An ion of velocity $\mathbf{v}_i$ sees the field of a wave of (angular) frequency ω and wave vector $\mathbf{k}$ at the frequency $\omega - \mathbf{k} \cdot \mathbf{v}_i$, so that the resonance condition reads

$$\omega - \mathbf{k} \cdot \mathbf{v}_i = \omega_{gi} \tag{4.37}$$

where ω_{gi} is the ion frequency of gyration. At this resonance, ions surf on the waves and can spin up, at the expense of them [12].

With a magnetic field of the order of 10^{-3} T in the low corona, the (angular) frequency of gyration is $\omega_{gp} = eB/m_p \sim 10^5$ Hz for a proton; it is somewhat smaller for heavy ions, since $\omega_{gi} = Z\omega_{gp}/A$ for ions of charge Ze and mass $m_i = Am_p$. Hence, though the magnetic field decreases with altitude, making the gyrofrequencies decrease accordingly, the resonant waves have still frequencies greater by many orders of magnitude than the typical values observed at the photosphere, where the observed periods are in the range of minutes. The wave generation mechanism has thus to be very subtle to bridge this gap, and this process should heat preferentially heavy ions, whose frequency of gyration is smaller.

The popularity of models based on this mechanism is due to their preferential heating of heavy ions perpendicular to the magnetic field – because the resonance involves the particle gyration – which agrees with observational inferences. Indeed, heavy ions of mass m_i are inferred to have temperatures (in the direction normal to the magnetic field) that may be greater than the proton one by a factor even larger than the mass ratio m_i/m_p. Furthermore, as we shall see in Section 6.4, Alfvén waves are routinely observed in the solar wind, suggesting that they may also be present in the corona. The major drawback of these models, however, is that despite much effort, adequate waves have not been observed at the place where they are needed for the models to work, and the mechanism responsible for their generation remains unknown (see [13]).

4.6.4 Filtration of a non-Maxwellian velocity distribution

As is the heating process discussed above, this one is also kinetic, involving the individual speeds of the particles. The key point is the realisation that there are not enough collisions in the outer solar atmosphere for the particle velocity distributions to be Maxwellian distributed. This implies a radical revision of ideas, since this assumption is a pillar of analyses in astrophysics, from ionisation and radiation loss calculations to spectroscopic diagnostic procedures and energy balance equations. Let us explain how this comes about.

Beware of the assumption of local thermal equilibrium

The key point is that the corona is a plasma, and we have seen in Section 2.1 that the collisional free path of charged particles increases strongly with energy. The *mean* free path of particles in the transition region is a few per cent of the scale of variation, and somewhat greater farther out. Naively, one might think that this should ensure local thermal equilibrium. But no! As we already noted, charged particles whose speed is, say, three times greater than average, have a free path greater by a factor of 3^4, and is therefore of the order of magnitude of the scale of variation. These particles thus traverse one or more scale heights before thermalising, so that local thermal equilibrium fails for them.

Put another way, even though particles of average speed $\sim v_{th}$ have a free path $l_f < H$, this is not so for faster particles, so that the usual expansion in power series of l_f/H, that work so well for neutral gases, now fails to converge uniformly. As a result, the local Maxwellian approximation is badly in error for speeds v satisfying $(v/v_{th})^4 > H/l_f$. With $H/l_f \sim 10^2$ in the transition region (and still smaller farther out), the velocity distributions should have non-Maxwellian tails at speeds greater than a few times the thermal speed [52].

This raises grave doubts as to the validity of a number of calculations. In particular, the heat flux (see Section 2.3) might be completely different from the usual collisional one; it may even flow from the cold side to the warm side – in sharp contrast with our deeply ingrained prejudices. It is fair to say that there is much controversy on this subject and that different calculations give conflicting results (see [28] and references therein).

There are two ways to solve the problem. One way, which is the simplest and the most popular, is to cling to the fluid description and use an empirical value of either the heat flux or the temperature variation. The other way, which (in theory) is the most satisfactory, is to give up the fluid description and to adopt a kinetic picture; the problem, however, is so difficult that nobody has yet come up with a complete solution.

In order to get a feeling of what happens, let us consider a simplified version of the problem: no collisions at all.

This is the basis of an elegant reasoning invented by J. Scudder [49], [50],[29] which builds on the following basic points:

---

[29] A more digestible version, in a different context, may be found in [36].

- the numerous non-thermal processes taking place in the solar atmosphere should prevent the (nearly collisionless) fast particles from relaxing towards Maxwellians, producing tails in the velocity distributions;

- in this case, the solar gravitational potential filters the particles by overpopulating these tails, making the temperature increase with distance.

How can the solar gravitational potential filter the particles? To understand this, let us first consider how particles behave when subjected to an attracting force in the absence of collisions.

A simple toy model

Consider particles that are attracted by a central object, so that their potential energy $\psi(r)$ is negative and increases with the distance r. How does the temperature vary with r? Let us first assume that the distribution at r_0 is a Maxwellian (of density n_0 and temperature T_0). We saw in Section 2.3 that when the potential energy varies, the particle energy distribution is translated by the same amount, which yields Boltzmann's law, even though there are no collisions; this holds only for attracted particles, for which there is no accessibility impeachment between r_0 and r whatever the energy.

Hence at distance r, where the potential has varied by $\Delta\psi = \psi(r) - \psi(r_0)$, the distribution is still a Maxwellian of temperature T_0, but the density has decreased to

$$n(r) = n_0 \times e^{-\Delta\psi/k_B T_0}. \tag{4.38}$$

What happens when the distribution at r_0 is no longer a Maxwellian (Problem 2.5.4)? The simplest case is a sum of two Maxwellians: one of number density n_C and temperature T_C, and a second one which is a minor constituent – of number density $n_H \ll n_C$, but of greater temperature $T_H \gg T_C$. This is a simple representation of a distribution having a suprathermal tail (Fig. 4.18). In that case, the total number density is $n_C + n_H \simeq n_C$, and the average temperature is

$$T(r_0) = \frac{n_C T_C + n_H T_H}{n_C + n_H} \tag{4.39}$$
$$\simeq T_C \quad \text{if} \quad n_H T_H \ll n_C T_C.$$

Between r_0 and r, each Maxwellian evolves according to (4.38) with its own temperature (which remains constant), so that the number densities of each Maxwellian decrease with distance as

$$n_C(r) = n_C \times e^{-\Delta\psi/k_B T_C} \quad ; \quad n_H(r) = n_H \times e^{-\Delta\psi/k_B T_H} \tag{4.40}$$

(Fig. 4.18). The average temperature at distance r is

$$T(r) = \frac{n_C(r) T_C + n_H(r) T_H}{n_C(r) + n_H(r)}. \tag{4.41}$$

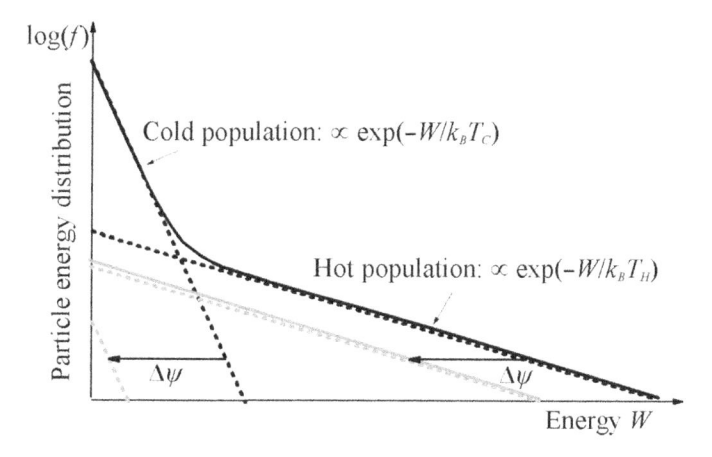

Figure 4.18 Filtration of particles by a potential well, for a distribution (continuous line) made of a sum of Maxwellian distributions (dotted lines). Black and grey lines denote respectively the original and filtered distributions. The variation in potential $\Delta\psi$ produces a translation towards smaller energies, making the cold component nearly disappear (the particles cannot surmount the potential well), whereas the hot component is barely affected and is the only surviving population at high altitudes.

Since $T_C \ll T_H$, there is some distance at which the variation in potential $\Delta\psi$ satisfies $\Delta\psi/k_B T_H \ll 1$ – so that the hot component is barely affected by the variation in potential, whereas on the contrary, $\Delta\psi/k_B T_C \gg 1$ – so that the cold component barely survives the variation in potential. This arises when the distance is large enough to ensure $e^{\Delta\psi/k_B T_C} > n_C T_C/n_H T_H$, while $e^{-\Delta\psi/k_B T_H} \simeq 1$. Then, substituting (4.40) into (4.41) yields the average temperature at distance r: $T(r) \simeq T_H$.

The remarkable thing is that the temperature has increased with altitude from T_C to T_H, without any heat being furnished. How can this be so? The answer is that the particles are hotter, but are much less numerous; they have not been heated, but rather filtered by the potential well, which lets only the most energetic ones go upwards. This 'velocity filtration' effect [49] is generic, provided that the distribution has a suprathermal tail and is subjected to a monotonous confining potential, without collisions. The result obtained with our unrealistic distribution may be generalised to a sum of any number of Maxwellian distributions (Problem 2.5.4), and to distributions having power law tails, as kappa distributions.

Elementary application to the Sun

Let us calculate the confining potential in the case of the Sun. Since the atmosphere is not neutral, the gravitational force is not the only one acting on

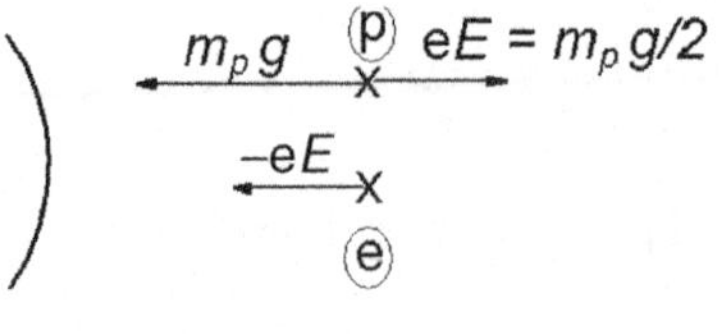

Figure 4.19 In a hydrostatic electron–proton atmosphere with equal temperatures, the tendency of the (light) electrons to escape from the (heavy) protons sets up a restraining electric field which reduces the net attractive force on a proton to half the solar gravitational force.

the particles. To keep the problem simple, consider a static atmosphere made of electrons and protons having similar velocity distributions. In the absence of electric field, the proton pressure gradient should equilibrate the gravitational attraction $F_g = m_p g$ per proton. But since the gravitational attraction on the electrons is negligible, there is nothing to equilibrate the electron pressure gradient (which equals that of the protons in this simple case), so that the electrons tend to be displaced outwards with respect to the protons. This produces a minute space charge which induces a radial electrostatic field $\mathbf{E}$ directed outwards. If the atmosphere is to remain static, this electrostatic field must adjust itself to ensure that the total attraction on a proton $(F_g - eE)$ equals the electrostatic attraction on an electron (eE), where e is the elementary charge. This occurs when $eE \simeq F_g/2$, so that the force on the protons is halved and equals the force on the electrons. This enables the pressure gradients to be equilibrated for each species (Fig. 4.19).

The total potential to which the particles are subjected at distance r is thus approximately half the gravitational potential, i.e. $\psi(r) \simeq -m_p M_\odot G/2r$. Hence, over a small distance Δr near the Sun, this potential varies by[30]

$$\Delta\psi \simeq \frac{m_p}{2} g \Delta r \quad ; \quad g = \frac{M_\odot G}{R_\odot^2}. \tag{4.42}$$

Let us use our toy model, and assume that the velocity distribution near the Sun (for both electrons and protons) is made of a Maxwellian of temperature $T_C \simeq 10^4$ K plus a small (Maxwellian) tail of temperature and density satisfying, say, $T_H/T_C \simeq 10^2$, $n_H/n_C \simeq 10^{-3}$. In this case, the average temperature near the Sun is roughly T_C, and we have seen that it increases to roughly T_H over a distance satisfying the conditions $e^{-\Delta\psi/k_B T_H} \simeq 1$ and $e^{\Delta\psi/k_B T_C} > n_C T_C/n_H T_H$. Substituting (4.42), this yields

$$3H_C < \Delta r < H_C T_H/T_C$$

where $H_C = 2k_B T_C/m_p g$ is the scale height at the temperature T_C.

---

[30] A simple generalisation to the case when there are other ions than protons, so that the mean mass per particle is different from $\mu \simeq m_p/2$, and when different particle species have different velocity distributions, can be found in [36].

Hence, the temperature increases from $T_C \simeq 10^4$ K to $T_H \simeq 10^6$ K over a distance of a few scale heights H_C. The large temperature achieved is the remnant of the high-energy tail of the original distribution, and is produced by filtration of the particles by the confining potential. We have devised this toy model for enlightening the physics; a realistic distribution should rather have a power law energy tail, as in [49], which produces a more regular temperature increase.

As other heating schemes based on kinetic processes, this one produces ion temperatures that increase roughly proportionally to their mass, since the filtering potential acting on a minor ion species is roughly proportional to its mass.[31] The heavier the ion, the more effective the filtration making the temperature rise.

Difficulties

This elegant scheme, however, has several worrisome points. For one thing, this process replaces the problem of heating the corona in bulk by another one: producing a small quantity of suprathermal particles. Indeed, it is not proven that the velocity distributions in the solar atmosphere have suprathermal tails adequate for producing the observed temperature increase. The arguments for the presence of such tails are rather similar to those invoked for Alfvén waves. First, the numerous acceleration processes acting in the solar atmosphere are expected to produce suprathermal tails. Whereas large flares are known to produce very high-energy particles, of which the tail needed by the filtration mechanism might be the low-energy side, the by-products of small flares are largely unknown. Second, just as the solar wind exhibits Alfvén waves, it also shows up velocity distributions having high-energy tails. And finally, there are some (still controversial) observational hints as to the presence of such velocity distributions in the corona.

A more serious cause for concern is that collisions between particles are not negligible in the transition region and in the low corona. Even though collisions are not sufficiently numerous for the usual energy equation with the (collisional) heat flux to hold, collisions cannot be neglected either, and are expected to modify significantly the velocity filtration in the transition region (see [28] and references therein). And finally, the above derivation ignores the energy loss by radiation.

To conclude this section on coronal heating, we must admit that neither of the mechanisms proposed is without difficulties; the process at work is elusive, to say the least, and it seems finally that all clues are pointing to a *boojum*.[32]

---

[31] More precisely, the potential acting on a minor ion of mass m_i and number of elementary charges Z_i varies as $m_i/\langle m \rangle - Z_i/(1 + \langle Z \rangle)$, where $\langle m \rangle$ and $\langle Z \rangle$ are respectively the average ion mass and number of elementary charges.

[32] Many thanks to David Mermin for his efforts in making Lewis Carroll's boojum an internationally accepted technical term (see Mermin, N. D. 2003, *Am. J. Phys.* **71** 296).

4.7 Hydrostatic instability of the corona

Let us take for granted the high temperature of the corona and examine the consequences. Consider first the simplest case of a radial magnetic field, which does not produce forces along the radial direction.

4.7.1 Simplified picture of a static atmosphere

We assume a spherical symmetry, all quantities depending only on the radial distance r to the Sun, and a static atmosphere made of electrons and protons having equal number density n and temperature T, immersed in the solar gravitational field. Since the gas is electrically neutral, there is no electromagnetic force in average, so that for the atmosphere to be static, the radial pressure gradient must balance the gravitational attraction, i.e.

$$\frac{dP}{dr} = -\frac{\rho M_\odot G}{r^2} \qquad (4.43)$$

where the gas mass density is $\rho \simeq nm_p$ (since the electron mass is negligible), and the pressure is $P = 2nk_BT = \rho k_BT/\mu$, where $\mu \simeq m_p/2$ is the average mass per particle. Eliminating ρ, this yields

$$\frac{1}{P}\frac{dP}{dr} = -\frac{\mu M_\odot G}{k_B T r^2}$$

whose integration gives the pressure at distance r

$$P = P_0 \exp\left[-\frac{\mu M_\odot G}{k_B} \int_{r_0}^{r} \frac{dr}{Tr^2}\right]. \qquad (4.44)$$

Since the coronal temperature increases only weakly from the low corona to the temperature maximum (Fig. 4.1), and decreases slowly outwards, we first assume a constant temperature. Substituting into (4.44) $T = T_0$, which may be factored out of the integral, we obtain

$$P = P_0 \exp\left[\frac{\mu M_\odot G}{k_B T_0}\left(\frac{1}{r} - \frac{1}{r_0}\right)\right]. \qquad (4.45)$$

Close to r_0, we put $r = r_0 + z$, so that $r_0/r \simeq (1 + z/r_0)^{-1} \simeq 1 - z/r_0$ for $z \ll r_0$, and (4.45) reduces to the familiar exponential decrease $P \propto e^{-z/H_0}$, with the scale height at r_0: $H_0 = k_B T_0/\mu g_0$, where $g_0 = M_\odot G/r_0^2$. On the other hand, at large distances satisfying $r/r_0 \gg r_0/H_0$, (4.45) yields

$$P \to P_\infty = P_0 \, e^{-r_0/H_0} \quad ; \quad \frac{r_0}{H_0} = \frac{\mu M_\odot G}{k_B T_0 r_0}. \qquad (4.46)$$

Substituting the solar parameters at the coronal temperature maximum, i.e. $r_0 \simeq R_\odot$, $T_0 \simeq 1.5 \times 10^6$ K and $P_0 \simeq 4 \times 10^{-4}$ Pa (from Fig. 4.1), we find $r_0/H_0 \simeq 8$, so that farther than a few tens $R_\odot$, the pressure is roughly equal to $P_\infty \sim 10^{-7}$ Pa.

Is this reasonable? We know that at a distance of a hundred AU or so, the solar atmosphere encounters the interstellar medium which exerts a pressure of order of magnitude $P_{ins} \sim 10^{-13}$ Pa (see Section 8.1). Hence $P_{ins} \ll P_\infty$ (even taking into account the large uncertainty on P_∞), so that the interstellar pressure is far too small to be able to confine the solar atmosphere. We conclude that the solar atmosphere cannot remain static and tends to blow out.

Is this conclusion inescapable? We made a bold approximation in assuming a constant temperature, since we know that this temperature actually decreases. How does this change our conclusion? You can show in Problem 4.8.3 that taking into account the temperature decrease still yields a pressure P_∞ that is far too large to be matched by the one of the interstellar medium.

How fast should the temperature decrease for the atmosphere not to blow out? For this to be so, the atmospheric pressure at large distances P_∞ should be of the order of magnitude of P_{ins}, which is so small that it is nearly zero compared to P_0. Now, look at (4.44). For the pressure to tend to zero at large distances, the integral $\int_{r_0}^{r} dr/\left(Tr^2\right)$ should tend to infinity as $r \to \infty$, i.e. $1/\left(Tr^2\right)$ should decrease more slowly than $1/r$. In other words, for the atmosphere not to blow up, T must decrease faster than $1/r$ as $r \to \infty$.

We know that this is not so. The conclusion is that the solar atmosphere blows up. Here comes the wind!

4.7.2 Magnetic field effects

Our simple derivation neglects the magnetic field. It only holds either when the magnetic forces are negligible – which is not true in the outer corona, or along the magnetic field direction, i.e. when **B** is approximately radial. In other directions, magnetic pressure enhances the plasma's tendency to expand outwards, whereas magnetic tension tends to retain the plasma if the field lines have both feet anchored at the photosphere (Fig. 4.20). As a result, and very roughly, we expect the outer corona to be in two basic states:

- pockets of plasma retained by magnetic field lines whose both ends are anchored at the photosphere, over active regions,

- plasma expanding along open (nearly radial) magnetic field lines, over regions where the photosphere magnetic field has a dominant polarity.

This is not, however, the whole story. Look at the closed magnetic field lines that retain pockets of plasma (left-hand side of Fig. 4.20), and imagine what should happen farther out. As the magnetic forces decline rapidly with distance, they are no longer able to confine the plasma; hence this plasma will expand. In turn, this expansion drags the field lines into an open configuration, so that finally the wind blows everywhere (right-hand side of Fig. 4.20). We shall refine this picture in Section 6.1.

During the solar activity minimum, when the solar magnetic field is not too far from that of a dipole, the pockets of trapped plasma extend only near

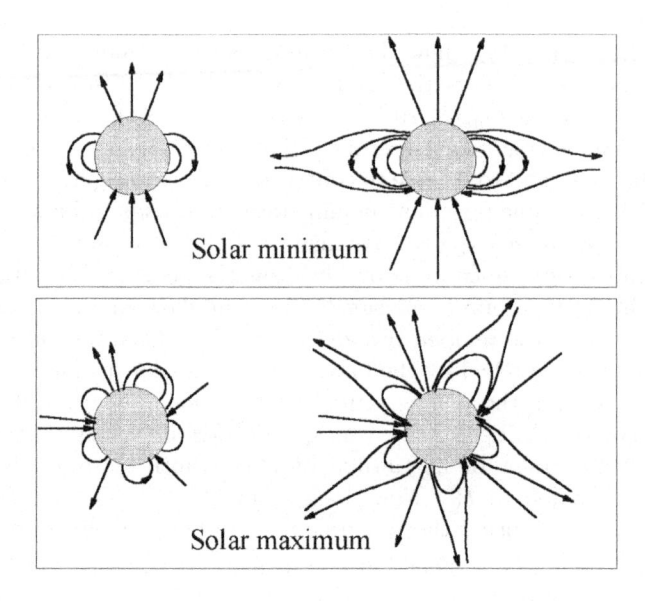

Figure 4.20 Schematic picture of the low (left) and high (right) corona in solar minimum and maximum of activity, showing two fundamental states.

the equatorial magnetic plane, whereas the unhampered flow occupies most of the heliosphere (Fig. 4.20, top). As solar activity increases during the cycle, huge loops of magnetic field emerge nearly everywhere on the Sun and the magnetic field becomes non-dipolar; one then finds both states: pockets of trapped plasma, with expanding plasma above, and regions of unhindered flow, scattered all over the Sun (Fig. 4.20, bottom).

This sheds some light on the different states of the corona revealed by observation, and on the wind picture outlined in Section 1.3. The large homogeneous regions of open magnetic field having a dominant polarity are the so-called coronal holes, which extend around magnetic poles at solar activity minimum, but are smaller and dispersed over the Sun when solar activity is near maximum and the solar magnetic field is no longer dipole-like. It is not surprising that these 'open' coronal regions have a smaller density and electron temperature than 'closed' regions. The wind coming from there is unimpeded, and has a simple geometry: this is the fast and regular solar wind.

On the other hand, the large pockets of trapped plasma are related to the so-called active regions (of greater density and electron temperature); therefore the expanding plasma above them depends strongly on the changing geometry and is variable: this is the slow and unsteady wind. The region in between is the not-so-quiet quiet corona, and the kind of wind it emits is not fully clear for everyone.

We shall refine this picture in Chapter 6. But before doing so, let us study the most basic aspects of how the solar atmosphere produces a wind.

4.8 Problems

4.8.1 Elementary temperature profile

Assume that energy balance in the solar atmosphere is governed by (4.22). Integrate this equation from the upper chromosphere to the outer corona, and show that energy balance is not satisfied. Nevertheless, go ahead and try to find a solution. Since this is an homogeneous second-order differential equation, it tells nothing about absolute values of distance and temperature, and you need adequate boundary conditions. Find reasonable ones and deduce a solution for the temperature as a function of distance.

Hints

Since the temperature both close to the Sun and at large distances is much smaller than the value of its maximum, we can assume

$$T \to 0 \quad \text{for} \quad r \to R_\odot \quad \text{and} \quad r \to \infty.$$

Since (4.22) is homogeneous, we need an additional condition, say

$$T = T_M \quad \text{for} \quad r = r_M.$$

With these boundary conditions, the solution is given by

$$T = T_M \left[\frac{1 - R_\odot/r}{1 - R_\odot/r_M} \right]^{2/7} \quad \text{for} \quad r < r_M \quad (4.47)$$

$$T = T_M \left(r_M/r \right)^{2/7} \quad \text{for} \quad r > r_M. \quad (4.48)$$

(This solution is depicted in Figure 9.15 of [47] cited in Chapter 3.) Show that dT/dr has a discontinuity at r_M. What happens there? Why is (4.22) unreasonable in this region?

4.8.2 Helicity of a string wrapped around a doughnut

Wrap a string around a doughnut by making three warps around the small radius of the torus for two warps around the large radius, and join the two ends of the string. Then eat the doughnut, leaving a knotted string that you can chew (but not break). Now, imagine that the string is a magnetic flux tube carrying the flux Φ (and that the wrapping you have done is, say, right-handed), and calculate the helicity. (To keep the problem simple, do not take into account a possible twist of the tube.) [44].[33]

Then, deform the knot to explore how the helicity distribution can be changed along the string without changing the total value. Try to put all the knot into the smallest region possible. How can the same knot be made more simply,

---

[33] The dynamics of such knots in ideal fluids has interesting properties (see Keener, J. P. 1990, J. Fluid Mech. **211** 629).

without the aid of a doughnut? Imagine again that this is a tube of magnetic flux. What is likely to happen when the knot is squeezed too much? Finally, cut the string and imagine several different ways of increasing the total helicity.

Hint

To calculate the helicity, deform the knot to give it a trefoil shape and then decompose it into two linked tubes having the topology of Fig. 4.12 (right) by inserting two self-cancelling flux elements.

4.8.3 A static solar atmosphere?

Consider a static atmosphere whose temperature decreases with distance as expected from the standard plasma heat conductivity.

Calculate the variation of pressure with distance. What is the value of the pressure at infinity? Compare with the interstellar pressure in the solar surroundings.

Show that the density has a minimum at some distance.

Show that the atmosphere becomes superadiabatic at some distance, so that it is convectively unstable.

Hints

Substitute (4.24) into (4.44) to find $\frac{P}{P_0} = \exp\left[\frac{7r_0}{5H_0}\left(\left(\frac{r_0}{r}\right)^{5/7} - 1\right)\right]$ so that with $r_0 = R_\odot$ and the above parameters we have $P_\infty \simeq P_0 e^{-11} \simeq 10^{-8}$ Pa.

The density $n \propto P/T$. Close to the Sun, the exponential term makes P decrease more rapidly than T, so that n decreases. On the other hand, far away, P is nearly constant, whereas T given by (4.24) still decreases, so that P/T increases.

Compare dT/dr from (4.24) to the adiabatic gradient (3.33), and find that it becomes superadiabatic at a distance $r = r_0 \left(7r_0/5H_0\right)^{7/5}$. With the above parameters this yields a distance of about $28 R_\odot$.[34]

References

[1] Alexander, D. and L. W. Acton 2001, in *The Century of Space Science*, ed. J. A. M. Bleeker *et al.*, New York, Kluwer, p. 1089.

[2] Antonucci, E., M. A. Dodero and S. Giordano 2000, Fast solar wind velocity in a polar coronal hole during solar minimum, *Solar Phys.* **197** 115.

[3] Anzer, U. 2002, in *Solar Variability: From Core to Outer Frontiers*, ed. A. Wilson, ESA SP-506, p. 389.

---

[34] Details may be found in Lemaire, J. 1969, Equilibre mécanique et thermique de la couronne solaire, Thèse de Doctorat, Université de Liège.

References

[4] Aschwanden, M. J. 2004, *Physics of the Solar Corona: An Introduction*, New York, Springer.

[5] Aschwanden, M. J. *et al.* 2000, Time variability of the quiet Sun observed with TRACE. II. Physical parameters, temperature evolution, and energetics of extreme-ultraviolet nanoflares, *Ap. J.* **535** 1047.

[6] Aschwanden, M. J. and C. E. Parnell 2002, Nanoflares statistics from first principles: fractal geometry and temperature synthesis, *Ap. J.* **572** 1048.

[7] Aulanier, G. *et al.* 2002, Prominence magnetic dips in three-dimensional magnetic sheared arcades, *Ap. J.* **567** L97.

[8] Bak, P. 1996, *How Nature Works: The Science of Self-Organized Criticality*, New York, Springer.

[9] Berger, M. A. 1990, in *Physics of Magnetic Flux Ropes*, Geophys. Monogr. Ser. **58**, ed. C. T. Russell, E. R. Priest and L. C. Lee, Washington DC, American Geophysical Union, p. 251.

[10] Brown, M. R., R. C. Canfield and A. A. Pevtsov eds. 1999, *Magnetic Helicity in Space and Laboratory Plasmas*, Geophys. Monogr. Ser. **111**, Washington DC, American Geophysical Union.

[11] Buchlin, E., S. Galtier and M. Velli 2005, Influence of the definition of dissipative events on their statistics, *Astron. Astrophys.* **436** 355.

[12] Cranmer, S. R. 2002, Coronal holes and the high-speed solar wind, *Space Sci. Rev.* **101** 229.

[13] Cranmer, S. R. 2002b, in *From solar min to max: half a solar cycle with SOHO*, ed. by A. Wilson, ESA SP-508, p. 361.

[14] Crooker, N., J.-A. Joselyn and J. Feynman eds. 1997, *Coronal Mass Ejections*, Geophys. Monogr. Ser. **99**, Washington DC; American Geophysical Union.

[15] Crosby, N. B., M. J. Aschwanden and B. R. Dennis 1993, Frequency distributions and correlations of solar X-ray flare parameters, *Solar Phys.* **143** 275.

[16] Démoulin, P. *et al.* 2003, The long-term evolution of AR 7978: testing coronal heating models, *Ap. J.* **586** 592.

[17] Dere, K. P., J.-D. F. Bartoe and G. E. Brueckner 1989, Explosive events in the solar transition zone, *Solar Phys.* **123** 41.

[18] Esser, R. and D. Sasselov 1999, On the disagreement between atmospheric and coronal electron densities, *Ap. J.* **521** L145.

[19] Feldman, U., I. E. Dammasch and K. Wilhelm 2000, The morphology of the solar upper atmosphere during the sunspot minimum, *Space Sci. Rev.* **93** 411.

[20] Fontenla, J. M., E. H. Avrett and R. Loeser 1993, Energy balance in the solar transition region. III. Helium emission in hydrostatic, constant-abundance models with diffusion, *Ap. J.* **406** 319.

[21] Gabriel, A. H. 1976, A magnetic model of the solar transition region, *Phil. Trans. Roy. Soc. London A* **281** 339.

[22] Golub, L. and J. M. Pasachoff 1997, *The Solar Corona*, Cambridge University Press.

[23] Habbal, S. R. 1992, Coronal energy distribution and X-ray activity in the small scale magnetic field of the quiet Sun, *Ann. Geophysicae* **10** 34.

[24] Hood, A. W. 1985, in *Solar System Magnetic Fields*, ed. E. R. Priest, Dordrecht, The Netherlands, D. Reidel, p. 80.

[25] Kane, S. R. et al. 1995, Energy release and dissipation during giant solar flares, *Ap. J.* **446** L47.

[26] Klimchuk, J. A. 2001, in *Space Weather, Geophys. Monogr. Ser.* **125**, ed. P. Song et al., Washington DC, American Geophysical Union, p. 143.

[27] Krucker, S. and Benz, A. O. 1998, Energy distribution of heating processes in the quiet solar corona, *Ap. J.* **501** L213.

[28] Landi, S. and F. G. Pantellini 2001, On the temperature profile and heat flux in the solar corona: kinetic simulations, *Astron. Astrophys.* **372** 686.

[29] Leroy, J.-L. et al. 1989, in *Dynamics and Structure of Quiescent Solar Prominences*, ed. E. R. Priest, New York, Kluwer, p. 77.

[30] Lin, H., M. J. Penn and S. Tomczyk 2000, A new precise measurement of the coronal magnetic field strength, *Ap. J.* **541** L83.

[31] Longair, M. S. 1994, *High Energy Astrophysics*, Cambridge University Press, vol. 2, p. 344.

[32] Low, B. C. 1990, Equilibrium and dynamics of coronal magnetic fields, *Annu. Rev. Astron. Astrophys.* **28** 491.

[33] Low, B. C. 1996, Solar activity and the corona, *Solar Phys.* **167** 217.

[34] Lu, E. T. and R. J. Hamilton 1991, Avalanches and the distribution of solar flares 1978, *Ap. J.* **380** L89.

[35] McWhirter, R. W. P., P. C. Thonemann and R. Wilson 1975, The heating of the solar corona. II. A model based on energy balance, *Astron. Astrophys.* **40** 63.

[36] Meyer-Vernet, N. 2001, Large-scale structure of planetary environments: the importance of not being Maxwellian, *Planet. Space Sci.* **49** 247.

[37] Miller, J. A. *et al.* 1997, Critical issues for understanding particle acceleration in impulsive solar flares, *J. Geophys. Res.* **102** 14631.

[38] Moffatt, H. K. and R. L. Ricca 1992, Helicity and the Călugăreanu invariant, *Proc. Roy. Soc. London A* **439** 411.

[39] Newman, M. E. J. 2005, Power laws, Pareto distributions and Zipf's law, *Contemp. Phys.* **46** 323.

[40] Parker, E. N. 1988, Nanoflares and the solar X-ray corona, *Ap. J.* **330** 474.

[41] Priest, E. R. 1982, *Solar Magnetohydrodynamics*, Dordrecht, The Netherlands, D. Reidel.

[42] Ramirez, R. and R. Soto 2003, Temperature inversion in granular fluids under gravity, *Physica A* **322** 73.

[43] Raouafi, N.-E., S. Sahal-Bréchot and P. Lemaire 2002, Linear polarization of the OVI λ1031.92 coronal line II, *Astron. Astrophys.* **396** 1019.

[44] Ricca, R. L. and M. A. Berger 1996, Topological ideas and fluid mechanics, *Phys. Today* **49** 12, 28.

[45] Rosner, R., W. H. Tucker and G. S. Vaiana 1978, Dynamics of the quiescent solar corona, *Ap. J.* **220** 643.

[46] Rosner, R. and G. S. Vaiana 1978, Cosmic flare transients: constraints upon models for energy storage and release derived from the event frequency distribution, *Ap. J.* **222** 1104.

[47] Ruzmaikin, A. and M. A. Berger 1998, On a source of Alfvén waves heating the solar corona, *Astron. Astrophys.* **337** L963.

[48] Schmelz, J. T. and J. C. Brown eds. 1992, *The Sun: A Laboratory for Astrophysics*, Dordrecht, The Netherlands, Kluwer.

[49] Scudder, J. D. 1992, On the causes of temperature change in inhomogeneous low-density astrophysical plasmas, *Ap. J.* **398** 299.

[50] Scudder, J. D. 1992, Why all stars should possess circumstellar temperature inversions, *Ap. J.* **398** 319.

[51] Shimizu, T. 1995, Energetics and occurrence rate of active region transient brightenings and implications for the heating of the active-region corona, *Publ. Astron. Soc. Japan* **47** 251.

[52] Shoub, E. C. 1983, Invalidity of local thermodynamic equilibrium for electrons in the solar transition region, *Ap. J.* **266** 339.

[53] Sittler, E. C. Jr and M. Guhathakurta 1999, Semiempirical two-dimensional magnetohydrodynamic model of the solar corona and interplanetary medium, *Ap. J.* **523** 812.

[54] Sterling, A. C. 2000, Solar spicules: a review of recent models and targets for future observations, *Solar Phys.* **196** 79.

[55] Taylor, J. B. 1974, Relaxation of toroidal plasma and generation of reverse magnetic fields, *Phys. Rev. Lett.* **33** 1139.

[56] Vourlidas, A. et al. 2000, Large-angle spectrometric coronagraph measurements of the energetics of coronal mass ejections, *Ap. J.* **534** 456.

[57] Walsh, R. W. 2002, in *From Solar Min to Max: Half a Solar Cycle with SoHO*, ed. A. Wilson, ESA SP-508, p. 253.

[58] Wilhelm, K. et al. 1998, The solar corona above polar coronal holes as seen by SUMER on *SoHO*, *Ap. J.* **500** 1023.

[59] Wilhelm, K. 2000, Solar spicules and macrospicules observed by SUMER, *Astron. Astrophys.* **360** 351.

[60] Winebarger, A. R. 2002, Energetics of explosive events observed with SUMER, *Ap. J.* **565** 1298.

[61] Withbroe, G. L. 1988, The temperature structure, mass, and energy flow in the corona and inner solar wind, *Ap. J.* **325** 442.

[62] Zirin, H. 1990, in *Physics of Magnetic Flux Ropes, Geophys. Monogr. Ser.* **58**, ed. C. T. Russell, E. R. Priest and L. C. Lee, Washington DC, American Geophysical Union, p. 33.

[63] Zirker, J. B. 1993, Coronal heating, *Solar Phys.* **148** 43.

5

How does the solar wind blow?

> ...such were the facts accumulated by the Scientific Researcher. And now, what deep, far-reaching *Theory* was he to construct from them?
>
> Lewis Carroll, *Sylvie and Bruno*

We have seen that the solar atmosphere is so hot, out to so large a distance, that it cannot be held back by the Sun's gravity nor confined by the pressure of the interstellar medium. The interstellar medium therefore sucks it, somewhat as a vacuum cleaner – a nice illustration of Aristotle's *horror vacui* doctrine, but a pack of puzzles for the modern physicist.

There are essentially two ways of addressing the problem. One way is to view the solar atmosphere as a fluid, flowing out under the action of the pressure imbalance between the Sun and the interstellar medium. Historically, this was the first theory of the solar wind, and even now it remains the one on which most theoretical attempts are based.

The fluid description, however, requires the medium to be close to local thermodynamic equilibrium, which does not hold in the outer solar atmosphere. As we already noted, the basic difficulty is that the fluid picture requires an assumption on heat transport. However, the flow is far from adiabatic, and indeed we shall see that if it were adiabatic, there would be no solar wind. Therefore, the heat conductivity plays an important role, but its classical expression (see Section 2.3) is not valid because there are not enough collisions to ensure an approximate equilibrium. The essence of the problem is that heat is transported by the fast electrons, and because the collisional free path of charged particles is proportional to the fourth power of their speed, these electrons undergo virtually no collisions, so that they are outside the scope of the usual fluid picture; in that case heat may flow in a completely unexpected way (see Section 4.6). Parker's pioneering theory assumed a uniform temperature, which requires an infinite heat conductivity. However, even with this extreme assumption, the

Figure 5.1 How does the solar wind blow? (Drawing by F. Meyer.)

fluid theory cannot explain the fast solar wind without assuming that some additional energy is furnished.

The other way of addressing the problem is a kinetic picture, which considers particles instead of fluids, and therefore bypasses the heat transport problem since the heat flux is calculated within the theory instead of being postulated a priori or calculated with an invalid approximation. The solar wind is then viewed as the evaporation of a hot atmosphere in the near vacuum of the interstellar medium. There is, however, a fundamental difference with a neutral atmosphere: the medium is made essentially of electrons and protons, of very different masses. Whereas the protons' thermal speed is smaller than their speed of escape from the solar gravitational attraction, the reverse is true for the electrons, which are much lighter. Electrons therefore tend to escape and leave ions behind, producing an outward electric field which adjusts itself in order to ensure electric quasi-neutrality (see Section 4.6). This produces an outward electric force on the protons, which outweighs the gravitational attraction at some distance from the Sun, producing a supersonic wind.

Historically, the kinetic picture was disregarded because the electric field was estimated incorrectly, with too small a value, producing a slow breeze [4] – an error that had large consequences on the subsequent evolution of the ideas because it took a decade to be corrected [26], [29]. The kinetic approach has the major advantage of enabling one to calculate the heat flux and to address non-equilibrium plasmas, which are perhaps the clue to the problem as we shall see later – but the disadvantage of more complexity.

The basic problem 225

This chapter is an attempt to address the physics, starting from first principles. In order to concentrate on basics, I shall consider a simple geometry and assume time independence, keeping geometrical and temporal aspects for the next chapter. Section 5.1 considers the simplest theory: an isothermal flow, and raises some basic questions. The following sections deal with these questions, introducing the corresponding complications step by step. To simplify the notations and conform to practical usage, the *fluid velocity* (the average velocity of particles) is noted **v** (modulus v), up to Section 5.5, where it is noted $\mathbf{v}_w$, to avoid confusion with the velocity of individual particles.

5.1 The basic problem

Let us first try to understand the acceleration of the solar wind by treating the solar atmosphere as a fluid, as did E. N. Parker in the late 1950s, when remarkably few facts were known about the solar wind.

We assume that all quantities are independent of time and depend only on the radial distance r (Fig. 5.2); the magnetic force is neglected (it vanishes if both the magnetic field and the velocity are radial). The medium is pictured as a single fluid that is electrically quasi-neutral so that the total electric force vanishes, is non-viscous, and behaves as a perfect gas of mean mass per particle $m = \mu$ and (local) temperature T, so that the pressure P and mass density ρ are related by $P = \rho k_B T/\mu$.

5.1.1 The solar wind on the back of an envelope

Let us start from Bernoulli's theorem, assuming the pressure and mass density to follow a polytrope relation $P \propto \rho^\gamma$ (Section 2.3). The simplest cases to which this relation applies are the two extreme cases: adiabatic and isothermal. The flow is adiabatic when changes are too fast for heat conduction to act,

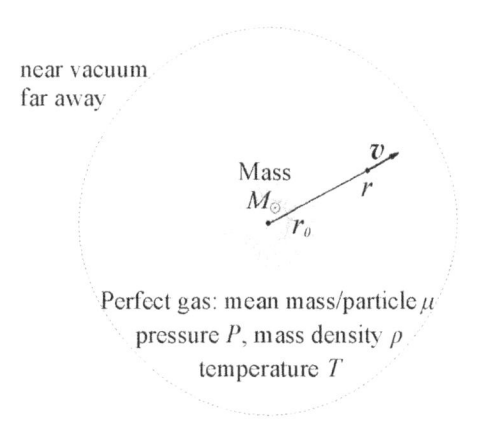

Figure 5.2 Simple problem addressed in Sections 5.1 and 5.2.

so that $\gamma = 5/3$ for a gas made of classical non-relativistic particles having 3 (space) degrees of freedom, which holds for electrons and protons. The flow is isothermal when, on the contrary, changes are sufficiently slow that heat conduction suppresses the temperature gradient, so that the perfect gas law yields $\gamma = 1$.

Let us first assume the flow to be adiabatic. Bernoulli's theorem then tells us that along flow lines:

$$w = \frac{v^2}{2} + \frac{\gamma}{\gamma-1}\frac{k_B T}{\mu} + \Phi_G = \text{constant} \quad ; \quad \Phi_G = -\frac{M_\odot G}{r}. \quad (5.1)$$

This is just conservation of energy (and mass), where the energy w per unit mass is made of:

- the bulk kinetic energy $v^2/2$,
- the enthalpy, that is the thermal energy $3k_B T/2$ per particle (with 3 degrees of freedom), plus $k_B T$ per particle (the potential for work through adiabatic expansion),
- the gravitational energy Φ_G.

At the base of the corona (distance $r_0 \simeq R_\odot$), we have $v_0 \simeq 0$, whereas at large distances both the gravitational energy Φ_G and the enthalpy are negligible. An upper limit to the kinetic energy at large distances $v_\infty^2/2$ is therefore the sum of:

- bulk kinetic energy at r_0: $v_0^2/2 \simeq 0$,
- enthalpy at r_0: $5k_B T_0/2\mu \simeq 0.5 \times 10^{11}\,\text{J kg}^{-1}$,
- gravitational energy at r_0: $\Phi_{G0} = -M_\odot G/r_0 \simeq -1.9 \times 10^{11}\,\text{J kg}^{-1}$,

where we have substituted the Sun's parameters, an average temperature of particles at the base of the corona of $T_0 \simeq 1.5 \times 10^6$ K and the average particle mass $\mu \simeq 0.6 m_p$.[1] The sum is negative, so that there is no wind! And this remains true with any realistic value of the parameters put into (5.1).

What is wrong? Producing a wind requires a much greater value of the energy w, which may be achieved with a much greater enthalpy at r_0, namely a value of γ much closer to unity than the adiabatic value of $5/3$.

Consider the isothermal case $\gamma = 1$. The enthalpy is then $H = V_S^2 \ln \rho$ (Section 2.3), where $V_S = (dP/d\rho)^{1/2}_{isothermal}$ is the *isothermal* sound speed[2]

$$V_S = (k_B T/\mu)^{1/2} \quad ; \quad T = \text{constant}. \quad (5.2)$$

---

[1] This temperature is some average of the badly known electron and ion temperatures at the base of the corona. The mass is a small refinement with respect to the simple proton–electron value of $\mu = (m_p + m_e)/2 \simeq 0.5 m_p$, to account for the small quantity of helium present in the mixture.

[2] Beware that this is not the speed of sound waves propagating adiabatically, which is $(\partial P/\partial \rho)^{1/2}_{adiabatic} = (\gamma k_B T/\mu)^{1/2}$ with $\gamma = 5/3$ for a classical gas made of particles having 3 degrees of freedom (see Section 2.3).

The basic problem

In this isothermal case, Bernoulli's theorem yields instead of (5.1)

$$\frac{v^2}{2} + V_S^2 \ln \rho + \Phi_G = \text{constant}. \tag{5.3}$$

The density ρ may be eliminated by using conservation of mass through a sphere of surface $4\pi r^2$, i.e.

$$\rho v r^2 = \text{constant} \tag{5.4}$$

so that (5.3) yields

$$\frac{v^2}{2} - V_S^2 \ln(vr^2) + \Phi_G = \text{constant} \; ; \quad \Phi_G = -\frac{M_\odot G}{r}. \tag{5.5}$$

The constant is the value at r_0, which is finite, whereas at large distances the dominant term is $\sim v^2/2 - V_S^2 \ln(r^2)$ (since v varies much less rapidly than r^2). Hence the speed at large distances increases indefinitely with r as $v \simeq 2V_S \left(\ln r\right)^{1/2}$. A wind blows!

5.1.2 Nasty questions, or why it is complicated

This simple derivation embodies a large part of the solar wind acceleration problem. No wind blows in the adiabatic case, which means that producing a wind requires heat transport, or a slow temperature decrease. In contrast, a wind blows if the medium is isothermal, and we shall consider in detail this case in Section 5.2, including the subtle question of how the Sun manages to choose between a wind, a breeze or an accretion.

The isothermal case, however, is unphysical since the speed increases without limit with distance, which requires an infinite amount of energy. We shall address this question in Section 5.3, considering the simplest generalisation of an isothermal fluid: a polytrope of arbitrary index $1 < \gamma < 5/3$; we will also examine the effects of a non-spherical geometry, energy addition and viscosity.

The polytrope approximation, however, is not satisfying. It hides the physics within a single parameter: the polytrope index, which is set to a constant ad hoc value; it also ignores a fundamental point: the solar wind is essentially made of electrons and protons, whose large difference in mass makes them behave very differently; as we said, the (massive) protons tend to be trapped by the Sun's gravitational attraction, whereas the (light) electrons tend to escape, producing an electrostatic field to preserve electric neutrality. Furthermore, (massive) protons carry momentum, whereas (fast) electrons carry heat; since exchanges between protons and electrons are insufficient to equalise their temperatures, they cannot be treated as a single fluid. How does this mixture behave? This problem is addressed in Section 5.4.

We shall see that this raises a further question. The medium is weakly collisional, and the particle velocity distributions are not close to Maxwellians. As a result, not only does the medium not behave as one fluid, but each particle species itself may not behave as a fluid, so that a kinetic picture is required. We shall address this problem in Section 5.5, but it is fair to say that it is not fully solved.

5.2 Simple fluid theory

This theory and its early refinements can be found in the book [39] and in a series of papers cited in [40].

5.2.1 The isothermal approximation

Consider in more detail the isothermal case, assuming time stationarity and spherical symmetry, with a radial velocity **v** (Fig. 5.2). The amplitudes along the radial direction are counted positive (respectively negative) when directed away from (respectively towards) the Sun. Taking the derivative with respect to r of Bernoulli's equation (5.5) and rearranging yields

$$\frac{1}{v}\frac{dv}{dr}\left(\frac{v^2}{V_S^2} - 1\right) = \frac{2}{r}\left(1 - \frac{M_\odot G}{2V_S^2 r}\right). \tag{5.6}$$

This is equivalent to writing the radial equation of motion, driven by the pressure gradient working against the solar gravitational attraction, i.e.

$$v\frac{dv}{dr} = -\frac{1}{\rho}\frac{dP}{dr} - \frac{M_\odot G}{r^2} \tag{5.7}$$

and substituting the (isothermal) perfect gas law $P = \rho k_B T/\mu$ and mass conservation (5.4).

Equation (5.6) has a special form: the left-hand side member vanishes when either $dv/dr = 0$, or $v = V_S$, whereas the right-hand side member vanishes at the distance

$$r_C = \frac{M_\odot G}{2V_S^2} \tag{5.8}$$

where the particle escape speed $v_{esc} = (2M_\odot G/r)^{1/2}$ equals twice the sound speed $V_S = (k_B T/\mu)^{1/2}$. With the solar parameters introduced above we have $r_C \simeq 4.5 R_\odot$.

With this notation, (5.6) takes the form

$$\frac{1}{v}\frac{dv}{dr}\left(\frac{v^2}{V_S^2} - 1\right) = \frac{2}{r}\left(1 - \frac{r_C}{r}\right). \tag{5.9}$$

At low heights v and r are small, so that $(v^2/V_S^2 - 1)$ and $(1 - r_C/r)$ are both negative, yielding $dv/dr > 0$. This acceleration stems from two effects: fast density decrease and mass conservation. The density decreases fast because the speed is so small near the Sun that the atmosphere has nearly the hydrostatic profile found in Section 4.7.1:

$$\rho \propto P \propto e^{\mu M_\odot G/(k_B T r)} \quad ; \quad v = 0. \tag{5.10}$$

With ρ therefore decreasing faster than $1/r^2$, v must increase in order to conserve $\rho v r^2$. How far does it continue to do so? As r and v increase together, either member of (5.9) may vanish, and the behaviour of the solution depends upon which vanishes first. There are three possibilities:

Simple fluid theory

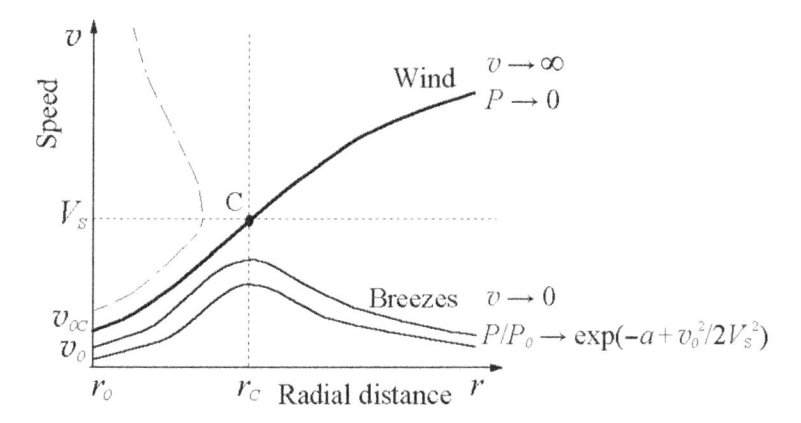

[Figure 5.3: graph showing Speed vs Radial distance with Wind, Breezes curves, critical point C at (r_C, V_S), starting at r_0 with v_0, v_∞.]

Figure 5.3 Sketch of the radial expansion speed for an isothermal flow starting subsonic at r_0. The solution crossing C is a wind that accelerates continuously, becoming supersonic at r_C. The other physical solutions are breezes whose speed is everywhere subsonic and vanishes far away.

- $v = V_S$ arises for $r < r_C$, so that (5.9) yields $dv/dr \to \infty$ as $v \to V_S$; this infinite derivative of the speed is outside the scope of the theory;

- $r = r_C$ is reached when $v < V_S$, so that $dv/dr = 0$; for $r > r_C$ the right-hand side of (5.9) becomes positive, whereas $(v^2/V_S^2 - 1)$ is still negative; hence $dv/dr < 0$, i.e. the flow slows down, producing a *subsonic breeze*;

- $v = V_S$ arises at $r = r_C$, so that both sides of (5.9) change sign together, yielding $dv/dr > 0$; the flow continues to accelerate, producing a *supersonic wind* beyond r_C.

These solutions are sketched in Fig. 5.3, where r_0 is some start-off distance where the flow speed is $v_0 \ll V_S$. To calculate the speed v as a function of the distance r, we write Bernoulli's equation (5.5). Since the constant is the value at r_0 where the speed is v_0, we have

$$\frac{v^2 - v_0^2}{2} = -V_S^2 \ln\frac{r_0^2 v_0}{r^2 v} + (\Phi_{G0} - \Phi_G). \tag{5.11}$$

Let

$$a \equiv \frac{-\Phi_{G0}}{V_S^2}$$
$$\equiv \frac{M_\odot G/r_0}{k_B T/\mu} \equiv \frac{2r_C}{r_0} \equiv \frac{r_0}{H_0} \equiv \frac{v_{esc}^2(r_0)}{2V_S^2} \tag{5.12}$$

where $H_0 = k_B T r_0^2 / \mu M_\odot G$ is the pressure scale height and $v_{esc} = (2M_\odot G/r_0)^{1/2}$ is the particle escape speed at r_0. Rearranging (5.11), we obtain the speed v at distance r in implicit form

$$\frac{v}{v_0}e^{-v^2/2V_S^2} = \frac{r_0^2}{r^2}\exp\left[a\left(1 - \frac{r_0}{r}\right) - \frac{v_0^2}{2V_S^2}\right] \tag{5.13}$$

whereas conservation of mass yields $\rho_0 v_0 r_0^2 = \rho v r^2$, so that the pressure $P = \rho V_S^2$ is given by

$$\frac{P}{P_0} = \frac{\rho}{\rho_0} = \exp\left[a\left(\frac{r_0}{r} - 1\right) + \frac{v_0^2 - v^2}{2V_S^2}\right]. \tag{5.14}$$

What is the value of the start-off distance r_0? For the theory to be consistent, r_0 must be chosen so that the temperature does not vary too much, whereas the bulk speed is small. Since some observations suggest that the speed may already be significant at a few tenths of solar radii above the photosphere, we must put r_0 within the corona, somewhere between a hundredth (or less) and a few tenths of solar radius above the photosphere.[3] Hence we may take $r_0 \simeq R_\odot$ (to within 50%). With the value of r_C found above, this yields $a \gg 1$, which means that the escape speed at r_0 is much greater than the thermal speed, namely the fluid is tightly bound.

With the start-off speed $v_0 < V_S$ at r_0, (5.13)–(5.14) yield the two classes of physical solutions already mentioned: breezes and supersonic wind.

Breezes

With $v < V_S$ everywhere, the flow equation (5.9) shows that the speed increases with distance below r_C and decreases beyond. There are an infinity of breezes, each starting with a different speed v_0.

With $v < V_S$, the pressure (5.14) is close to the one of a static atmosphere. In particular, since $v \to 0$ for $r \to \infty$, the large distance pressure is

$$P \to P_\infty = P_0 \exp\left(-a + v_0^2/2V_S^2\right) \quad ; \quad r \to \infty \tag{5.15}$$

and mass conservation ensures that since $\rho \propto P = $ constant, $v \propto 1/r^2$ for $r \to \infty$.

One sees that the large distance pressure (5.15) is slightly greater than the one of a static atmosphere by the factor $e^{v_0^2/2V_S^2}$. This reveals a very counterintuitive behaviour: the greater the initial speed v_0, and therefore (from Fig. 5.3) the greater the speed everywhere, the greater the terminal pressure – which should match that of the interstellar medium (if such solutions do exist). Hence if one increases the external pressure applied to the breeze, expecting to slow it down, pressure equilibrium requires it to accelerate instead! We shall return to this puzzling point later.[4]

---

[3] Parker's original derivation assumed $r_0 \simeq 1.4 R_\odot$ and $\mu = m_p/2$.
[4] Another difficulty is that with $\rho = $ constant, the total mass within a sphere of radius r surrounding the Sun increases without limit for large r, which would change the gravitational potential. As an exercise, you may calculate the distance at which this occurs, and find that, with the parameters used in this section, it is much larger than the size of the heliosphere.

Simple fluid theory 231

Transonic wind

This solution accelerates everywhere – becoming supersonic at the critical distance r_C. Note that no information on the state of the fluid can propagate upstream from any point beyond r_C, since the speed there is greater than the sound speed – the speed at which pressure disturbances propagate (within the frame of this simple theory). This means that the upstream part of this flow cannot be altered by what happens beyond r_C.

With $v \to \infty$ for $r \to \infty$, (5.14) shows that $P \to 0$ at large distances, a behaviour that departs significantly from that of a breeze or of a static atmosphere. So, at large distances, the flow speed increases indefinitely whereas the density and pressure vanish.

Another important difference from the breezes is that there is not an infinity of such wind solutions, but only one. This is because the initial speed v_{0C} of this 'critical' solution is not arbitrary, but is determined by the condition: $v = V_S$ at $r = r_C \equiv ar_0/2$, i.e. from (5.13)

$$\frac{v_{0C}}{V_S} e^{-v_{0C}^2/2V_S^2} = \frac{a^2}{4} e^{-a+3/2}. \tag{5.16}$$

Note that for the flow to pass through the critical distance r_C, this distance must be above the base of the flow, i.e. $r_C > r_0$, or $a > 2$; this condition is amply satisfied for the Sun. If the opposite inequality $a \leq 2$ were to hold instead, the atmosphere would not be tightly bound at the base and would 'explode', with a start-off speed of the order of magnitude of the sound speed, and would not necessarily follow the critical solution since in that case $r_C < r_0$.

It is easily seen that (5.16) has two solutions for v_{0C}, one subsonic, the other supersonic (reducing to a single one in the limiting case $a = 2$). For $a \gg 1$ the subsonic solution satisfies $v_{0C} \ll V_S$ and is approximately given by

$$\frac{v_{0C}}{V_S} \simeq \frac{a^2}{4} e^{-a+3/2}. \tag{5.17}$$

Substituting this value of v_0 into (5.11) yields the flow speed of the transonic wind at distance r

$$\frac{v^2}{V_S^2} = 2\ln\left(\frac{r^2 v}{r_C^2 V_S}\right) - 3 + 4\frac{r_C}{r}. \tag{5.18}$$

Since at large distances $v/V_S \gg 1$ and $\ln\left(r^2/r_C^2\right) \gg \ln(v/V_S)$, the speed is roughly

$$v \sim 2V_S \left(\ln r/r_C\right)^{1/2} \quad ; \quad r \to \infty \tag{5.19}$$

(in agreement with our 'back-of-the-envelope' estimate).

The speed increases with V_S, i.e. with the temperature, and does not depend on r_0 – a satisfying result, since in this simple theory r_0 is not precisely defined.

What physical effects determine the continuous speed increase? We have seen that below r_C, the fast density decrease forces the speed to increase in order to conserve mass. On the other hand, above r_C, gravity becomes less important, so that the gas expands nearly freely into a (near) vacuum. This behaviour is akin to the flow in a *de Laval nozzle* – a key ingredient of rocket engines (Problem 5.7.1).

The mass loss rate produced by the wind is $M' = 4\pi r_0^2 \rho_0 v_{0C}$; substituting v_{0C} from (5.17), this yields

$$M' \simeq \pi r_0^2 \rho_0 V_S a^2 e^{-a+3/2}$$
$$= (\pi e^{3/2}) \frac{(M_\odot G)^2}{(k_B T/\mu)^{3/2}} \left[\rho_0 e^{-\mu M_\odot G/(k_B T r_0)}\right]. \quad (5.20)$$

Note that the term in brackets is, from (5.10), equal to the density the (isothermal) atmosphere would have at large distances if it were in hydrostatic equilibrium. Hence the mass flux does not depend on the exact location r_0 of the base of the wind, provided the density there is nearly that of a static (isothermal) atmosphere; indeed in this case, the bracket in (5.20) is independent of r_0 since changing the value of r_0 changes $\rho_0 = \rho(r_0)$ according to (5.10).

Tentative application to the Sun

Let us set the base of the wind at a low height above the photosphere, i.e. $r_0 \simeq R_\odot$. We saw that the results do not depend sensitively on this choice. With the parameters of Section 5.1.1, we have the escape speed at the base: $(2M_\odot G/R_\odot)^{1/2} \simeq 6.2 \times 10^5 \text{ m s}^{-1}$, the sound speed (5.2): $V_S \simeq 1.4 \times 10^5 \text{ m s}^{-1}$, the critical distance: $r_C \simeq 4.5 R_\odot$, whence $a = 2r_C/r_0 \simeq 9$, and the initial speed (5.17) of the transonic wind is $V_{0C} \simeq 6 \times 10^{-4} V_S$. We deduce from (5.18) the flow speed at 1 AU from the Sun, that is at $r \simeq 214 R_\odot \simeq 48 r_C$: $v \simeq 3.9 V_S \simeq 5.5 \times 10^5 \text{ m s}^{-1}$. To estimate the mass flux, we need the mass density ρ_0 at r_0; from Fig. 4.14, we have $\rho_0 \simeq 10^{14} \times m_p \text{ kg m}^{-3}$ at $r_0 = 1.01 R_\odot$; the mass loss rate (5.20) is thus $M' \simeq 1.6 \times 10^9 \text{ kg s}^{-1}$.

These figures are within 50% of the observed values (Table 1.3) – an astoundingly good result for so simple a model. Increasing the temperature T by 30% would increase the speed at 1 AU by merely 20%; however, if ρ_0 were not changed, this would also increase the mass flux by a factor of six – putting it outside the range of observed values. We shall return to this point later.

Before trying to improve this model, we must clarify an important point: how does the Sun choose among the different mathematical solutions of (5.9)? Why a supersonic wind rather than one of the gentle breezes sketched in Fig. 5.3?

5.2.2 Breeze, wind or accretion?

A simple answer is that the breezes have a pressure at large distances that is too large to be matched by the pressure of the interstellar medium, so that the transonic wind is the only acceptable solution starting at a low speed near the Sun (Fig. 5.3). Indeed, the terminal pressure of breezes (5.15) is even greater than the one of a static atmosphere, which we saw in Section 4.7 to be far too

Simple fluid theory 233

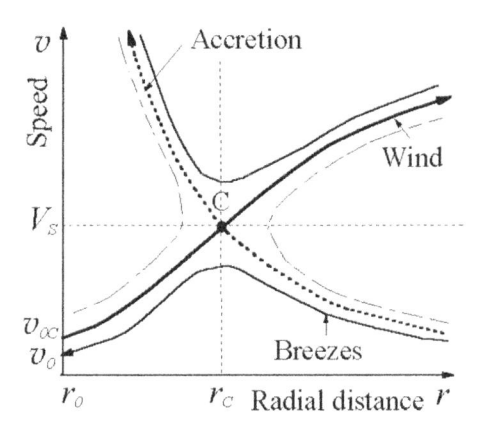

Figure 5.4 Topology of the solutions of (5.9). Parker's transonic wind (bold) and Bondi's transonic accretion (bold dotted; cf. Problem 5.7.3) cross at the sonic point C. In addition to these (unique) solutions, there are an infinity of breezes (one of which is drawn as a thin line) which are everywhere subsonic, and three other families of solutions discussed in the text.

great to be matched by the interstellar medium. In contrast, the wind solution has a vanishing pressure at large distances, so that it can easily accommodate to the small pressure of the interstellar medium with the help of a shock transition (see Section 2.3 and [22]).

A little reflection shows that this answer is not fully satisfying. First of all, it does not explain how the transonic wind is established, and whether it is stable. Furthermore, the argument for choosing the transonic solution relies on the terminal pressure of breezes (5.15), which depends strongly on the temperature, through a. A static corona of temperature a few times smaller than the actual value would have a terminal pressure of the order of the one of the interstellar medium; in that case, would it remain static, or would it blow a breeze? Are these solutions stable? Note also that (5.9) is invariant by changing v into $-v$, and should thus describe accretion of matter as well as ejection. How then does the Sun choose between ejection and accretion? And finally, what would happen if the Sun entered a denser or thinner region of the galaxy, so that the external pressure would change? The answer to these questions is subtler than generally thought, as discussed in [53] and references therein.

To understand what happens, we must first complete Fig. 5.3, which sketches only the physical solutions starting subsonic close to the Sun. It is easily seen from (5.9) that the transonic wind is not the only solution crossing the sonic point C. There is another solution, for which $v \to 0$ at large distances, whereas $v \gg V_S$ at small distances (Fig. 5.4). This solution, discovered by Bondi [2], represents a transonic accretion (Problem 5.7.3).

At C, the wind and accretion curves cross with opposite slopes

$$\left[\frac{dv}{dr}\right]_{r_C} = \pm \frac{V_S}{r_C} \qquad (5.21)$$

– a result we shall demonstrate in Section 5.3.2 for a more general case. Two other classes of solutions (dashed) have a slope that is infinite at $v = V_S$, so that the fluid theory is expected to break down there; a part of the lower branch of these solutions, however, can be connected to respectively the transonic wind and to the transonic accretion, if a shock transition is allowed; we shall return to this point later. Finally, there are also solutions that are everywhere supersonic, just as the breezes are everywhere subsonic (Fig. 5.4).

Note also that all the breezes lie between the horizontal axis ($v = 0$) and the fastest breeze, which is made of a superposition of the subsonic parts of the wind (bold continuous curve) and of the accretion curves (dotted), which meet at C; this critical breeze has a slope that is discontinuous at C.

Stability and time reversal

To complete the picture, we must distinguish between outward and inward solutions and investigate their stability, which requires reintroducing the time into the problem. Indeed, we have assumed the problem to be time invariant, so that the term $\partial v/\partial t$ has been dropped in the fluid equation of motion (Section 2.3) to obtain (5.7); changing v into $-v$ in this term does not change it *only if* we also change t into $-t$. This reveals that the symmetry between v and $-v$ exhibited by (5.6) is an artificial feature that results from our choosing a stationary solution.

This has a major consequence on stability. Since stability and instability mean respectively that a small perturbation decreases or increases with time, the transformation $t \to -t$ generally changes stability into instability and vice versa. Hence, if a solution is stable for an outward velocity, the symmetrical inward solution is expected to be unstable, and vice versa. Now, it is known that a continuous transition from supersonic to subsonic speed is unstable to the formation of shocks. Hence, the transonic solution that is supersonic at large distances is expected to be stable only for an outward flow.[5] From the same argument, the transonic solution that is subsonic at large distances is stable only for inward flow, i.e. accretion.

What about the stability of breezes? Assume that an outward breeze has somehow managed to be established; since it is subsonic, there is no shock and it must be in pressure equilibrium with the exterior medium exerting some pressure. Let this applied pressure decrease. The resulting pressure gradient makes the gas expand more rapidly, which in turn increases the breeze terminal pressure, according to (5.15). This increases the pressure mismatch, worsening the problem. A similar instability arises if we increase the applied pressure instead of decreasing it. Hence, outward breezes are expected to be unstable. This is not so with an *inward* breeze, because a *decrease* (or respectively, an increase) in the applied pressure *decreases* (respectively, increases) the speed, which in turn according to (5.15) *decreases* (respectively, increases) its terminal

---

[5] A simple stability analysis may be found in [40].

Simple fluid theory

pressure and cures the original pressure mismatch. Finally, therefore, breezes are stable only if the speed is inwards.

We conclude that, even when the applied pressure matches the terminal pressure of a breeze, an outward breeze is not expected to flow because it is unstable.

Choosing how to blow

When does the pressure applied at large distances match that of a breeze? One sees in Fig. 5.4 that the breezes lie between the horizontal axis and the *critical breeze* which is made of the subsonic parts of both the transonic wind and the transonic accretion, and whose start-off (outward) speed is therefore the same as that of the transonic wind, i.e. v_{0C} given by (5.17). Hence, breezes have speeds at r_0 in the range $0 < v_0 \leq v_{0C}$, and their large distance pressure P_∞ given by (5.15) therefore lies in the range

$$P_{\infty static} < P_\infty \leq P_{\infty C} \qquad \text{(breezes)} \qquad (5.22)$$

where

$$P_{\infty static} = P_0 e^{-a} \qquad \text{(static atmosphere)}$$
$$P_{\infty C} = P_0 \exp\left(-a + v_{0C}^2/2V_S^2\right) \qquad \text{(critical breeze)}. \qquad (5.23)$$

Outside this range of applied pressure, there is no matching breeze, so that there is only one solution: a transonic wind if the applied pressure $P_\infty < P_{\infty static}$, a transonic accretion if the applied pressure $P_\infty > P_{\infty C}$. But what happens when the applied pressure lies in the range (5.22) where matching breezes do exist?

Let us assume that a supersonic wind is established, with a very small pressure at large distances, and let the applied pressure P_∞ increase (Fig. 5.5). The wind accommodates to this increase with a terminal shock, as usually occurs when a supersonic flow encounters an obstacle (in this case, the interstellar medium). At the shock, the flow becomes subsonic and continues along a part of the subsonic downward branch of one of the double-valued solutions appearing beyond r_C (one of these is shown as a dotted line on the right-hand side of Fig. 5.4). The location of the shock is determined by the requirement that the increase in pressure across the shock plus the one downstream of the shock allows to match the applied pressure P_∞ (Fig. 5.5(2)). We shall see in Section 8.1 that this happens at the distance where the ram pressure of the wind (ρv^2) is roughly equal to the total interstellar pressure. As the applied pressure P_∞ continues to increase, the shock is pushed inwards and decreases in amplitude, until it reaches the critical point C.

Then, the flow remains unchanged below r_C, i.e. it still follows the subsonic part of the wind solution, but above r_C the flow must follow the only solution going through C and having a large terminal pressure: the critical breeze (Fig. 5.5(3)). The shock has disappeared, the discontinuity in speed being replaced by a discontinuity in the derivative of the profile at C. The pressure at large distances is $P_{\infty C}$ and it must match the applied pressure since no shock

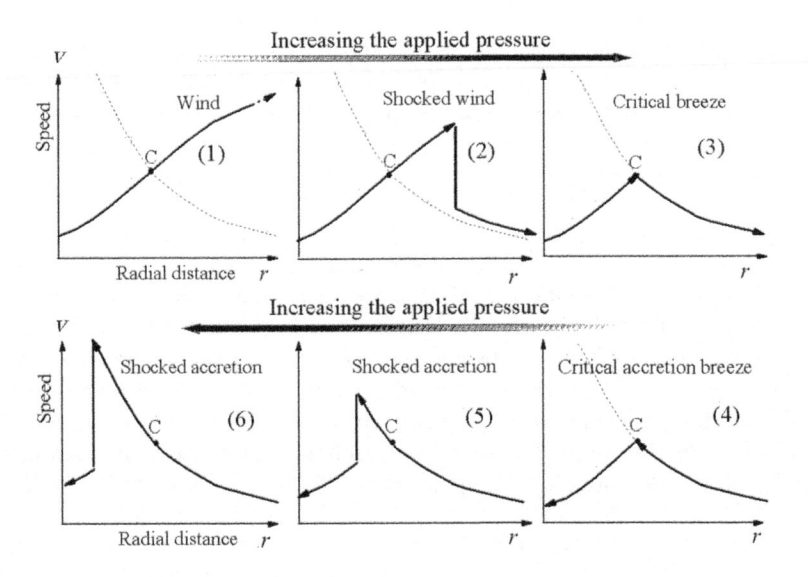

Figure 5.5 How the isothermal wind accommodates to an increase in the applied pressure P_∞ (from 1 to 6), as first predicted in [52]. As P_∞ increases, the terminal shock moves inwards (1–2). When the shock reaches r_C, the wind turns into the outward critical breeze (3). As P_∞ continues to increase, the flow collapses into the critical accretion breeze (4), and to a transonic accretion having a shock that moves inwards (5–6) as P_∞ increases.

can exist in a subsonic flow. Now, what happens if we increase the applied pressure above $P_{\infty C}$? Since the breezes have $P_\infty \leq P_{\infty C}$, no breeze will do, even the accretion breeze (4) – whose terminal pressure is still $P_{\infty C}$, so that the flow must collapse into a transonic accretion (5) having a shock below r_C. As the applied pressure continues to increase, the shock moves inwards (6) [53].

In this case, therefore, a transonic wind blows when the applied pressure $P_\infty < P_{\infty C}$, and it collapses into a transonic (shocked) accretion when the applied pressure $P_\infty > P_{\infty C}$.

What happens if we apply the transformation in reverse? Namely, we begin by applying a large pressure, producing a transonic accretion, and we decrease continuously this applied pressure. The only stable neighbouring solution when P_∞ reaches $P_{\infty C}$ is an inward breeze – a state which persists as P_∞ is decreased, until $P_\infty = P_{\infty static}$, at which point the flow turns into the only possible solution compatible with $P_\infty < P_{\infty static}$: the transonic wind (with a shock, in order to accomodate the pressure). In that case, therefore, there is an inward breeze in the range of applied pressure (5.22).

The conclusion of all this is that in the pressure range (5.22) where several solutions are possible, two of which are stable, the state of the system depends on its history; it is either an outward transonic (shocked) wind or an inward breeze (Problem 5.7.2). This behaviour has been confirmed by numerical simulations ([53] and reference therein).

5.3 Letting the temperature vary

Let us now consider a question that has been swept under the rug: the variation in temperature. Not only does the isothermal approximation neglect the outward temperature decrease, but it produces a speed that diverges at a large distance, implying infinite addition of energy to the flow, which is clearly unphysical. Furthermore, a proper theory should derive the temperature radial profile (as well as the bulk speed), from the physical processes at work, and should not impose it a priori.

Since the variation in temperature is determined by the energy flow in the gas, let us consider in detail the energy balance.

5.3.1 Energy balance

The energy of the gas

We saw in Section 5.1.1 that (for particles having 3 degrees of freedom) the energy per unit mass is

$$w = \frac{v^2}{2} + \frac{5}{2}\frac{k_B T}{\mu} + \Phi_G. \tag{5.24}$$

At the base of the flow, where the speed is small, the energy consists mainly of the large (negative) gravitational energy Φ_G, so that $w < 0$. On the other hand, at large distances where Φ_G is negligible, the energy of a supersonic wind is of the order of the bulk kinetic energy $v^2/2$, and is therefore positive. This difference in energy is the central problem of the solar wind acceleration.

Energy flux

In the absence of radiation,[6] the total energy flux (per metre squared per second) is the sum of:

- the energy carried by the flow, equal to the energy per unit mass (w) times the mass flux (ρv),

- the heat flux Q produced by heat conduction (due to individual particle motions in the frame where the gas is at rest).

The total energy crossing a sphere of radius r per second is thus

$$W' = M'w + 4\pi r^2 Q \tag{5.25}$$

where

$$M' = 4\pi r^2 \rho v \quad \text{(mass loss rate)}. \tag{5.26}$$

---
[6] We have seen in Section 4.3 that the typical timescale of cooling by radiation in the corona is of the order of 10 h outside active regions. This is far greater than other timescales – a property that is true at larger distances, too. Radiation can thus be safely neglected in both the corona and the wind.

It is a constant in the absence of additional heating or loss process. Substituting the energy per unit mass (5.24), we get

$$\frac{v^2}{2} + \frac{5}{2}\frac{k_B T}{\mu} + \Phi_G + \frac{Q}{\rho v} = \frac{W'}{M'} = \text{constant} \quad ; \quad \Phi_G = -M_\odot G/r. \quad (5.27)$$

Note that derivating this equation and subtracting the momentum equation (5.7), we obtain

$$\frac{k_B T^{5/2}}{\mu \rho} \frac{d}{dr}\left(\rho T^{-3/2}\right) = \frac{d}{dr}\left(\frac{Q}{\rho v}\right) \quad (5.28)$$

which is the usual relation between the variation in entropy and that of the heat flux normalised to the mass flux.

Terminal speed

Writing the energy balance (5.27) between the base r_0, where the speed is negligible, and a large distance, where both the enthalpy and the gravitational energy are negligible compared to the values at r_0, we find the kinetic energy at large distances

$$\frac{v_\infty^2}{2} \simeq \frac{5}{2}\frac{k_B T_0}{\mu} + \Phi_{G_0} + \left[\frac{Q}{\rho v}\right]_0 - \left[\frac{Q}{\rho v}\right]_\infty \quad (5.29)$$

where the indices 0 and ∞ refer respectively to the base and to large distances. Let us apply this equation to the solar wind. The values at $r_0 \simeq R_\odot$ have been estimated in Section 5.1.1 and we have (per unit mass):

- enthalpy at r_0: $5k_B T_0/2\mu \simeq 0.5 \times 10^{11}\,\text{J kg}^{-1}$,
- gravitational energy at r_0: $\Phi_{G_0} = -M_\odot G/r_0 \simeq -1.9 \times 10^{11}\,\text{J kg}^{-1}$,
- kinetic energy at large distances: $v_\infty^2/2 \simeq (1 \leftrightarrow 3) \times 10^{11}\,\text{J kg}^{-1}$, for typical solar wind speeds of 400↔800 km s^{-1}.

Required heat flux

Accelerating the solar wind therefore requires the heat flux per particle at r_0:[7]

$$\left[\frac{Q}{\rho v}\right]_0 \geq [-0.5 + 1.9 + 1 \leftrightarrow 3] \times 10^{11} \simeq (2.5 \leftrightarrow 4.5) \times 10^{11}\ \text{J kg}^{-1} \quad (5.30)$$

the lower and larger limits corresponding respectively to the slow and fast wind. With a total mass loss rate $M' \simeq 10^9$ kg s^{-1} (typical of the fast wind), the mass flux averaged over the solar surface is

$$[\rho v]_0 \simeq \frac{M'}{4\pi r_0^2} \simeq 1.6 \times 10^{-10}\ \text{kg m}^{-2}\,\text{s}^{-1}. \quad (5.31)$$

---

[7] The inequality accounts for the possibility that $[Q/\rho v]_\infty$ may not be negligible in (5.29).

Letting the temperature vary 239

So that the total heat flux required at the base of the corona to accelerate the fast solar wind is, from (5.30) with the larger value

$$Q_0 \geq 70 \text{ W m}^{-2} \quad (5.32)$$

if the expansion is radial.[8] Where does this heat flux come from?

In the fluid scheme, the heat flux stems from the heat conductivity produced by collisions (Section 2.3), i.e.

$$Q_c = -q_0 T_e^{5/2} dT_e/dr \quad (5.33)$$

with $q_0 \simeq 10^{-11}$. This is the heat flux that would exist *if* the medium were sufficiently collisional. For an order of magnitude estimate, we assume $T_e \propto r^{-2/7}$ (see Section 4.3), which yields

$$Q_{c0} \simeq (2/7) q_0 T_{e0}^{7/2}/r_0.$$

With $r_0 \simeq R_\odot$ and $T_{e0} \simeq 1.5 \times 10^6$ K, we have $Q_{c0} \simeq 17$ W m^{-2}, which is far too small compared to (5.32). This conclusion is especially robust for the fast wind, despite the great sensitivity of Q_c to T_e, because the electron temperature in coronal holes (where the fast wind comes from) may be smaller – but not greater – than the value used in the above estimate, which is therefore an upper limit.

We conclude that the collisional heat flux is not large enough to power the solar wind. From our modern vantage point [49], this conclusion may appear irrelevant since we now know that the corona is not sufficiently collisional for this estimate of the heat flux to be correct (see Section 4.6). However, this inability of the collisional heat flux to power the solar wind had major consequences on early theoretical studies, because it focused the efforts on finding external energy supplies instead of finding a proper way of calculating the heat flux.

This problem is connected to a major unsolved problem encountered in Section 4.6: the origin of the large coronal temperature. The energy required to accelerate the fast wind is of the same order of magnitude as that required to heat the corona outside active regions.

What about the slow wind? Equation (5.30) yields a smaller value for the normalised heat flux required at the base, but the mass loss rate is greater (see Table 1.3), so that the required heat flux Q_0 is the same (assuming again radial expansion). Interestingly, we shall see in Section 8.3 that a comparison of winds blown by a large sample of stars suggests that they require a similar value of the heat flux Q_0 at the base of the wind.

5.3.2 Polytrope approximation

A convenient heat flux

What is the actual value of the heat flux in the corona and in the solar wind? We have seen that the usual collisional value might be inapplicable, and that,

---

[8]We shall see in Section 5.3.3 that local deviations from spherical expansion that might change this estimate (by changing the particle flux at r_0) would still increase the heat flux requirements.

anyway, it is too small. If we insist on building a simple fluid theory, we may try using an empirical heat flux proportional to the number density of particles, to their temperature, and to the bulk speed, as

$$Q = \alpha \rho v k_B T / \mu. \tag{5.34}$$

We shall see later that such a heat flux with α in the range $[1 \leftrightarrow 10]$ has some basis on both empirical and theoretical grounds.

Substituting this expression of Q into the energy equation (5.28) yields

$$\alpha \rho \frac{dT}{dr} = T^{5/2} \frac{d}{dr} \left(\rho T^{-3/2} \right)$$

whose integration yields $T^\alpha \propto \rho T^{-3/2}$. Since the pressure $P \propto \rho T$, this yields the polytrope law:

$$P \propto \rho^\gamma \quad ; \quad \gamma = \frac{5 + 2\alpha}{3 + 2\alpha} \quad \text{or:} \quad \alpha = \frac{\gamma}{\gamma - 1} - \frac{5}{2}. \tag{5.35}$$

A positive value of (Q/v) requires α to be positive, i.e. $1 < \gamma < 5/3$, with the limiting cases:

- isothermal: $\gamma = 1$ or $\alpha \to \infty$ (infinite heat flux),
- adiabatic: $\gamma = 5/3$ or $\alpha = 0$ (zero heat flux).

Note that the equation of energy balance (5.1) may be obtained by taking the integral of the momentum equation (5.7) in which one substitutes $P \propto \rho^\gamma \propto T^{\gamma/\gamma-1}$ with $P = \rho k_B T / \mu$. With γ given in (5.35), the energy balance (5.1) may also be obtained by substituting the expression (5.34) of Q into the energy balance (5.27), which yields

$$\frac{v^2}{2} + \left(\frac{5}{2} + \alpha \right) \frac{k_B T}{\mu} + \Phi_G = \text{constant}. \tag{5.36}$$

Returning to the energy equation (5.27), we see that the expression (5.34) of Q is formally equivalent to assuming that (1) the enthalpy $H = (5/2 + \alpha) k_B T / \mu$ instead of $(5/2) k_B T / \mu$, and (2) the heat flux is zero (i.e. the flow is adiabatic). For particles having N degrees of freedom, the thermal energy per particle is $N k_B T / 2$, whence the enthalpy $H = (1 + N/2) k_B T / \mu$ per unit mass. Hence the heat flux (5.34) is formally equivalent to assuming an adiabatic flow of particles having an effective number of degrees of freedom N satisfying $5/2 + \alpha = 1 + N/2$, i.e.

$$N = 3 + 2\alpha = 2/(\gamma - 1). \tag{5.37}$$

Thus $\gamma < 5/3$ corresponds formally to $N > 3$, which requires the particles to have internal degrees of freedom in addition to the three space co-ordinates of their centre of mass. This is so for non-colinear molecules, which have three

Letting the temperature vary 241

free angles of rotation and may also vibrate and have electron transitions if the temperature is adequate. We shall see an application to the expansion of cometary atmospheres in Section 7.5. A similar effect occurs in a weakly ionised plasma, through electron ionisation and recombination,[9] because when the heat furnished may serve to ionise atoms instead of raising the temperature, the specific heat is greater. Note that $\gamma = 1$ corresponds formally to an infinite number of degrees of freedom.

What is the significance of a polytrope fluid having $\gamma < 5/3$, though it is fully ionised and made of point-like particles? In some sense, the particles of such a plasma might be viewed as being coupled to a number of wave modes and turbulent fluctuations, that produce additional effective degrees of freedom. In some sense only, as the number of modes available for storing energy depends on the temperature, so that the index γ should not be a constant. An application may be found in [45]. Another interpretation, to which we shall return later, involves the absence of equilibrium.

Polytrope flow

It is easy to generalise the isothermal calculations performed in Section 5.1 to a polytrope flow. With $P \propto \rho^\gamma$, the sound speed[10] is defined as

$$V_S = (dP/d\rho)^{1/2} = (\gamma P/\rho)^{1/2} = (\gamma k_B T/\mu)^{1/2}. \tag{5.38}$$

Writing $dP/dr = V_S^2 d\rho/dr$ and substituting $d\rho/dr$ from the conservation of mass (5.4), we have

$$\frac{1}{\rho}\frac{dP}{dr} = \frac{V_S^2}{\rho}\frac{d\rho}{dr} = -V_S^2\left(\frac{1}{v}\frac{dv}{dr} + \frac{2}{r}\right). \tag{5.39}$$

Substituting into the equation of motion (5.7), we get

$$\frac{1}{v}\frac{dv}{dr}\left(\frac{v^2}{V_S^2} - 1\right) = \frac{2}{r}\left(1 - \frac{r_C}{r}\right) \quad ; \quad r_C = \frac{M_\odot G}{2V_S^2} = \frac{\mu M_\odot G}{2\gamma k_B T}. \tag{5.40}$$

At first sight, this looks like the equation (5.9) obtained in the isothermal case, so that the solutions might be expected to have a similar topology, with a critical point where $v = V_S$ at distance $r_C = M_\odot G/2V_S^2$. This similarity, however, is only formal, because V_S and r_C are no longer constants, being instead auxiliary variables that depend on the temperature, as defined in (5.38)–(5.40).

Let the flow start at r_0 with the speed v_0, and use similar notations as previously, i.e.

$$a \equiv -\frac{\mu M_\odot G}{k_B T_0 r_0} \quad ; \quad V_{S0} = \left(\frac{\gamma k_B T_0}{\mu}\right)^{1/2}. \tag{5.41}$$

---

[9] We have seen in Section 3.2.2 an application of this effect in the weakly ionised outer part of the solar interior.
[10] Beware that for classical particles having 3 degrees of freedom, only for $\gamma = 5/3$ does this definition correspond to the speed of (adiabatic) sound waves $(dP/d\rho)^{1/2}_{adiabatic}$.

With a subsonic speed at r_0, we have $(v^2/V_S^2 - 1) < 0$, so that $dv/dr > 0$ only if the right-hand side member of (5.40) is negative, i.e.

$$1 - \frac{\mu M_\odot G}{2\gamma k_B T_0 r_0} < 0 \quad \text{or:} \quad a > 2\gamma, \tag{5.42}$$

namely, if the gas is strongly bound at the base. This is similar to the condition we found in the isothermal case ($\gamma = 1$), with, however, a major difference. Now, as r and v increase together, a critical point in (5.40) may be reached only if $(M_\odot G/2V_S^2 r)$ decreases with distance, i.e. if T decreases more slowly than $1/r$. Such a slow decrease in temperature requires a significant heat flux, i.e. a value of γ significantly smaller than $5/3$. We shall return to this point later.

To proceed further, we use:

- energy balance (5.1) (or the integral of the equation of motion),
- mass conservation (5.4) with the polytrope law $T \propto \rho^{\gamma-1}$, so that

$$T \propto (vr^2)^{1-\gamma}. \tag{5.43}$$

These two equations, with the unknowns v and T, enable one to calculate v and T at any distance as a function of their values at r_0.

Transonic wind

Contrary to the isothermal case, energy conservation now ensures that the polytrope wind has a finite asymptotic speed v_∞. From the polytrope law (with mass conservation), the asymptotic parameters therefore vary with the distance r as

$$P_\infty \propto \rho_\infty^\gamma \propto (v_\infty r^2)^{-\gamma} \propto r^{-2\gamma} \quad ; \quad T_\infty \propto \rho_\infty^{\gamma-1} \propto r^{2(1-\gamma)}$$

so that the pressure, density and temperature of the wind all vanish at infinite distance. Hence, the energy balance (5.1) between r_0 and large distances where both T and Φ_G are negligible yields

$$\frac{v_\infty^2}{2} = \frac{v_0^2}{2} + \frac{\gamma}{\gamma-1} \frac{k_B T_0}{\mu} + \Phi_{G0} \tag{5.44}$$

or:

$$\frac{v_\infty^2 - v_0^2}{V_{S0}^2} = \frac{2}{\gamma}\left[\frac{\gamma}{\gamma-1} - a\right]$$

since from (5.41) $a = -\gamma \Phi_{G0}/V_{S0}^2$. For the terminal speed to be greater than the start-off speed, one must therefore have $\gamma/(\gamma-1) > a$. This ensures that the heat and enthalpy fluxes are sufficient to lift the gas out of the Sun's gravitational well. This inequality together with (5.42) may be written

$$2\gamma < a < \frac{\gamma}{\gamma-1} \tag{5.45}$$

which are compatible only if $\gamma < 3/2$.

Letting the temperature vary 243

We conclude that a transonic polytrope flow whose speed starts subsonic and increases with distance requires inequalities (5.45) to hold, which require that $\gamma < 3/2$.[11] Let us calculate the corresponding mass flux. Since this solution crosses the sonic point, the start-off speed v_{0C} may be calculated by writing the conservation equations (5.1) and (5.4), with the polytrope relation (5.43), between the base r_0 (where the temperature is T_0) and the critical distance (where, from (5.40), $v = V_S$ and $r = \frac{M_\odot G}{2V_S^2}$). A little manipulation yields

$$\frac{v_{0C}^2}{2V_{S0}^2} + \frac{1}{\gamma - 1} - \frac{a}{\gamma} = \left(\frac{1}{\gamma - 1} - \frac{3}{2}\right)\left[\frac{4\gamma^2}{a^2}\frac{v_{0C}}{V_{S0}}\right]^{\left(\frac{1}{\gamma-1} - \frac{3}{2}\right)^{-1}}. \tag{5.46}$$

If v_{0C} is small enough, the first term on the left may be neglected, whence

$$\frac{v_{0C}}{V_{S0}} \simeq \frac{a^2}{4\gamma^2}\left[\frac{2}{\gamma} \times \frac{\gamma - a(\gamma - 1)}{5 - 3\gamma}\right]^{\frac{1}{\gamma - 1} - \frac{3}{2}}$$

so that the mass flux $M' = 4\pi r_0^2 \rho_0 v_{0C}$ is given by

$$M' \simeq 4\pi r_0^2 \rho_0 V_{S0} \frac{a^2}{4\gamma^2}\left[\frac{2}{\gamma}\frac{\gamma - a(\gamma - 1)}{5 - 3\gamma}\right]^{\frac{1}{\gamma - 1} - \frac{3}{2}}. \tag{5.47}$$

A closer examination of the critical point

How does the speed behave at the critical point? To address this question, we may use the *de l'Hopital* rule which gives the slope of the solution of a differential equation near a critical point. The flow equation (5.40) is of the form:

$$f(r)\frac{dv}{dr} = g(r) \quad ; \quad f(r_C) = g(r_C) = 0.$$

If the derivatives f' and g' of respectively f and g do not vanish at r_C, the slope of the velocity profile at r_C is given by

$$\left[\frac{dv}{dr}\right]_{r_C} = \frac{g'(r_C)}{f'(r_C)}. \tag{5.48}$$

Multiplying both sides of (5.40) by V_S^2, we have

$$f = v - \frac{V_S^2}{v} \quad ; \quad g = \frac{2V_S^2}{r} - \frac{M_\odot G}{r^2} \quad ; \quad V_S^2 \propto (vr^2)^{1-\gamma} \tag{5.49}$$

---

[11] With similar arguments, it is easily shown that in the remaining range $3/2 < \gamma < 5/3$, a transonic flow does exist for $\gamma/(\gamma - 1) < a < 2\gamma$, but its speed now declines with distance, (albeit less rapidly than does the sound speed, so that v/V_S still increases) [7]. In the limiting case $\gamma = 3/2$, there is a solution for which $T \propto r^{-1}$ so that, from (5.43), $v =$ constant. Putting $dv/dr = 0$ into (5.40) yields $M_\odot G/2V_S^2 = r$ everywhere, which implies in particular $a = 2\gamma = 3$. In the adiabatic limit $\gamma = 5/3$, the energy balance (5.1) yields a special solution for which the (constant) fluid energy is zero [5]. With the constant equal to zero, (5.1) and (5.43) imply $v^2 \propto T \propto r^{-1}$, so that the ratio v/V_S is a constant, and the pressure $P \propto T^{\gamma/(\gamma - 1)}$ vanishes at large distances. With the energy of the gas equal to zero at r_0, $v_0^2/V_{S0}^2 = 6a/5 - 3$ so that $v_0 < V_{S0}$ requires $a < 10/3$.

where the latter equation stems from the polytrope law with mass conservation. At the critical point $r = r_C$, we have

$$v^2 = V_{SC}^2 = \frac{M_\odot G}{2r_C} \quad \text{with} \quad V_{SC} = V_S(r_C). \tag{5.50}$$

Let

$$y = \frac{r_C}{V_{SC}} \left[\frac{dv}{dr}\right]_{r_C}. \tag{5.51}$$

Using (5.48), and derivating f and g with respect to r with (5.49)–(5.50), we have

$$y = \frac{r_C}{V_{SC}} \frac{g'(r_C)}{f'(r_C)} \tag{5.52}$$

$$f'(r_C) = \frac{V_{SC}}{r_C} \left[2(\gamma - 1) + (1 + \gamma) y\right] \tag{5.53}$$

$$g'(r_C) = 2 \left(\frac{V_{SC}}{r_C}\right)^2 [3 - 2\gamma + (1 - \gamma)y]. \tag{5.54}$$

Substituting (5.53)–(5.54) into (5.52) we obtain a (non-dimensional) quadratic equation for y. From the solution y, we deduce from (5.51) the slope of the speed profile at r_C:

$$\left[\frac{dv}{dr}\right]_{r_C} = \frac{V_{SC}}{r_C} \left[\frac{\pm\sqrt{2(5 - 3\gamma)} - 2(\gamma - 1)}{1 + \gamma}\right]. \tag{5.55}$$

In the range $1 \leq \gamma < 3/2$, the $\pm$ signs yield respectively a positive and a negative slope, corresponding respectively to the transonic wind and to the transonic accretion with the topology shown in Fig. 5.4. As an exercise,[12] one may verify that in this range of γ, the temperature falls off less rapidly than $1/r$. In the limiting case $\gamma = 1$ (isothermal), the slopes (5.55) reduce to the values (5.21).

Breezes

For breeze solutions whose terminal speed vanishes, energy balance (5.1) between r_0 and large distances yields

$$\frac{V_{S\infty}^2}{V_{S0}^2} = \frac{\gamma - 1}{2} \frac{v_0^2}{V_{S0}^2} + 1 - a\frac{\gamma - 1}{\gamma} \quad \text{(breezes)} \tag{5.56}$$

$$= \left(\frac{P_\infty}{P_0}\right)^{\frac{\gamma-1}{\gamma}} \tag{5.57}$$

---

[12] Using $T \propto (vr^2)^{1-\gamma}$, we calculate the derivative of T from the one of v, and deduce that $(dT/dr)_{r_C} = -T_C/r_C$ times a factor that is smaller than unity in this range of γ.

Letting the temperature vary 245

where the latter equation stems from the polytrope law. This shows that, for $a < \gamma/(\gamma-1)$, the terminal pressure of breezes is finite and increases with the base speed, as it does in the isothermal case. Hence, both the large pressure and the instability of outward breezes are expected to hold true also for a polytrope fluid.

Tentative application to the Sun

Let us apply these results to the Sun. With $r_0 \simeq R_\odot$ and $T_0 \simeq 1.5 \times 10^6$ K, we have $a \gg 1$, so that a transonic wind requires, from the inequalities (5.45), a value of γ rather close to unity, equivalent to a large heat flux. With, say, $\gamma = 1.08$, we find $v_\infty \simeq 400$ km s^{-1}, $v_{0C}/V_{S0} \simeq 4 \times 10^{-4}$, whence a mass flux $M' \simeq 10^9$ kg s^{-1}. The temperature at large distances is given by $T = T_0 (\rho/\rho_0)^{\gamma-1}$ with $\rho/\rho_0 \simeq (v_{0C} r_0^2/v_\infty r^2)$, which yields $T/T_0 \simeq 0.2$ at 1 AU from the Sun. Obtaining a speed twice as great would require $\gamma = 1.06$, which yields a mass flux roughly four times greater, i.e. outside the range of measured values, and a temperature at 1 AU greater by roughly 30%. The agreement with observation is not too bad – given the simplicity of the model – but (as with the isothermal approximation) it is difficult to reproduce simultaneously the fast wind speed and mass flux observed.

Conclusion on polytrope winds

A polytrope model $P \propto \rho^\gamma$ is the simplest extension of the (unphysical) isothermal assumption. This is formally equivalent to a heat flux

$$Q = \left(\frac{\gamma}{\gamma-1} - \frac{5}{2}\right) \rho v \frac{k_B T}{\mu}.$$

The flow behaves qualitatively as if it were isothermal ($\gamma = 1$), except that the terminal speed is finite and that the condition for having a transonic wind is stricter. Indeed, a transonic isothermal wind only requires that the parameter a, which measures the tightness of the particle binding at the base of the wind, satisfies $a > 2$ (if a is smaller, the sonic point lies below the base). In contrast, producing a transonic polytrope wind whose speed increases outwards requires a to lie in a narrow range given by (5.45), and γ to be small enough. Stated another way, the (finite) heat flux powering the wind must be large enough to lift the gas out of the gravitational well and to provide it with a large bulk speed.

The major interest of this model is its simplicity, which enables one to study analytically how the flow depends on the parameters. However, it does not solve the major problem of heat transfer, since the relevant physics is hidden in the adiabatic index γ, whose value is not explained, and, in any case, is not expected to be independent of distance.

5.3.3 Changing the geometry

Does the flow depend critically on the assumed spherical geometry? This is an important question because the expansion may not be strictly radial near the Sun. We shall return to this question in Section 6.2. In particular, at solar activity minimum, the expansion may be faster than radial in the polar regions, whereas near the equator it should vary in a non-monotonous way.

Consider the simple case of a stream line that is parallel to a locally radial magnetic field, and imagine that the neighbouring stream lines diverge non-radially, so that the cross-sectional area A of a flow tube is no longer proportional to r^2. Let

$$A \propto f(r) r^2$$

where the function $f(r)$ denotes the deviation of the cross-section A from a spherical expansion. Since the expansion is known to be spherical at large distances, $f(r)$ tends to a constant: f_∞, as $r \to \infty$.

Neither the equation of motion (5.7) nor the energy equation in the form (5.27) are changed, but mass conservation now reads

$$\rho v A = \text{constant}. \tag{5.58}$$

Let us consider briefly the consequences. A detailed study may be found for example in [20].

Consider first the simple isothermal case (infinite heat flux). One sees easily that the large distance speed varies as $\left[\ln(f^{1/2}(r)r)\right]^{1/2}$ rather than $[\ln r]^{1/2}$, and is therefore greater (smaller) if the expansion is faster (slower) than radial.

Consider now the case when a given heat flux is provided at the base, so that the terminal speed is given by (5.29). A change in the cross-sectional area changes the base mass flux $[\rho v]_0$ deduced from the mass loss observed at large distances $M' = \left[\rho v \times 4\pi r^2\right]_\infty$, since mass conservation now yields

$$[\rho v]_0 = \frac{[\rho v A]_\infty}{A_0} = \frac{M'}{4\pi r_0^2} \times \frac{f_\infty}{f_0}. \tag{5.59}$$

With an expansion faster than spherical, we have $f_\infty > f_0$, so that $[\rho v]_0$ is greater than with spherical expansion. Indeed, if most of the solar mass loss comes from only a fraction of the solar surface, the mass flux per unit surface at r_0 must be greater. In that case, a greater heat flux at the base is required to produce the observed solar wind speed, which worsens the problem of furnishing enough heat flux at the base to accelerate the wind.

Another question is how the speed profile and the topology of the solution are affected. The variation in f introduces an additional term $f^{-1} df/dr$ in the right-hand side of the flow equation (5.40). At first sight it might seem that this changes only the location of the critical distance, and perhaps the topology of the solution in this vicinity. More subtle consequences may arise, however, since the additional term may make the right-hand side of the equation

Letting the temperature vary

vanish at several locations, so that additional critical points may arise. The *continuous* transonic solution passes through one critical point only. However, if *discontinuous* shock transitions may occur between solutions passing through different sonic points, then transonic flows involving several shock transitions may have the same parameters as the normal continuous transonic flow, at both the base and large distances. This opens the interesting possibility of having multiple wind states (see [15]).

These geometry-dependent effects are only a part of the numerous physical effects determined by the structure of the magnetic field. We shall take up this subject in Chapter 6.

5.3.4 Further pushing or heating the wind

We have seen that the acceleration of the solar wind requires the gas to conduct heat very efficiently at the base of the wind. However, if some additional process did manage to furnish an adequate amount of energy, this requirement would be relaxed. Hence it may be interesting to imagine that some (yet unspecified) process pushes and/or heats the gas, and to examine the consequences.

Since a large majority of the literature on the solar wind is based on such an assumption, let us assume that some boojum conspiration not only heats the corona but also heats and/or pushes the solar wind, and examine the consequences.

Balance equations with deposition of energy

Let us assume that:

- the fluid is subjected to an additional outward force $F_{ext}(r)$ per unit mass,

- some additional process puts heat into the fluid at the rate $Q_{ext}(r)$ per unit volume.

In that case, the mass balance is still given by (5.4), but the momentum (5.7) and energy (5.27) balance equations per unit mass become:

- momentum balance:

$$v \frac{dv}{dr} = -\frac{1}{\rho}\frac{dP}{dr} - \frac{M_\odot G}{r^2} + F_{ext} \qquad (5.60)$$

- energy balance:

$$\frac{d}{dr}\left[\frac{v^2}{2} + \frac{5}{2}\frac{k_B T}{\mu} + \Phi_G + \frac{Q}{\rho v}\right] = F_{ext} + \frac{Q_{ext}}{\rho v}. \qquad (5.61)$$

The first term on the right-hand side of (5.61) stems from the work done by the force (which adds energy to the fluid at the rate $F_{ext}\rho v$ per unit volume); the second term corresponds to the energy added via heating.

Mass loss rate

Let us first consider the simple case of an atmosphere that is approximately isothermal below the sonic point. The outward force F_{ext} introduces an additional (positive) term in the right-hand side member of the flow equation (5.9), making it positive at the distance r_C defined in (5.8) instead of vanishing. This means that the critical distance where this member vanishes (and therefore $v = V_S$) is no longer given by (5.8) but lies closer to the Sun. Denoting by r_C this new critical distance, the mass loss rate M' may be calculated as:

$$M' = 4\pi r_C^2 V_S \rho(r_C). \tag{5.62}$$

To calculate $\rho(r_C)$, we substitute the (isothermal) perfect gas law into the momentum balance equation (5.60), to obtain

$$\frac{d}{dr}\left[\frac{v^2}{2} + V_S^2 \ln \rho - \frac{M_\odot G}{r}\right] = F_{ext} \quad \text{(isothermal)}.$$

We deduce $\rho(r_C)$ by integrating this equation between r_0 and r_C (where $v = V_S$) and rearranging, to obtain the mass loss rate

$$M' \simeq 4\pi r_C^2 V_S \rho_0 \exp\left[-\frac{M_\odot G}{r_0 V_S^2}\left(1 - \frac{r_0}{r_C}\right) + \int_{r_0}^{r_C} dr \frac{F_{ext}}{V_S^2} - 1/2\right]. \tag{5.63}$$

Hence the action of the *external force* F_{ext} below r_C increases the mass loss rate by the factor: $\exp\left(\int_{r_0}^{r_C} dr \, F_{ext}/V_S^2\right)$; the inward displacement of the critical point produces a further increase in M' since the exponential term $\exp(M_\odot G/r_C V_S^2)$ varies faster than the factor r_C^2. Therefore an outward force acting in the subsonic region increases M', as does an increase in temperature. This is easily understood, as the outward force counteracts gravity, and thus both raises the effective scale height and moves the sonic point inward, thereby increasing the density at the critical distance. These results remain qualitatively true if the atmosphere is not isothermal. In that case, the mass flux can also be increased by *heat addition* (Q_{ext}) in the subsonic region since this raises the temperature. Finally, therefore, adding energy in whatever form in the subsonic region increases the mass loss rate.

On the other hand, since the mass flux M' is determined by the structure of the flow below the critical point, adding energy (in whatever form) in the supersonic region does not change M'. This is easily understood from basic considerations. Since the loss rate is already determined at the critical point, it cannot be changed by what happens farther out, since this information cannot travel upstream.

Terminal speed

Now let us consider the speed at large distances, where the gravitational energy and enthalpy are both negligible. The energy balance (5.61) yields (neglecting

Letting the temperature vary 249

the start-off speed):
$$\frac{v_\infty^2}{2} \simeq \frac{5}{2}\frac{k_B T_0}{\mu} + \Phi_{G0} + \left[\frac{Q}{\rho v}\right]_0 - \left[\frac{Q}{\rho v}\right]_\infty + \int_{r_0}^{r_\infty} dr \left[F_{ext} + \frac{Q_{ext}}{\rho v}\right]. \quad (5.64)$$

At first sight it might seem that, because of the right-hand side term, addition of energy in whatever form always increases the terminal speed. However we have seen that if this energy is furnished below the critical point, the mass loss rate increases; one sees on (5.64) that this increase in ρv may balance or even overbalance the speed increase produced by energy deposition. This can be understood on simple grounds: if there is more energy available but also more particles, then there is not necessarily more energy per particle. On the other hand, we have seen that depositing energy above the critical point barely changes the mass loss rate, so that in that case the terminal speed always increases.

Conclusion on energy addition

We conclude that in order to increase the terminal speed, one should add energy to the flow (by pushing or heating) in the supersonic region. Depositing energy in the subsonic region only increases the mass loss rate (and might even decrease the terminal speed if energy is furnished via a force). Basic discussions of these points may be found in [21] and [27].

Note, finally, that the external force introduces an additional term F_{ext}/V_S^2 into the equation of motion (5.40), just as the deviation of the expansion from spherical introduces a term $f^{-1}df/dr$. One may thus expect similar effects to arise in both cases. Some forms of the function $F_{ext}(r)$ may indeed produce multiple critical points, and thus – as do some shapes of the flow tubes – open the possibility of multiple wind solutions and bi-stability, if either the zeros are close to each other or if shock transitions are possible (see for example [15]).

Before returning to the energy problem, let us consider another point.

5.3.5 What about viscosity?

We have ignored the viscosity of the medium, as do virtually all modern solar wind models – a question that has raised many disputes (see [44] and references therein). A rough justification is that the ratio of the diffusion coefficients associated respectively to viscosity and to heat transport[13] (Section 2.3) is significantly smaller than unity, so that thermal conductivity acts much faster than does viscosity.

Basically, this is because viscosity is associated with momentum, which is transported by protons, whereas heat is transported by electrons – which move much faster. The kinematic viscosity is of order of magnitude $\nu \sim v_{thp} l_f$, where v_{thp} is the proton thermal speed and l_f the mean free path. On the other hand, the thermal diffusivity is of order of magnitude $\chi \sim v_{the} l_f$, where $v_{the} \gg v_{thp}$ is the electron thermal speed, while the free path has the same order of magnitude as the one of protons. Hence $\chi \gg \nu$.

---
[13] The so-called *Prandtl number*.

This rough argument, however, must be used with caution for two reasons. First, viscosity yields an additional term in the equation of motion, as

$$v\frac{dv}{dr} = -\frac{1}{\rho}\frac{dP}{dr} - \frac{M_\odot G}{r^2} + \frac{1}{\rho r^3}\frac{d}{dr}\left[r^3 \times \frac{4}{3}\nu\rho r \frac{d}{dr}\left(\frac{v}{r}\right)\right]. \qquad (5.65)$$

The term v/r arises because of lateral momentum transfer in the radially diverging flow. We see that viscosity introduces a second order derivative in the flow equation, which removes the singularity at the critical point. Let us evaluate its importance. Making the approximations $d/dr \sim 1/L$, L being a characteristic scale, $v_{thp} \sim V_S$, and $P \sim \rho V_S^2$, we find that the ratio of the viscous force to the pressure force is of order of magnitude

$$\frac{v}{V_S} \times \frac{l_f}{L} \qquad (5.66)$$

which is a priori not negligible since we shall see that the mean free path l_f is of the order of magnitude of the characteristic scale L in the solar wind.

Second, since viscosity acts through second derivatives in the equation of motion, it may have important effects even when it is small on large scales, because velocity gradients may produce small scales at which the second derivative becomes large. This is basically why water produces on moving boats a drag force that does not depend on viscosity, whereas this force vanishes in the absence of viscosity; in other words, the limit $\nu \to 0$ may not coincide with the case $\nu = 0$. We will find another illustration of this point in Section 6.4 when studying turbulence.

Nevertheless, viscosity is expected to make a relatively small contribution to the energy balance, due to the small Prandtl number. Finally, a sobering remark is that, since the solar wind is not collisional enough for the usual (collisional) transport coefficients to be valid (Section 2.3.2), the viscosity term in (5.65) is extremely dubious. One should use a kinetic description instead of a fluid one, making the viscosity question irrelevant. We shall return to this point in Section 5.5.

5.4 A mixture of fluids

The simple fluid picture considered above has a major drawback. Since the medium (1) is made of electrons and protons (plus heavier ions in lesser concentration), and (2) is weakly collisional, it should not be pictured as a single fluid but as (at least) two fluids. This question was first addressed long ago [16], and early reviews may be found in [3] and [25].

To make things simple, we neglect the ions heavier than protons because their concentration is too small for them to affect the overall dynamics much. We therefore consider a plasma made of electrons and protons pictured as two different fluids, rather than a single fluid made of 'average' particles. How does this change the physics? Or rather does this bring about qualitatively new results?

A mixture of fluids 251

Electrons and protons have opposite charges, but their masses differ by the factor $m_p/m_e \simeq 1837$, so that electrons have a thermal speed greater than protons by a factor of order of magnitude $(m_p/m_e)^{1/2} \simeq 43$ (because their temperatures have generally the same order of magnitude). This has several consequences:

- whereas protons are strongly bound close to the Sun, electrons barely feel gravity; indeed, at a temperature of 10^6 K, their thermal speed $\simeq 5.5 \times 10^6 \, \mathrm{m\,s^{-1}}$ is nearly 10 times greater than the escape speed;

- the greater thermal speed of electrons is expected to make them carry heat much faster than do protons;

- collisions between electrons and protons exchange energy at a rate $\sim (m_p/m_e)$ slower than the rate of momentum exchange, which is itself slow since the medium is weakly collisional; hence electrons and protons may have different temperatures;

- since electrons and protons are subjected to very different forces (and may have different temperatures), an electric field sets up to preserve electric quasi-neutrality.

Since electrons and protons have opposite charges, electric quasi-neutrality requires them to have roughly the same number density n. Furthermore, since the radial electric current must vanish otherwise electric charge would accumulate indefinitely on the Sun, electrons and protons should have also the same radial bulk speed. The simplest generalisation of the one-fluid picture is therefore to consider two fluids having the same bulk velocity but different particle masses, temperatures and heat fluxes.

5.4.1 Simple balance equations

We have seen in Section 4.6 that in a static isothermal atmosphere with equal proton and electron pressures, the gravitational attraction – acting essentially on protons – tends to displace them inwards with respect to electrons. The corresponding space charge induces a radial electric field E directed outwards, which adjusts itself so that the total attraction on a proton $m_p M_\odot G/r^2 - eE$ is equal to the attraction on an electron eE, whence $E = m_p M_\odot G/2er^2$. We shall see later that when the plasma is moving and the proton and electron pressures are not equal, the electric field has a somewhat different value.

Let $\Phi_G = -M_\odot G/r$ and Φ_E be respectively the gravitational and electrostatic potential, set equal to zero at infinite distance. As a fluid, protons are subjected to the Sun's gravitational force $-m_p d\Phi_G/dr$ and to the (outward) electric force $-ed\Phi_E/dr$ (per particle), in addition to the pressure force. On the other hand, because of the small mass of electrons, both inertia and gravity are negligible for them, so that they are only subjected to the (inward) electric force $ed\Phi_E/dr$ (per particle) in addition to the pressure force.

The collisions do not produce any mutual force on the proton and electron fluids because they have the same velocity **v**, so that there is no direct momentum transfer between them. We deduce the fluid momentum equations in the radial direction (in the absence of viscosity):

$$m_p v \frac{dv}{dr} + \frac{1}{n}\frac{d}{dr}(nk_B T_p) = -\frac{d}{dr}(e\Phi_E + m_p \Phi_G) \quad \text{(protons)} \quad (5.67)$$

$$\frac{1}{n}\frac{d}{dr}(nk_B T_e) = e\frac{d\Phi_E}{dr} \quad \text{(electrons)}. \quad (5.68)$$

Let us neglect the (small) collisional transfer of energy between protons and electrons; we shall estimate its role later. The energy equations are then a trivial generalisation of the one-fluid equation (5.27) applied to each species, where we add the electrostatic energy and neglect the small electron mass (in the absence of exterior heating):[14]

$$m_p \frac{v^2}{2} + \frac{5}{2}k_B T_p + m_p \Phi_G + e\Phi_E + \frac{Q_p}{nv} = \text{constant} \quad \text{(protons)} \quad (5.69)$$

$$\frac{5}{2}k_B T_e - e\Phi_E + \frac{Q_e}{nv} = \text{constant} \quad \text{(electrons)}. \quad (5.70)$$

where the indices p and e refer respectively to protons and electrons.

Since the proton heat flux is expected to be much smaller than the electron one, let us neglect it. Subtracting the derivative of the energy equations from the momentum equations, we then find the equivalent form of the energy equation, for respectively protons and electrons

$$\frac{d}{dr}\left(nT_p^{-3/2}\right) = 0 \quad (5.71)$$

$$\frac{k_B T_e^{5/2}}{n}\frac{d}{dr}\left(nT_e^{-3/2}\right) = \frac{d}{dr}\left(\frac{Q_e}{nv}\right), \quad (5.72)$$

a generalisation of the one-fluid equation (5.28) applied to each species.

The proton energy equation expresses that $T_p \propto n^{2/3}$, i.e. the proton fluid is adiabatic (because we have neglected the proton heat flux and the collisional energy transfer between protons and electrons). In this case, at large distances where the flow speed is constant so that conservation of particles implies $n \propto r^{-2}$, the proton temperature falls off as

$$T_p \propto r^{-4/3} \quad ; \quad n \propto r^{-2} \quad \text{(adiabatic protons)}. \quad (5.73)$$

In contrast, if the electron heat flux falls off more rapidly than the particle flux nv, the electron temperature falls off less rapidly than in the adiabatic case.

---

[14] One may verify that when the proton and electron equations are added together, the electric field contribution cancels, and one recovers the one-fluid momentum and energy equations (5.7) and (5.27), with the total mass density $\rho \simeq nm_p$, pressure $P = nk_B(T_e + T_p)$, and heat flux $Q = Q_e + Q_p$; the single fluid has the average particle mass $\mu \simeq m_p/2$ and the average temperature $T = (T_e + T_p)/2$, so that $P = \rho k_B T/\mu$.

A mixture of fluids 253

For example, if the electron heat flux is of the polytrope form

$$Q_e = \alpha_e n v k_B T_e \equiv Q_{epoly} \tag{5.74}$$

then, as we found previously for one fluid, (5.72) implies the polytrope law $P \propto n^{\gamma_e}$ for electrons with

$$\gamma_e = \frac{5 + 2\alpha_e}{3 + 2\alpha_e}. \tag{5.75}$$

In this case, at large distances where the speed is constant so that $n \propto r^{-2}$, the electron temperature falls off as

$$T_e \propto r^{-2(\gamma_e - 1)}. \tag{5.76}$$

5.4.2 Observed proton and electron temperatures

How do the temperatures actually vary in the solar wind? This is a difficult question because the temperature is one of the most difficult to measure quantities in space plasmas. The main reason is the already mentioned absence of local thermodynamic equilibrium. Not only do different species have different temperatures, but these temperatures are difficult to define because the velocity distributions are not close to Maxwellians. Indeed, not only are the pressures not exactly isotropic – but in fact tensorial – and not only is there an excess of fast particles over the Maxwellian amount, but each particle species is made up of several populations having different properties, as we shall see below. Worse still, the notion of thermodynamic equilibrium is so deeply ingrained in the mind that it is implicit in most measuring schemes, and different measurements of the temperature often yield results that depend on the measurement method (see Problem 5.7.6).

Table 5.1 shows the number density and mean temperatures of protons and electrons in the low corona (see for example [9] and references therein) and at 1 AU from the Sun, together with the polytrope indices measured at that distance, for the fast wind, as deduced from a number of studies[15] (see [14], [50], [43], the reviews [46], [34], and that by M. Neugebauer cited in Chapter 1). The slow wind is more difficult to define because it is very variable in both space and time, and its observation generally deals with a mixture of different states; on the whole, it is denser by roughly a factor of three than the fast wind, its electrons are roughly twice as hot in the corona, and about 50% hotter at 1 AU, with a polytrope index somewhat greater, whereas the protons are cooler by a factor of six at 1 AU and closer to adiabatic, compared to the fast wind values in Table 5.1. Owing to the above-mentioned limitations, the temperatures are

---

[15] To be consistent with the fluid description, which considers all particles of a given species as a single entity, we only give the *mean* temperature measured for each species, even though most measurements concern the temperatures of parts of the velocity distributions; we shall return to this point later. The polytrope index γ is generally obtained by measuring the variation of the density n and temperature T as a function of a parameter; for example if $n \propto r^{-2}$ and $T \propto r^{-\beta}$, then $nT \propto r^{-(\beta+2)} \propto n^{1+\beta/2}$ so that $\gamma = 1 + \beta/2$.

Table 5.1 *Density and temperatures of protons and electrons measured in the low corona and the fast wind at 1 AU from the Sun, with the corresponding polytrope indices*

Fast wind	n_0 (corona)	T_0 (corona)	n (1 AU)	T (1 AU)	γ (1 AU)
Protons	10^{14} m^{-3}	2×10^6 K	3×10^6 m^{-3}	3×10^5 K	$1.5 \leftrightarrow 1.7$
Electrons	10^{14} m^{-3}	10^6 K	3×10^6 m^{-3}	10^5 K	1.2

known hardly better than to within a factor of two (see the careful discussion of errors in [43]); furthermore, some observations (but not all) suggest that the slope of the electron temperature profile – and therefore the polytrope index – declines with distance, in agreement with kinetic calculations [37].

One sees in Table 5.1 that the protons have a polytrope index slightly smaller than the adiabatic value of 5/3, while the one of electrons is much closer to unity. This raises these basic questions:

- what is the importance of collisions and do they significantly affect the temperatures?

- what produces the heat flux responsible for the observed strong departure of electrons from the adiabatic law?

Let us first examine the role of collisions. Their primary role is to tend to make the velocity distributions relax to Maxwellians. If they were frequent enough, they would also tend to equalise the proton and electron temperatures. Furthermore, their rate determines whether the heat flux is given by the usual (collisional) Fourier law, as already mentioned.

5.4.3 The role of collisions

What is the importance of the exchange of energy by collisions between protons and electrons? Collisional exchange of energy acts at the rate $\nu_E = (2m_e/m_p)\,\nu_e$ where ν_e is the collision frequency estimated in Section 2.1; this adds an energy exchange term in the energy equations (5.71)–(5.72), which become

$$\frac{1}{nT_p^{-3/2}} \frac{d}{dr}\left(nT_p^{-3/2}\right) = \frac{3}{2}\frac{\nu_E}{v}\frac{T_p - T_e}{T_p} \tag{5.77}$$

$$\frac{1}{nT_e^{-3/2}} \frac{d}{dr}\left(nT_e^{-3/2}\right) = \frac{3}{2}\frac{\nu_E}{v}\frac{T_e - T_p}{T_e} + \frac{1}{nvr^2 k_B T_e}\frac{d}{dr}\left(r^2 Q_e\right). \tag{5.78}$$

This shows that the distance over which collisions significantly affect the temperatures is of the order of magnitude of $(v/\nu_E)(T/\Delta T)$ where T is the temperature of the species considered, and ΔT is their difference in temperature. With $\Delta T \sim T$, this distance is of the order of magnitude of $(m_p/m_e)(v/v_{the})\,l_f$, where $l_f \sim v_{the}/\nu_e$ is the mean free path defined in Section 2.1 and $v_{the} =$

A mixture of fluids 255

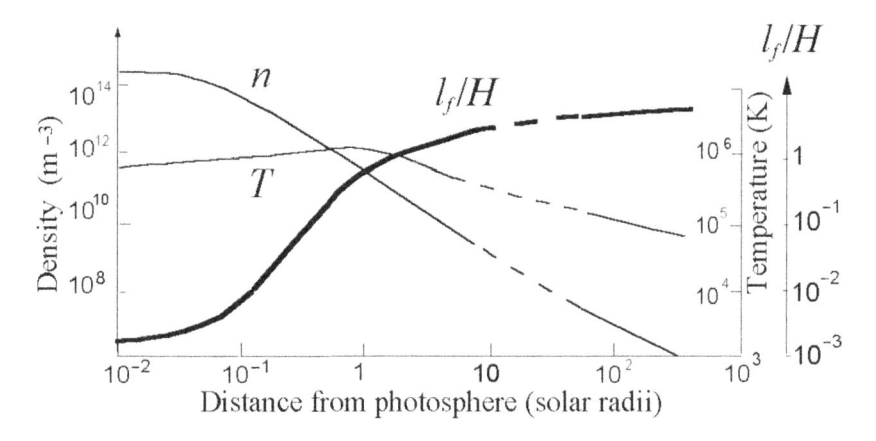

Figure 5.6 Electron density, temperature and mean free path (normalised to the scale height), as a function of distance from the solar surface, in the fast wind (the region where no measurements are available is shown by dashed lines).

$\sqrt{2k_B T_e/m_e}$. Since the mean free path at distance r is of the order of magnitude of r in most of the solar wind (Fig. 5.6), and the terminal speed is not very much smaller than v_{the}, we conclude that collisional exchange is not expected to affect the energy balance very much, except at small distances.[16]

Now, let us examine whether collisions are sufficiently frequent for the heat flux to be given by the usual collisional value. This question is addressed through the *Knudsen number*, that is the ratio of the particle free path to the scale height of variation in temperature. Since, however, the temperature gradient is even less well known than the temperature itself, we shall rather compare the mean collisional free path of particles with the density scale height

$$H = \left| \left[d\left(\ln n \right) / dr \right]^{-1} \right|$$

which is known reasonably well. Figure 5.6 shows the ratio of the electron mean free path l_f to H as a function of distance, estimated with the parameters of Fig. 4.1 close to the Sun, and of Table 5.1 at large distances where $n \propto r^{-2}$ so that the scale height is $H = r/2$. The slow wind is more collisional, but weakly so.

One sees that l_f/H is greater than 10^{-3} in the corona and is of the order of unity or greater in the solar wind. As we already noted, the medium is not sufficiently collisional in this case for the classical heat conduction formula to be justified (cf. [49]). Indeed, with $l_f \propto v^4$ and $l_f/H > 10^{-3}$, the electrons of speed $v > 10^{3/4} v_{the}$ have $l_f/H > 1$. There are grave doubts on the applicability of the usual collisional heat flux in the corona, and it should certainly not apply in the solar wind, where $l_f/H > 1$.

---

[16] However, plasma waves and instabilities (Section 6.4) might affect the energy balance.

5.4.4 Heat flux

What, then, determines the electron heat flux?

Polytrope electron heat flux

Let us estimate the value of the electron heat flux required to explain the large observed deviation of electrons from the adiabatic law. The polytrope index $\gamma_e \simeq 1.2$ observed near 1 AU corresponds to a heat flux given by (5.74), with, from (5.75)

$$\alpha_e = \frac{\gamma_e}{\gamma_e - 1} - \frac{5}{2}. \tag{5.79}$$

With $\gamma_e \simeq 1.2$ (from Table 5.1) this yields $\alpha_e \simeq 3.5$, with a large uncertainty since γ_e is badly known and changing it by merely 10% would change α_e by a factor of two.

The polytrope electron heat flux (5.74) is therefore

$$Q_{epoly} \equiv \alpha_e n v k_B T_e \simeq \alpha_e \times 0.3 \times 10^{-5} \text{ W m}^{-2} \tag{5.80}$$

where we have substituted the parameters of the fast wind at 1 AU given in Table 5.1 and $v = 750\,\text{km s}^{-1}$. With the above value of α_e, this amounts to about 10^{-5} W m^{-2}; if v and α_e are roughly constant, the polytrope heat flux Q_{epoly} varies as $nT_e \propto r^{-2\gamma_e} = r^{-2.4}$.

Note that if this empirical heat flux is interpreted in terms of degrees of freedom as given in (5.37), the solar wind electrons have $N \sim 10$ degrees of freedom.

Collisional electron heat flux

Since many solar wind fluid models use the collisional heat flux, whose full expression is (Section 2.3)

$$Q_c = -\frac{3}{2} n k_B v_{the} l_f \frac{dT_e}{dr}, \tag{5.81}$$

despite the strong doubts on its applicability, it is interesting to compare Q_c and Q_{epoly}. With a temperature varying as $T_e \propto r^{-\beta}$, so that $dT_e/dr = -\beta T_e/r$, we have

$$Q_c/Q_{epoly} = (3\beta/2\alpha_e) \times (l_f/r) \times (v_{the}/v). \tag{5.82}$$

In the fast solar wind, with the parameters from Fig. 5.6 and Table 5.1, this ratio is close to unity near 1 AU, and varies weakly with distance. In the more variable slow wind, this ratio is somewhat greater but not much so. We conclude that it is not easy to distinguish between the two expressions on empirical grounds; this numerical coincidence (and the observational problems indicated below) may explain the conflicting views that pervade the literature on the subject.

Observed electron heat flux

What does observation tell us about the heat flux? The electron heat flux is obtained by calculating the moment of order three of the electron velocity distribution – i.e. the total energy flux – measured in its rest frame. This requires *in situ* measurements, which up to now have only been performed farther than about 0.3 AU from the Sun. Since the heat flux is a vector (aligned with the local magnetic field and varying relatively rapidly, as do the fast electrons carrying it), its precise measurement requires a three-dimensional detector having a good resolution – a luxurious item of equipment that is often not available on space probes. As a result, heat flux measurements are in general extremely inaccurate.

The electron heat flux is observed to lie in the range:

$Q_{eobs} \sim (0.5 \leftrightarrow 1) \times 10^{-5}$ W m^{-2} ; observed electron heat flux at 1 AU

with a variation with distance $Q_{eobs} \propto r^{-\delta}$ with $\delta \sim 1.5 \leftrightarrow 3$, depending on the time, distance, wind properties, etc. and measurement method [43], [47].

This is close to the polytrope heat flux estimated above from the observed variation in electron temperature in the regular fast wind. This might indicate that the electron energy balance (5.72) is approximately correct in the solar wind in the range covered by these observations, without introducing any unidentified heating term. Given the large uncertainty on the measured value of γ_e and on its variation with distance, however, this does not prove that the heat flux has the polytrope form (5.74) with a constant value of α_e.

On the contrary, a flow having such a polytrope index all the way out from the base of the corona cannot produce the wind observed if it is left to its own devices. Indeed, we have seen in Section 5.3.2 that the value of γ that a single polytrope should have in order to produce the observed speeds is much closer to unity than the value observed in the solar wind for electrons near 1 AU (Table 5.1). This result remains qualitatively true with two polytrope fluids – a question that is addressed explicitly in Problem 5.7.4. This may also be seen directly from the value of the total normalised heat flux $Q/\rho v$ (5.30) required in the low corona to produce the observed wind speed; dividing by $k_B T_e$ with the observed coronal electron temperature (Table 5.1) and substituting $\rho \simeq nm_p$, we find that the value of $Q/(nvk_B T_e)$ required at the base of the corona to produce the fast wind is roughly $\alpha_{e0} \simeq 50$, i.e. greater by one order of magnitude than the value observed near 1 AU.

5.4.5 The electric field

With protons and electrons behaving as two different (charged) fluids, the electric field plays a major role.

Estimates of the electrostatic field and potential

The electric field serves to balance the electron pressure force (cf. (5.68)), and it is directly related to the electron heat flux since applying the electron energy balance (5.70) between r and infinite distance yields

$$e\Phi_E(r) = \left[\frac{Q_e}{nv} + \frac{5}{2}k_BT_e\right]_r - \left[\frac{Q_e}{nv} + \frac{5}{2}k_BT_e\right]_\infty. \qquad (5.83)$$

Neglecting in (5.83) the value of the bracket at infinite distance, we find that the electrostatic potential at r is roughly determined by the sum of the enthalpy and of the heat flux per particle:

$$e\Phi_E \simeq \frac{Q_e}{nv} + \frac{5}{2}k_BT_e \quad \text{at any distance } r. \qquad (5.84)$$

Applying the proton energy equation (5.69) between the base r_0 (where the speed is negligible) and infinite distance (where the temperature is negligible), and neglecting the proton heat flux (which is small compared to the electron heat flux and therefore compared to the contribution of the electric potential), the terminal speed is approximately given by

$$\frac{v_\infty^2}{2} \simeq \frac{5}{2}\frac{k_BT_{p0}}{m_p} + \Phi_{G0} + \frac{e}{m_p}\Phi_{E0}. \qquad (5.85)$$

Estimating the electric potential at r_0 from (5.84), where the heat flux dominates the enthalpy, we see that the terminal wind speed may be thought of as being produced by either the electric potential or the electron heat flux.

What is the electric potential required at the coronal base for producing a terminal speed of 400↔800 km s^{-1}, with $T_{p0} \simeq 2\times 10^6$ K, and $r_0 \simeq R_\odot$? Putting these numbers in (5.85), we find $\Phi_{E0} \simeq (2 \leftrightarrow 5) \times 10^3$ V. Note that this value is greater than that of a static atmosphere, which we have seen to correspond to an electric potential energy equal to half the modulus of the gravitational proton energy, i.e. $\Phi_{E0} = m_pM_\odot G/2r_0e \simeq 10^3$ V at r_0. If the base of the corona is viewed as a surface of radius $R_\odot$, it can be thought of as carrying a positive electric charge $Q_0 \simeq 4\pi\epsilon_0 R_\odot \Phi_{E0}$, which amounts to a few hundred coulombs. Given the huge size of the object, this charge is extremely small. A spacecraft (of radius smaller by nine orders of magnitude) or a dust grain (of radius smaller by 15 orders of magnitude) in the Earth's magnetosphere can be at a similar electrostatic potential (though of opposite sign), and the electrostatic field in their vicinity is greater by the same huge factors.

How does the electrostatic potential vary with distance? In the case of polytrope electrons, the heat flux is given by (5.74)–(5.75), so that (5.84) yields

$$e\Phi_{Epoly} = \frac{\gamma_e}{\gamma_e - 1}k_BT_e. \qquad (5.86)$$

In that case the electrostatic potential varies proportionally to the electron temperature, and the electrostatic field amplitude satisfies, in order of magnitude $eE \sim e\Phi/r \sim k_BT_e/r$. It is interesting to verify that this electric field is compatible with the plasma quasi-neutrality. Let Δn be the difference between the proton and electron number densities. Writing Poisson equation $\nabla.\mathbf{E} = \rho/\epsilon_0$ and substituting the charge density $\rho = \Delta n \times e$ yields in order of magnitude $\Delta n \sim \epsilon_0 k_BT_e/(er)^2$. It is easy to see that this amounts to an extremely minute fraction of n, so that the plasma remains quasi-neutral.

A mixture of fluids 259

Particle viewpoint

Even though it is extremely small from the point of view of electric quasi-neutrality, the electric field is huge from the point of view of the particles. This can be seen from a simple order of magnitude estimate. Since the electron temperature varies much more slowly than the density in both the corona and the wind, we may estimate the electric field from the electron momentum equation (5.68) by putting the temperature outside the derivative to obtain

$$eE \sim -\frac{k_B T_e}{n}\frac{dn}{dr} \sim k_B T_e/H \qquad (5.87)$$

where H is the density scale height. A particle moving radially is accelerated by the field, whereas collisions tend to brake it on a scale length equal to its free path. The energy gained by a particle across one mean free path is

$$eE \times l_f \sim k_B T_e \times l_f/H. \qquad (5.88)$$

Since l_f/H is of order unity in the wind (including its acceleration region), the energy gained by a particle across one mean free path is of the order of the mean particle energy. In other words, the electric field is of the order of magnitude of the so-called *Dreicer electric field* [8], which accelerates particles moving faster than thermally more rapidly than they are braked by collisions. Since the free path of a particle of speed v increases as v^4, fast particles have a huge free path and are therefore negligibly braked by collisions, so that the electric field can accelerate them to very high speeds.

The great importance of the electric field for the particles may be seen another way. The time for the electrostatic force to change significantly the mean electron speed is equal to their thermal speed $v_{the} = \sqrt{2k_B T_e/m_e}$ divided by their acceleration eE/m_e, i.e.

$$\tau_{dyn} \sim \frac{m_e v_{the}}{eE} \quad ; \quad \text{electron dynamic time.}$$

This can be compared to the collision timescale:

$$\tau_c \sim l_f/v_{the} \quad ; \quad \text{electron collision time.}$$

Equation (5.88) with $l_f/H \sim 1$ expresses the fact that the two timescales are roughly equal, which means that while the coherent electric field and collisions are of similar importance in the dynamics of thermal electrons, the field hugely dominates for faster-moving particles.

This shows another difficulty of the fluid approximations, which ignore the dynamics of single particles, in addition to being unable to determine the heat flux in this weakly collisional medium (see [48] and references therein). We shall return to this point in Section 5.5.

Total force on individual protons

To investigate further the particle dynamics, it is interesting to examine how the force on individual protons varies with distance from the Sun. Individual

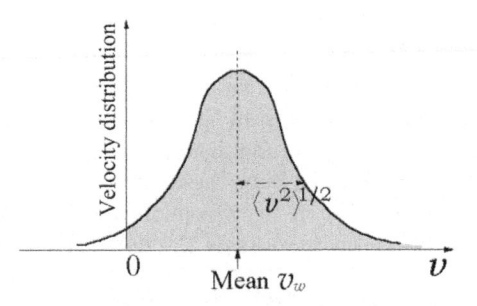

Figure 5.7 One-dimensional sketch showing that, for most of the particles of a typical distribution to be moving in the same direction (shaded in grey), the mean must be greater than the width.

protons are subjected to the total force $m_p M_\odot G/r^2 - eE$. Close to the Sun, where they are strongly bound, the gravitational attraction dominates. However, we see from either (5.86) or (5.87) that since the temperature decreases more slowly than r^{-1} in a transonic wind, the electric field E decreases more slowly than r^{-2}, i.e. than the gravitational force. Hence, there is some distance where both forces balance, so that the total potential energy of a proton $\Phi_E - m_p M_\odot G/r$ has a maximum. Farther out, the slowly decreasing outward electric force on protons dominates, accelerating them outwards. Hence, the electric field furnishes an interesting interpretation to the wind acceleration, in terms of individual particles: at large distances, protons – which carry the mass of the wind – are propelled outwards by the electric field, which dominates gravity.

At what distance r_M is this maximum of energy located, compared to the critical distance r_C where the wind becomes supersonic? Above r_M, the decreasing potential energy means that the total force on a proton is directed outwards; since collisions are not dominant, most of the protons are therefore escaping from the Sun there, i.e. have a velocity directed outwards as seen from the Sun. Now, consider a velocity distribution whose mean velocity is v_w, directed outwards. If $v_w = 0$, only half the protons are moving outwards. For most of them to be moving outwards, their mean bulk velocity v_w must be greater than the width of the distribution, or the mean square speed $\langle v^2 \rangle^{1/2}$ of protons, which is of the order of magnitude of v_{thp} (Fig. 5.7). Hence $v_w > v_{thp}$ above the distance r_M, so that v_w is greater than the sound speed V_S there (because $V_S \sim v_{thp}$). This means that the distance r_M where the proton potential energy has a maximum is located in the supersonic region, i.e. *above* the critical distance where the wind becomes supersonic; this is proved explicitly in Problem 5.7.4 for a polytrope fluid (see [37]).[17] We shall see the consequences of this property in Section 5.5.

---

[17] Where it is also proved that in a transonic wind, the potential energy of protons has necessarily a maximum.

5.4.6 Fluid picture balance sheet and refinements

The two-fluid description improves the fluid picture by taking into account the huge difference in mass between protons and electrons, which makes them feel different forces, carry heat differently and have few mutual energy exchanges. Basically, ions carry momentum (because of their large mass), whereas electrons carry heat (because of their large individual speeds).

To preserve zero electric charge and flux, an electric field sets up, whose amplitude is greater than implicitly assumed in the fluid theories that close the infinite ladder of moment equations (Section 2.3) by using the usual (collisional) heat flux. Note that, contrary to a frequent misconception, the electric field does not make ions and electrons behave as if they were glued together; rather they distribute themselves so that there is no significant *average* charge accumulation.

Observations of the electron temperature and heat flux yield $Q_e/nvk_BT_e \sim 5 \leftrightarrow 10$ in the solar wind. This value is compatible with the observed variation of T_e with distance in the wind, without having to invoke unidentified heating processes. However, not only is this value unexplained in the frame of the fluid picture, but it is much smaller than the normalised heat flux Q_e/nvk_BT_e required in the corona to power the wind (in the absence of additional energy sources).

Put differently, electrons behave in the fluid picture as if they had $N \sim 2/(\gamma_e - 1) \sim 10$ degrees of freedom in the fast wind, but roughly 10 times more at the base of the wind. Why is this so?

There are two ways out of this dilemma. The easiest way is to circumvent the difficult problem of calculating the heat flux by assuming that the energy is instead mainly furnished by plasma waves; the fluid models (equipped with the usual collisional heat flux) are then rescued by postulating ad hoc heating and pushing functions supposed to mimic a local deposit of wave energy; the corresponding terms are added to the balance equations as indicated in Section 5.3.4. The alternative – and more difficult – way is to return to the basics and adopt a kinetic description that calculates properly the heat flux; we shall discuss such attempts in the next section.

The easiest refinement of the simple two-fluid model considered above is to allow the pressures to be anisotropic – with different temperatures in the directions parallel and perpendicular to the magnetic field – and to consider many particle species – pictured as separate fluids (see for example [23], [24]). Note that, just as the classical fluid models implicitly assume the particle velocity distributions to be nearly Maxwellians, these extensions assume them to be nearly bi-Maxwellians.[18]

Since it is relatively straightforward to build such multi-fluid constructions that reproduce the observations if the heating and pushing functions are suitably chosen and if enough arbitrary parameters are put into the recipe, there is a host of such models. These models have been adorned with more and more refinements as more computational power has become available, in order to

---

[18] I.e. Maxwellians having different temperatures in the directions parallel and perpendicular to the magnetic field.

rescue them in the face of contradicting and more accurate observations. In some sense, these ingenious schemes are reminiscent of the elaborate theory of planetary motions known as the Ptolemaic system, which was ingeniously adorned with epicycles that became more and more complicated to rescue it in the face of contradicting observations, until Kepler's brilliant construction pulled it to pieces.

These fluid models use the conservation of mass, momentum and energy – including sources terms – to calculate the density, mean velocity and temperature, with a (local) hypothesis on the heat flux. A further refinement is to consider higher-order moments (see for example [31]). The price to be paid is a great increase in complexity, and the benefits are not evident because these schemes do not alleviate the major problem encountered by fluid models: their doubtful validity when the particle distributions are not close to Maxwellians (or bi-Maxwellians), because in that case, only the infinite set of moment equations is correct. Ironically enough, just as adding more terms in a divergent asymptotic series may worsen the approximation, considering higher-order moments of the velocity distributions may not improve the theory because the higher the order of the moment, the greater the role of the fast particles which are the least collisional and are thus the most easily driven out of equilibrium.

5.5 Kinetic descriptions

We discuss in this section the other alternative: the kinetic description. Basically, a kinetic description deals with particles and their equations of motion. As we saw in Section 2.3, a convenient method, which is in some sense intermediate between the particle and the fluid concepts, deals with the velocity distributions of the particles; more exactly, one *first* calculates the evolution of the velocity distributions – a very difficult task, and *then* deduces their moments – which is straightforward. The difficulty of calculating the evolution of the velocity distribution is linked to the special character of collisions in a plasma and to the lack of local thermal equilibrium.

5.5.1 Some notations

When collisions are rare, the velocity distributions may be very different from Maxwellians. The number density n and mean velocity $\mathbf{v}_w$ of a particle species are given by the first two moments of the velocity distribution $f(\mathbf{v})$ as

$$n = \int d^3v \, f(\mathbf{v}) \quad ; \quad n\mathbf{v}_w = \int d^3v \, \mathbf{v} \, f(\mathbf{v}).$$

The mean velocity $\mathbf{v}_w$ coincides with the fluid velocity of the fluid picture.[19] The mean temperature represents the width of the distribution as

$$T = m \int d^3v \, |\mathbf{v} - \mathbf{v}_w|^2 \, f(\mathbf{v}) / 3nk_B. \tag{5.89}$$

---

[19] We noted its modulus v in the previous sections.

Kinetic descriptions

The heat flux vector quantifies the skewness of the distribution, through the moment of order three in the frame where the mean velocity is zero

$$\mathbf{Q} = \frac{m}{2} \int d^3v \, f(\mathbf{v}) \, (\mathbf{v} - \mathbf{v}_w) \, |\mathbf{v} - \mathbf{v}_w|^2. \tag{5.90}$$

The particle gyration generally ensures some symmetry around the magnetic field. In the simple case when the magnetic field is parallel to the mean velocity $\mathbf{v}_w$, the temperatures in the parallel and perpendicular directions are defined from the widths of the distribution in these directions as

$$nk_B T_\parallel = m \int d^3v \, v_\parallel^2 \, f(\mathbf{v}) - nmv_w^2 \; ; \quad nk_B T_\perp = \frac{m}{2} \int d^3v \, v_\perp^2 \, f(\mathbf{v})$$

where the factor $1/2$ takes into account that there are two perpendicular directions (but only one parallel direction). The mean temperature is

$$T = (T_\parallel + 2T_\perp)/3.$$

In this case, the mean velocity and heat flux amplitudes may be written

$$nv_w = \int d^3v \, v_\parallel f(\mathbf{v})$$

$$Q = \frac{m}{2} \int d^3v \, v_\parallel v^2 f(\mathbf{v}) - nv_w \left[3k_B T_\parallel / 2 + k_B T_\perp + mv_w^2/2 \right]. \tag{5.91}$$

In the expression of the heat flux Q, the first term is the total energy flux in the frame where $f(\mathbf{v})$ is defined, whereas the rest represents the flux of enthalpy and of bulk kinetic energy, in accord with energy balance (see (5.27)).

5.5.2 Observed proton and electron velocity distributions

Velocity distributions are at the basis of kinetic descriptions. They have been measured in detail over a large range of distances in the solar wind, farther than about 0.3 AU. Their main property is their strong deviation from local thermal equilibrium, especially in the regular fast wind.

Solar wind protons

For protons, the main deviation from a Maxwellian in the plasma frame is a lack of isotropy, with a beam and different temperatures in the directions parallel and perpendicular to the magnetic field. Figure 5.8 shows a typical proton velocity distribution measured in the fast wind at 1 AU from the Sun (see for example [13][20]). The non-thermal character appears to decrease with distance

---

[20] This is one among the large number of publications reproducing this distribution (see also [12]), measured aboard the venerable spacecraft IMP 7 on 22 March 1973. This distribution was modelled as a sum of two bi-Maxwellian distributions having different mean speeds and temperatures, namely a cold distribution moving at about 700 km s^{-1}, plus a hotter and faster beam of smaller density (we do not reproduce this modelisation because it is not unique). In contrast, in the direction perpendicular to the magnetic field, the distribution is roughly symmetrical, and the main component is generally hotter.

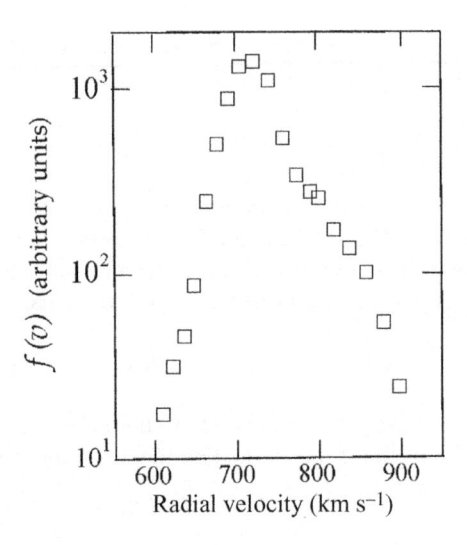

Figure 5.8 Typical proton velocity distribution measured in the fast solar wind at 1 AU from the Sun (projection on the radial direction). (Adapted from [13].)

to the Sun, and is much less conspicuous in the slow wind, which may be due to the smaller value of the particle mean free path (see [34]).

Solar wind electrons

Electron velocity distributions have different properties. First of all, they are never close to Maxwellians or bi-Maxwellians, even in the (more collisional) slow wind, since there is always a large excess of electrons moving faster than two to three times the thermal speed. At speeds smaller than the thermal speed, the distribution is generally close to a Maxwellian (or rather a bi-Maxwellian), but at greater speeds, the distribution decreases much more slowly with energy – decreasing as W^{-p} (with $p \sim 3-6$ at 1 AU) rather than as an exponential (see [13]).

In the high-speed wind, the distribution is still farther from equilibrium, as it carries also an excess of particles moving outwards roughly parallel to the magnetic field, faster than the mean of the distribution; the temperature of this component – called the *strahl* – is nearly 10^6 K (see [42] and references therein). This is illustrated in Fig. 5.9, which shows a typical electron velocity distribution measured in the fast wind,[21] which is close to a Maxwellian at low speeds, but has a power law shape at large speeds and a flux of escaping fast electrons [32].

As we already noted, this excess of high-speed particles is not surprising, especially in the fast wind where collisions are rare, since the faster the particles, the smaller their collisional cross-section, hence the slower they relax to a Maxwellian, and therefore the more easily they are driven out of equilibrium.

---

[21] Measured aboard the spacecraft WIND, at 1 AU from the Sun.

Kinetic descriptions

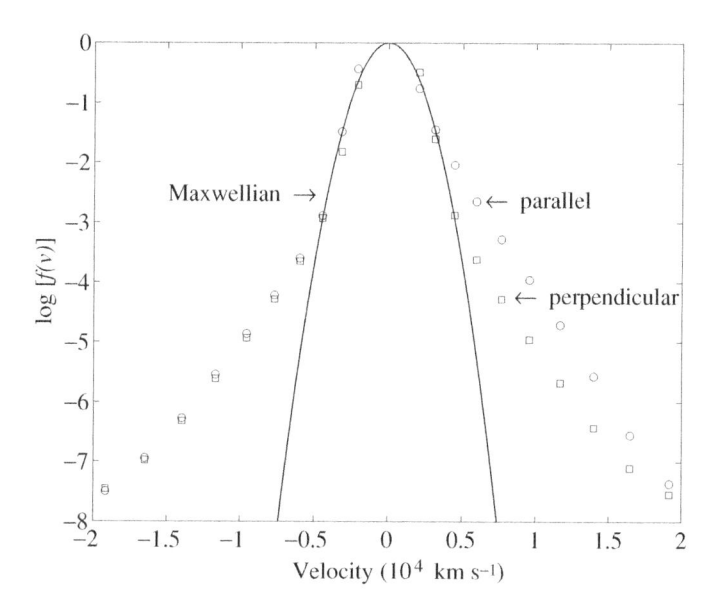

Figure 5.9 Typical electron velocity distribution measured in the fast solar wind at 1 AU from the Sun (projections parallel (circles) and perpendicular (squares) to the magnetic field), compared to a Maxwellian (continuous line). Along both directions, the distribution approaches a Maxwellian at low speeds but has more high-speed particles than a Maxwellian. Furthermore, there is a flux of fast electrons escaping along the magnetic field. (Data courtesy of I. Zouganelis.)

The kappa distribution

How to model a distribution that is nearly Maxwellian at low speeds, but decreases as a power law at high speeds? In the menagerie of mathematical functions, a convenient choice is the generalised Lorentzian, or kappa function:

$$f_\kappa(\mathbf{v}) \propto \left[1 + \frac{v^2}{\kappa v_{th}^2}\right]^{-(\kappa+1)} \tag{5.92}$$

with a normalisation factor ensuring that $\int d^3v \ f(\mathbf{v})$ equals the number density.[22]

Whereas the shape of the Maxwellian distribution is characterised by only one quantity: the temperature (the density acting as a normalisation factor), the kappa distribution has one more parameter, κ, which measures how it deviates

---

[22] The normalisation factor in (5.92) is the product of the number density by $A_\kappa \pi^{-3/2} v_{th}^{-3}$ with $A_\kappa = \kappa^{-3/2} \Gamma(\kappa+1) / \Gamma(\kappa - 1/2)$. The gamma function is defined in the *Handbook of Mathematical Functions*, ed. M. Abramowitz and I. A. Stegun (New York, Dover, 1970), p. 253. For κ integer, it may be easily calculated by using $\Gamma(\kappa+1) = \kappa!$, $\Gamma(1/2) = \sqrt{\pi}$, and $\Gamma(\kappa+1/2) = \Gamma(1/2) \times 1 \times 3 \times 5 \times \cdots \times (2\kappa - 1) / 2^\kappa$. When $\kappa \to \infty$, $A_\kappa \to 1$.

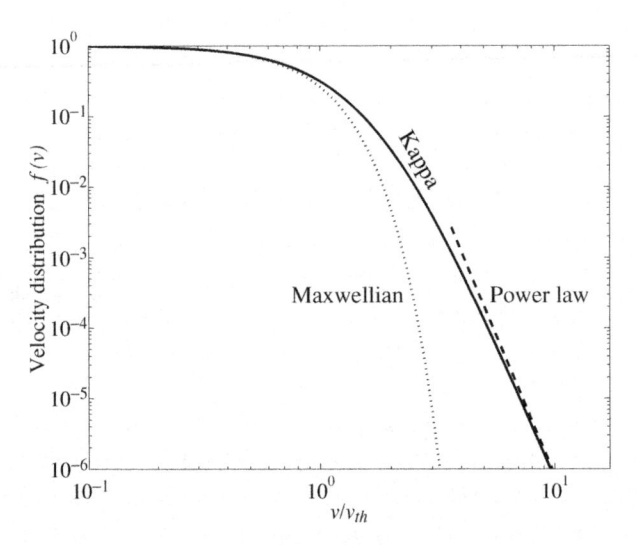

Figure 5.10 A kappa distribution (normalised so that $f(0) = 1$) with $\kappa = 3$, compared to its Maxwellian low-speed approximation: $\exp - \left[(\kappa + 1) v^2 / \kappa v_{th}^2 \right]$ and to its high-speed limit: $\left(v^2 / \kappa v_{th}^2 \right)^{-\kappa - 1}$.

from a Maxwellian; the larger the value of κ, the fewer suprathermal particles, and the closer the distribution to a Maxwellian; in the limit $\kappa \to \infty$

$$\left[1 + \frac{v^2}{\kappa v_{th}^2} \right]^{-(\kappa+1)} \to e^{-v^2/v_{th}^2} \quad ; \quad \kappa \to \infty$$

i.e. the kappa distribution tends to a Maxwellian of temperature $T = m v_{th}^2 / 2 k_B$. Furthermore:

- at speeds $v \ll \sqrt{\kappa} v_{th}$, f_κ has the same series expansion – to first order – as a Maxwellian of temperature $T = (\kappa/(\kappa+1)) \times (m v_{th}^2 / 2 k_B)$

- in the opposite limit $v \gg \sqrt{\kappa} v_{th}$, $f_\kappa \propto \left(v^2 / \kappa v_{th}^2 \right)^{-\kappa - 1}$.

Figure 5.10 shows a kappa distribution together with the Maxwellian and the power law approximating it at respectively low and high speeds.

The probability that the speed will lie between v and $v + dv$ is $f_\kappa(\mathbf{v}) \times 4\pi v^2 dv$, so that the most probable speed (the speed at which the derivative of $v^2 f_\kappa(\mathbf{v})$ vanishes) is v_{th}, whatever the value of κ – including the Maxwellian limit $\kappa \to \infty$. However, the kinetic temperature[23]

$$T \equiv \frac{m \langle v^2 \rangle}{3 k_B} = \frac{\kappa}{\kappa - 3/2} \frac{m v_{th}^2}{2 k_B} \tag{5.93}$$

is greater than the one in the Maxwellian limit, the more so as κ decreases – due to the increasing importance of the tail. Note that κ is constrained by the inequality $\kappa > 3/2$ for the temperature to remain finite. For any finite value of

---

[23] The angular brackets denote an average over the distribution, i.e. $\langle v^2 \rangle = \int d^3 v \, v^2 f(\mathbf{v}) / \int d^3 v \, f(\mathbf{v})$.

Kinetic descriptions 267

κ, there are only a limited number of finite moments, so that this distribution must be handled with care.

At speeds greater than the most probable speed v_{th}, the kappa distribution has more particles than a Maxwellian – an excess that increases with speed (Fig. 5.10[24]). It is not surprising that such a property is often observed of particles in space. Indeed, we have already noted that the faster the particles, the more easily they are driven out of equilibrium; furthermore, many acceleration processes produce a distribution of particles decreasing as a power law at high speeds. We have seen in Section 4.5 a simple way of generating a kappa distribution, and a number of more ingenuous methods have been devised (see [51], [6] and references therein). Problem 5.7.5 invites you to torture a kappa distribution, and Problem 5.7.6 examines different empirical determinations of the temperature for non-equilibrium distributions.

5.5.3 Non-collisional electron heat flux

We saw that the concept of temperature, which lies at the heart of classical macroscopic physics, must be applied with extreme caution in a weakly collisional plasma. The same holds for the heat flux.

What determines the heat flux in the absence of collisions? Since the heat flux is an odd moment of the velocity distribution, the question may be reformulated as: what determines the observed skewness of the electron distribution?

A hint may be found by studying the trajectories of individual electrons. In the absence of collisions, the particle energy is conserved. At distance r where the speed is v and the electrostatic potential is $\Phi_E(r)$, the total electron energy is

$$m_e v^2/2 - e\Phi_E(r) = \text{constant}$$

along the trajectory. Let the electrostatic potential be positive, decreasing monotonously with distance and vanishing at infinity, and define the speed

$$v_E = (2e\Phi_E/m_e)^{1/2}.$$

We deduce that at distance r:

- an electron of speed $v < v_E$ has not enough energy to go to infinity, and is therefore trapped by the Sun,

- an electron of speed $v > v_E$ moving outwards is able to reach infinity.

A little reflection tells us more. An electron trapped by the Sun may have a speed directed either inwards or outwards. An electron of speed $v > v_E$, however, is necessarily escaping from the Sun because otherwise, it would be coming *from infinity* – a possibility that we may ignore because the Sun is immersed in a very dilute medium.

---

[24] In Fig. 5.10, the kappa and the Maxwellian distributions are normalised so as to be both equal to unity at $v = 0$. When normalised so as to have the same number density, the kappa has *fewer* particles than the Maxwellian at low speeds and *more* at high speeds.

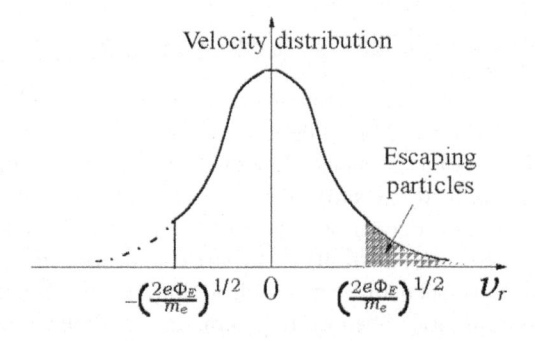

Figure 5.11 One-dimensional sketch of a velocity distribution in an attractive potential. Slow particles are trapped and can therefore be moving inwards or outwards, whereas because no particles come *from* infinity, fast particles are all escaping, and produce both the mean outward velocity and the heat flux.

Finally, therefore, the velocity distribution is made of the trapped electrons, of speed $v < v_E$, whose distribution is symmetrical with zero mean velocity and heat flux, and the escaping ones, of speed $v > v_E$, which are responsible for both the particle flux and the heat flux. Figure 5.11 sketches a cut of the distribution along the radial. The particle flux and the heat flux are respectively the moments of order one and three of the part of the velocity distribution shown in grey. These values are calculated in Problem 5.7.7 when the escaping electrons are assumed to consist of *all* the electrons having an outward velocity $v > v_E$,[25] and the distribution is either a Maxwellian or a kappa.[26]

In this kinetic collisionless picture, the wind speed and the heat flux per particle have a similar origin: the electrons escaping from the electrostatic potential. This is in sharp contrast with the usual (collisional) view of the heat flux in the fluid picture.

5.5.4 Exospheric models

Particle orbits and how to deal with them

We have seen that in the absence of collisions the bulk speed and the heat flux are produced by the escaping tail of the electron velocity distribution, which is determined by the electrostatic potential. However, contrary to a frequent misconception, the above calculation does not yield the true collisionless flux. This is because it includes all the electrons having an outward velocity of modulus $v > v_E$, i.e. all velocity directions in a whole half-space $0 < \theta < \pi/2$, where θ is the inclination of the velocity to the radial. Without collisions, however, escaping electrons cannot have any outward velocity direction because their

---
[25] An assumption that is not consistent with the absence of collisions, as we shall see in the next section.

[26] In the particular case when the electrostatic potential energy $e\Phi_E \gg k_B T_e$, this net flux is $Q_e \sim n v_w e \Phi_E$, as already found from energy balance (see (5.83)).

Kinetic descriptions

trajectories are constrained by the invariants of motion. Namely, both their energy $m_e v^2/2 - e\Phi_E$ and their magnetic moment $\propto v^2 \sin^2\theta/B$ (θ is the pitch angle)[27] are conserved along the trajectory from the base of the wind r_0 (Section 2.2). With a radial magnetic field satisfying from flux conservation $B \propto r^{-2}$, we thus have

$$m_e v^2/2 - e\Phi_E = m_e v_0^2/2 - e\Phi_{E0} \tag{5.94}$$

$$v^2 r^2 \sin^2\theta = v_0^2 r_0^2 \sin^2\theta_0. \tag{5.95}$$

The pitch angle θ of an electron at r is therefore determined by the value θ_0 at r_0, the speed v and the difference of electric potential between r and r_0. Since an escaping electron at r_0 may have any pitch angle between 0 and $\pi/2$, (5.95) shows that the pitch angle of escaping electrons at distance r must satisfy $\sin\theta < (v_0 r_0/vr)$ with v_0 given in (5.94). Hence $\theta < \theta_M$ with

$$\sin\theta_M = \frac{r_0}{r}\left[1 + \frac{2e(\Phi_{E0} - \Phi_E)}{m_e v^2}\right]^{1/2} \tag{5.96}$$

which is very small for fast electrons far from the Sun.

The particle and heat fluxes at distance r are thus produced by particles whose velocities lie in a small cone of half-angle θ_M, instead of the whole half-space of outward velocities that was considered in deriving (5.119).

Let the electron velocity distribution at r be $f(\mathbf{v})$; the escaping flux is therefore

$$nv_w = 2\pi \int_{(2e\Phi_E/m_e)^{1/2}}^{\infty} dv\,v^3 \int_0^{\theta_M} d\theta \sin\theta \cos\theta\, f(\mathbf{v}). \tag{5.97}$$

Conservation of energy and magnetic moment enables one to calculate the whole velocity distribution at r from Liouville's theorem (Section 2.3). The distribution $f(\mathbf{v})$ at distance r of particles coming from r_0 where the distribution is $f_0(\mathbf{v}_0)$ is given by $f_0(\mathbf{v}_0) = f(\mathbf{v})$ with the velocities related by the conservation equations (5.94)–(5.95). Substituting the above value of θ_M, making the change of variables $v_0 = \left[v^2 + 2e(\Phi_{E0} - \Phi_E)/m_e\right]^{1/2}$, (5.97) yields[28]

$$nv_w = \pi\left(\frac{r_0}{r}\right)^2 \int_{(2e\Phi_{E0}/m_e)^{1/2}}^{\infty} dv\,v^3 f_0(v). \tag{5.98}$$

Since the integral depends only on the values at r_0, this equation expresses flux conservation between r_0 and r through flux tubes whose cross-section increases as r^2.

What about the electrons that are not escaping? With no particles coming from infinity, they may be divided in two classes. Those coming from the base r_0 with a speed too small to reach infinity are reflected on a ballistic trajectory. The second category are those trapped between two reflection points: one due

---

[27] The angle between $\mathbf{v}$ and $\mathbf{B}$.
[28] For simplicity, we assume $f_0(\mathbf{v}_0)$ to be isotropic in the velocity range where it is not zero.

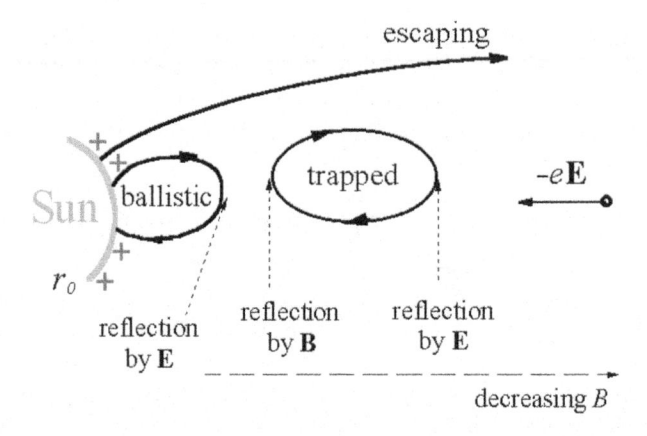

Figure 5.12 Sketch of electron orbits around the Sun. The outward electric field **E** is produced by the small electron to proton mass ratio, which tends to make electrons escape more easily than protons. Some electrons coming from the corona are reflected by **E** on a ballistic trajectory. Some are trapped by **E** and the magnetic mirror force. Those having a large enough speed are escaping.

to inward reflection by the electrostatic field, the other due to outward reflection by the magnetic mirror force (see Section 2.2) produced by the increase in B towards the Sun (Fig. 5.12). Since, without collisions, no electrons can be put on or leave such trapped orbits, this population is arbitrary. However, even a very small number of collisions can populate these orbits after a sufficiently long time; hence one may assume them to be in equilibrium with the ballistic electrons, so that the distribution at r is symmetrical for energies smaller than $e\Phi_E(r)$. These trapped particles represent the vast majority of electrons and are therefore essential to ensure electric quasi-neutrality [57].

How about protons? A similar scheme holds, except that the gravitational attraction is not negligible. We saw in Section 5.3 that their total potential energy has a maximum somewhere in the supersonic region, so that all the protons present at greater distances are escaping. Hence, in contrast to the electrons, reflected protons cannot be found at any distance, but only closer to the Sun than this maximum.

Following this scheme, one can calculate the particle velocity distributions everywhere from Liouville's theorem as a function of the electrostatic potential, deduce the number densities and fluxes, and finally deduce the potential itself at any distance by imposing equal densities and fluxes of electrons and protons; once the electrostatic potential is calculated, the wind properties are deduced.

Consequences of the electric potential

The total potential energy of an electron and of a proton are given respectively by

$$\psi_e = -e\Phi_E \tag{5.99}$$
$$\psi_p = e\Phi_E - m_p M_\odot G/r \tag{5.100}$$

Kinetic descriptions 271

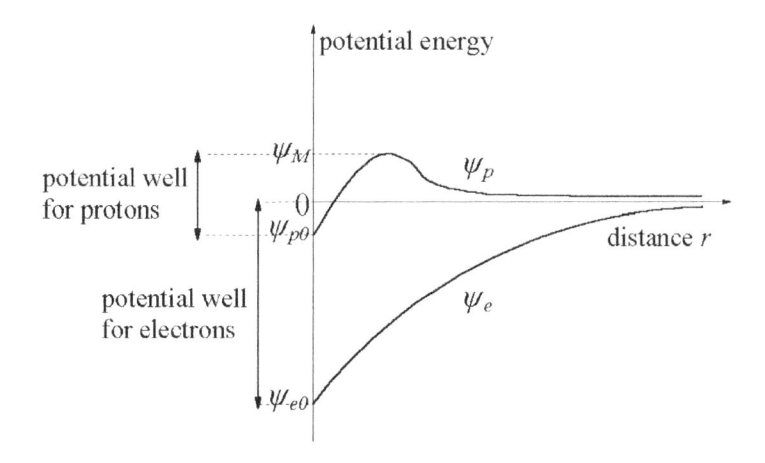

Figure 5.13 Sketch of the total potential energy of electrons and protons as a function of heliocentric distance.

and sketched in Fig. 5.13. One can easily understand how the electrostatic potential arises to prevent permanent electric charge accumulation. Let ψ_{e0} and ψ_{p0} be the values of ψ_e and ψ_p at the base r_0. There, the flux of escaping electrons is that of all outward-going electrons of energy greater than the depth of the potential well $|\psi_{e0}|$. With a Maxwellian distribution, this flux is given by (5.119) with $b = e\Phi_{E0}/k_B T_e$ (Problem 5.7.7).

The case of escaping protons is subtler, because their potential energy does not vary monotonously with distance but has a maximum if the wind is transonic, as we saw in Section 5.3.5.[29] This maximum of potential satisfies $\psi_M > (0, \psi_{p0})$, so that the escaping protons are those of energy greater than the well depth $\psi_M - \psi_{p0}$. With a Maxwellian distribution, this flux is given by (5.119) with $b = (\psi_M - \psi_{p0})/k_B T_p$, and v_{the} replaced by v_{thp}. Both fluxes must be equal. (The electron and proton densities must be equal, too, which means that their values of n_0 in (5.119) are not strictly equal.) Since electrons have a much greater thermal speed than protons, the potential well retaining them must therefore be much greater than the one retaining protons, in order to ensure density and flux balance. This means that the electrostatic potential at r_0 is significantly greater than the value of a static atmosphere[30] – in agreement with what we found from energy balance considerations in Section 5.3.5. It is the failure to recognise this fact that was responsible for the failure of early exospheric theories [4]. The exospheric point of view resurfaced when self-consistent calculations showed that the actual electrostatic field is greater than in a static atmosphere, and can produce a supersonic (albeit slow) solar wind if the wind

---

[29] Indeed, if ψ_p is monotonously decreasing (and therefore positive) then all protons present at r_0 are escaping (Fig. 5.7), so that their mean velocity there is close to their thermal speed, i.e. the wind starts nearly supersonic.

[30] Because with the same value at r_0 as in a static atmosphere, one would have $\psi_{p0} = \psi_{e0} < 0$, whence $\psi_M - \psi_{p0} = \psi_M - \psi_{e0} = \psi_M + |\psi_{e0}| > |\psi_{e0}|$, so that the potential well retaining electrons would be smaller than the one retaining protons.

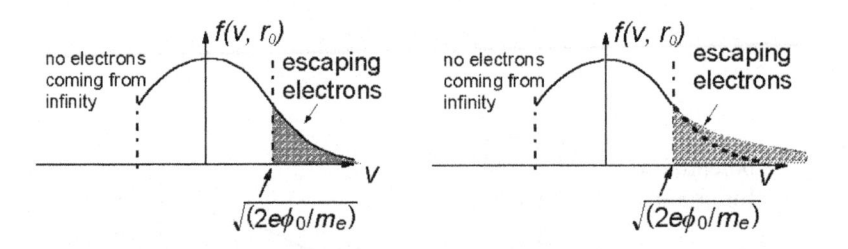

Figure 5.14 In the exospheric picture, the electrons that escape from the corona are the fast ones (left). If the velocity distribution there has an excess of such electrons, the escaping electron flux tends to increase (right), making the electric potential rise in order to keep that flux equal to the proton one; the greater electric field accelerates the protons and thus increases the wind speed.

starts at a distance r_0 of a few solar radii (see [26], [29], [30]). In retrospect, it is amusing to note, from our twenty-first-century vantage point, that George Fitzgerald was right: this electric field produces an outward force on protons equal in magnitude to a few times the gravitational force to the Sun.

In this view, the solar wind is the evaporation of the high-speed tail of a Maxwellian particle distribution in the corona, with, however, a major difference with respect to the evaporation of a neutral exosphere. The protons are accelerated outwards by the electric field induced by the tendency of electrons to escape because of their small mass. This approach does not contradict the fluid picture, since the wind is pushed by the electrons, whereas the fluid picture has to assume by brute force a large heat conductivity which ultimately must be carried by the electrons.

Now, assume that the electron velocity distribution in the corona has a greater proportion of high-speed electrons than a Maxwellian, being for example close to a kappa distribution. If the electrostatic potential remains the same as with a Maxwellian, the escaping flux of electrons increases, whereas the flux of protons does not change. This produces an accumulation of positive charge at the base of the corona, making the electrostatic potential increase, until it traps enough electrons to keep their escaping flux equal to the proton one. And in turn, this greater electrostatic field accelerates the protons outward, thereby increasing the wind speed (Fig. 5.14). Another way of understanding this effect is to note that increasing the number of high-speed electrons increases the heat flux, which in turn produces a greater terminal wind speed because of the global energy balance (see [57] and references therein; a simple analytic estimate is given in [36] in a simpler case[31]).

---

[31]The problem is simpler when the proton thermal speed at the base is of the order of (or greater than) the escape speed. In that case, the proton (positive) electrostatic energy dominates everywhere the (negative) gravitational one, so that protons are all escaping; hence their mean velocity is roughly equal to the thermal one, so that the wind starts nearly at the sonic speed. This may occur under conditions relevant for the slow wind, which is more collisional than the fast one, so that the collisionless region starts at a few solar radii from the Sun.

Balance sheet of the exospheric picture

The exospheric picture provides insights into two physical processes that are not accounted for by fluid models. First, studying the individual particle trajectories, governed by the electrostatic field, reveals the existence of different classes of particles that lead different lives, although they belong to the same species. This suggests that treating all particles of a given species as a fluid obliterates a major aspect of the physics. Second, the role of suprathermal electrons in increasing the escaping electron flux and the heat flux, and therefore the wind speed, reveals a major weakness of the fluid picture which ignores these particles. Instead of being determined by collisions and being proportional to the temperature gradient, the fast wind heat flux is due in the exospheric picture to the nearly collisionless fast electrons, and governed by the electrostatic potential.

The exospheric picture may provide a key to explaining the fast wind speed, and provides a direct interpretation for the observed electron velocity distributions that have a nearly Maxwellian cold component – presumably made of the electrons trapped by the electrostatic potential well, and a suprathermal tail that is strongly skewed in the direction away from the Sun – presumably made of the escaping electrons.

However, the observed distributions differ from the theoretical predictions in several aspects (see for example [42]), so that a better picture is required.[32]

5.5.5 Kinetic models with collisions and wave–particle interactions

Even though the exospheric picture provides insights into the behaviour of the velocity distributions and may enable one to evaluate bulk properties of the wind, it does not account properly for the measured velocity distributions.

Explicitly incorporating collisions into the theory does not change drastically the wind speed and heat flux [57], and still produces velocity distributions that disagree with observation and may drive plasma instabilities (see [13] and references therein). This suggests that both particle collisions and plasma waves produced by instabilities should be incorporated into the theory, which is a

---

[32] First, according to the collisionless picture, the trapped electrons should be at rest with respect to the Sun; this is not observed. Second, the observed pitch angle of escaping electrons is broader than predicted by (5.96). Third, since ions are all escaping far from the Sun in the collisionless picture, their velocity distribution should be strongly anisotropic. This is because, from conservation of the magnetic moment ($v_\perp^2 \propto B$), the perpendicular temperature of escaping particles should decrease as $T_\perp \propto B \propto r^{-2}$ (with a radial magnetic field). In contrast, their $T_\parallel$ should not vary with distance, because both their parallel pressure and their number density vary as the inverse of the flux tube cross-section, so that the ratio is constant. Hence, the theory predicts $T_{p\parallel} \gg T_{p\perp}$. That such a large anisotropy is not observed is not surprising, because it should drive the plasma unstable. Indeed, when the pressure parallel to the magnetic field exceeds the perpendicular one by an amount greater than the restoring magnetic curvature force B^2/μ_0, the plasma becomes unstable – the so-called fire-hose instability. This produces waves, which interact with the particles, decreasing the anisotropy and keeping the distribution stable (see [35] and references therein).

difficult task. Because of the complexity of the calculations, and because nobody has yet come up with a satisfying solution, we refer the reader to the literature (see for example [11], [13], [41], [28] and references therein).

5.6 Building a 'full' theory?

The difficulty of understanding energy transport in both the corona and the solar wind, whereas energy must be transmitted to the wind in some way or another in order to explain the observed acceleration, is reminiscent of the unsolved coronal heating problem. This suggests that both problems should be treated simultaneously.

Treating coronal heating and the solar wind as a single problem requires placing the base of the wind not in the corona but in the chromosphere. This involves in particular:

- the radiative losses, which are important at low heights and involve a huge number of particle species out of equilibrium,

- the heat conductivity, which is unknown in a weakly collisional medium,

- a source of energy presumably involving the magnetic field, of which there is no predictive theory nor unambiguous measurements, and which may produce waves and/or non-thermal velocity distributions,

- a hideously complicated geometry at low heights (see Section 4.1), as well as temporal variations.

Given the complexity of the problem, building a proper theory for the origin of the solar wind requires understanding the physical processes having a dominant effect, and doing observations capable of constraining the theory.

5.6.1 More and better observations (beware of hidden assumptions)

As we saw in Sections 4.1–4.2, the density and temperature are difficult to measure in the wind acceleration region, that is the corona (and the magnetic field is even more so), where no spacecraft has yet been sent. The available data rely on a multitude of unsecured assumptions, and different measurements give conflicting results. Among the dubious assumptions are:

- Maxwellian (or bi-Maxwellian) particle velocity distributions, which are implicit in most spectral measurements,

- the homogeneity – or lack of homogeneity – of the medium, generally accounted for by an arbitrary 'filling factor',

- the factor by which the flow lines deviate from the radial, generally accounted for by an arbitrary 'expansion factor'. This factor is generally

Figure 5.15 A scale model of a NASA project for a solar probe, which should make a near-Sun fly-by: an age-old dream of space scientists, which is within the reach of twenty-first-century technology.

inferred to be large on the basis of extrapolations of the magnetic field measured at the photosphere (see [12] and references therein), whereas other studies show on the contrary that the fast wind flows mainly radially (see [56] and references therein). We shall return to this point in Section 6.2.

Observation of helium and minor ions might give some further indications as to the physics at work. The temperature of these ions is found to be at least proportional to their mass in the corona – a fact that favours most kinetic theories. Furthermore, they flow faster than protons and may exhibit a large anisotropy, with greater temperatures in the direction perpendicular to the magnetic field – a fact that supports cyclotron heating, but is still controversial (see [9] and references therein). We shall return to these ions in Section 6.5.

In this context, a major step will be achieved when a space agency launches a solar probe, to measure the coronal physical parameters *in situ* (Fig. 5.15).

5.6.2 Difficult theoretical questions

Understanding the physical processes involved requires answering several fundamental questions.

What are the shapes of the velocity distributions in the corona, and might the wind be 'suprathermally driven'?

Three arguments suggest that the particle velocity distributions in the corona (and even below) may have high-speed tails. First, it is relatively easy to produce such tails from the magnetic energy present in the solar atmosphere (see [54], [55] and references therein). Second, some observations suggest that this might be so [10] (though other observations disagree). Third, we have seen that this is so in the solar wind. If the particle velocity distributions in the corona are found to have high-speed tails, then not only will a large part of the observations have to be re-examined, but this tail might accelerate the solar wind through the increased heat flux at the base (see [57] and references therein), as suggested long ago [38]. This question is still open.

If waves or turbulence do push and heat, how and where do they act?

Waves and magnetic field fluctuations are fashionable candidates for heating the corona and accelerating the wind, and the literature on this subject would fill several books (see [19] and references therein). There are several reasons for that. First, it is relatively easy to produce such waves in the solar atmosphere, and to incorporate them into fluid models through ad hoc macroscopic parameters. Second, some observations can be interpreted as a signature of such waves (see however [33] and references therein). Third, such waves and fluctuations *are* observed in the solar wind, as we shall see in Section 6.4.

However, the question of their role in heating the corona and accelerating the wind is still open, and one does not understand what form the magnetic fluctuations take (waves, shocks, turbulence, or all forms together?), how and where they are generated, which modes are important, and finally, how and where their energy is used to heat the medium and produce the wind. A recent review may be found in [18].

Are spatial and temporal inhomogeneities a 'detail'?

We have seen in the previous chapter that the solar atmosphere is far from homogeneous and stationary, exhibiting jets and explosive events in permanence. Furthermore, some observations suggest the nascent wind to be made of micro streams. One may seriously ask, therefore, whether it is permissible to ignore these irregularities in wind theories (see [12] and references therein). We shall return to this point in the next chapter.

How is energy transported, and should fluid theories be pensioned off?

Energy transport is perhaps the most important unsolved problem in weakly collisional space plasmas, where, as we have already said, the usual (collisional) heat conductivity is invalid because heat is mainly transported by the fast (collisionless) electrons. Put another way, these electrons make the conductivity non-local, whereas fluid models involve local parameters and their local derivatives. The solution of this problem may require concepts of non-equilibrium thermodynamics, a subject still in its infancy. Preliminary results suggest that in a weakly collisional plasma, the heat flux is indeed extremely different from the classical one (see [28] and references therein).

Does this mean that the old fluid theories should be pensioned off? Not necessarily so, since – hopefully – kinetic studies may enable one to

5.7 Problems

5.7.1 Transonic flows in ducts: the de Laval nozzle

Consider a stationary flow along some co-ordinate z, through a tube whose cross-section S varies with z. Assume that the parameters are uniform across S so that the problem is one-dimensional and mass conservation yields $\rho v S =$ constant along z. Assume that the pressure P of the flowing gas is a function of ρ only, so that variations in pressure and density follow $dP = V_S^2 d\rho$, with V_S the sound speed. Neglect all forces except the pressure gradient. Show that the flow speed varies in the duct according to

$$\frac{dv}{v}\left(1 - \frac{v^2}{V_S^2}\right) = -\frac{dS}{S}. \tag{5.101}$$

Show that when the gas is subsonic, it accelerates only if the duct is convergent. Can you explain this property without writing any equation?

Show that when the gas is supersonic, it accelerates only if the duct is divergent. Again, try to explain this from physical arguments.

Show that for the gas to change smoothly from subsonic to supersonic, the duct must have a throat where S reaches a minimum value (Fig. 5.16). Is this condition sufficient to obtain a supersonic exhaust?

Assume that the gas is adiabatic, i.e. $P \propto \rho^\gamma$. Show that the maximum speed of the gas v_{max} at the exit of the duct is given by

$$\frac{\mu v_{max}^2}{2} = \frac{\gamma}{\gamma - 1} k_B T_0 \tag{5.102}$$

where T_0 is the temperature at the entrance of the duct and μ is the mean mass per particle.

Comment on the analogy between this problem and the solar wind. What is the major difference?

Could the duct be used in reverse, for changing from a supersonic to a subsonic flow?

Hints

Convergent duct
The Mach number v/V_S provides a measure of the importance of compressibility. In a subsonic flow ρ varies less than v. Hence an increase in v requires a decrease in S to maintain a constant flux of mass. This corresponds to usual experience: for example, the speeding up of a river as the channel narrows, or the fact that the wind blows faster at the top of hills, where the flow lines are convergent.

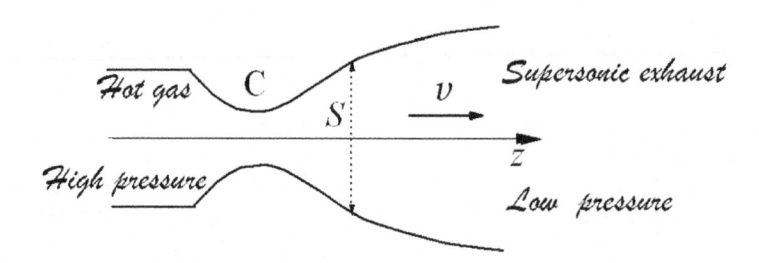

Figure 5.16 The de Laval nozzle used in rocket engines. The juxtaposition of a convergent duct and a divergent duct enables the flow to accelerate continuously, becoming supersonic at the throat.

Divergent duct
The counter-intuitive result that at supersonic speeds, an increase in velocity requires an increase in the cross-section S is also due to mass conservation. An increase in S decreases the product ρv. In a supersonic flow, however, the density varies faster than the speed and in the opposite sense (see Section 2.3). As a result, a decrease in ρv requires an increase in v.

Nozzle
From (5.101), dv/dz remains finite when $v = V_S$ only if $dS/dz = 0$. To obtain a supersonic exhaust, therefore, the gas should flow through a converging–diverging nozzle. This condition, however, is not sufficient. If you own such a nozzle and place it on a table, nothing happens, unless you provide adequate boundary conditions, in particular an adequate pressure difference between both ends. These principles are at the basis of the design of jet engines and rockets.

Maximum speed
Write the Bernoulli equation (or conservation of energy)

$$\frac{\mu v^2}{2} + \frac{\gamma}{\gamma - 1} k_B T = \text{constant} \tag{5.103}$$

between the entrance of the duct where the flow is subsonic and the exit where it is supersonic, and note that the maximum speed is obtained when both the entrance speed and the exit temperature are very small.

Solar wind analogy
The converging part of the tube replaces gravity for making the speed increase in order to conserve mass. The flaring out part of the tube replaces the radial expansion of the solar wind at large distances where gravity is no longer dominant. However, in the solar wind the sonic transition would not occur if the gas were adiabatic, because energy is needed to lift it out of the solar gravitational well.

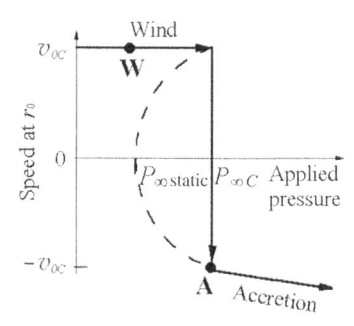

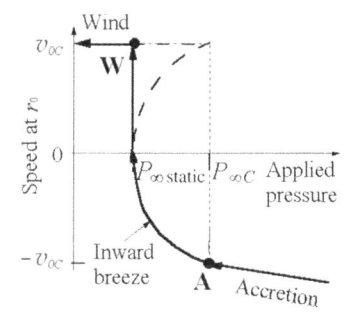

Figure 5.17 The hysteresis cycle of an isothermal flow in response to changes in the applied pressure P_∞ (the dashed lines show unstable solutions). The transition from a transonic wind (**W**) to a transonic accretion (**A→**) when P_∞ is *increased* occurs at $P_{\infty C}$ (left). In contrast when P_∞ is *decreased*, the transition from a transonic (shocked) accretion (**A←**) to a transonic wind (**W**) involves an inward breeze in the range $P_{\infty static} < P_\infty < P_{\infty C}$ (right).

A nozzle in reverse

In practice, shocks form, which require a very subtle design of the nozzle. This is related to the extreme sensitivity of the speed on variations in S at the sonic point, exhibited by (5.101).

A very good survey of compressible flows, including an extensive discussion of the *de Laval nozzle*[33] is [1].

5.7.2 The hysteresis cycle of an isothermal flow

Assume that a transonic isothermal wind flows from a star immersed in a medium whose pressure is extremely small, and that this applied pressure P_∞ increases continuously. Sketch the flow speed at distance r_0 close to the star as a function of P_∞, and the transition through different types of flow. Conversely, assume that the external pressure is large, producing an accretion, and that it decreases continuously. Sketch again the flow speed at r_0 versus P_∞, and explain the transition through different types of flow.

Hint

Figure 5.17 (left) sketches the speed v_0 at r_0 versus the applied pressure P_∞, with the transonic wind starting at v_{0C}, the transonic (shocked) accretion flowing faster than $-v_{0C}$ at r_0, and the unstable breezes (dashed) in the intermediate range (see [53]).

---

[33] Carl G. P. de Laval was a very inventive engineer and a successful businessman. In the late nineteenth century, he designed a steam turbine which incorporated a convergent–divergent nozzle, upstream of the turbine blades. He is better known in Europe, however, for the design and manufacture of centrifugal machines for the separation of cream in milk, of which he sold more than a million. It is interesting to note that de Laval and other contemporary engineers were not certain that the flow was actually supersonic in the de Laval nozzle.

5.7.3 Spherical accretion by a star: the Bondi problem

Consider a star of mass M immersed in a medium of density ρ_∞, and study how the star accretes the surrounding matter. Assume the problem to have spherical symmetry and to be stationary, so that the parameters depend only on the distance r from the star. Assume a polytrope flow $P \propto \rho^\gamma$ (with $\gamma \neq 1$) and let $V_{S\infty}$ be the sound speed in the distant medium. A characteristic length scale is $r_B = MG/V_{S\infty}^2$ [2].

Show that energy balance between large distances (where the flow speed vanishes), and distance r (where the flow speed and the sound speed are respectively v and V_S) yields

$$\frac{v^2}{2V_{S\infty}^2} + \frac{1}{\gamma - 1}\left[\frac{V_S^2}{V_{S\infty}^2} - 1\right] - \frac{r_B}{r} = 0. \tag{5.104}$$

From (5.40), the transonic solution satisfies $v = V_S$ at the critical distance $r = MG/2V_S^2$. Show that for this solution, the sound speed at the critical distance is equal to $V_{S\infty}\sqrt{2/(5-3\gamma)}$. Deduce that the critical distance is $r_C = r_B(5-3\gamma)/4$. (In particular, the sonic transition occurs at the origin in the limit $\gamma = 5/3$.)

Use mass conservation and the adiabatic law to show that the accretion rate is

$$M' = \pi r_B^2 \rho_\infty V_{S\infty} \left(\frac{2}{5-3\gamma}\right)^{\frac{5-3\gamma}{2(\gamma-1)}}. \tag{5.105}$$

Deduce that the order of magnitude of the accretion rate is $M' = \pi r_B^2 \rho_\infty V_{S\infty}$, and that this value is exact in the limit $\gamma = 5/3$.

Now consider a breeze accretion, whose speed is very small at close distances. Show that the density at a close distance r satisfies $\rho/\rho_\infty = [1 + (\gamma - 1)r_B/r]^{1/(\gamma-1)}$.

Estimate this ratio near the surface of the star, assuming that it has the same radius and mass as the Sun, and is immersed in an interstellar medium made of hydrogen of number density 10^6 m^{-3} and temperature 10^2 K, if $\gamma = 1.1$.

Deduce that such a solution is unphysical, so that in this case, the accretion will be transonic.

Hint

The density near the star satisfies $\rho/\rho_\infty = [1 + (\gamma - 1)r_B/R_\odot]^{1/(\gamma-1)}$ for the accretion breeze. We have $r_B/R_\odot = 2.3 \times 10^6/\gamma$, so that $\rho/\rho_\infty \simeq 2 \times 10^{40}$, which would give a number density of 2×10^{46} m^{-3}! With this huge density, there would be about 10^{74} atoms in a shell of width of $R_\odot$ around the star; this number is much more than the number of particles in a whole galaxy! (Anyway, at such a density, the physics would be different; in particular, the gas would be degenerate and would have a different value of γ.)

5.7.4 A wind with polytrope protons and electrons

Consider a plasma made of protons and electrons having polytrope indices equal respectively to γ_p and γ_e, i.e. the proton and electron temperatures vary with the number density n as

$$T_p \propto n^{\gamma_p - 1} \quad ; \quad T_e \propto n^{\gamma_e - 1}.$$

Show that the flow speed follows the same formal equation (5.40) as a single polytrope, but with

$$V_S = [k_B (\gamma_p T_p + \gamma_e T_e)/m_p]^{1/2}. \tag{5.106}$$

Deduce from this equation that, for the flow to be accelerating at the base r_0, one must have there

$$2(\gamma_p T_{p0} + \gamma_e T_{e0}) < m_p |\Phi_{G0}|/k_B. \tag{5.107}$$

Deduce from the energy balance that, for the terminal speed to be greater than the start-off speed, one must have

$$\frac{\gamma_p}{\gamma_p - 1} T_{p0} + \frac{\gamma_e}{\gamma_e - 1} T_{e0} > m_p |\Phi_{G0}|/k_B. \tag{5.108}$$

Show that in the particular case when $T_{p0} = T_{e0}$ and the protons are adiabatic, a transonic accelerating wind must have an electron polytrope index satisfying $-0.3 < \gamma_e < 1.38$.

Compare this result with the case of a single polytrope and comment.

Show that the total potential energy of a proton in the electric and gravitational fields is at distance r:

$$\psi = \frac{\gamma_e}{\gamma_e - 1} k_B T_e - \frac{m_p M_\odot G}{r}. \tag{5.109}$$

In the particular case when electrons and protons have the same temperature T and polytrope index γ, show that if a transonic wind does exist, then ψ has a maximum at some finite distance r_M. In the usual case when $\gamma > 1$, show that this maximum is located above the critical distance r_C where the wind becomes supersonic.

Hints

First, prove that if $1 < \gamma < 3/2$, $d\psi/dr$ is negative at large distances. Second, prove that $d\psi/dr$ is positive at r_C; for doing so, you may calculate dT/dr as a function of dv/dr by using the polytrope law and the conservation of particles, and use the value of dv/dr at r_C calculated in Section 5.3.2 (note that in this case, $\mu = m_p/2$). This shows that ψ has a maximum somewhere between r_C and infinite distance [37].

5.7.5 Playing with the kappa distribution

Poisson, Maxwell and ... kappa

Show that the kappa distribution $f_\kappa(\mathbf{v})$ is related to the Maxwellian distribution through the Poisson distribution[34] $P_\kappa(t) = t^\kappa e^{-t}/\Gamma(\kappa+1)$ by

$$f_\kappa(\mathbf{v}) = A_\kappa \int_0^\infty dt\, P_\kappa(t) \times \frac{1}{\pi^{3/2} v_{th}^3} e^{-tv^2/\kappa v_{th}^2}. \tag{5.110}$$

(A_κ is defined in (5.118).)

Hint

Write the Laplace transform[35] of t^κ

$$\int_0^\infty dt\, e^{-st} t^\kappa = \Gamma(\kappa+1)/s^{\kappa+1}$$

and substitute $s = 1 + v^2/\kappa v_{th}^2$.

A modified kappa distribution

One often encounters a modified kappa distribution defined as

$$f_{\kappa md}(\mathbf{v}) \propto \left[1 + \frac{v^2}{\kappa v_{thmd}^2}\right]^{-\kappa} \tag{5.111}$$

which tends to the Maxwellian e^{-v^2/v_{thmd}^2} in the limit $\kappa \to \infty$, and has the same series expansion to first order as this Maxwellian.

Show that the most probable speed depends on κ, being equal to $v_{thmd} \times (\kappa/\kappa - 1)^{1/2}$.

Show that the kinetic temperature is

$$T = \frac{\kappa}{\kappa - 5/2} \frac{m v_{th}^2}{2 k_B}$$

Show that the normalisation factor in (5.111) is $A_{\kappa md} \pi^{-3/2} v_{thmd}^{-3}$ with $A_{\kappa md} = \kappa^{-3/2} \Gamma(\kappa)/\Gamma(\kappa - 3/2)$.

Hint

This distribution is formally equivalent to the normal kappa distribution considered in Section 5.4, changing the values of κ and v_{th} as $f_{\kappa md}(\mathbf{v}, v_{thmd}) = f_{\kappa-1}(\mathbf{v}, (\kappa/\kappa - 1)^{1/2} v_{thmd})$.

---

[34] The probability of κ events during time t if the average is one event per unit time.
[35] The required Laplace transform may be found in the *Handbook of Mathematical Functions*, ed. M. Abramowitz and I. A. Stegun (New York, Dover, 1970), p. 1022.

Problems

5.7.6 'Temperature' or 'temperatures'?

At local thermodynamic equilibrium, the velocity distributions are Maxwellian with $f(\mathbf{v}) \propto e^{-v^2/v_{th}^2}$, where the temperature is $T = mv_{th}^2/2k_B$ for particles of mass m. In this case, temperature measurements generally yield (hopefully) the actual value T. When the velocity distributions are not Maxwellian, however, the measured temperatures depend on the measuring scheme if the observers are not careful enough. Some examples are given below for an isotropic distribution.

Kinetic temperature and other moments

The kinetic temperature is $T = m\langle v^2 \rangle/3k_B$ (where the angular brackets denote an average on the distribution). Its determination involves an integration over the measured distribution, which gives the correct result if the accuracy and resolution of the measurements are sufficient over the full velocity range that is relevant in the integration. Aboard space probes, measurements at small speeds are generally spoiled by the electrostatic potential of the space probe (which affects the detected particle speeds) and by the particles it ejects (which mix with the genuine ones). Some devices are biased towards either small or large speeds, and give instead a generalised temperature deduced from a different moment of the distribution. One may define a 'generalised temperature' as

$$k_B T_q/m = (\langle v^q \rangle/c_q)^{2/q} \tag{5.112}$$

with

$$c_q = (q+1)!! \quad ; \quad q \text{ even} \tag{5.113}$$

$$c_q = \frac{2^{1+q/2}}{\sqrt{\pi}} \times \left(\frac{q+1}{2}\right)! \quad ; \quad q \text{ odd} \tag{5.114}$$

where $q > -3$. The greater the index q the faster the particles responsible for this 'temperature'; T_2 is the usual kinetic temperature.

Show that with a Maxwellian distribution, all T_q are equal to the true temperature. How about a kappa distribution?

Show that the mean random speed is related to T_1 by $\langle v \rangle = (8k_B T_1/\pi m)^{1/2}$, whereas the mean square speed is $\langle v^2 \rangle = 3k_B T_2/m$.

Show that the Debye length relative to the species considered is given by $L_D = (\epsilon_0 k_B T_{-2}/ne^2)^{1/2}$, where n is the particle number density.

What error results if one deduces the temperature from the mean square speed, thinking that the distribution is a Maxwellian, whereas it is in fact a kappa distribution?

Differential temperature

The temperature is sometimes determined from the log derivative of the measured velocity distribution as

$$T_{diff} = -k_B \left[d\left(\ln f\right)/dW\right]^{-1} \tag{5.115}$$

where $W = mv^2/2$ (see [42]).

Show that if $f(\mathbf{v})$ is a Maxwellian, T_{diff} coincides with the true temperature. Show that for a Kappa distribution

$$T_{diff} = \frac{mv_{th}^2}{2k_B} \frac{\kappa + v^2/v_{th}^2}{\kappa + 1} \tag{5.116}$$

which coincides with the value of its Maxwellian low-speed limit at low speeds, but is no longer a constant at high speeds and increases with energy.

5.7.7 Non-collisional heat flux

Calculate the particle flux and the heat flux due to the electrons of speed $v > v_E = (2e\Phi_E/m_e)^{1/2}$ whose velocity is pointing away from the Sun, Φ_E being the electrostatic potential at distance r. Consider first a Maxwellian velocity distribution

$$f(\mathbf{v}) = \frac{n_0}{\pi^{3/2} v_{the}^3} e^{-(v/v_{the})^2} \tag{5.117}$$

having density n_0, temperature $T_e = m_e v_{the}^2/2k_B$, and zero mean velocity and heat flux.

Because of the fast decrease of the Maxwellian distribution at high speeds, the bulk speed and heat flux produced by the escaping electrons are small. This is not so, however, with a distribution decreasing less rapidly at high speeds. To see this, replace the Maxwellian by a kappa distribution

$$f_\kappa(\mathbf{v}) = \frac{n_0 A_\kappa}{\pi^{3/2} v_{the}^3} \left[1 + \frac{v^2}{\kappa v_{the}^2}\right]^{-(\kappa+1)} \quad ; \quad A_\kappa = \frac{\Gamma(\kappa+1)}{\kappa^{3/2} \Gamma(\kappa - 1/2)} \tag{5.118}$$

where the gamma (factorial) function Γ ensures the normalisation.[36] Calculate again the electron flux.

Hints

Remove from the Maxwellian distribution the particles having both a negative radial velocity and a speed modulus $v > v_E$. This removal distorts the distribution in the outward direction (see Fig. 5.11), so that the mean velocity and the heat flux no longer vanish. Calculate the outward particle and energy fluxes by integrating in spherical co-ordinates $[v, \theta, \phi]$, over $v > v_E$ and $0 < \theta < \pi/2$ (i.e. outward directions) with $v_r = v \cos\theta$, $d^3v = v^2 \sin\theta d\theta d\phi$, to obtain

$$nv_w = \int d^3v v_r f(v) = \pi \int_{v_E}^{\infty} dv v^3 f(v) = \frac{n_0 v_{the}}{2\sqrt{\pi}} (b+1) e^{-b} \tag{5.119}$$

$$Q_{SF} = \frac{m_e}{2} \int d^3v v_r v^2 f(v) = \frac{\pi m_e}{2} \int_{v_E}^{\infty} dv v^5 f(v) = \frac{n_0 m_e v_{the}^3}{4\sqrt{\pi}} \left[(b+1)^2 + 1\right] e^{-b}$$

---
[36] In the limit $\kappa \to \infty$, (5.118) reduces to (5.117).

Problems

where n is the density, v_w the modulus of the mean (radial) speed, Q_{SF} the energy flux in the solar frame and

$$b = (v_E/v_{the})^2 = e\Phi_E/k_BT_e.$$

The heat flux – which should be calculated in the frame where the mean velocity vanishes – is deduced from the one in the solar frame (Q_{SF}) by subtracting the enthalpy and bulk energy fluxes (see (5.91)). Note that our removing of some electrons from the initial Maxwellian has changed slightly the particle density and temperature, so that n_0 and T_e do not represent the exact density and temperature. If, however, the potential is large enough so that $v_E \gg v_{the}$, the number of removed electrons is a small fraction of the total, so that the number density n and the temperature – determined mainly by the symmetrical part of the distribution – are roughly equal respectively to n_0 and T_e; furthermore, in this case the bulk speed v_w – determined by the non-symmetrical part – is small compared to v_{the}, and the heat flux Q_e – determined by the non-symmetrical part, too – is approximately given by its value in the solar frame Q_{SF}, minus the enthalpy flux $5nv_wk_BT_e/2$ ((5.91) where v_w is negligible and $T_\| \simeq T_\perp$). In this case, we deduce from (5.119) that the electron heat flux may be written as

$$Q_e \simeq \alpha_e nv_w k_B T_e \; ; \quad \alpha_e = \left[b + 1 + \frac{1}{b+1} - \frac{5}{2} \right]. \tag{5.120}$$

Note that this yields, for $b \gg 1$, $Q_e \simeq nv_{sw}\left(e\Phi_E - \frac{5}{2}k_BT_e\right)$ as already found from energy balance arguments. In this picture, a large value of α_e requires a large electrostatic potential. An exact calculation may be found in [17].

With a kappa distribution, removing the particles coming inwards with $v > v_E$ yields the electron flux

$$nv_w = \frac{n_0 A_\kappa \kappa v_{the}}{2\sqrt{\pi}(\kappa-1)}(1+b)\left(1+\frac{b}{\kappa}\right)^{-\kappa} \; ; \quad b = \left(\frac{v_E}{v_{the}}\right)^2. \tag{5.121}$$

When b is large, this yields greater electron flux and heat flux than with a Maxwellian. Note that since the heat flux is a moment of order three, a finite heat flux requires that $v^3 \times v^{2-2(\kappa+1)}$ decreases faster than v^{-1}, i.e. that $\kappa > 2$.

Note that this calculation does not yield the true collisionless heat flux, except at the base r_0, because conservation of energy and magnetic momentum decrease the inclination of the particle velocity to the radial as the particle moves outwards, so that in the absence of collisions the velocity of escaping electrons at large distances points in a small cone instead of the whole half-space (see Section 5.5.4).

5.7.8 An imaginary wind with charges of equal masses

Assume a star made of positive and negative charges of equal mass m (and temperatures) ... just for the purpose of this problem, and don't bother about the other consequences. The fluid and kinetic points of view yield in that case very different winds – thereby enlightening some aspects of the physics.

Consider the simple Parker's picture of an isothermal wind. Show that if the ratio of the escape speed to the thermal speed of an 'average' particle is the same as for the Sun (and the large distance pressure is small enough), the wind is similar to that found in Section 5.1. Answer the same question for a polytrope wind with a sufficiently small γ, comparing to Section 5.2.

Now, adopt a kinetic point of view. Show that the (large-scale) radial electrostatic field is zero, and that charges of both signs are strongly bound to the star. Show qualitatively from energy arguments that there is in that case no supersonic wind.[37]

Why do the fluid and kinetic points of view give so different results?

Hint

The key point is the isothermal or polytrope assumption made in the fluid models, which implicitly assumes the heat flux to be large (even infinite in the isothermal case). It is because of the small electron mass that this approximation is not bad for the solar wind.

References

[1] Anderson, J. D. 1990, *Modern Compressible Flow*, New York, McGraw-Hill.

[2] Bondi, H. 1952, On spherically symmetrical accretion, *Mont. Not. Roy. Astron. Soc.* **112** 195.

[3] Brandt, J. C. 1970, *Introduction to the Solar Wind*, San Francisco CA, W. H. Freeman.

[4] Chamberlain, J. W. 1960, Interplanetary gas. II. Expansion of a model corona, *Ap. J.* **131** 47.

[5] Chamberlain, J. W. 1961, Interplanetary gas. III. A hydrodynamic model of the corona, *Ap. J.* **133** 675.

[6] Collier, M. R. 2004, Are magnetospheric suprathermal particle distributions (κ) functions inconsistent with maximum entropy considerations? *Adv. Space Res.* **33** 2108.

[7] Dahlberg, E. 1964, On the stellar-wind equations, *Ap. J.* **140** 268.

[8] Dreicer, H. 1960, Electron and ion runaway in a fully ionised plasma. II, *Phys. Rev.* **117** 329.

[9] Esser, R. *et al.* 1999, Plasma properties in coronal holes derived from measurements of minor ion spectral lines and polarized white light intensity, *Ap. J.* **510** L63.

---

[37]This problem has some similarity with the one considered by [4], who did not find a supersonic wind, but a breeze.

[10] Esser, R. and R. J. Edgar 2000, Reconciling spectroscopic electron temperature measurements in the solar corona with in situ charge states observations, *Ap. J.* **532** L71.

[11] Feldman, W. C. *et al.* 1979, A possible closure relation for heat transport in the solar wind, *J. Geophys. Res.* **84** 6621.

[12] Feldman, W. C. *et al.* 1996, Constraints on high-speed solar wind structure near its coronal base: a Ulysses perspective, *Astron. Astrophys.* **316** 355.

[13] Feldman, W. C. and E. Marsch 1997, in *Cosmic Winds and the Heliosphere*, ed. J. R. Jokipii *et al.*, Tucson AZ, University of Arizona Press, p. 617.

[14] Feldman, W. C. *et al.* 1998, Ion energy equation for the high speed solar wind: Ulysses observations, *J. Geophys. Res.* **103** 14547.

[15] Habbal, S. R. and K. Tsinganos 1983, Multiple transonic solutions with a new class of shock transitions in steady isothermal solar and stellar winds, *J. Geophys. Res.* **88** 1965.

[16] Hartle, R. E. and P. A. Sturrock 1968, Two-fluid model of the solar wind, *Ap. J.* **151** 1155.

[17] Hollweg, J. V. 1974, On electron heat conduction in the solar wind, *J. Geophys. Res.* **79** 3845.

[18] Hollweg, J. V. 2003, In *Solar Wind*, vol. 10, ed. M. Velli, R. Bruno and F. Malava, American Institute of Physics, p. 14.

[19] Hollweg, J. V. and P. A. Isenberg 2002, Generation of the fast wind: A review with emphasis on the resonant cyclotron interaction, *J. Geophys. Res.* **107** 101029.

[20] Holzer, T. E. 1977, Effects of rapidly diverging flow, heat addition, and momentum addition in the solar wind and stellar winds, *J. Geophys. Res.* **82** 23.

[21] Holzer, T. E. 1988, in *Solar Wind*, vol. 6, ed. V. J. Pizzo *et al.*, National Center for Atmospheric Research Tech. Notes, Boulder CO.

[22] Holzer, T. E. and W. I. Axford 1970, The theory of stellar winds and related flows, *Annu. Rev. Astron. Astrophys.* **8** 31.

[23] Hu, Y. Q. *et al.* 1997, A fast solar wind model with anisotropic proton temperature, *J. Geophys. Res.* **102** 14661.

[24] Hu, Y. Q. *et al.* 2000, A four-fluid turbulence-driven solar wind model for preferential acceleration and heating of heavy ions, *J. Geophys. Res.* **105** 5093.

[25] Hundhausen, A. J. 1972, *Coronal Expansion and Solar Wind*, New York, Springer.

[26] Jockers, K. 1970, Solar wind models based on exospheric theory, *Astron. Astrophys.* **6** 219.

[27] Lamers, H. J. G. L. M. and J. P. Cassinelli 1999, *Introduction to Stellar Winds*, Cambridge University Press.

[28] Landi, S. and F. Pantellini 2003, Kinetic simulations of the solar wind from the subsonic to the supersonic regime, *Astron. Astrophys.* **400** 769.

[29] Lemaire, J. and M. Scherer 1971, Kinetic models of the solar wind, *J. Geophys. Res.* **76** 7479.

[30] Lemaire, J. and M. Scherer 1973, Kinetic models of the solar and polar winds, *Rev. Geophys. Space Phys.* **11** 427.

[31] Lie-Svendsen, O. et al. 2001, A 16-moment solar wind model: from the chromosphere to 1 AU, *J. Geophys. Res.* **106** 8217.

[32] Maksimovic, M. et al. 2005, Radial evolution of the electron distribution functions in the fast solar wind between 0.3 and 1.5 AU, *J. Geophys. Res.* **110** 9104.

[33] Mancuso, S. and S. R. Spangler 1999, Coronal Faraday rotation observations: measurements and limits on plasma inhomogeneities, *Ap. J.* **525** 195.

[34] Marsch, E. 1991, In *Physics of the Inner Heliosphere*, vol. 2, ed. R. Schwenn and E. Marsch, New York, Springer, p. 45.

[35] Matteini, L. et al. 2006, Parallel proton fire-hose instability in the expanding solar wind: hybrid simulations, *J. Geophys. Res.* in press.

[36] Meyer-Vernet, N. 1999, How does the solar wind blow? A simple kinetic model, *Eur. J. Phys.* **20** 167.

[37] Meyer-Vernet, N. et al. 2003, Some basic aspects of the solar wind acceleration, in *Solar Wind*, vol. 10, ed. M. Velli, R. Bruno and F. Malara, American Institute of Physics, p. 263.

[38] Olbert, S. 1981, In *Plasma Astrophysics*, ed. T. D. Guyenne and G. Levy, ESA SP-161, p. 135.

[39] Parker, E. N. 1963, *Interplanetary Dynamical Processes*, New York, Wiley.

[40] Parker, E. N. 1965, Dynamical theory of the solar wind, *Space Sci. Rev.* **4** 666.

[41] Pierrard, V. et al. 2001, Self-consistent model of solar wind electrons, *J. Geophys. Res.* **106** 29305.

[42] Pilipp, W. G. *et al.* 1987, Characteristics of electron velocity distribution functions in the solar wind derived from the Helios plasma experiment, *J. Geophys. Res.* **92** 1075.

[43] Pilipp, W. G. *et al.* 1990, Large-scale variations of thermal electron parameters in the solar wind between 0.3 and 1 AU, *J. Geophys. Res.* **95** 6305.

[44] Price, J. C. *et al.* 1975, Interplanetary gas. XXI. Validity of the Chapman–Enskog description of the solar wind for protons, *Ap. J.* **199** 756.

[45] Roussev, I. I. *et al.* 2003, A three-dimensional model of the solar wind incorporating solar magnetogram observations, *Ap. J.* **595** L57.

[46] Schwenn, R. 1991, In *Physics of the Inner Heliosphere*, vol. 1, ed. R. Schwenn and E. Marsch, New York, Springer, p. 99.

[47] Scime, E. E. *et al.* 2001, Solar cycle variations in the electron heat flux: Ulysses observations, *Geophys. Res. Lett.* **28** 2169.

[48] Scudder, J. D. 1996, Dreicer order ambipolar electric fields at Parker's steady state solar wind sonic critical point, *J. Geophys. Res.* **101** 13461.

[49] Shoub, E. C. 1988, in *Solar Wind*, vol. 6, ed. V. J. Pizzo *et al.*, NCAR TN-306, Boulder CO, p. 59.

[50] Totten, T. L. *et al.* 1995, An empirical determination of the polytropic index for the free-streaming solar wind using Helios 1 data, *J. Geophys. Res.* **100** 13.

[51] Treumann, R. A. 2001, Statistical mechanics of stable states far from equilibrium: thermodynamics of turbulent plasmas, *Astrophys. Space Sci.* **277** 81.

[52] Velli, M. 1994, From supersonic winds to accretion: comments on the stability of stellar winds and related flows, *Ap. J.* **43** L55.

[53] Velli, M. 2001, Hydrodynamics of the solar wind expansion, *Astrophys. Space Sci.* **277** 157.

[54] Vinas, A. F. *et al.* 2000, Generation of electron suprathermal tails in the upper solar atmosphere: implications for coronal heating, *Ap. J.* **528** 509.

[55] Vocks, C. and G. Mann 2003, Generation of suprathermal electrons by resonant wave-particle interaction in the solar corona and wind, *Ap. J.* **593** 1134.

[56] Woo, R. and S. R. Habbal 2002, The origin of the solar wind, *Am. Scientist* **90** 532.

[57] Zouganelis, I. *et al.* 2005, Acceleration of weakly collisional solar-type winds, *Ap. J.* **626** L117.

6

Structure and perturbations

> The researcher looked round him once more:
> and now the Facts accumulated in such bewildering profusion, that
> the Theory was lost among them.
> Lewis Carroll, *Sylvie and Bruno*

In order to concentrate on the basic physics of the solar wind acceleration, we considered in Chapter 5 a spherically symmetric and stationary problem, with a radial magnetic field. Unfortunately, the solar wind is more complicated. We now introduce some of these complications, trying however to keep a bias towards basics. More details may be found in the books [29], [1], [3], with some updates in [38] and [7]. Most of this chapter considers the wind having already been accelerated to a large velocity, which we note $\mathbf{v}_w$.

6.1 Basic large-scale magnetic field

6.1.1 Parker's spiral

Figure 6.1 (left) reminds us of the geometry considered in the previous chapter: a radially expanding solar wind with a radial magnetic field. This is an application of the frozen-in magnetic field concept (see Section 2.3). Since the magnetic Reynolds number is extremely large,[1] any magnetic flux tube in the steady plasma flow will hold the same fluid parcels later on, so that the magnetic field lines are dragged by the flow and tend to be aligned with the radial flow lines. Conservation of the magnetic flux then yields a radial magnetic field varying as $B \propto r^{-2}$.

---

[1] With an electron temperature of 10^6 K in the corona, the conductivity is $\sigma \sim 3 \times 10^5$ mhos, so that with a scale $L \sim 1 R_\odot$ and a bulk speed $v \sim 100$ km s^{-1} near the sonic point, the magnetic Reynolds number $R_M \sim \mu_0 \sigma v L \sim 10^{13}$. In the solar wind proper, the temperature is smaller, but both the typical scale and the speed are greater, so that R_M is still greater.

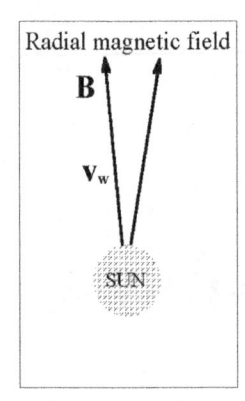

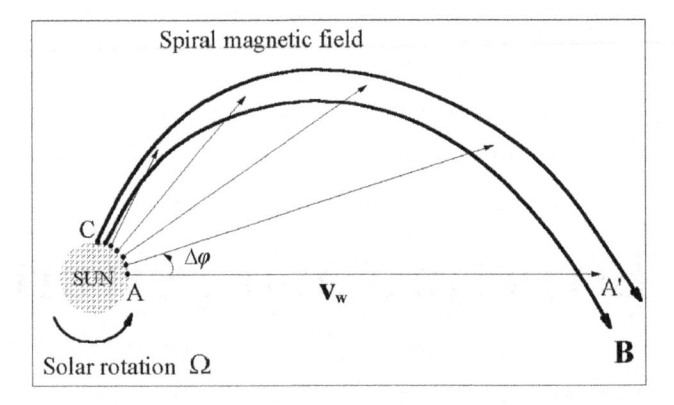

Figure 6.1 Magnetic field lines drawn by the radially expanding solar wind. On the left, the Sun is not rotating – a reasonable approximation in polar regions and/or at relatively close distances. On the right, the solar rotation draws the field lines into a spiral shape. The arrows materialise the trajectory of fluid parcels ejected at radial velocity $\mathbf{v}_w$ by a fixed source on the rotating Sun, which was at A (respectively C) when the first (respectively last) parcel was ejected.

Simple derivation

There is, however, a complication to this nicely simple picture: we have seen in Section 3.2 that the Sun is rotating. How does this change the magnetic field structure? Basically, the change is associated with the velocity due to the solar angular rotation Ω. At radial distance r and latitude θ, the distance to the rotation axis is $r\cos\theta$, so that the rotation speed is $\Omega r \cos\theta$. Therefore, the change in the magnetic field is expected to vary in order of magnitude as the ratio of this speed to the solar wind speed $v_w = |\mathbf{v}_w|$, i.e. as $\Omega r \cos\theta / v_w$. With $\Omega \simeq 2.7 \times 10^{-6}$ rad s^{-1} (Fig. 3.6) and $v_w \simeq 400$ km s^{-1}, we have

$$\Omega r \cos\theta / v_w \simeq r_{AU} \times \cos\theta. \qquad (6.1)$$

We thus expect the effect of the solar rotation on the magnetic field to be negligible closer than about 1 AU from the Sun, and/or at high latitudes where $\cos\theta \ll 1$.

What is the structure of the magnetic field elsewhere? Figure 6.1 (right) illustrates what happens. Consider a fixed source on the rotating Sun that is ejecting fluid parcels of radial velocity v_w, and is initially, say, at point A. As the ejected parcel travels along the arrow starting from A, the source rotates, so that it ejects the following parcels from different locations; the arrows indicate the trajectories of parcels ejected at regular time intervals Δt. When the parcel ejected at A has travelled to A′, the parcels ejected later have had less time to travel, and are thus closer to the Sun but along different arrows, whereas the last parcel has just been ejected and is still at C; the location of these parcels materialises a magnetic flux tube at this time (bold lines). Consider the parcel

Basic large-scale magnetic field

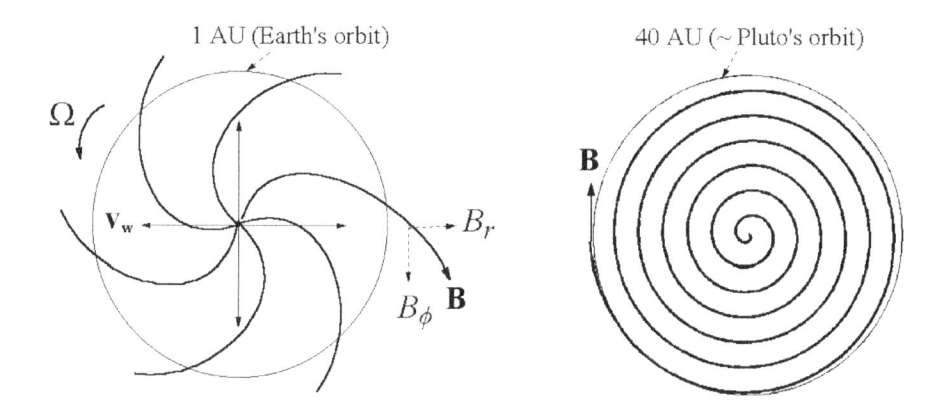

Figure 6.2 Spiral magnetic field lines frozen in the radially expanding solar wind in the equatorial plane, drawn at two different scales. Left, six magnetic field lines, drawn up to a distance of about 1 AU. Right, one field line, drawn up to about 40 AU. At 1 AU, the field makes an angle of less than about 45° to the radial, but it is tightly wound at large distances.

ejected Δt later than the one ejected from A; during Δt, the Sun has rotated by the angle $\Delta \varphi = \Omega \Delta t$, so that the parcel follows a (straight) trajectory making the angle $\Delta \varphi$ to AA', but – having started later – it has travelled less by the amount $\Delta r = -v_w \Delta t$. The field lines thus follow the equation

$$d\varphi/dr = -\Omega/v_w \qquad (6.2)$$

so that they take the shape of Archimedes spirals of equation $r = -v_w \varphi / \Omega$. Since by definition **B** is along the field lines, its radial and azimuthal components satisfy $B_\varphi / B_r = r d\varphi/dr$ in the equatorial plane, so that from (6.2)

$$B_\varphi / B_r = -\Omega r / v_w. \qquad (6.3)$$

The magnetic field inclination to the radial: $\arctan(-\Omega r/v_w)$ thus increases with distance. Substituting the numerical values of Ω and v_w, we find $\Omega r/v_w \simeq r_{AU}$ when the wind is slow and twice less when it is fast. Hence the magnetic field inclination to the radial is about 25–45° at 1 AU (in the equatorial plane), but nearly 90° beyond 10 AU, so that at large distances the field lines roughly follow circles around the Sun (Fig. 6.2).

The solar rotation therefore winds up the magnetic field, the more so as distance increases, rather as streams of water ejected by a rotating water sprinkler.

What is the magnetic structure outside the equatorial plane? The field lines satisfy $B_r/B_\varphi = dr/(r \cos\theta d\varphi)$ with the latitude $\theta =$ constant (Fig. 6.3, left), so that (6.2) yields

$$B_\varphi / B_r = -\Omega r \cos\theta / v_w \qquad (6.4)$$

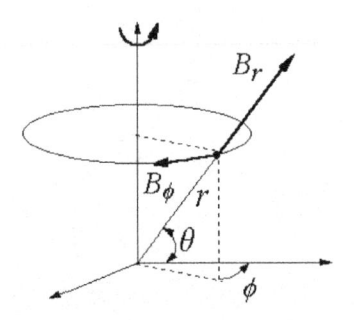

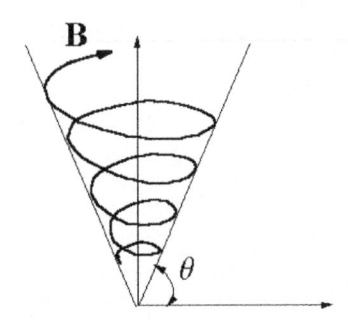

Figure 6.3 Spherical co-ordinate system (left); θ is the latitude; because of the sense of the solar rotation, when B_r is positive (pointing outwards), B_φ is negative (opposite to direct sense). Right, the magnetic field lines are spirals wrapped on cones of half-angle the colatitude.

and the component B_θ vanishes. The field lines are spirals wound up on the surface of cones whose half-angle is the colatitude $\pi/2 - \theta$ (Fig. 6.3, right). From magnetic flux conservation, the radial component of **B** satisfies

$$B_r \propto r^{-2}. \tag{6.5}$$

Rotating frame

Another way of understanding this magnetic structure – the so-called Parker's spiral – is to consider a reference frame rotating with the angular velocity of the Sun, as in Parker's original (1958) paper.[2] With respect to the stationary frame, or more precisely to the inertial frame of the Sun, the frame rotating with the Sun moves at the azimuthal speed

$$v_\Omega = \Omega r \cos\theta \quad \text{(azimuthal)} \tag{6.6}$$

at distance r and latitude θ.

In this rotating frame, the source ejecting the plasma is no longer rotating, so that the path followed by fluid parcels from a given source area – which defines a magnetic field line – is simply a flow stream line. The plasma velocity is equal to the (radial) solar wind velocity $\mathbf{v}_w$ transformed to the rotating frame, so that its radial and azimuthal components are

$$\begin{aligned} v_r &= v_w \\ v_\varphi &= -v_\Omega \end{aligned} \tag{6.7}$$

at distance r and latitude θ. Since the magnetic field follows the path of the plasma, which coincides with the stream lines in this rotating frame since the source is fixed, we have $B_\varphi/B_r = v_\varphi/v_r$. Substituting (6.6) and (6.7) yields

---

[2] Cited in Chapter 1.

Basic large-scale magnetic field

(6.4). The magnetic field lines thus follow Parker's spiral – a result that is not changed by transforming back to the stationary frame because **B** is invariant for non-relativistic velocity transformations (see Section 2.2).

To summarise, in the frame rotating with the Sun, the flow stream lines follow Archimedean spirals, and so do the dragged magnetic field lines; transforming back to the stationary frame does not change the magnetic field (whereas the speed becomes radial). Two comments are in order. First, we have swept under the carpet the difficult problem of the structure near the 'source' of the wind, and assumed implicitly that its distance r_0 to the Sun's centre is very small compared to the heliocentric distances considered, so that Ωr_0 is much smaller than v_w whereas Ωr is not. We shall return to this problem later.

Second, what about the electric field? Consider the solar wind frame, that is moving radially at $\mathbf{v}_w$ but non-rotating. Standard MHD (see Section 2.3) tells us that the electric field vanishes in this frame because of the large electric conductivity of the plasma. Consider now the frame that is rotating with the Sun (it has zero radial velocity and the azimuthal speed v_Ω). Transforming the electric field from the plasma frame (where $\mathbf{E} = 0$) to that frame yields $\mathbf{E} = \mathbf{v}_{rel} \times \mathbf{B}$, where $\mathbf{v}_{rel} = -\mathbf{v}_w + \mathbf{v}_\Omega$ is the relative velocity, so that $\mathbf{E}$ has only a θ component: $E_\theta = v_w B_\varphi + v_\Omega B_r = 0$, from (6.4) and (6.6).

Therefore the (MHD) electric field vanishes not only in the solar wind frame but also in the frame rotating at the angular velocity of the Sun. The spiral magnetic pattern might also be viewed as rotating with the Sun (even though the plasma does not) – just as the grooves in an old gramophone record are rotating, whereas the gramophone needle moves (roughly) in the radial direction – as does the solar wind plasma.

Consider finally the stationary frame (non-rotating, no radial motion). Transforming the zero electric field of the plasma frame or of the rotating frame to that stationary frame yields an electric field $\mathbf{E} = -\mathbf{v}_w \times \mathbf{B} = -\mathbf{v}_\Omega \times \mathbf{B}$. Since $\mathbf{v}_w$ is radial and $\mathbf{v}_\Omega$ azimuthal, this yields $E_\theta = v_w B_\varphi = -v_\Omega B_r$. Because of the sense of the solar rotation ($v_\Omega > 0$), B_φ and B_r have opposite signs, so that the electric field in the stationary frame points northwards (respectively southwards) if $B_r > 0$ (respectively $B_r < 0$).

To summarise:

- solar wind frame (velocity $\mathbf{v}_w$, radial): $\mathbf{E} = 0$, because of the high conductivity,
- rotating frame (velocity $\mathbf{v}_\Omega$, azimuthal): $\mathbf{E} = (-\mathbf{v}_w + \mathbf{v}_\Omega) \times \mathbf{B} = 0$,
- stationary frame (velocity = 0): $\mathbf{E} = -\mathbf{v}_w \times \mathbf{B} = -\mathbf{v}_\Omega \times \mathbf{B} \neq 0$.

Estimating the magnetic field

These results enable one to estimate the solar wind magnetic field as a function of distance and latitude as

$$B_r = B_0 \times (r_0/r)^2 \tag{6.8}$$
$$B_\varphi = -B_0 \times \Omega \cos\theta \times r_0^2/v_w r \tag{6.9}$$

where B_0 is the magnetic field at the source's location r_0, assumed to be radial there and independent of latitude. To get a rough estimate, let us start in the low corona at $r_0 \simeq R_\odot$ with an average magnetic field $B_0 \sim 2 \times 10^{-4}$ T (see Table 4.1). This yields

$$|B_r| \simeq 4 \times 10^{-9}/r_{AU}^2 \text{ T} \qquad (6.10)$$

$$|B_\varphi| \simeq 4 \times 10^{-9} \times \cos\theta/r_{AU} \text{ T} \qquad (6.11)$$

where r_{AU} is the distance in AU (1 AU $\simeq$ 214 $R_\odot$), and we have used the estimate (6.1), so that the value of B_φ holds for a slow solar wind speed of $v_w \simeq 400$ km s^{-1} and is roughly twice less in the fast wind. The corresponding electric field in the stationary frame is

$$E_\theta = v_w B_\varphi \sim 1.6 \times 10^{-3} \times \cos\theta/r_{AU} \text{ V m}^{-1}. \qquad (6.12)$$

We shall see later that this simple model is a rather good approximation to the observations, up to a very large distance.

A small aside is in order. When calculating the electric field in the stationary frame $E_\theta = v_w B_\varphi$, we used the standard MHD approximation that the large-scale electric field is roughly zero in the plasma frame. But what about the large-scale electric field discussed in Sections 5.3–5.4, which compensates for the difference in electron and proton masses? From (5.87), its order of magnitude is $E_\parallel \sim k_B T_e/er$ at radial distance r, so that in the equatorial plane at $r \sim 1$ AU, we have $E_\parallel/v_w B_\varphi \sim 3 \times 10^{-8}$ for $T_e \sim 10^5$ K. This means that the parallel electric field is indeed very small compared to the MHD value E_θ, even though it has important physical consequences.

This simple picture of the magnetic field structure has two important limitations: one in distance, the other in latitude. The limitation in distance arises because we have assumed that the solar wind velocity is constant and radial. This hypothesis is incorrect near the Sun, for two reasons. First, we saw in Chapter 5 that it is only somewhat farther than the sonic point – located at a few solar radii – that the solar wind speed approaches sufficiently its asymptotic value to remain approximately constant. Second, since the magnetic field is anchored in the rotating Sun, the plasma and field near the Sun tend to rotate with it as they do at its surface. For the radially moving solar wind to be able to drag the field lines out of corotation, it must overpower the magnetic field. This requires that the kinetic energy density surpasses the magnetic one, i.e. that (in the supersonic wind) $\rho v_w^2/2 > B^2/2\mu_0$, or $v_w > v_A$ – the Alfvén speed. We shall see later that this inequality holds true farther from about 10 solar radii.

6.1.2 Basic heliospheric current sheet and other currents

The limitation in latitude of the above picture is related to our hypothesis that the magnetic field B_0 at the source's location is independent of latitude. We shall see in Section 6.2.2 that this approximation is reasonable, except for one important point: in the simplest configuration (which holds approximately true

Basic large-scale magnetic field 297

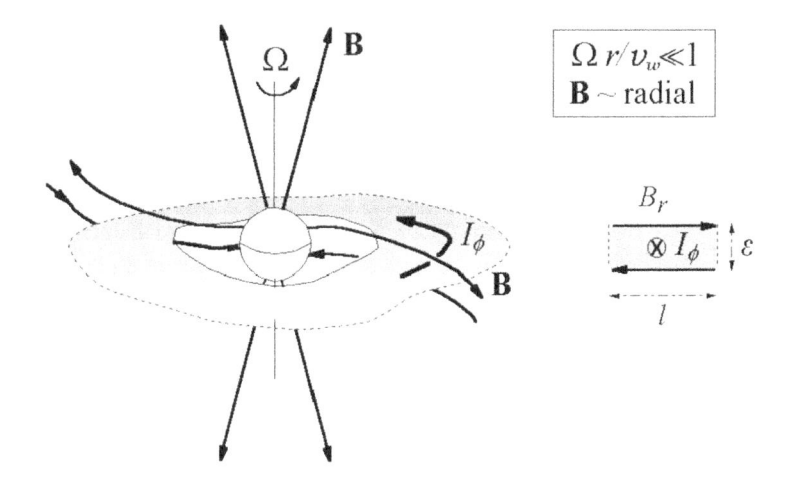

Figure 6.4 Basic structure of the heliospheric magnetic field at small distances, assuming that the Sun's rotation and magnetic axes coincide. The reversal of **B** near the equatorial plane corresponds to a thin sheet of current that is nearly azimuthal at relatively small distances, where **B** is nearly radial. The right panel shows a rectangular cut across this sheet.

near solar activity minimum), the sense of the solar radial magnetic field is opposite in both solar hemispheres – reversing at the equator, as we saw in Section 1.3.3.

To derive Parker's spiral, we sketched in Fig. 6.1 a magnetic flux tube pointing radially outwards at the solar surface, and viewed from above the Sun's north pole. Figure 6.4 completes the picture in three dimensions, near solar activity minimum – when the solar magnetic dipole is nearly aligned with the rotation axis. The magnetic field in both hemispheres follows spirals that are wound up in the same sense, but have opposite **B**. The figure corresponds to a case when **B** points outwards in the northern hemisphere, as occurred near the last solar activity minimum of 1996 – a direction that reverses every 11 years.

This magnetic field distribution corresponds to volume currents flowing in both hemispheres according to Maxwell's equation

$$\nabla \times \mathbf{B} = \mu_0 \mathbf{J} \tag{6.13}$$

that may be calculated from the expressions (6.10)–(6.11) of **B**. With the basic geometry sketched in Fig. 6.4, something peculiar takes place in the equatorial plane: the reversal of the magnetic field direction there corresponds to a thin sheet of electric current flowing in this plane, as already mentioned in Section 1.3.3. Let us calculate this current.

Consider first relatively small distances, so that $\Omega r/v_w \ll 1$, making the magnetic field nearly radial. Imagine a rectangle (perpendicular to the equatorial plane) of length l along the radial and thin (vertical) width ϵ (Fig. 6.4,

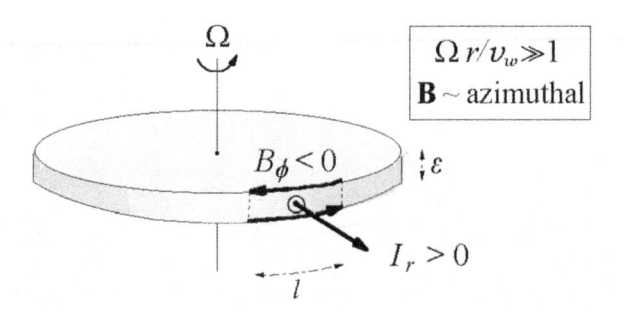

Figure 6.5 At large distances, the magnetic field is nearly azimuthal; in that case, the reversal of **B** near the equatorial plane corresponds to a thin sheet of nearly radial current.

right), and integrate Maxwell's equation (6.13) across the surface of this rectangle. Using Stokes' theorem to transform the surface integral of the left member of (6.13) into a line integral, we find $2B_r l = \mu_0 J_\varphi \times l\epsilon$. The current density J_ϕ corresponds to a thin sheet of azimuthal current $I_\varphi = J_\varphi l\epsilon$ in the equatorial plane,[3] i.e.

$$I_\varphi = 2B_r/\mu_0 \quad \text{per unit of radial length.} \quad (6.14)$$

At greater distances, **B** is no longer roughly radial, so that the current is no longer roughly azimuthal, but its azimuthal component is still given by (6.14). Let us calculate its amplitude at 1 AU. Substituting the magnetic field (6.10) into (6.14), we find $I_\varphi \sim 6 \times 10^{-3}$ A m^{-1}, which yields roughly 10^9 A across a radial length of 1 AU.

Consider now the other extreme: large distances, so that $\Omega r/v_w \gg 1$, making **B** nearly azimuthal.

To calculate the current flowing in the equatorial sheet, we draw a rectangle still perpendicular to the equatorial plane and of thin (vertical) width ϵ, but now of length l along the azimuthal direction (Fig. 6.5). Integrating Maxwell's equation (6.13) over this rectangle as above and using Stokes' theorem, we now find $-2B_\varphi l = \mu_0 J_r \times l\epsilon$, whence the radial current[4] $I_r = J_r l\epsilon$, i.e.

$$I_r = -2B_\varphi/\mu_0 \quad \text{per unit circumference length.} \quad (6.15)$$

Since at radial distance r the length of the circumference is $2\pi r$, the total radial current flowing through the sheet is

$$I_r = -4\pi r B_\varphi/\mu_0. \quad (6.16)$$

---

[3] With $B_r > 0$ in the northern hemisphere, (6.14) yields $I_\varphi > 0$, i.e. the current flows in the direct sense.

[4] Here B_φ is the magnetic field in the northern hemisphere, which is negative in the case considered, so that $I_r > 0$, i.e. the current flows outwards.

Basic large-scale magnetic field

At smaller distances, **B** is no longer roughly azimuthal, so that the current is no longer roughly radial, but its radial component is still given by (6.16). To calculate its amplitude, we substitute the magnetic field (6.11) into (6.16), to find a total radial current $I_r \sim 5 \times 10^{-9}$ A, independent of distance. This current flows outwards in the case considered in the figures, where the radial magnetic field points outwards in the northern hemisphere, so that B_φ is negative there – a configuration that reverses every 11 years.

We conclude that in this simple picture the vicinity of the equatorial plane carries a current that is nearly azimuthal at small distances (where **B** is nearly radial), and nearly radial at large distances (where **B** is nearly azimuthal). At intermediate distances, this current is still perpendicular to **B**, so that it also follows a spiral shape.

Now, a question arises. The current flowing radially outwards in the equatorial sheet must return to the Sun somewhere, otherwise the Sun would charge indefinitely; how does it manage to do so? The answer is simple. The current flowing radially along the sheet balances the sum of all the volume currents that flow radially inward in both hemispheres (Problem 6.6.2).

6.1.3 Magnetic field effects on the wind

We have seen how the solar wind affects the magnetic field, drawing the field lines along spirals near the solar equatorial plane, and roughly radially at high latitudes (where the rotation plays a smaller role). But how does the magnetic field affect the wind? At the microscopic level, the particles are forced to follow helical paths along the field lines. At the macroscopic level, this yields two main effects: a macroscopic magnetic force acting on the plasma, and an inhibition of the transport processes in the direction perpendicular to the field, so that the heat flux is directed along the magnetic field.

The magnetic forces were ignored in the previous chapter because we assumed spherical symmetry with both the magnetic field and the velocity oriented radially, so that $\mathbf{v}_w \times \mathbf{B} = 0$. The above results indicate that this approximation is acceptable at distances smaller than about 1 AU, and/or at high latitudes, where **B** is roughly radial. Elsewhere, the magnetic forces may be neglected only if the magnetic energy $B^2/2\mu_0$ is negligible compared to either the kinetic energy ρv_w^2 or the thermal energy, i.e. if $v_w \gg v_A$ (the Alfvén speed), or if

$$\beta \equiv \frac{n k_B (T_p + T_e)}{B^2/2\mu_0} \gg 1. \tag{6.17}$$

Table 6.1 shows the values of v_A and β estimated with the parameters of Fig. 4.1 and of Table 5.1, and the magnetic field given by (6.10)–(6.11), at the distances of 10 $R_\odot$ and 1 AU. Much farther out, both the density n and B^2 decrease roughly as r^{-2}, so that v_A is roughly constant with distance; furthermore, since from Section 5.3 the sum of temperatures decreases relatively slowly, the parameter β decreases slowly with distance beyond 1 AU.

One sees in Table 6.1 that the parameter β is generally of order unity or smaller, but the ratio of the flow speed to the Alfvén speed – which is far

Table 6.1 *Orders of magnitude of the Alfvén speed and of the plasma β (ratio of thermal to magnetic energy) at the heliocentric distances of 10 $R_\odot$ and 1 AU*

Distance r	10 $R_\odot$	1 AU
v_A (km s^{-1})	$\sim 10^3$	$\sim 10^2$
β	$\sim 10^{-2} - 10^{-1}$	~ 1

smaller than unity in the low corona – increases strongly with distance, so that v_w becomes greater than v_A beyond 10–20 $R_\odot$. In fact, since only the magnetic field component perpendicular to the flow speed acts, and (outside solar active regions) the magnetic field is nearly radial close to the Sun, the magnetic forces may be neglected already at distances from the Sun significantly smaller than the one where $v_w = v_A$.

At large distances, the magnetic field is nearly perpendicular to the flow velocity, but the magnetic forces are nevertheless of minor importance there because the solar wind speed far surpasses the Alfvén speed. In contrast, the magnetic forces are expected to play a major role closer than the distance where $v_w = v_A$ if **B** is not close to the radial there.

The solar wind ejection by the rotating Sun has another consequence: the Sun ejects not only mass but also angular momentum. Indeed, a unit mass of matter rotating at the angular speed Ω has the angular momentum Ωr^2 at distance r. If matter were ejected directly from the solar surface at the rate of M' kg s^{-1}, the Sun would lose angular momentum at the rate $M' \times \Omega R_\odot^2$. However, we have seen that the matter around the Sun tends to rotate rigidly with it when magnetic forces dominate, i.e. roughly out to a distance r_A at most equal to the one where $v_w = v_A$. Hence the rate of angular momentum loss by the Sun is instead about

$$J' \sim M' \times \Omega r_A^2. \qquad (6.18)$$

More correct calculations yield roughly the same result (see [40]).

With the mass loss rate given in Table 1.3, $\Omega \simeq 2.7 \times 10^{-6}$ rad s^{-1}, and $r_A \leq 20 R_\odot$, we find $J' \leq 10^{24}$ kg m^2 s^{-2}. Since the angular momentum of the Sun is roughly $J \sim 2 M_\odot \Omega R_\odot^2 / 5 \sim 10^{42}$ kg m^2 s^{-1}, the solar wind will brake the solar rotation in a time $J/J' \geq 10^{18}$ s, or about 3×10^{10} years – significantly more than the solar nuclear life time (see Section 3.1.3).

6.2 Three-dimensional structure during the solar cycle

The simple model of Section 6.1 must be refined in several aspects, even though it embodies a large part of the physics.

6.2.1 Warped heliospheric current sheet

To begin with, even in the simplest configuration, which holds near solar activity minimum, the solar rotation axis is not exactly aligned with the magnetic axis, making a small angle α with it, as we mentioned in Section 1.3. As a result, the latitude θ_0 of a fixed point on the solar magnetic equator varies with the azimuthal angle φ_0 as $\sin\theta_0 = \sin\alpha \, \sin\varphi_0$. As the Sun rotates, this yields

$$\sin\theta_0 = \sin\alpha \, \sin(\varphi_0 - \Omega t) \tag{6.19}$$

so that the fixed point on the magnetic equator is alternately above and below the rotational equator. The magnetic equator gives the position of the current sheet close to the Sun, before being drawn outwards by the solar wind flow. This flow transforms the initial angular co-ordinates (θ_0, φ_0) related by (6.19) into values at a large distance r (Fig. 6.6, left) given by

$$\theta = \theta_0$$
$$\varphi = \varphi_0 - \Omega r/v_w$$

from Section 6.1. Substituting these values into (6.19), we deduce that when $\alpha \ll 1$ so that $\sin\theta_0 \simeq \theta_0$, the current sheet is a surface of equation

$$\theta \simeq \alpha \, \sin(\varphi + \Omega r/v_w - \Omega t) \tag{6.20}$$

where θ is the latitude, φ the azimuthal angle and r the radial distance (assumed much greater than the radius of the magnetic equator from which the sheet is drawn). Equation (6.20) represents a surface whose intersection with the rotational equatorial plane ($\theta = 0$) is an Archimedean spiral rotating at angular speed Ω, whereas the intersection with a meridian plane ($\varphi = $ constant) has a wavy shape, varying in time at the frequency Ω and in distance at the wavelength $2\pi v_w/\Omega \sim 6$ AU for $v_w \sim 400$ km s^{-1}.

The solar magnetic equatorial plane is thus deformed as it is carried outwards by the solar wind, and takes the form of a wavy current sheet resembling a spinning ballerina's skirt, whose latitudinal extension (see 6.20) is equal to the angle between the solar rotation and magnetic axes (Fig. 6.6, right).

This wavy current sheet serves as a magnetic equator separating two regions of inward and outward Archimedean spiral fields, and organising the large-scale structure of the solar wind. This simple picture should be refined in many ways, to take into account further dynamic effects in the wind and complexities of the solar magnetic field; furthermore, as solar activity increases, the tilt of the magnetic axis increases, and the Sun's magnetic structure becomes more complex (see [23]). Finally, the whole pattern reverses every 11 years.

6.2.2 Observed large-scale structure

How does this picture compare with observation? On the whole, reasonably well. Detailed reviews may be found in [21] and [30].

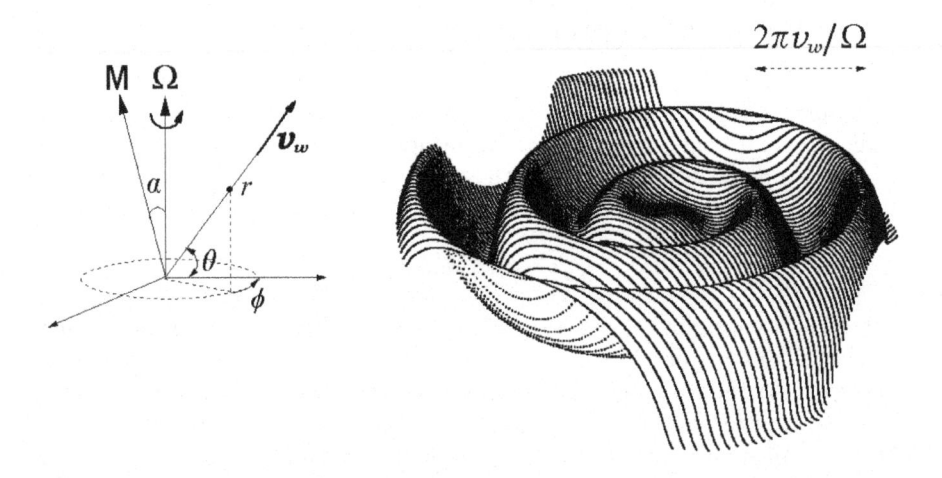

Figure 6.6 Deformation of the current sheet by a (small) tilt α of the Sun's magnetic axis (**M**) with respect to the rotation axis ($\mathbf{\Omega}$), with a constant solar wind radial velocity. The picture shown (right, from [16]) is calculated according to (6.20) with $\alpha = 15°$; the co-ordinate system is indicated on the left.

Observations at moderate distances

As we said in Section 1.3, the only spacecraft having reached high heliographic latitudes is Ulysses, which made a snapshot of the solar wind latitudinal structure at a distance of 1–2 AU, at both solar activity minimum and maximum (Fig. 1.20). When it passed from high southern latitudes to high northern latitudes near solar activity minimum, it found two hemispheres of opposite magnetic polarities with a magnetic field direction very close to Parker's spiral, and a high-speed and dilute wind (see Figs. 1.21–1.22), except in a latitude band of $\pm 20°$ around the equator, where it crossed the current sheet several times (see [6]).

This is illustrated in Fig. 6.7, which shows magnetic measurements from the Ulysses magnetometer.[5] Except in the low-latitude band, the radial component of **B** times r^2 (top panel) is found to be remarkably constant, not only with distance (as expected from conservation of magnetic flux), but also with latitude, with $B_r \simeq 3.1 \times 10^{-9}/r_{AU}^2$ T – a result rather close to our rough estimate (6.10). The tilt of **B** to the radial (bottom panel) is found to be rather close to the one of the expected spiral, whose tilt is from (6.8)–(6.9): $\arctan(\Omega r \cos\theta/v_w)$, with $v_w \simeq 750$ km s^{-1} – the mean solar wind speed measured during this period (curved dashed lines) [8].

On this large-scale structure are superimposed other variations. To begin with, in the low-latitude band where Ulysses encountered the heliospheric current sheet and a variable wind, the changes in direction of the magnetic field were accompanied by large increases in amplitude. These increases in magnetic

---
[5]Described in Balogh, A. et al. 1992, *Astron. Astrophys. Supp.* **92** 221.

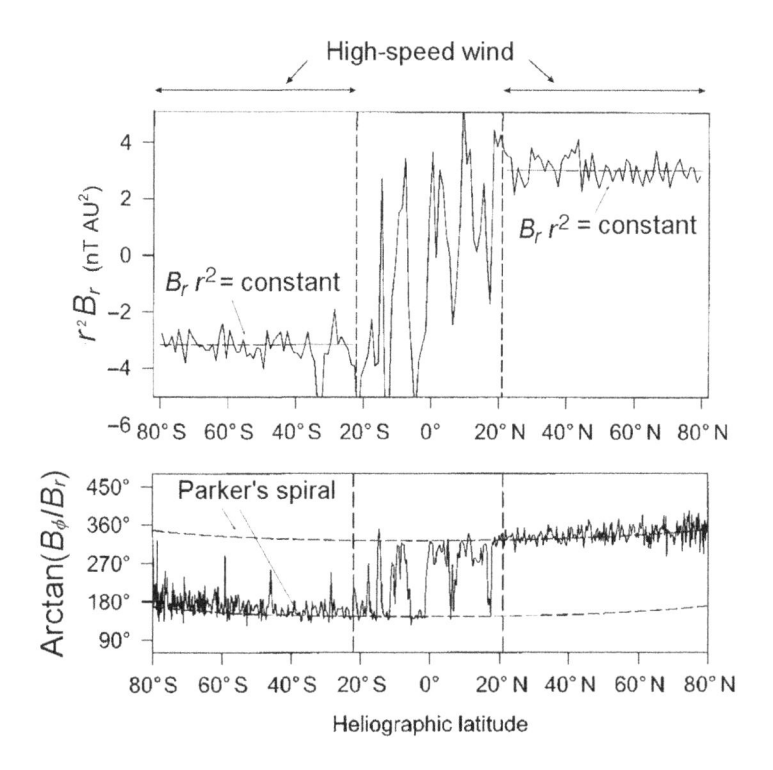

Figure 6.7 Ulysses magnetic field observations during the pole-to-pole transit at solar activity minimum: radial component of **B** normalised to the inverse distance squared (top) and angle between **B** and the radial direction (bottom); the vertical dashed lines indicate the low-latitude boundaries of the high-speed wind. (Adapted from [8].)

field are accompanied by increases in particle density (Fig. 1.18) and are produced by plasma compression as streams of different speeds interact; we shall return to these effects in Section 6.3. The other major feature seen in Fig. 6.7 is a large level of fluctuation in the magnetic field direction, especially near the poles; we shall return to these fluctuations in Section 6.4.

This simple structure becomes more complex as solar activity increases and the solar magnetic structure changes. Not only does the magnetic dipolar axis shift in direction, but multi-polar terms become important (see [26]). Near the maximum of activity, the heliospheric current sheet is highly warped and tilted to the equator. As a result, near solar activity maximum, even though the product $|B_r| \times r^2$ was still observed to be (on average) roughly independent of position and the (average) magnetic field direction was close to Parker's spiral, Ulysses crossed a current sheet many times at virtually all latitudes [32].

These observations concern moderate distances from the Sun. What is observed farther away?

Observations at large distances

A fleet of four spacecraft has explored the solar wind at large distances. Pioneer 10 and 11, launched respectively in 1972 and 1973 by NASA, were the first missions to navigate the asteroid belts, visit outer planets and explore the distant solar wind. They can no longer be counted upon to return data, although they will continue their journey towards deep space for an indeterminate period of time, at speeds of a few AU per year.[6]

The next members of the remote fleet are the two Voyager probes, sent in 1977. Designed and operated by the Caltech Jet Propulsion Laboratory and NASA, they are among the best scientific achievements of unmanned space exploration. Equipped with a battery of sophisticated instruments collecting radiation and particles over a wide energy range and managed by flexible inboard systems, they provided considerable knowledge and understanding on the four outer planets, their atmospheres, magnetospheres, satellites and rings, and a lot of data on the outer heliosphere near the fringes of the solar system, to which we shall return in the next chapter. Using planetary gravity assist[7] – a trick already used to enable Pioneer 11 to meet Saturn after having encountered Jupiter – the two Voyagers made sophisticated jumps from planet to planet.[8] In this way, Voyager 1 was able to return wonderful images and data from Jupiter's and Saturn's environments, whereas Voyager 2 did so for Jupiter, Saturn, Uranus and Neptune, taking advantage of a rare alignment of these planets and the Earth.

Having completed their planetary service with honours, both Voyagers are now travelling away from the Sun at speeds of about 3 AU per year. They are heading towards the incoming interstellar wind, that we shall study in Section 8.1. As of January 2006, Voyager 1 was at roughly 98 AU from the Sun and 34° heliographic latitude, and Voyager 2 at 78 AU and −26° heliographic latitude. Hopefully, they have enough electrical power and thruster fuel in reserve to operate until the late 2010s. The gravity assist manoeuvres have pushed them outside the ecliptic plane, so that their heliographic latitudes are larger than that of in-ecliptic spacecraft, but not very much so.

Figure 6.8 (right) shows the magnetic field amplitude measured by Voyager 1 up to very large distances, along the trajectory plotted in the left panel from 1978 to 1997. It agrees well with the value calculated from Parker's spiral, taking into account the variations in solar magnetic field and in wind speed measured during this period (solid line); to illustrate the effect of variations in wind speed, the dotted lines are calculated with a constant solar wind speed of 400 and 800 km s^{-1} respectively [4].

---

[6]The venerable Pioneer 10 sent its last signal in January 2003 when it was at about 82 AU from the Sun, presumably because it could no longer send transmissions to Earth, due to degradation of its radioisotope power source. The last signal from Pioneer 11 was recorded in November 1995, when it was at about 44 AU from the Sun; shortly after, the Earth went out of the view of its telemetry antenna in spite of the efforts at manoeuvring the spacecraft.

[7]Gurzadyan, G. A. 2002, *Space Dynamics*, New York, Taylor & Francis.

[8]Kohlhase, C. ed. 1989, *The Voyager Neptune Travel Guide, JPL Pub. 89-24*, Pasadena CA, NASA.

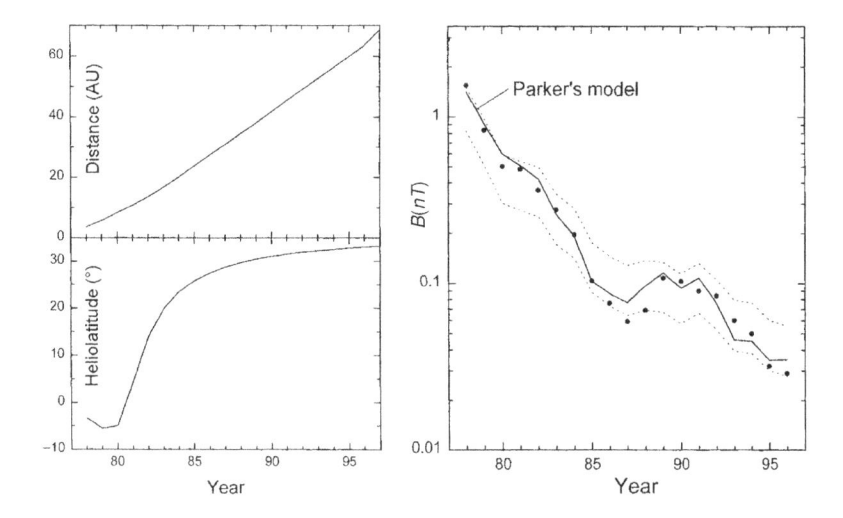

Figure 6.8 Magnetic field strength observed by Voyager 1 versus time (right panel, solid dots) compared with Parker's model (solid curve) taking into account variations in solar magnetic field and wind speed; the dotted curves show Parker's model assuming constant solar wind speeds of 400 and 800 km s^{-1} respectively. The spacecraft trajectory (distance and heliographic latitude) is shown in the left panel. (Adapted from [4].)

Voyager 1 has suffered from its close encounter with Saturn's environment and its plasma instrument has been damaged. But Voyager 2 is still in relatively good shape. Figure 6.9 shows the mean speed (roughly equal to the mean radial velocity) and the mean number density of protons measured by the plasma science instrument[9] on Voyager 2 from 1977 to 2003. The data, covering more than 25 years, are plotted as daily averages. There is not much systematic variation in wind speed with distance, except that small-scale fluctuations decrease with increasing distance; the mean wind speed is 440 km s^{-1}. With a radial speed roughly constant on a large scale, conservation of mass implies that the proton number density should vary roughly as the inverse squared distance; hence we have plotted the density multiplied by Voyager's distance (in AU) squared. As expected, this normalised density does not vary much with distance on large scales; the mean value is 6.6 cm^{-3}.

6.2.3 Connecting the Sun and the solar wind, or: where do the fast and slow winds come from?

To derive the estimates (6.8)–(6.9), we assumed the magnetic field to be drawn at constant radial speed from a rotating source located at some distance r_0. For

---

[9]Described in Bridge, H. S. et al. 1977, The plasma experiment on the 1977 Voyager mission, *Space Sci. Rev.* **21** 259.

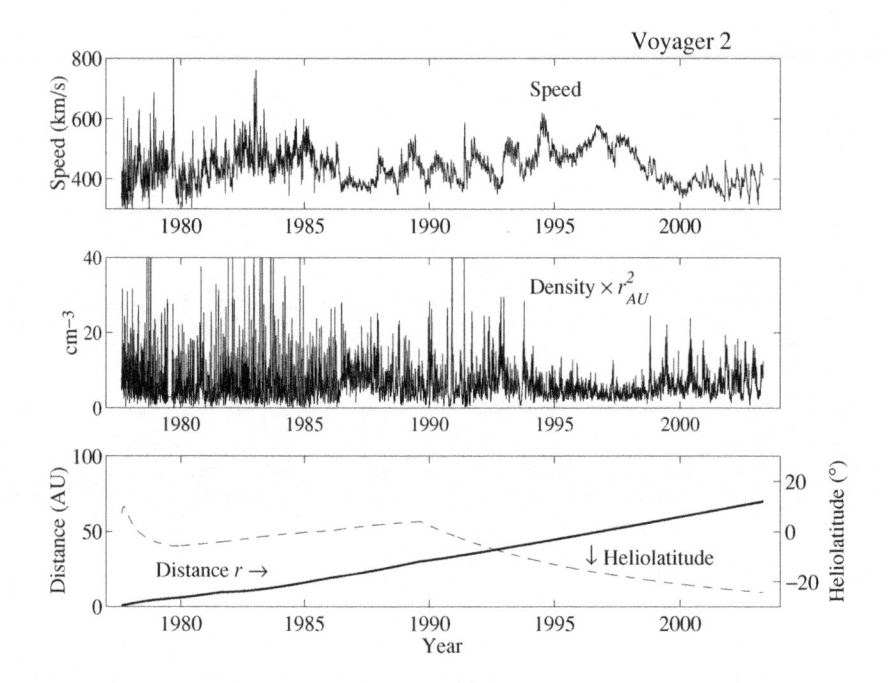

Figure 6.9 Mean proton velocity (top) and density (middle) measured on the spacecraft Voyager 2 from 1977 to 2003. Densities are normalised to 1 AU by multiplying by Voyager's distance squared. The bottom panel shows the distance (thick line) and latitude (dashed line) in solar co-ordinates. (Courtesy of the MIT Space Plasma Group.)

this assumption to be reasonable, r_0 must be somewhat greater than the solar radius $R_\odot$ for two reasons: first, near the solar surface, the atmosphere corotates with the Sun, and, second, it has not yet reached its asymptotic speed there. But how does the magnetic field vary between $R_\odot$ and r_0? And how is the solar wind structure connected to the magnetic structure on the Sun?

This is a difficult problem, because (except for solar active regions[10]) the magnetic field is measured essentially at the solar surface, albeit with not enough resolution, and at large distances in the solar wind. In between, our knowledge of the magnetic field in the source of the wind is based on a few uncertain measurements and on models of unproven validity.

A standard, but far from satisfying, method for relating the magnetic field at the Sun and in the solar wind consists of two steps. First, one maps any point in the wind inward back to some 'source' distance taken somewhat arbitrarily near 2.5 $R_\odot$ to determine a foot point, by assuming radial flow at constant

---

[10] The large magnetic field in solar active regions is measured best, but of little use for modelling the regular fast wind.

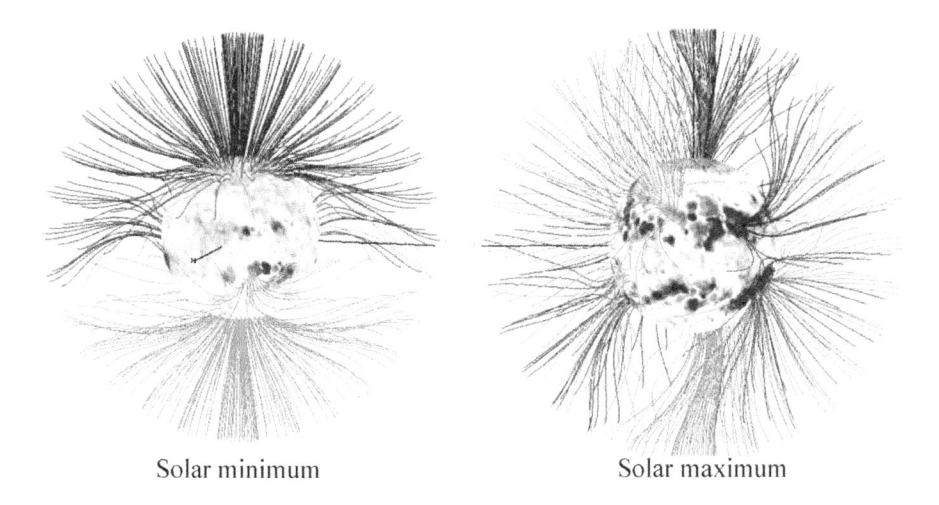

Three-dimensional structure during the solar cycle 307

Solar minimum   Solar maximum

Figure 6.10 Heliospheric magnetic field calculated by extrapolating the measured photospheric and solar wind fields at solar activity minimum (left) and maximum (right). Fields pointing outwards (respectively, inwards) are plotted as black lines (respectively, grey). (Adapted from [20].)

speed as outlined in Section 6.1. Second, one calculates the field between this distance and the solar surface, by assuming that it derives from a potential, with its radial component equal to that measured at the photosphere, and forcing it to be radial at the 'source' surface (see [28], [39] and references therein). Figure 6.10 shows the resulting open magnetic field lines calculated near solar activity minimum and maximum [20].

Near solar activity minimum, the heliospheric magnetic field exhibits the simple bipolar structure already discussed, with oppositely directed fields in both hemispheres (see Fig. 6.4), in contrast to the complex structure exhibited near activity maximum.

An important feature of this kind of modelling is that the field lines diverge faster than radial above the coronal holes, so that most of the fast wind originates from relatively small source regions, as seen in Fig. 6.10 (left). This feature has important consequences on solar wind models, as it affects the solar wind speed and the energy balance (see Sections 5.2.3 and 5.5.1), but it is still controversial. Indeed, as we noted in Section 5.5, some studies claim instead that the magnetic field lines extend radially as sketched in Fig. 6.11 except above active regions, and that comparison of the density measured in the corona and in the solar wind show that the fast wind expands radially, coming from a large part of the solar surface near activity minimum, and not solely from the coronal holes (see [43] and references therein), whereas the slow wind only comes from the vicinity of active regions (which are near the magnetic equator at solar activity minimum).

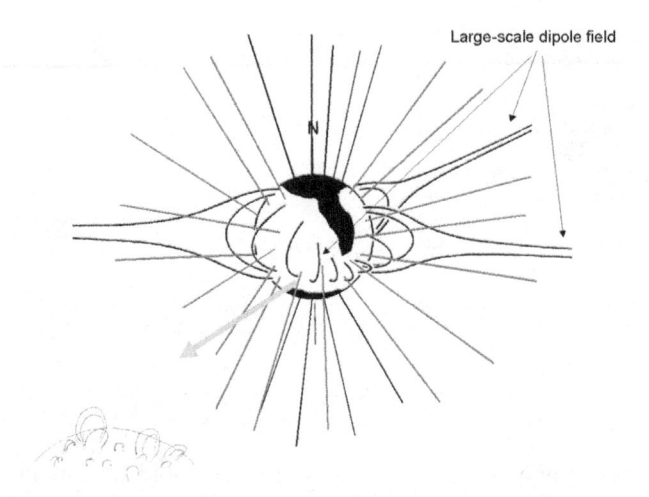

Figure 6.11 An alternative view of the geometry of magnetic field lines around the Sun. The small-scale magnetic field at short distances is essentially radial and the fast wind flows along it, except in the vicinity of active regions [43]. (Courtesy of S. R. Habbal.)

6.3 Major perturbations

6.3.1 Interaction between the fast and slow winds

Consider our observer lying in the wind, equipped with the necessary instruments (Fig. 1.17). Assume first that it lies close to the ecliptic near Earth's distance, at a time where solar activity is close to minimum. Once every week or so, it observes large perturbations as the one that can be seen in Fig. 1.18, which shows the speed, density and magnetic field as a function of time, during nearly a month.

These large perturbations occur when fast and slow wind meet. When does this happen? We have seen that near solar activity minimum the slow wind – coming mainly from the vicinity of (equatorial) active regions – flows along the vicinity of the equatorial sheet, whereas the fast wind flows elsewhere. Because of the inclination and warping of the sheet, the observer encounters it – and therefore the corresponding slow wind – from time to time as the Sun rotates. This means that the observer encounters now and again an interface between slow wind and fast wind, that lies approximately along an Archimedean spiral rotating with the Sun as does its source. This interface is thus called a *corotating interaction region* in the jargon of solar wind specialists. There, the leading edge of the fast wind collides with the slower wind ahead of it. Because each kind of wind originates on different field lines which cannot easily cross, and tends to be frozen to them, the two kinds of wind do not mix freely. The advance of the fast wind towards the slow wind therefore compresses the medium and the frozen in magnetic field.

What is the structure of the interaction region? On the forward side lies the undisturbed slow wind, whereas on the rear side lies the undisturbed fast wind. Consider first the forward side. As time passes and the wind flows, the pressure increases within a small region that moves outwards and pushes the slow wind ahead. In contrast, on the trailing side, a pressure perturbation propagates in the reverse sense – sunward with respect to the fast wind, tending to decelerate it. The net effect is to tend to decrease the speed difference between the fast and slow flows.

How does the interaction region evolve farther from the Sun? As distance increases, the Alfvén speed decreases (because the magnetic energy decreases faster than the particle density), and the sound speed decreases, too (because the temperature decreases), so that the speed of pressure waves (the so-called fast speed calculated in Section 2.3) decreases. When this wave speed becomes too low with respect to the fast/slow speed difference, the pressure waves steepen into shocks. This happens typically farther than a few AU from the Sun. As distance increases still more, the compression regions expand and tend to merge with other ones, while the shocks tend to damp out the fast/slow speed difference. We shall return to shocks in Section 6.3.3.

How does this picture change during the solar cycle? Near activity minimum, the heliospheric current sheet has a small inclination to the rotational equator and is weakly warped, so that corotating interaction regions occur only at very small latitudes. When activity is greater, so is the sheet inclination to the equator, producing interaction regions in a greater latitude range. Finally, close to activity maximum, there are few large fast wind streams and they do not persist for long, so that there are few conspicuous corotating interaction regions.

These structures and their associated shocks have three important consequences: first, they may form barriers to cosmic rays, which tend to inhibit them from entering the inner heliosphere, second, they accelerate particles to high energies, and third, they are responsible for perturbations at Earth. We shall return to these points in the following chapters.

6.3.2 Coronal mass ejections in the solar wind

The other kind of large scale perturbations is produced by *coronal mass ejections* from the Sun (see Fig. 4.10). We have seen that now and then the Sun ejects a huge plasma pocket of $10^{12} - 10^{13}$ kg, moving as fast as a few 10^3 km s^{-1} in the corona. On average over the whole Sun, this happens roughly once a week at solar activity minimum, and three times per day at maximum. These huge bullets produce large perturbations along their trajectory in the solar wind.

These perturbations have features in common with those studied above: the interaction of two different plasmas moving at different velocities, and the production of shocks when the difference in speeds is large compared to the speed of pressure waves. However, the very transitory nature of interplanetary coronal mass ejections makes them difficult to study. Furthermore, close to the Sun, we have only a global knowledge of these objects, inferred from photographs and measurements from a large distance. On the other hand, in the solar wind,

Figure 6.12 There is still no agreement as to the structure of interplanetary coronal mass ejections. (Drawing by F. Meyer.)

we can make detailed *in situ* measurements at a few locations, but we have not enough of them to deduce the whole structure. As a result, not only do we ignore the physics of their initiation, as we saw in Section 4.5, but we do not even know with precision:

- their basic geometry,

- their signature and their effects on the wind.

Even though ideas on coronal mass ejections in the interplanetary medium pre-date the ones on the solar wind, because of their major consequences on our close environment (as we saw in Section 1.1), there is still no agreement on these basic questions.

These events are frequently observed in the ecliptic, where the Earth and most spacecraft lie. This is because their coronal sources, the solar active regions, lie close to the solar magnetic equator near activity minimum, whereas at maximum, where they are much more frequent, they lie virtually everywhere on the Sun.

Their structure is threaded by magnetic field lines and generally carries magnetic helicity, appearing as a twisted flux tube that is nearly force-free (see Section 4.2.3). Near the Sun, they generally appear as closed magnetic field lines attached to the Sun at both ends, but as the structure moves forward, the lines may presumably detach from the Sun. As a consequence of this geometry and of the expansion, these structures exhibit in general (but not always) suprathermal electrons that move towards the Sun along the magnetic field, and low temperatures.

When these structures move faster than the ambient wind, they push it and compress the medium and its magnetic field. Indeed, a large magnetic field often (but not always) characterises them, and when the field is large enough, they are called *magnetic clouds*. On the other hand, when they move slower than the ambient wind, they are pushed ahead by it. When the speed difference is sufficiently large, these pressure perturbations produce shocks, as in the case of corotating interaction regions. Therefore, disturbances moving faster than

Major perturbations 311

the ambient solar wind tend to be decelerated, those moving slower tend to be accelerated, so that in most cases their speed lies in the range of those of the ambient solar wind.

Finally, what are their global effects on the heliosphere? We noted in Chapter 4 that they may enable the Sun to get rid of the excess of magnetic helicity produced in each hemisphere by the solar differential rotation. On a large scale, coronal mass ejections in the interplanetary medium form barriers to cosmic rays, accelerate particles, and perturb the Earth's environment, rather as do corotating interaction regions – and perhaps to a greater extent. We shall return to these points in the next chapters. On average over the solar cycle and in order of magnitude, they contribute about a tenth of the total solar wind mass flux in the ecliptic – a proportion that varies by a large factor during the activity cycle.

6.3.3 Associated shocks

As we noted, the large-scale perturbations of the solar wind often involve compressions at speeds greater than the phase speed of the associated waves, producing shocks. Basic references on this kind of shocks, although somewhat outdated, are [34] and [2]. More specialised accounts may be found in the monographs [35] and [36], with updates in the volume [25], while recent advances based on numerical simulations are reviewed in [17] and [14].

Heliospheric shocks, as do most shocks in space plasmas, are fundamentally different from the classical shocks of gas dynamics (Section 2.3) because they are magnetised, and – most importantly – they are collisionless.

Magnetised collisionless shocks

Since the particle and magnetic pressures are of the same order of magnitude, as are the sound and Alfvén speeds (since from Table 6.1, $\beta \sim 1$ in order of magnitude), the magnetic field affects considerably the compressional waves. There are two wave modes involving compression that can steepen into shocks, and none of them is the usual sound wave of ordinary fluid dynamics. The mode generally involved in heliospheric shocks is the so-called *fast wave* (Section 2.3). This wave makes the particle and magnetic pressures vary in phase, and propagates at a phase speed that depends on the direction of propagation: $(V_A^2 + V_S^2)^{1/2}$ normal to **B**, and the larger of V_A and V_S along **B** (Section 2.3).[11]

The other compressional wave, called the *slow wave* (Section 2.3), makes the particle and magnetic pressures vary out of phase; it gives rise to shocks relatively rarely. It does not propagate across **B**; in this direction it yields merely a tangential discontinuity, in which both the velocity and the magnetic field are parallel to the boundary; without plasma flow across the boundary, the total pressure must be conserved across it in order to ensure dynamic equilibrium;

---

[11] Across such a shock, **B** increases downstream, so that since the normal component is constant, the downstream field turns away from the shock normal.

Fast shock — v (v_n ≠ 0), B_u, B_d, Compression, Dissipation. Particle and magnetic pressures increase.

Tangential discontinuity — v, B. No dissipation. Total (particle + magnetic) pressure conserved. v_t and B_t may vary.

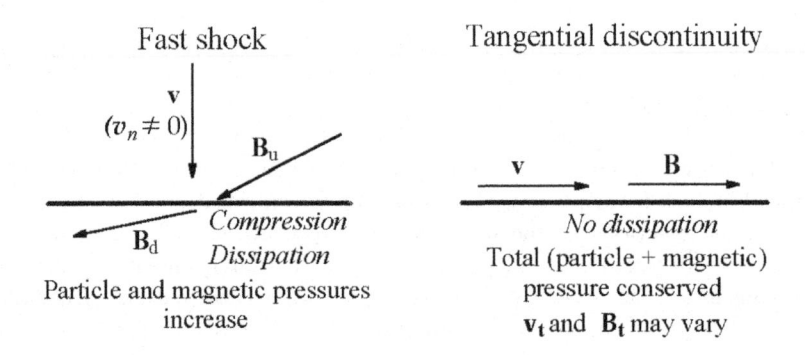

Figure 6.13 Basic geometry of the boundaries that are the most frequent in the heliosphere: fast shocks and tangential discontinuities. The subscripts n (t) denote the components normal (parallel) to the boundary, and u (d) denote the upstream (downstream) values.

hence the particle and magnetic pressure vary in opposite senses. Tangential discontinuities are ubiquitous in the heliosphere, where the (total) pressure has ample time to reach equilibrium.

We shall see in the next chapter that fast shocks and tangential discontinuities (Fig. 6.13) form systematically as the solar wind encounters planets and other large bodies.

Since the Alfvén wave does not change the particle pressure, it does not steepen into shocks.[12] Its steepening yields instead a rotational discontinuity where **B** and **v** change in direction but not in magnitude, with a flow across the boundary but no compression or dissipation; this structure propagates with a speed $B_n/\sqrt{\mu_0 \rho}$ along **B**.

The second – and the greatest – difference with classical shocks is that the free path is larger than the other scales by many orders of magnitude, making these structures virtually collisionless. A basic property of shocks, however, is that some dissipation acts to transform the bulk kinetic energy into particle random motions. Since binary collisions cannot do the job, something else must come to the rescue. This involves collective particle interactions with the electric and magnetic field, including wave instabilities and turbulence, in a way that is still not fully understood, in spite of several decades of research involving observation, theory and simulation.

The particle motion in the mean fields of the shock modifies the velocity distributions and increases the kinetic temperature across the shock, but it does not formally provide dissipation since the equations of motion are reversible. Dissipation is produced by kinetic micro-instabilities, which yield some anomalous resistivity producing Joule heating. However, this process becomes insufficient when the speed exceeds some critical speed (typically $v/V_A > 1 - 3$), and an additional dissipation mechanism is then needed. This additional dissipation

---
[12] Except in very special cases.

Table 6.2 *Typical orders of magnitude of the basic scales in the solar wind at 1 AU from the Sun*

Debye length L_D	10 m
Electron gyroradius r_{ge}	3 km
Proton inertial length $c/\omega_{pi} \equiv V_A/\omega_{gi}$	10^2 km
Proton convected gyroradius v_w/ω_{gi}	10^3 km
Collisional free path l_f	1 AU

mechanism is thought to be provided by the reflection of protons by the shock electric and magnetic fields, which converts upstream bulk motion into gyration motion, and ultimately into random kinetic energy through scattering by waves and turbulence. The irreversibility may be thought of as coming from the development of finer and finer structures, which after some time exceed the resolution of any measurement, so that some spatial and temporal average of the velocity distribution is observed instead of the local instantaneous value.

The rarity of binary collisions has too other consequences. First, the shock drives the plasma far from equilibrium; not only do the electrons and protons have different temperatures, but the velocity distributions are far from Maxwellians – much farther than in the undisturbed solar wind. Not only is the bulk of the velocity distribution driven out of equilibrium, but a few particles are accelerated to very high energies, contributing to the dissipation in the shock via the plasma wave instabilities they generate. This acceleration of particles to high energies makes shocks basic acceleration structures in the Universe, as we shall see in Section 8.2.

Second, the dissipation takes place in a rather large region, so that the shock thickness is not clearly defined. Whereas in classical shocks the shock thickness is of the order of magnitude of the collisional free path, in collisionless shocks it is determined instead by other basic scales of the plasma, defined by the dissipative mechanism at work. Table 6.2 shows typical orders of magnitude of some scales: the Debye length, and the electron and proton gyroradii (note that since $\beta \sim 1$, the gyroradius of a particle moving at the thermal speed is roughly equal to its so-called *skin depth* or *inertial length*[13]).

Because of the role of the magnetic field, the shock structure – and in particular its thickness – is strongly dependent on the geometry (Fig. 6.14). When the magnetic field is roughly perpendicular to the shock surface (the shock is called *quasi-parallel*, in reference to the shock normal **n**), the motion of the particles along **B** carries them through the shock (and away from it) very easily; this makes the width of the shock fairly large. In contrast, when the magnetic field is roughly parallel to the shock surface (the shock is then called

---

[13] The electron gyroradius is $r_{ge} \sim v_{the}/\omega_{ge}$, and with a little manipulation we find $\omega_{ge}/\omega_p \simeq (m_p/m_e)^{1/2} V_A/c$. Hence $r_{ge} \sim (v_{thp}/V_A) \times c/\omega_p$, so that $r_{ge} \sim c/\omega_p$ if $\beta \sim 1$. Similarly, $c/\omega_{pi} \equiv V_A/\omega_{gp}$ – the gyroradius of a proton moving at the Alfvén speed (equal in order of magnitude to the proton thermal speed in the solar wind).

Quasi-parallel   **Quasi-perpendicular**

Shock surface   Shock surface

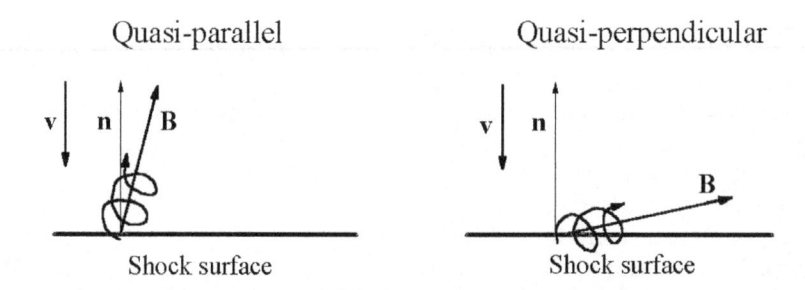

Figure 6.14 One-dimensional sketch showing the extreme cases when the magnetic field is roughly normal to the shock (left) and roughly parallel to it (right). When **B** is normal to the shock, the particles can easily cross its surface, producing a large shock width, whereas when **B** is parallel to the shock, the particle gyration keeps them close to its surface, producing a small width.

quasi-perpendicular), the motion of the particles along **B** keeps them close to the shock surface, making the shock thickness fairly small – typically of the order of magnitude of the ion gyroradius.

These various parameters produce a wide range of shock structures. Whatever the details, however, the basic conservation relations of mass, momentum and energy hold.

Conservation relations at MHD discontinuities

In the ideal case when the shock is stationary, when the particle distributions are not too far from Maxwellians, and when the waves do not carry away an appreciable part of the energy, the MHD generalisation of the fluid equations yields conservation relations that generalise the Rankine–Hugoniot relations (Section 2.3).

Consider the simple case of a one-dimensional steady shock, as in Section 2.3, in the frame where the shock is stationary. Mass conservation yields, as for an ordinary shock

$$[\rho v_n] = 0. \tag{6.21}$$

Here the subscripts n (and t respectively) denote the normal (and tangential respectively) components, and the symbol $[\ ]$ denotes the discontinuity at the shock. From Maxwell equations, we have

$$[B_n] = [\mathbf{E}_t] = 0. \tag{6.22}$$

With $\mathbf{E} = -\mathbf{v} \times \mathbf{B}$, this yields $[v_n \mathbf{B}_t - B_n \mathbf{v}_t] = 0$, whence

$$[v_n \mathbf{B}_t] = B_n [\mathbf{v}_t]. \tag{6.23}$$

Momentum balance yields in the normal direction

$$[\rho v_n^2 + P + B^2/2\mu_0] = 0 \tag{6.24}$$

and in the transverse direction: $[\rho v_n \mathbf{v}_t - B_n \mathbf{B}_t/\mu_0] = 0$, whence, using (6.21)–(6.22)

$$\rho v_n [\mathbf{v}_t] = B_n [\mathbf{B}_t]/\mu_0. \tag{6.25}$$

Finally, energy conservation yields

$$\left[\rho v_n \left(\frac{v^2}{2} + \frac{\gamma}{\gamma-1}\frac{P}{\rho}\right) + v_n \frac{B^2}{\mu_0} - \mathbf{v}\cdot\mathbf{B}\frac{B_n}{\mu_0}\right] = 0 \tag{6.26}$$

where the first term is the flux of bulk kinetic energy and enthalpy, and the last two terms stem from the energy flux $\mathbf{E}\times\mathbf{B}/\mu_0$, with $\mathbf{E} = -\mathbf{v}\times\mathbf{B}$.

These relations hold for any MHD discontinuity, of which shocks are a special case which requires a flow through the surface ($v_n \neq 0$) and some compression and dissipation. In the case of tangential discontinuities (Fig. 6.13), one sees immediately that the conservation relations reduce to the continuity of the total (particle plus magnetic) pressure.

Applying these conservation relations to a shock, one can show that the upstream and downstream magnetic field and the shock normal all lie in the same plane (Problem 6.6.3).

In the vicinity of shocks, the menagerie of plasma waves and instabilities is extremely active, as is turbulence, and the electric field accelerates particles in an efficient way.

We shall consider the acceleration of particles in Section 8.2, and make below a short incursion into the menagerie of plasma waves, which are ubiquitous in the solar wind, with a population that is extremely variable – depending as expected on the energy resources.

6.4 Waves and turbulence

Let us return to our solar wind observer (Fig. 1.17). In addition to the large-scale structures discussed above, and to some other ones that we shall spare the reader, it observes in permanence small-scale variations. There is a whole continuum of variations of virtually all scales and amplitudes, from the ubiquitous quasi-thermal fluctuations to large disturbances. Because of the coupling between waves and particles, these perturbations may be investigated with emphasis either on particles, waves or turbulence, depending on their properties ... and on the bias of the investigator. These perturbations are especially conspicuous in the vicinity of the large-scale perturbations studied in the previous section.

6.4.1 Waves

A huge number of wave modes can propagate in a magnetised plasma. Many of these modes may become unstable under some circumstances, and an interesting zoo has yet to be interpreted [27]. In particular, we have seen in Section 5.4

that even in the stationary solar wind the electron velocity distribution exhibits a large skewness – responsible for a large heat flux – and that the proton distribution tends to be anisotropic. This can drive many plasma wave instabilities; although these instabilities may be viewed as providing additional interacting degrees of freedom which may come to the rescue of particle collisions for providing dissipation, they only prevent the velocity distributions from becoming too crazy, but do not keep them close to Maxwellian.

In the solar wind, the plasma frequency f_p is much greater than the electron cyclotron frequency f_{ge}. For example at 1 AU from the Sun, with a density $\sim 5\times 10^6$ m^{-3}, and a magnetic field amplitude $\sim 5\times 10^{-9}$ T, we have $f_p \sim 2\times 10^4$ Hz and $f_{ge} \sim 150$ Hz. Hence the waves of frequency near and above f_p are virtually unaffected by the magnetic field. We have seen that high-frequency electromagnetic waves are virtually undamped, so that they may go to a large distance from their region of production, and may therefore be observed very far from it. In contrast, electrostatic waves are in general heavily damped, so that they must be observed close to their place of emission. Since unstable electrostatic waves are generally intermittent, they are therefore difficult to observe directly, and are more frequently inferred from the electromagnetic waves they generate, which fill a far larger space.

The solar wind velocity $\mathbf{v}_w$ introduces a complication: the wave frequency is measured on a spacecraft that is generally moving at a small velocity (compared to v_w) with respect to the Sun, and therefore at a velocity of about $-\mathbf{v}_w$ relative to the medium. As a result, a wave of angular frequency ω and wave vector $\mathbf{k}$ is observed at a frequency $\omega - \mathbf{k}\cdot\mathbf{v}_w$, so that the relative Doppler frequency shift is $\Delta\omega/\omega \sim v_w/v_\phi$, where v_ϕ is the phase speed. This shift is especially important for acoustic and MHD waves whose phase speeds – of the order of magnitude of the sound and Alfvén speeds – are much smaller than v_w.

The waves observed most frequently in the solar wind are:

- electromagnetic waves, of $k = \sqrt{\omega^2 - \omega_p^2}/c$, that propagate nearly as in vacuum at $f \gg f_p$,

- electrostatic Langmuir waves, of $k = \sqrt{\omega^2 - \omega_p^2}/(\sqrt{3/2}v_{th})$, so that $\omega/k \sim v_{th}$ in order of magnitude – producing Landau damping – except at $\omega \simeq \omega_p$, where $\omega/k \gg v_{th}$ so that they interact with suprathermal electrons,

- acoustic waves, whose Doppler shift carries them up to frequencies much above the proton plasma frequency,

- whistler waves (below the electron gyrofrequency in the plasma frame),

- electromagnetic MHD waves (Alfvén and magnetosonic) below the proton gyrofrequency in the plasma frame, of which only the Alfvén waves are not strongly damped.

Figure 6.15 (top) is a typical wave spectrogram at frequencies of the order of magnitude of f_p in the solar wind near solar activity minimum. It is displayed as

Waves and turbulence 317

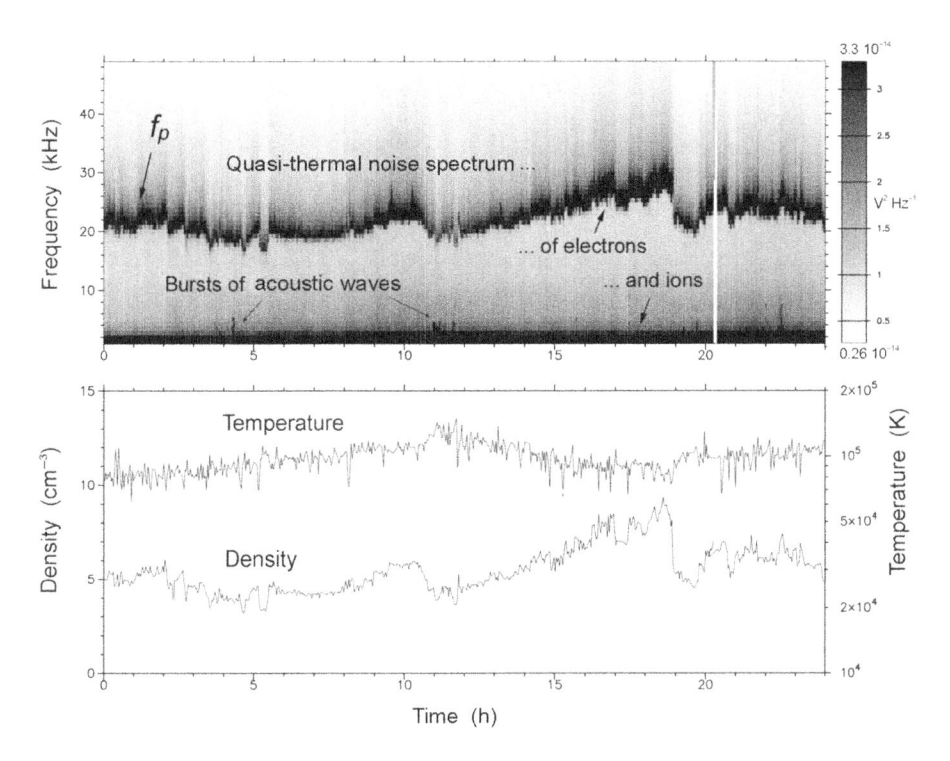

Figure 6.15 Typical electric spectrogram near solar activity minimum in the ecliptic, with the ubiquitous quasi-thermal spectrum of the plasma, plus a few bursts of acoustic waves. It was measured on Ulysses with the 70-m tip-to-tip electric antennae of the URAP instrument in March 1995, near the ecliptic at $\simeq$ 1.3 AU. The bottom panel shows the electron density and temperature during this day, deduced in routine from analysis of the quasi-thermal spectrogram. (Courtesy of M. Moncuquet.)

frequency versus time with intensity indicated by the grey bar chart on the right; hence each vertical cut represents the spectral density of the squared voltage on the electric antenna at a given time. The spectrogram essentially shows a broad line peaking close to the local plasma frequency, which represents the quasi-thermal (electrostatic) fluctuations of the local plasma. The shape of the spectrum is due to the fact that the electric antenna selects the wavelengths of the order of magnitude of its proper length; therefore if this length is greater than the Debye length $L_D \sim v_{th}/\omega_p$, the antenna selects the Langmuir waves near ω_p, whose wavelength is greater than L_D and which are excited in permanence by the quasi-thermal motions of the electrons. This makes the spectrum peak at the local plasma frequency. The spectral shape can be calculated from the velocity distributions of the particles, and its inversion furnishes a diagnostics of these particles (Fig. 6.15, bottom) [18]. Indeed, apart from bursts of acoustic waves, the spectra observed in the solar wind can be quantitatively explained with an

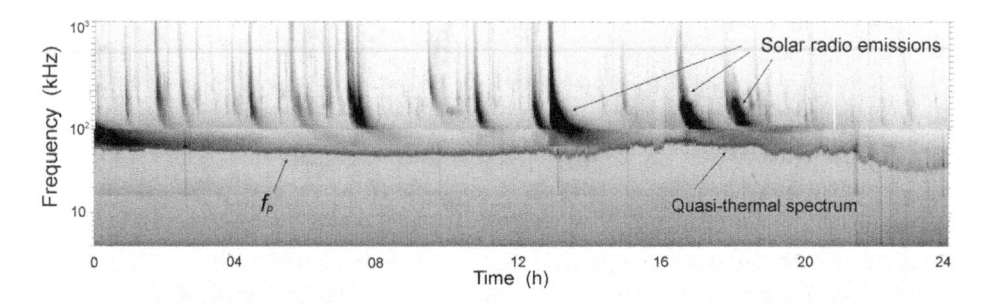

Figure 6.16 Typical electric spectrogram near solar activity maximum in the ecliptic, showing a number of radio emissions produced by electrons accelerated in solar flares, in addition to the ubiquitous plasma quasi-thermal spectrum around f_p. This spectrum was measured on the spacecraft WIND with the 100-m tip-to-tip electric antennae of the WAVES instrument in December 2001, in the ecliptic at 1 AU. (Data courtesy of C. Perche.)

accuracy of a few per cent by the theory of plasma quasi-thermal fluctuations [15].

The relatively large fluctuations observed in the plasma parameters (Fig. 6.15, bottom) are typical of those observed near solar activity minimum in the ecliptic, where we have seen that different kinds of winds interact (compare with Fig. 1.21, obtained at high latitudes where the spacecraft is permanently immersed in fast wind).

We have seen in Section 4.5 that solar flares are frequent near solar activity maximum. The corona then ejects packets of electrons travelling at high speeds, which produce a wave instability at the local plasma frequency along their trajectory in the solar wind. This produces electromagnetic waves at both f_p and its harmonic along their trajectory. Such waves can be seen in Fig. 6.16, which is a typical spectrogram acquired near solar activity maximum in the ecliptic at 1 AU. Particles accelerated in the vicinity of shocks can also produce electromagnetic waves. A review of these emissions may be found in [12].

At low frequencies, the figures also show a few bursts of acoustic waves. Bursts of various low frequency waves are ubiquitously observed in the solar wind. In particular, whistler waves are observed nearly in permanence below the electron gyrofrequency, and (Doppler-shifted) acoustic waves are observed below the plasma frequency and are strongly intermittent.

6.4.2 Turbulence

Magnetohydrodynamic waves are ubiquitous in the solar wind, in the form of Alfvén waves, and their distribution has much in common with fluid turbulence. But what is turbulence?

Oddly enough, there is no precise answer to this question. Turbulent flows are all around us – and even inside us – in both air and water, and have fascinated minds for ages (Fig. 6.17). Yet, despite more than a century of active

Figure 6.17 A sketch by Leonardo da Vinci, of what he called 'turbolenza'.

research on many fronts, including mathematics, modern statistical and nonlinear physics, engineering and computer simulations, turbulence is still not properly understood [33]. Rather surprisingly, 'less is known about the fine scale of turbulence – for example, the scale of 1 mm in the Earth atmosphere – than about the structure of atomic nuclei' [9]. Turbulence is often considered as the last great problem of modern physics that remains unsolved.

Yet turbulence has innumerable practical applications. It enhances enormously the transport of matter, momentum and heat. Life as we know it would not be possible without turbulence, and we have seen in Chapter 3 the importance of turbulence for heat transport and magnetism in the Sun. It has also undesirable consequences, and its control would bring considerable progress in our technology.

Elements of fluid turbulence

Leonardo's sketch illustrates beautifully a cheap experiment that can be performed in your kitchen. Open a water tap slightly, and you get a regular stationary water column; open it a little more, and you may see regular structures or periodicities, which become irregular as you open more the tap, to yield a complex 'turbulent' structure, both in the jet and in the sink. Turbulence has a very special character: it is disordered, but not completely so, showing both order and disorder; namely, small fluctuations and macroscopic structure do coexist.

At first sight, the basic problem appears simple, being based on fluid equations (Section 2.3) that have been known for much more than a century, and seem benign enough. The physics neither involves small scales requiring quantum mechanical concepts, nor large scales requiring general relativity concepts, and the experiments do not involve great energies requiring giant particle accelerators. The equations are, however, far from benign, having singularities, and containing a non-dimensional parameter – the Reynolds number R (the ratio of inertial to viscous forces) that is generally much greater than unity. The problem is simple only at low Reynolds numbers; as R increases, the flow undergoes a sequence of instabilities and becomes turbulent. Due to the large value of the non-dimensional parameter R, there is a wide range of relevant scales, so that numerical simulations require desperately long computer times.

The problem is non-linear and dissipative, has chaotic solutions (it exhibits great sensitivity to the initial conditions) and shows intermittency. A review of these concepts may be found in [9], the subtleties of which are nicely discussed at a simple level in a book by David Ruelle [24],[14] and in more detail, in the book [44].

Chaos means that you cannot predict the long-term evolution (although the governing equations are deterministic), because any small variation in the initial conditions (or any small perturbation) produces a change that increases exponentially with time. A popular image is that a butterfly may change drastically the weather at some future time and distant place – a valid excuse to the inability of meteorologists to predict the weather over more than a few days, first noted by Edward Lorenz.[15]

The intermittency of turbulence manifests itself in the fact that turbulent activity is more and more localised as one observes smaller scales. Large fluctuations are far more probable than expected from Gaussian statistics, so that the probability distributions of the fluctuation amplitude have tails (as do the kappa functions introduced in Section 5.5), the more so as the scale considered in the statistics decreases. This behaviour can be described with the tools used to study fractals[16] – a popular example of which is the length of a coast: as you measure it with a tool of smaller size, you see smaller details, so that the total measured length increases.

Let us return to our tap. The viscosity of water in normal conditions is $\nu \sim 10^{-6}$ m^2 s^{-1}. With a diameter $L \sim 1$ cm, the Reynolds number $R = vL/\nu \sim 10^4 \times v$, where v is the flow speed in m s^{-1}; this yields $R \gg 1$ even at modest speeds. In astronomy, the Reynolds numbers are often still greater because of the huge scales. Since virtually every three-dimensional flow of large Reynolds number exhibits turbulence, the ubiquity of turbulence is not surprising.

We have seen in Section 2.3 that fluids behave as incompressible at small speeds. In this case, the continuity equation reduces to $\nabla \cdot \mathbf{v} = 0$, so that the equation of motion (Section 2.3) reduces to

$$\partial \mathbf{v}/\partial t + \mathbf{v} \cdot \nabla \mathbf{v} = -\nabla p/\rho_0 + \nu \nabla^2 \mathbf{v} \quad \text{with} \quad \nabla \cdot \mathbf{v} = 0, \qquad (6.27)$$

the incompressible *Navier–Stokes* equations.

---

[14]Who first introduced chaos in the theory of the transition to turbulence, with the paper: Ruelle, D. and F. Takens 1971, On the nature of turbulence, *Comm. Math. Phys.* **20** 167. It is amusing to note that, as often occurs for seminal papers and as nearly occurred for E. Parker's seminal paper on the solar wind, this paper was rejected by the editor to which it was first submitted.

[15]Lorenz, E. N. 1963, Deterministic non periodic flow, *J. Atmos. Sci.* **20** 130.

[16]More precisely, multifractals (i.e. the scaling to smaller scales is not self-similar).

Waves and turbulence

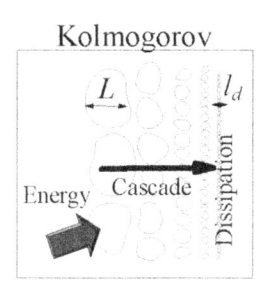

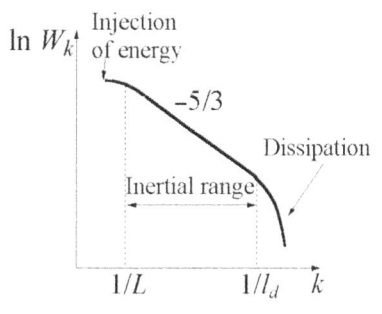

Figure 6.18 Kolmogorov's scheme. Left, sketch of the cascade in which energy is injected at large scales and is transferred through a series of steps to small scales where it is dissipated (assuming that turbulence fills all space, i.e. no intermittency). Right, corresponding wave number spectrum of the turbulent energy.

Kolmogorov's law

To compensate for our inability to solve these equations, one may use their form to derive universal similarity laws. Assume that mechanical energy is injected at some large scale L across which the velocity difference is V and let $R = VL/\nu \gg 1$ be the corresponding Reynolds number. Differential motions create shear and eddies, and in steady state, the energy fed into the fluid cannot accumulate; it cannot dissipate either because $R \gg 1$ makes viscosity negligible at the scale L.

There is, however, a way for energy to dissipate: producing scales so small that viscosity becomes efficient. What is the limiting scale l_d at which this happens? Let v_d be the velocity difference at this scale (i.e. the difference in speed between two points distant by l_d). With the order of magnitude estimate $\nabla \sim 1/l_d$, the corresponding inertial term in (6.27) is $\sim v_d^2/l_d$, whereas the viscous force is $\sim \nu v_d/l_d^2$. Viscosity becomes efficient when the latter dominates, i.e. for scales smaller than

$$l_d \sim \nu/v_d. \tag{6.28}$$

Since energy is injected at scale L and cannot dissipate except at scales smaller than l_d, it is progressively transferred via non-linear interactions to eddies of smaller and smaller scale (Fig. 6.18, left); this continues down to the scale l_d where the cascade ends because viscous forces become important. In the range between L and l_d (the so-called inertial range, for which inertia dominates viscosity), viscosity is negligible, so that the system is ruled by Euler equations, which contain no parameter. The solution is therefore scale invariant, and if it is steady, homogeneous and isotropic, one can deduce how the velocity difference v_l across an eddy of size l varies with l.

Indeed, when viscosity may be neglected, the only relevant physical parameter for an eddy is the energy ϵ that cascades through it per unit time per unit mass. From energy conservation, ϵ is the rate of energy injection at large scales, the rate of energy dissipation by viscosity at scales below l_d, and the rate of energy cascading through eddies of intermediate size. Dimensional analysis yields $\epsilon \sim v_l^3/l$; indeed, the energy per unit mass $\sim v_l^2$ is transferred during a time $\sim l/v_l$, so that the energy per unit time per unit mass is $\epsilon \sim v_l^2/(l/v_l)$. Hence the velocity fluctuation at scale l varies as

$$v_l \sim (l\epsilon)^{1/3} \propto l^{1/3} \qquad (6.29)$$

where we have dropped ϵ out of the parenthesis since it does not depend on the size l in the absence of viscous dissipation (i.e. for $l_d < l < L$).

This is the classical Kolmogorov's law: the velocity fluctuation varies with the scale as $l^{1/3}$ – a universal scaling that does not depend on the type of flow, nor on the Reynolds number. By the same argument, the moments of order n of the velocity differences v_l at scale l (the so-called structure functions) obey the relation

$$\langle v_l^n \rangle \propto (l\epsilon)^{n/3}. \qquad (6.30)$$

One can deduce how the energy fluctuations at a given location are distributed over spatial scales. This is generally plotted as the spectral density of the energy fluctuations as a function of wave number: $W_k(k)$. Since the energy in the fluctuations at scale l varies as $v_l^2 \propto l^{2/3}$ from (6.29), the fluctuation energy per unit wave vector (in one direction) $k \sim l^{-1}$ varies as $l^{2/3} \times l \propto l^{5/3}$, i.e. as $k^{-5/3}$. Hence the (one-dimensional) spectrum of the fluctuation energy varies as

$$W_k \propto k^{-5/3} \quad \text{for} \quad L^{-1} < k < l_d^{-1} \quad \text{(inertial range)}. \qquad (6.31)$$

Note that applying Kolmogorov's law (6.29) to both l_d and the largest scale L yields $(l_d/L)^{1/3} = v_d/V$. Substituting into (6.28), we get[17]

$$l_d \sim L \times (\nu/VL)^{3/4} \sim L \times R^{-3/4}. \qquad (6.32)$$

Consequences and limitations

These scaling laws are universal (within the framework of the hypotheses made) and are indeed found to hold in a host of systems in which the Reynolds number is large. This brings about three important points. First, in order to solve numerically a three-dimensional problem, the computational work must vary as the third power of the ratio of the outer to the inner scale: L/l_d; from (6.32), this means that the required computational work increases as $R^{9/4}$, for a single time step. In order to study the temporal evolution, computational times are thus expected to be extremely long for large Reynolds numbers.

---

[17] Beware that this estimate may be out by a large factor, given the drastic hypotheses made.

Second, since the dissipation rate $\epsilon \sim v_l^3/l$ is independent of viscosity, it should remain finite as $\nu \to 0$. Therefore, dissipation acts even in the limit of infinitely small viscosity, whereas there is no dissipation in the absence of viscosity. This apparent paradox is related to the well-known crucial difference between an ideal inviscid fluid (a fluid without viscosity, which John von Neumann nicely called *dry water* to emphasise its unrealistic character) and a real fluid having a small viscosity.[18]

This may be illustrated with a simple example. Consider the flow past an object (Fig. 6.17, top). This is equivalent to the same object moving through the fluid, and we know that if the Reynolds number is large, the object is subjected to a drag force $F_D \sim \rho v^2/2$ per unit cross-section, where ρ is the fluid mass density and v the velocity; the formula is correct to a factor of order unity that depends very weakly on the Reynolds number. For a three-dimensional object of size L, the total force is $F_D L^2$, so that the power dissipated by the flow past the object is $F_D L^2 \times v \sim \rho v^3 L^2/2$. Since the mass involved is ρL^3, the power dissipated by unit mass of the flow is $\sim \rho v^3/2L$ – roughly independent of the Reynolds number provided it is large. In other words, the dissipation is indeed finite in the limit of zero viscosity. Basically, this is because as viscosity decreases, vortices of smaller and smaller size appear, so that the viscous term in the dynamic equation (6.27) does not tend to zero when viscosity does.

This brings about a third consequence. Turbulence is expected to enhance considerably diffusion and the other transport coefficients.

A detailed review of Kolmogorov's ideas and modern ones may be found in the book [10].

This old Kolmogorov picture is based on several assumptions; in particular it assumes turbulence to be steady, homogeneous and isotropic, which is not fully true. Turbulence is highly intermittent, becoming less and less space-filling as the scale decreases, which produces deviations from the Kolmogorov scaling law that increase with the order of the moment considered in the statistics. This intermittency often involves highly anisotropic vortex tubes, whose dynamics is not well understood; stretching vortex lines in a three-dimensional flow enhances vorticity, which, however, requires some viscosity to be created.

Turbulence in the solar wind

The application of these concepts in the solar wind is far from straightforward, for several reasons. First of all, the solar wind is compressible. In supersonic compressible turbulence, the fluctuations tend to steepen into shock waves, in which dissipation may occur in one jump, without having to go through an intermediary cascade as in the Kolmogorov scheme. Second, the solar wind viscosity (Section 2.3) yields a Reynolds number $R \sim v_w L/v_{thp} l_f$, which is not very large, contrary to ordinary turbulent fluids; the dissipation is expected to be driven by waves and instabilities rather than the ordinary viscosity. Third, the solar wind is magnetised; the magnetic field introduces an anisotropy and

---

[18] Feynman R. P., R. B. Leighton and M. Sands 1964, *The Feynman Lectures on Physics*, vol. 2, New York, Addison-Wesley, p. 3.

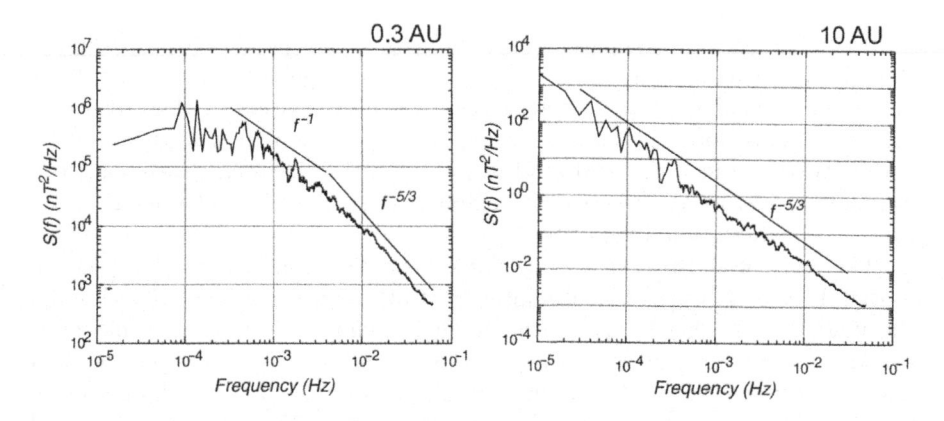

Figure 6.19 Power spectra of magnetic field fluctuations at 0.3 AU from the Sun (measured on the Helios spacecraft) and at 10 AU (measured on Voyager 2). (Adapted from [13].)

additional forces, so that the turbulent 'eddies' are MHD waves (more precisely Alfvén waves since the other ones are in general damped); since pure Alfvén waves are solutions of the ideal MHD equations, non-linear interactions yielding an energy cascade are not inevitable. Furthermore, in this MHD problem the cascade itself might be expected to yield a scaling relation different from the one in an ordinary fluid (Problem 6.6.4). And finally, we have seen in the previous chapters that the MHD description itself is very dubious, the more so as the scales of turbulence are ridiculously much smaller than the particle mean free path.

What does observation tell us? At frequencies below the local proton gyrofrequency, where fluid Alfvén waves propagate, one observes fluctuations of the magnetic field **B** and the velocity **v** that are highly correlated, as

$$\delta \mathbf{v} \sim \pm \delta \mathbf{B}/\sqrt{\mu_0 \rho} \tag{6.33}$$

with the individual components of **B** fluctuating much more than the modulus B. This reveals Alfvén waves, and for the dominant wave population the sign is found to correspond to waves propagating away from the Sun in the plasma frame; this strongly suggests a solar origin.

Whereas large-scale perturbations are more frequent in the slow wind, these small-scale turbulent fluctuations in both the magnetic field and the velocity are of greatest amplitude in the fast wind; there the magnetic field fluctuations are of the order of magnitude as the average field itself.

Figure 6.19 shows the energy spectrum of fluctuations measured at both relatively small and large distances from the Sun. The spectra were obtained as a function of frequency, because the magnetic field was measured aboard spacecraft as a function of time. Since the spacecraft are moving with respect to the medium at roughly $-\mathbf{v}_w$ (producing a Doppler shift $\mathbf{k} \cdot \mathbf{v}_w$), much greater than the phase speed of MHD waves, the spectra in frequency may be transformed into spectra in the wave vector component k along $\mathbf{v}_w$ by the transformation $k \simeq \omega/v_w \simeq 2\pi f/v_w$; this means that 1 second corresponds to a wavelength of the order of several hundreds kilometres – the solar wind speed. For example, the slow to fast wind changes that occur roughly once every 10 days (Fig. 1.18) contribute to the spectral density of fluctuations at frequencies around 10^{-6} Hz.

Figure 6.19 shows the power spectrum of magnetic fluctuations at greater frequencies, i.e. smaller scales. One sees that the spectrum has a power law shape with index $-5/3$, above a frequency that decreases with increasing heliocentric distance. This frequency varies from about 4×10^{-3} Hz at 0.3 AU to around 10^{-5} Hz at 10 AU [13]. At 1 AU, it is observed to lie around 10^{-4} Hz, which corresponds to a length scale $l_{out} \sim v_w/f \sim 2 \times 10^{-2}$ AU.

We shall see in Section 8.2 that this scale has important consequences on the propagation of cosmic rays through the heliosphere. This Kolmogorov-like spectrum, corresponding to the inertial range of fully developed fluid turbulence, holds up to very large frequencies or wave numbers (very small scales), corresponding roughly to the inverse of the proton gyroradius, where dissipation is presumed to take place. Detailed measurements also indicate that the turbulence is strongly intermittent and anisotropic, depending on the inclination of the wave vector to the magnetic field.

Several important questions arise. First, what produces the injection of energy at large scales where the non-linear cascade starts, and what determines the value l_{out} of the corresponding outer scale, to which corresponds the small frequency limit of the $-5/3$ spectrum? In order for this spectrum to derive from the non-linear energy cascade discussed above, the typical time for a turbulent eddy to interact non-linearly ($t_{NL} \sim l/v_l$) should be smaller than the solar wind expansion time ($\sim r/v_w$) at distance r, which yields an outer limit on the scale l: $l_{out} \sim rv_l/v_w$; with $v_l \sim V_A$ for $l \sim l_{out}$, this yields $l_{out} \sim rV_A/v_w$, which is of the same order of magnitude as the value observed.[19]

Second, what produces the dissipation whose scale sets the inner spatial scale (i.e. outer frequency limit) of the $-5/3$ power spectrum at about the proton gyroradius (i.e. about the proton gyrofrequency)? The mechanism should involve kinetic effects, since MHD is out of the question at such scales. And why does the spectrum generally vary as $f^{-5/3}$ as in *isotropic* fluid turbulence? These questions are far from being understood, and involve the shape and geometry of the structures, which are strongly intermittent. The observed intermittency is greater than in ordinary fluid turbulence and has a different character; instead of occurring mainly as filament vortices, the large perturbations are one-dimensional, in the form of current sheets and shocks (see [5] and references therein); furthermore again, the intermittency of magnetic field fluctuations is different from the velocity ones.

---

[19] A. Mangeney and C. Salem, personal communication 2005. Note that using (6.29), $l_{out} \sim rv_l/v_w$, and $v_l \sim V_A$ for $l \sim l_{out}$, we have $l_{out} \sim \epsilon^{1/2} (r/v_w)^{3/2}$ and $\epsilon \sim V_A^2 \times v_w/r$, i.e. the cascading energy per unit mass corresponds to the square of the Alfvén speed divided by the expansion time.

6.5 Minor constituents

The solar corona and the wind are made of protons and electrons, plus minor constituents that change the mean particle mass by at most 20%, so that they are not expected to change the basic behaviour of the medium.

Minor constituents, however, are not always of minor importance. For example, we saw above that viscosity cannot be ignored even when it is infinitesimally small; another example of the importance of minor constituents is catalysis in chemistry. Furthermore, as we already mentioned, the different charges and masses of minor ions in the solar corona and wind enable them to act as tracers for testing the validity of some theories.

Charge transfer collisions of heavy solar wind ions with neutral atoms of various origins leave the ions in a highly excited state, whose de-excitation produces emission in extreme ultraviolet and X-ray wavelengths (Section 2.4.4). This takes place in particular near comets (Section 7.5.6), planets (Section 7.2) and in the outer heliosphere (Section 8.1.2).

And finally, the knowledge of the composition of the Sun and the solar wind is essential for understanding the evolution of the stars and of the universe.

6.5.1 Abundances: from the Universe to the solar wind

When the solar system formed, some $4 - 5 \times 10^9$ years ago, it captured a bit of the interstellar medium, which subsequently evolved to produce the Sun, planets and smaller bodies, and their environments. How are these productions typical of the universe as a whole?

The question is not simple. To begin with, the bit of the interstellar medium that gave birth to the solar system was not necessarily 'typical'. Second, we have seen that the Sun processes elements in its core, and the interstellar medium is expected to have evolved chemically since the formation of the solar system. Indeed, while hydrogen, helium and other light elements were produced at the beginning of the universe, elements heavier than lithium are produced in stars. Third, gravitational and electrostatic forces in the solar atmosphere and wind act differently on elements of different charge and mass, which may modify their concentrations. And finally, the system is not closed, since particles are penetrating it from outside and leaving, too, as we shall see in Section 8.1 (see [41]). Nevertheless, for reasons that are not fully understood, the solar and galactic abundances appear to be relatively similar, apart from a few exceptions.

The solar composition [11] is essentially determined from the spectroscopy of solar radiation – which yields the composition of the photosphere, and also from analysis of a class of meteorites[20] – that are supposed to have retained most of the elements present in the primitive solar nebula, and to yield more accurate measurements. The agreement between both methods is remarkably good.

---

[20] The CI carbonaceous chondrites.

The composition of the solar wind [42] is somewhat different, being altered by several mechanisms that involve ionisation processes and dynamics in the solar atmosphere and the solar wind.

6.5.2 Helium and heavier solar wind ions

Helium

Apart from hydrogen, helium is the most abundant element in the Sun and the solar wind. Essentially all solar wind helium is in the doubly ionised state He^{++} – i.e. alpha particles. Its relative abundance relative to protons is $n_\alpha/n \simeq 0.04 - 0.05$; because its atomic mass is nearly $4m_p$, helium constitutes nearly 20% of the solar wind mass. The helium concentration is somewhat smaller and variable in the slow wind, and somewhat greater and rather constant in the fast wind. It is extremely small near the heliospheric current sheet, and is often considerably enhanced in the interplanetary coronal mass ejections.

Ionisation states

Heavier elements are in various ionisation states, that reveal the physical conditions in the corona, in the region where frequent collisions between particles ensure that ionisation and recombination balance each other; at larger distances, the ionisation state is expected to remain frozen in since there are not enough collisions to change it.

The freezing-in is expected to occur at a distance where the timescale for ionisation change equals the time for the particle to travel a scale height. Since ionisation depends on the electron temperature, the electron temperature in this region of the corona may be estimated from charge-state measurements, as about 1 million degrees in the fast wind and 50% more in the slow wind – consistent with measurements in the corona which yield smaller electron temperatures in the regions where the fast wind comes from.

Abundances of heavy ions

Elements heavier than helium together make up only 0.1% by number. Their concentrations are modified with respect to the photosphere according to their energy of ionisation and to the dynamics of the flow. Atoms having a low first ionisation potential, such as metals, are easily ionised. Those whose first ionisation potential is smaller than the ionisation energy of hydrogen (the Bohr energy) tend to be more abundant with respect to the photosphere than harder-to-ionise elements, in particular helium – a difference that is greater in the slow wind than in the fast wind.

A large part of this fractionation is expected to occur in the chromosphere, where most of the ionisation takes place, but the observations are not yet explained, although a large number of models have been proposed.

Temperatures and speeds of ions

All ions have temperatures roughly proportional to their mass in the fast wind, whereas they are rather similar in the more collisional low-speed wind.

This means that their thermal speeds, rather than their temperatures, are roughly equal in the fast wind. In particular, He^{++} ions are roughly four times hotter than protons.

Likewise, in the fast wind, alpha particles and heavier ions are observed to move faster than protons, with a speed difference that is roughly along the magnetic field and equal to the Alfvén speed (or smaller), and decreases with heliocentric distance. In contrast, in the more collisional slow wind, all ions move at approximately the same speed.

Measuring ions having temperatures roughly proportional to their mass and moving faster than protons is not a surprise if kinetic effects are at work in the absence of collisions. Explaining the observations in a self-consistent way, however, is a difficult task. The most popular interpretation involves wave particle interactions at cyclotron resonance (see [37] and references therein). However, velocity filtration (see [22] and references therein) may produce a behaviour qualitatively similar since the gravitational potential that makes the temperature increase (Section 4.6) is proportional to the particle mass (and heavier ions are subjected to a greater electric force since they carry a greater electric charge). However, none of the proposed theory is fully satisfying.

6.5.3 Pick-up ions

The above picture concerns the bulk of the ions. There is, however, another component of the ion distribution: the *pick-up* ions. Pick-up ions are produced when neutral particles are ionised locally in the solar wind, either by photoionisation or by charge exchange with a solar wind proton. These neutrals come from diverse sources: essentially the dust and the various bodies present in the wind (Chapter 7), and the interstellar medium (Section 8.1). The importance of pick-up ions lies in the fact that they can reveal their sources, and when they are sufficiently numerous – as happens near some bodies and in the outer heliosphere – they affect the dynamics. We shall return to this later. Far from their sources, however, the concentration of pick-up ions is extremely small – 10^{-3} (or less) smaller than the one of protons.

Pick-up ions are easily distinguished from genuine ions coming from the corona with the solar wind by three main properties. First, contrary to heavy ions of coronal origin, heavy pick-up ions are in general singly charged (except He, whose double ionisation cross-section is very small), because their sources are much colder than the corona. Second, their spatial distribution is very different from the one of normal solar wind ions, revealing their origin. And finally, their velocity distribution is extremely different.

The difference in velocity distribution may be understood qualitatively from simple analysis. Pick-up ions are generally produced from neutrals moving with respect to the Sun at a speed that is much smaller than the solar wind speed v_w.

These neutrals are thus roughly at rest in the solar frame. Once ionised, they are subjected to the solar wind magnetic field and to the electric field $-\mathbf{v}_w \times \mathbf{B}$ in the solar frame. Hence, whereas they keep their initial velocity parallel to $\mathbf{B}$ (which is roughly zero *in the solar frame*), they begin to gyrate in the magnetic field and acquire a drift velocity equal to $\mathbf{v}_{w\perp}$ – the projection of $\mathbf{v}_w$ normal to $\mathbf{B}$ (Section 2.2). This happens very quickly, in a time of the order of the inverse of their gyrofrequency. In other words, they are picked up by the solar wind in the direction perpendicular to $\mathbf{B}$.

Hence in the *solar wind frame*, they have a velocity $\simeq -\mathbf{v}_{w\|}$ (the projection of $\mathbf{v}_w$ on $\mathbf{B}$) parallel to $\mathbf{B}$, and a gyration at a speed equal to their initial perpendicular velocity in the solar wind frame, which is roughly $v_{w\perp}$ because their initial velocity in the solar frame is roughly zero. The resulting velocity distribution has both a beam and a ring, and is therefore highly unstable. It is thus quickly isotropised, producing in the solar wind frame a spherical shell distribution, of speed v_w (because of energy conservation).

Therefore, *in the solar frame* (moving at $-\mathbf{v}_w$ with respect to the solar wind), the speed of the particles varies from zero (particles whose velocity is $-\mathbf{v}_w$ in the solar wind frame) to $2v_w$ (particles whose velocity is $\mathbf{v}_w$ in the solar wind frame). The velocity distribution in the solar frame is thus relatively flat, up to a speed of twice the solar wind speed, and falls off quickly for greater speeds [19].

The maximum energy of pick-up ions is therefore $4 \times m_p v_w^2/2$ for protons with a solar wind speed of 400 km s^{-1} – which comes to about 3 keV, and tens of keV for heavy ions.

We shall see examples of pick-up ions produced from neutrals ejected by dust, comets and planets in Chapter 7, and produced from interstellar neutrals streaming through the heliosphere in Section 8.1.

6.6 Problems

6.6.1 Parker's spiral

Find an approximation of the heliospheric distance where an average magnetic field line near the solar equatorial plane has wrapped itself around the Sun once. How many times has the magnetic field wound around the Sun by a distance of 100 AU? To perform these estimates, choose a reasonable solar wind speed. Answer these questions for high latitudes, too; in that case, what solar wind speed will you assume near solar activity minimum?

6.6.2 Heliospheric currents

Show that the total radial current flowing in the heliosphere in both hemispheres is exactly the opposite of the radial current flowing along the current sheet [31].

Hint

Calculate the radial current density from Maxwell's equation (6.13) with the magnetic field given in (6.11), and integrate it over a sphere of radius r.

6.6.3 Coplanarity in MHD shocks

Prove that the upstream ($\mathbf{B}_u$) and downstream ($\mathbf{B}_d$) magnetic fields and the normal to the shock all lie in the same plane. Deduce that the normal to the shock is parallel to the vector $(\mathbf{B}_u - \mathbf{B}_d) \times (\mathbf{B}_u \times \mathbf{B}_d)$.

Hints

The conservation relations imply that if $B_n \neq 0$, then $[v_n \mathbf{B}_t]$ and $[\mathbf{B}_t]$ are parallel since they are both parallel to $[\mathbf{v}_t]$. From $[v_n \mathbf{B}_t] \times [\mathbf{B}_t] = 0$, write explicitly the discontinuities as the difference between upstream and downstream quantities, to find $[v_n] (\mathbf{B}_{tu} \times \mathbf{B}_{td}) = 0$. Hence with $[v_n] \neq 0$, $\mathbf{B}_{tu}$ and $\mathbf{B}_{td}$ are parallel, so that $\mathbf{B}_u$, $\mathbf{B}_d$ and $\mathbf{n}$ all lie in the same plane. Thus $(\mathbf{B}_u - \mathbf{B}_d)$ and $(\mathbf{B}_u \times \mathbf{B}_d)$ both lie in the plane of the shock, so that their scalar product is parallel to $\mathbf{n}$.

6.6.4 Kraichnan's spectrum in magnetofluid turbulence

The magnetic field destroys the isotropy of turbulence. Let us picture the fluid 'eddies' whose interaction produces the energy cascade as Alfvén waves moving at opposite Alfvén speeds v_A. In this case, they interact only during the time taken by an Alfvén wave to travel their size, i.e. $t_l \sim l/v_A$. The energy cascading through an eddy of size l during this time t_l is

$$\Delta E_l \sim \frac{v_l^2}{l/v_l} \times \frac{l}{v_A} \sim \frac{v_l^3}{v_A}. \tag{6.34}$$

The energy v_l^2 (per unit mass) exchanged with a number N of such interactions is (assuming a random walk process): $v_l^2 \sim \sqrt{N} \Delta E_l$, whence (substituting the above value of ΔE_l) $N \sim (v_A/v_l)^2$. Since N interactions of duration t_l require the time $N t_l$, the energy cascading per unit time is

$$\epsilon \sim v_l^2 / (N t_l) \sim v_l^4 / (l v_A). \tag{6.35}$$

Show that this yields $v_l \propto l^{1/4}$ instead of Kolmogorov's law (6.29), so that the spectral density of turbulent energy should vary as $k^{-3/2}$ instead of the Kolmogorov value $k^{-5/3}$.

References

[1] Balogh, A., R. G. Marsden and E. J. Smith 2001, *The Heliosphere near Solar Minimum*, New York, Springer.

[2] Biskamp, D. 1973, Collisionless shock waves in plasmas, *Nucl. Fus.* **13** 719.

[3] Burlaga, L. F. 1995, *Interplanetary Magnetohydrodynamics*, Oxford University Press.

[4] Burlaga, L. F. *et al.* 1998, Heliospheric magnetic field strength out to 66 AU: Voyager 1, 1978-1996, *J. Geophys. Res.* **103** 23727.

[5] Carbone, V. *et al.* 1995, Experimental evidences for differences in the extended self-similarity scaling laws between fluid and magnetohydrodynamic turbulent flows, *Phys. Rev. Lett.* **75** 3110.

[6] Crooker, N. U. *et al.* 1997, Coronal streamer belt asymmetries and seasonal solar wind variations deduced from Wind and Ulysses data, *J. Geophys. Res.* **102** 4673.

[7] Fleck, B. and T. H. Zurbuchen eds. 2005, *Solar Wind*, vol. 11/*SoHO 16: Connecting Sun and Heliosphere*, ESA SP-592, ESA/ESTEC, Noordwijk.

[8] Forsyth, R. J. *et al.* 1996, The heliospheric magnetic field at solar minimum: Ulysses observations from pole to pole, *Astron. Astrophys.* **316** 287.

[9] Frisch, U. and S. A. Orszag 1990, Turbulence: challenges for theory and experiment, *Physics Today* **43** 24.

[10] Frisch, U. 1996, *Turbulence: The Legacy of A. N. Kolmogorov*, Cambridge University Press.

[11] Grevesse, N. and A. J. Sauval 1998, Standard solar composition, *Space Sci. Rev.* **85** 161.

[12] Gurnett, D. A. 1995, Heliospheric radio emissions, *Space Sci. Rev.* **72** 243.

[13] Goldstein, M. L. 2001, Major unsolved problems in space plasma physics, *Astrophys. Space Sci.* **277** 349.

[14] Hellinger, P. 2003, Structure and stationarity of quasi-perpendicular shocks: numerical simulations, *Planet. Space Sci.* **51** 649.

[15] Issautier, K. *et al.* 1999, Quasi-thermal noise in a drifting plasma: theory and application to solar wind diagnostics on Ulysses, *J. Geophys. Res.* **104** 6691.

[16] Jokipii, J. R. and B. Thomas 1981, Effects of drift on the transport of cosmic rays. IV. Modulation by a wavy interplanetary current sheet, *Ap. J.* **243** 1115.

[17] Lembege, B. *et al.* 2004, Selected problems in collisionless shock physics, *Space Sci. Rev.* **110** 161.

[18] Meyer-Vernet, N. et al. 1998, in *Measurement Techniques in Space Plasmas: Fields*, Geophys. Monogr. Ser. **103**, ed. R. F. Pfaff et al., Washington DC, American Geophysical Union, p. 205.

[19] Möbius, E. et al. 1985, Direct observation of He$^+$ pick-up ions of interstellar origin in the solar wind, *Nature* **318** 426.

[20] Neugebauer, M. et al. 2002, Sources of the solar wind at solar activity maximum, *J. Geophys. Res.* **107** 1488.

[21] Ness, N. F. and L. F. Burlaga 2001, Spacecraft studies of the interplanetary magnetic field, *J. Geophys. Res.* **106** 15803.

[22] Pierrard, V. et al. 2004, Exospheric distributions of minor ions in the solar wind, *J. Geophys. Res.* **109** A02118.

[23] Riley, P. et al. 2002, Modelling the heliospheric current sheet: solar cycle variations, *J. Geophys. Res.* **107** 1136.

[24] Ruelle, D. 1991, *Chance and Chaos*, Princeton University Press.

[25] Russell, C. T. ed. 1995, *Physics of Collisionless Shocks*, Oxford UK, Pergamon.

[26] Sanderson, T. R. 2003, Observations of the Sun's magnetic field during the recent solar maximum, *J. Geophys. Res.* **108** 1035.

[27] Schwartz, S. J. and I. W. Roxburgh 1980, Instabilities in the solar wind, *Phil. Trans. Roy. Soc. London* A **297** 555.

[28] Schatten, K. H. 1999, In *Sun–Earth Plasma Connections*, Geophys. Monogr. Ser. **109**, ed. J. L. Burch et al., Washington DC, American Geophysical Union, p. 129.

[29] Schwenn, R. and E. Marsch eds. 1991, *Physics of the Inner Heliosphere*, vols. 1 and 2, New York, Springer.

[30] Smith, E. J. 2001, The heliospheric current sheet, *J. Geophys. Res.* **106** 15819.

[31] Smith, E. J. and B. Tsurutani 1978, Observations of the interplanetary sector structure up to heliographic latitudes of 16°: Pioneer 11, *J. Geophys. Res.* **83** 717.

[32] Smith, E. J. et al. 2001, Ulysses in the south polar cap at solar maximum: heliospheric magnetic field, *Geophys. Res. Lett.* **28** 4159.

[33] Sreenivasan, K. R. 1999, Fluid turbulence, *Rev. Mod. Phys.* **71** S383.

[34] Tidman, D. A. and N. A. Krall 1971, *Shock Waves in Collisionless Plasmas*, New York, Wiley.

[35] Tsurutani, B. T. and R. G. Stone eds. 1985, *Collisionless Shocks in the Heliosphere: A Tutorial Review*, Geophys. Monogr. Ser. **34**, Washington DC, American Geophysical Union.

[36] Tsurutani, B. T. and R. G. Stone eds. 1985, *Collisionless Shocks in the Heliosphere: Reviews of Current Research*, Geophys. Monogr. Ser. **35**, Washington DC, American Geophysical Union.

[37] Tu, C.-Y. and E. Marsch 1995, *MHD Structures, Waves and Turbulence in the Solar Wind*, New York, Kluwer.

[38] Velli, M., R. Bruno and F. Malara eds. 2003, *Solar Wind*, vol. 10, American Institute of Physics.

[39] Wang, Y.-M. et al. 1996, The magnetic nature of coronal holes, *Science* **271** 464.

[40] Weber, E. J. and L. Davis Jr 1967, The angular momentum of the solar wind, *Ap. J.* **148** 217.

[41] Wimmer-Schweingruber, R. F. ed. 2001, *Solar and Galactic Composition*, American Institute of Physics.

[42] Wimmer-Schweingruber, R. F. 2002, The composition of the solar wind, *Adv. Space Res.* **30** 23.

[43] Woo, R. and S. R. Habbal 2000, Connecting the Sun and the solar wind: source regions of the fast wind observed in interplanetary space, *J. Geophys. Res.* **105** 12667.

[44] Zeldovich, Ya. B. et al. 1990, *The Almighty Chance*, Singapore, World Scientific.

7

Bodies in the wind: dust, asteroids, planets and comets

> We are no other than a moving row
> Of Magic Shadow-shapes that come and go
> Round with the Sun-illumined Lantern held
> In Midnight by the Master of the Show;
>
> Omar Khayyam, *Rubaiyat*[1]

The light of the Sun heats the bodies of the Solar System, making them evaporate, and driving their atmospheres, while photons of adequate energy ionise the surfaces and atmospheres. But the wind of the Sun does much more: it carves these environments, producing elaborate structures and powering mighty engines.

The Solar System bodies are extremely diverse, from dust grains to asteroids, comets, planets – and space probes. Even though the interactions are extremely diverse, too, depending on the flow properties, on the nature of the object, and on its size relative to the basic plasma scales, they illustrate similar basic processes. We survey in this chapter the different kinds of bodies and the basic physics of their interaction with the solar wind.

The range of sizes is very large: from small dust grains made of minute assemblages of atoms, to the largest planet, Jupiter. The small dust grains respond to a number of forces: the solar gravitational and radiative forces, electrostatic and Lorentz forces (since they carry an electric charge), and the ram pressure of the solar wind. Because these objects are much smaller than the basic scales of the medium, they do not perturb it significantly, except through the exchange of particles that are absorbed on their surface or emitted.

Larger bodies, on the other hand, may produce significant perturbations, depending on their size, electric conductivity, magnetic field and atmosphere. Bodies like our Moon, which are made of insulating material and nearly devoid

---
[1] Trans. E. FitzGerald.

of atmosphere, interact weakly, essentially in absorbing the incident solar wind particles and in creating a wake depleted of plasma. In contrast, objects having either a conspicuous atmosphere (for example comets, Venus and Mars) and/or magnetic field (for example the Earth and the giant planets) are capable of halting the solar wind, producing a bow shock and a cavity containing matter of their own that is elongated in a long tail. While the plasma tail of comets may be seen as a blue straight tail oriented in the anti-solar direction, the tails of planets are too tenuous to be observed remotely, but they are responsible for accelerating the particles that produce the auroras.

Solar wind perturbations on planets have instigated the emergence of a new discipline: space meteorology – how the Sun affects the Earth environment, and how this can be predicted. And, finally, a growing number of space probes carrying various instruments travel in the solar wind; understanding how they interact is essential for interpreting the observations, and for designing propulsion and navigation.

7.1 Bodies in the wind

7.1.1 Various bodies

Solar wind dust particles are released by comets and other bodies, and produced by collisions between asteroids; another contribution comes from the interstellar medium. Reviews may be found in the books [52] and [34], and in [53] and [54].

Their properties are deduced from a combination of observation techniques. Small particles reflect and scatter sunlight, producing what is seen as the so-called *zodiacal light* when observation points away from the sun, and as the so-called *solar F-corona* when observation points close to the Sun.[2] Zodiacal light appears to the naked eye on dark clear nights as a faint band of light distributed about the ecliptic. When they penetrate into the Earth's atmosphere, dust grains and larger bodies may be observed as meteors, which can be analysed with various techniques – improving in many ways over the age-old observation of 'shooting stars'. Additional measurements come from statistics of craters produced by impacts on the Moon, as the size of the crater is related to the size of the projectile. And finally, accurate *in situ* measurements of small dust grains are acquired from space probes, which carry detectors of increasing sophistication [77].

The distribution of large bodies includes comets and asteroids [43]. It is watched with special attention because of potential hazards on the Earth. The great majority of asteroids lies in the region between Mars and Jupiter – the *asteroid belt* – but an increasing number of bodies are being discovered near the orbit of Neptune and beyond: the so-called *Kuiper belt* objects [51].

Finally, at the far end of the mass distribution lie the Moon and planets. While most of these bodies and their environments have been studied from long

---

[2]The latter is observed during solar eclipses, when the contribution of Thompson scattering by the coronal electrons (Section 1.2) is called the *K-corona*.

Bodies in the wind   337

Table 7.1 *Some basic characteristics of the Moon and the planets (rounded up to a few per cent)*

	$d/d_\oplus$	$M/M_\oplus$	$R/R_\oplus$	$\Omega/\Omega_\oplus$	$\mu/\mu_\oplus$	Atmosphere	$H/H_\oplus$
Moon	1.00	0.0122	0.27	0.036	~ 0	Na	20
Mercury	0.39	0.055	0.38	0.017	$3 - 6 \times 10^{-4}$	Na	8
Venus	0.72	0.81	0.95	0.0041	$< 10^{-5}$	CO_2	1.9
Earth	1.0	1.0	1.0	1.0	1.0	N_2	1.0
Mars	1.5	0.107	0.53	1	$< 10^{-6}$	CO_2	1.3
Jupiter	5.2	318.	11.2	2.4	2.0×10^4	H_2	2.9
Saturn	9.5	95	9.4	2.3	5.8×10^2	H_2	5.8
Uranus	19	14.5	4.0	1.4	48	H_2	3.1
Neptune	30	17.1	3.9	1.3	28	H_2	2.2
Pluto	39	0.0022	0.18	0.16		N_2	5

Notes: The mean heliocentric distance d, mass M, equatorial radius R, rotation rate Ω and magnetic moment $\mu \simeq (4\pi/\mu_0) B_0 R^3$ A m^2 (B_0 is the planet's magnetic field amplitude at equator) are normalised to the Earth's values, respectively equal to $d_\oplus \simeq 1.5 \times 10^{11}$ m (1 AU), $M_\oplus \simeq 6. \times 10^{24}$ kg, $R_\oplus \simeq 6.4 \times 10^6$ m, $\Omega_\oplus \simeq 7.3 \times 10^{-5}$ rad s^{-1} and $\mu_\oplus \simeq (4\pi/\mu_0) 7.9 \times 10^{15}$ A m^2. The last two columns indicate the main constituent of the atmosphere, and its approximate scale height $H \simeq k_B T/mg$ (with m the mass of the main constituent and $g = MG/R^2$), normalised to the Earth's value $H_\oplus \simeq 8 \times 10^3$ m.

ago through their radiation, with increasing sophistication and in an increasing range of wavelengths, from radio to X-rays and gamma rays, long-range space exploration has enabled us to study them *in situ* with a luxury of details. The properties relevant for interaction with the solar wind are listed in Table 7.1, and will be discussed in this section. Far more may be found in the books [17] and [6].

Basically, planets lie in two families (Fig. 7.1). The four inner planets are small and rocky, being made mainly of iron and silicates. The four outer ones (not including Pluto, which is now recognised as a large Kuiper-belt object but is nevertheless listed as a planet for historical reasons) are relatively large and consist mainly of gaseous material, essentially hydrogen.[3] Many planets have satellites, which generally lie in the planetary environment, but some of them – such as the Earth's Moon which orbits at $\simeq 60 R_\oplus$ from the Earth – lie in the solar wind proper. All planetary orbits are nearly circular and close to the ecliptic plane, with an eccentricity smaller than 0.1 and an inclination smaller than 4°, except the innermost (Mercury: eccentricity 0.21, inclination 7°) and the outermost (Pluto: eccentricity 0.25, inclination 17°).

---

[3] For the gaseous planets, the 'radius', i.e. the base of the atmosphere, is arbitrarily defined as the level where the atmospheric pressure is 1 bar, i.e. 10^5 Pa.

338   Dust, asteroids, planets and comets

Figure 7.1 From 'Un autre Monde' by Granville (Ed. H. Fournier, Paris, 1844).

The most fundamental properties of these bodies are their mass and size: how mass is distributed among them, and how the mass of a body is related to its size. This is examined below. We shall then examine their atmosphere and other properties that are crucial from the point of view of the solar wind.

7.1.2 Mass distribution

Figure 7.2 shows the cumulative flux of bodies from minute 10^{-2} µm grains to several kilometres-sized asteroids, moving in the interplanetary medium at about 1 AU from the Sun. The plot (thick dashed) is based on the cumulative impact rate per year on the Earth compiled in [15] from various data, that we have divided by the factor 1.63×10^{22} (the Earth's surface, times the duration of a year).[4] In these data, the values for $m < 10^{-9}$ kg are from [33], whose interplanetary dust distribution is superimposed as a thin continuous line.[5] The

---

[4] Not taking into account any concentration effect by the Earth's gravitation, which would yield a factor of about two – smaller than the uncertainties on the overall distribution.

[5] $F_{G85} = \left(a_1 m^{0.306} + 15\right)^{-4.38} + a_2 \left(m + 10^{14} m^2 + 10^{33} m^4\right)^{-0.36} + a_3 \left(m + 10^9 m^2\right)^{-0.85}$
with $a_1 = 1.8 \times 10^4$, $a_2 = 1.1 \times 10^{-10}$, $a_3 = 3.7 \times 10^{-19}$, for $10^{-21} < m < 0.1$ kg, based on

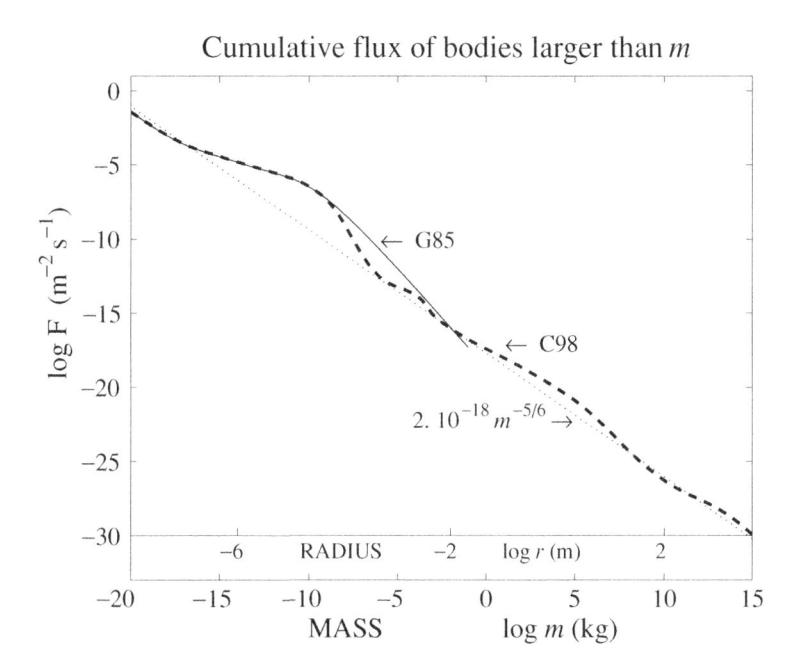

Figure 7.2 Cumulative flux of bodies of mass greater than m plotted versus m (log–log co-ordinates) at 1 AU from the Sun. The curve labelled C98 (thick dashed) is based on [15]; G85 (thin continuous) is the interplanetary flux by [33]. The superimposed power law $F \propto m^{-5/6}$ (thin dotted) corresponds to fragmentation equilibrium. The (cumulative) number density is $N \sim 2 \times 10^{-4} F$. The radius scale is for spheres of mass density 2.5×10^3 kg m^{-3}.

flux F plotted for mass m is the mean number of particles of mass greater than m that are impinging per unit time per unit area on one side of a flat detector if the particle velocity distribution is isotropic.[6] We have also indicated on the figure the corresponding radii, assuming spheres of mass density 2.5×10^3 kg m^{-3}, i.e. $m \simeq 10^4 \times r^3$ kg.

The flux F may be converted into a number density N by assuming an average impact speed $\langle v \rangle$, so that for an isotropic velocity distribution we have $F = N \langle v \rangle / 4$ (where $\langle v \rangle \sim 20$ km s^{-1} in order of magnitude near 1 AU). The factor $1/4$ is the product of two factors $1/2$: one because half the particles are coming from one side of the detector; the other because of averaging over the angle of incidence.[6]

These values are cumulative ones. Namely, the total number density of objects in the mass interval $[m_{min}, m_{max}]$ is $N(m_{min}) - N(m_{max})$, where

---

the formula A3 in [33] converted so that m is in kilograms.

[6]The effective solid angle is then $\Omega = \int_0^{\pi/2} \cos\theta \times 2\pi \sin\theta d\theta = \pi$. Indeed, 2π would represent a half-space (one side of a plane) in the absence of projection effects, but the flux is determined from the average of $v \cos \theta$ (the projection of the speed v on the normal to the plane) over directions, i.e. $\langle v \cos \theta \rangle = v/2$ if the distribution is isotropic.

$N(m) \sim 4F(m)/\langle v \rangle$, whereas the differential number density is dN/dm. For example, one sees on Fig. 7.2 that the number density of dust grains with radii larger than 1 µm ($m > 10^{-14}$ kg) is a few particles per km^3.

Although the distribution across this vast range of scales cannot be expressed by a single power law, a power law $F \propto m^{-5/6}$ (thin dotted line on Fig. 7.2) appears to account reasonably well for masses greater than 10^{-5} kg (and even for $m < 10^{-15}$ kg); however, it does not fit the conspicuous dust grain population in the mass range $10^{-13} < m < 10^{-7}$ kg.

A power law decreasing as $m^{-5/6}$ for the cumulated flux or density distribution (i.e. a differential distribution $dN/dm \propto m^{-11/6}$) has a special property: it should ensure fragmentation equilibrium, i.e. the number of particles created in unit mass interval by fragmentation of larger objects equals the number of particles in that mass interval which are destroyed by collisions [25]. The basic effect of collisions on the mass distribution may be understood by noting that with a typical relative speed $v \sim 20$ km s^{-1}, the kinetic energy per nucleon is $m_p v^2 / 2 \sim 3 \times 10^{-19}$ J, i.e. roughly 2 eV – an amount sufficient to vaporise matter. Collisions between two objects may therefore vaporise the smaller object and cause cratering and/or fragmentation of the larger one.

A simple dimensional analysis may be instructive. We use the basic fact that the cross-section for collision of two objects of masses m and m' is proportional to $\left(m^{1/3} + m'^{1/3}\right)^2 \sim m'^{2/3}$ if $m' > m$, and assume that a target of mass m will be catastrophically disrupted by a projectile of mass m' if $m' > km$, where k is some factor (in practice much smaller than unity). The number of objects destroyed by catastrophic collisions (per unit time per unit volume) in the mass range $[m, m + dm]$ is thus

$$d\nu \propto dm\, dN/dm \int_{km}^{m_{max}} dm'\, dN/dm'\, m'^{2/3}.$$

With $dN/dm \propto m^{-\alpha}$, this yields $d\nu \propto dm\, m^{5/3 - 2\alpha}$ if $\alpha > 5/3$. The total mass crushed from objects in a given mass range $[m_1, m_2]$ is thus $\int_{m_1}^{m_2} dm\, m\, d\nu/dm \propto \int_{m_1}^{m_2} dm\, m^{5/3 + 1 - 2\alpha}$. For the result to depend only (and weakly) on the ratio m_2/m_1, and not to be sensitively dependent on the extremes m_1 and m_2, the integrant must vary as m^{-1}. This requires that $\alpha = 11/6$, whence $N \propto F \propto m^{-5/6}$. With this distribution, the mass crushed from objects in a given logarithmic interval of mass is independent of the mass.

The whole mass distribution shown in Fig. 7.2 cannot be modelled by this process only, especially the dust grains, whose distribution involves a number of other processes which are not necessarily in equilibrium, making the distribution vary. In particular, we shall see in Section 7.4 that small dust grains are braked by the Poynting–Robertson force, making the size of their heliocentric orbit decrease until they pass close enough to the Sun to evaporate. Other dust grains are blown out of the solar system by the solar radiation pressure, and comets may replenish the losses.

Note that a differential *mass distribution* of the form $dN/dm \propto m^{-\alpha}$ is equivalent to a differential *size distribution* $dN/dr = dN/dm \times dm/dr \propto r^{-q}$

Bodies in the wind

with $q = 3\alpha - 2$; for example $\alpha = 11/6$ corresponds to $q = 3.5$.

To keep some perspective, note that the plot is far from complete. The smaller mass plotted contains about 6 million nucleons, whereas the larger mass plotted is less massive than Ceres – the most massive asteroid of the asteroid belt – by nearly six orders of magnitude, and than the biggest planet – Jupiter – by more than 12 orders of magnitude.

From Fig. 7.2, one sees that one body larger than 10 m in diameter (i.e. $m > 10^6$ kg) impacts the Earth in average every year, whereas bodies larger than 100 m imperil the Earth's biosphere on a timescale about 300 times greater.

The distribution shown in Fig. 7.2 is based on measurements at 1 AU from the Sun, close to the ecliptic plane, where the majority of the bodies lie. The amount of bodies decreases with heliocentric distance d, (very) roughly as d^{-1}; since the mean speed of bodies orbiting the Sun varies as $d^{-1/2}$, the flux falls off (very) roughly as $F \propto d^{-3/2}$.

7.1.3 Mass versus size

Dust grains are generally inferred to have a mass density of $(2-3) \times 10^3$ kg m^{-3}, with smaller values for icy dust and fragile porous aggregates. The mass density of planets lies in the range $(0.6 - 6) \times 10^3$ kg m^{-3}, the smaller values corresponding to the outer planets (mainly made of hydrogen). What is the origin of these densities? We shall derive a crude relation between radius and mass from simple basic arguments, in a similar way as we did in Section 3.1 for stars.

These solid bodies are so cold (and dense) that the electrons are fully degenerate, i.e. the thermal pressure is negligible. They are also sufficiently dense for the electrostatic attraction between nuclei and electrons to be important. In the general case, they are thus held together by three forces:

- the electrostatic attraction between electrons and nuclei,

- the Fermi pressure of electrons, which acts against compression, the corresponding energy per electron being $(3\pi^2 n)^{2/3} \hbar^2 / 5 m_e$ times $3/2$ (Appendix), where n is the electron number density (Section 2.1),

- the gravitational attraction, which, unlike electricity, is not neutralised by neighbouring particles, so that it is cumulative with increasing mass, and becomes important for large bodies.

This means that solid bodies are held together by the balance of electrostatic attraction and Fermi pressure if they are not too large, whereas bigger bodies tend to be further compressed by the gravitational force.

Consider a spherical object of mass M and radius R made of atoms having mass number A and Z bound electrons. Hence the body contains $M/(Am_p)$ nuclei and $N \simeq ZM/(Am_p)$ electrons. This amounts to $N = M/m_p$ electrons in the volume $V = 4\pi R^3/3$ if the body is made of hydrogen, and about half as much if it is made of other elements since in that case $A \sim 2Z$ in order of magnitude.[7] With the electron number density $n = N/V$, the total Fermi energy in the body is

$$W_F \simeq \frac{(3\pi^2)^{2/3} \hbar^2}{5 m_e} \frac{N^{5/3}}{V^{2/3}} \sim \frac{\hbar^2}{m_e m_p^{5/3}} \frac{M^{5/3}}{R^2} \quad (7.1)$$

where we have dropped a factor of order of magnitude unity.

Let us now evaluate the electrostatic energy. Each neighbouring attracting pair of charges roughly neutralises each other force on the remaining particles. Hence the electrostatic energy per nucleus is about $-Z^2 e^2 f_{A,Z}/4\pi\epsilon_0 r$, where $r \sim R/(N/Z)^{1/3}$ is the average separation between electrons and nuclei, and $f_{A,Z}$ is a function of A and Z roughly equal to unity for $A = Z = 1$ to account for the electron distribution. With N/Z nuclei, the total electrostatic energy is therefore

$$W_E \sim -N^{4/3} \frac{e^2 Z^{2/3} f_{A,Z}}{4\pi\epsilon_0 R} \sim -\frac{e^2 Z^{2/3} f_{A,Z}}{4\pi\epsilon_0 m_p^{4/3}} \frac{M^{4/3}}{R} \quad (7.2)$$

where we have dropped again a factor of order of magnitude unity. To this accuracy, this is equivalent to the result of Ref. [20] of Chapter 3.

The gravitational energy is

$$W_G \simeq -M^2 G/R \quad (7.3)$$

and Virial theorem (Section 3.1) tells us that $W_F = -(W_E + W_G)/2$. Substituting (7.1), (7.2), (7.3) and rearranging, we obtain a relation between the mass M and the radius R of the body (in order of magnitude)

$$r_{Bohr}/R \sim (M/m_p)^{-1/3} Z^{2/3} f_{A,Z} + (M/m_p)^{1/3} \left(4\pi\epsilon_0 m_p^2 G/e^2\right) \quad (7.4)$$

where $r_{Bohr} = 4\pi\epsilon_0 \hbar^2/m_e e^2 \simeq 0.53 \times 10^{-10}$ m is the Bohr radius. The dimensionless number

$$4\pi\epsilon_0 m_p^2 G/e^2 \simeq 0.8 \times 10^{-36}$$

which arises from the gravitational contribution, has a fundamental significance: it is the ratio of the strength of gravitational to electric forces between protons.

Let us examine the radius–mass relation $R(M)$ given by (7.4). One sees that it behaves very differently for small and large masses. When M is small, the first term (the electrostatic one) on the right-hand side of (7.4) dominates, yielding $R \propto M^{1/3}$; in contrast, when M is large enough, the second term (the gravitational one) dominates, yielding $R \propto M^{-1/3}$, a shrinking produced by gravitational compression; for intermediate masses, R has a maximum. Let us see this in more detail.

---

[7] Because nuclei have roughly the same number of protons and neutrons, and there are as many electrons as protons.

Bodies in the wind

When the mass is small enough that the gravitational term in (7.4) is negligible, this equation reduces to a balance between electrostatic attraction and electron degeneracy pressure, yielding

$$M \sim f_{A,Z}^3 \, Z^2 \frac{m_p}{(r_{Bohr})^3} R^3 \sim 10^4 f_{A,Z}^3 \, Z^2 R^3. \tag{7.5}$$

This gives a mass density of about 2×10^3 kg m^{-3} for hydrogen ($f_{A,Z} \sim 1$), which increases with Z. These figures are of the order of magnitude of the mass densities observed for these objects – a reasonably good agreement given the crudeness of this estimate.

When the mass is large enough that the gravitational term in (7.4) is no longer negligible, the radius no longer increases as $M^{1/3}$, and reaches a maximum when the two terms on the right-hand side of (7.4) are equal in order of magnitude. For hydrogen ($f_{A,Z} \sim 1$), this occurs at the mass

$$M \sim m_p \left(e^2 / 4\pi\epsilon_0 m_p^2 G \right)^{3/2} \simeq 2.3 \times 10^{27} \text{ kg} \equiv M_M \tag{7.6}$$

and therefore the radius

$$R \sim r_{Bohr} (M_M/m_p)^{1/3} \equiv r_{Bohr} \left(e^2 / 4\pi\epsilon_0 m_p^2 G \right)^{1/2} \simeq 6 \times 10^7 \text{ m} \tag{7.7}$$

which are close to those of Jupiter. So Jupiter's mass and radius are close to the values corresponding to the maximum radius for planets made essentially of hydrogen.

What happens for more massive objects? Since the gravitational term then dominates the right-hand side of (7.4), this equation yields

$$M/m_p \sim (r_{Bohr}/R)^3 (M_M/m_p)^2 \tag{7.8}$$

where we have substituted the definition (7.6) of M_M; this is the classic $R \propto M^{-1/3}$ relationship holding for low-mass white dwarfs, which results from the balance between gravitational attraction and electron degeneracy pressure. Let us calculate the distance between nuclei in that case. For a body made of hydrogen, it is about $R(m_p/M)^{1/3}$; substituting the relation (7.8) between M and R, this yields a distance of about $r_{Bohr} (M_M/M)^{2/3}$. Hence for $M > M_M$, the average distance between nuclei becomes smaller than r_{Bohr}, making the atomic structure break down. The mass M_M thus sets the order of magnitude of the mass above which compression of matter produces ionisation (Section 2.5),[8] but it is still below the minimum mass of stars (Section 3.2). Note, incidentally, that around the maximum of the function $R(M)$ given by the balance equation (7.4), lies a mass range where the radius is roughly constant; this corresponds to brown dwarfs (see the references cited in Section 3.1).

---

[8] This is the case for white dwarfs.

7.1.4 Atmospheres and how they are ionised

Atmospheres

The last two columns of Table 7.1 give the main detected constituent of the atmosphere of each planet, and its approximate pressure scale height $H = k_B T/mg$. Here, T is the temperature at the base of the atmosphere, $g = MG/R^2$, and m is the mass of the main atmospheric constituent.[3] The scale height H would represent the order of magnitude of the width of the atmosphere if the temperature would remain constant with altitude (see (1.8)); since $\sqrt{R/H}$ is the ratio of the escape speed to the thermal speed, it is an indication of the binding of the atmosphere.

The Moon and Mercury have atmospheres so tenuous that particles have practically no collisions and are thus very far from equilibrium, making the scale heights indicated of limited practical value;[9] these exospheres are thought to consist mainly of atoms liberated from the body's surface by solar radiation and solar wind particles. Pluto may have an escaping atmosphere that sublimates from the surface rather as in a comet, with seasonal changes.[10] The other planets have relatively dense and stable atmospheres.

Planetary atmospheres are subjected to solar radiation, which ionises them, producing ionospheres. Basically, solar ionising radiation is that part of the solar flux whose energy per photon is greater than about the Bohr energy W_{Bohr}, i.e. whose wavelength $\lambda < hc/W_{Bohr}$. This radiation comes from regions of the solar atmosphere where the particle thermal energy $\sim k_B T > W_{Bohr}$, i.e. the chromosphere and corona.

Although the structure of ionospheres is a complicated balance between ionisation, recombination, chemical reactions and transport [16], simple order-of-magnitude estimates may be obtained from the physics introduced in Section 2.4, when one may assume local equilibrium between ionisation and recombination of a single atmospheric constituent.

Chapman layers

When solar ionising radiation penetrates through a planetary atmosphere, it dissociates and ionises its constituents. In doing so, it is progressively absorbed, making the ionising flux decrease as altitude decreases, so that little ionising radiation reaches the bottom of the atmosphere. However, the number of neutrals available for ionisation varies in the opposite sense, being maximum at the bottom. Since the rate of ionisation is proportional to the product of both quantities, ionisation is weak at the bottom because there is little ionising radiation, and weak at the top, too, because there are few neutrals. This produces a maximum of ionisation at some altitude, if the ionosphere is determined by ionisation equilibrium.

---

[9] With a number density at the surface of the order of magnitude of $n_0 \sim 10^8$ m^{-3} for the Moon (see for example [75] and references therein), and $n_0 \sim 10^{11}$ m^{-3} for Mercury [44], on the sunlit side, the particle mean free path for collisions $\sim (n_0 \sigma_0)^{-1}$ (where σ_0 is a typical collisional cross-section given in Section 2.1) is much greater than the scale height.

[10] See [73], [27] and references therein.

Bodies in the wind

We saw in Section 1.2 that the flux of solar ultraviolet radiation amounts to about 3×10^{-3} W m^{-2} at Earth on average over the solar cycle, and varies as the inverse squared heliocentric distance d^{-2}. This corresponds to an average flux of ionising photons of order of magnitude

$$F_\odot \sim 3 \times 10^{14} \, (d_\oplus/d)^2 \text{ photons m}^{-2}\text{s}^{-1} \tag{7.9}$$

incident at the top of the atmosphere. The flux varies with solar activity, being typically twice as small at activity minimum, and twice as great at activity maximum.

How does this flux vary as radiation penetrates through the atmosphere? At altitude z (assumed much smaller than the radius of the planet so that the problem is one-dimensional), where the density of neutrals is n_n and the ionising flux is F, the ionisation rate per unit volume is $F n_n \sigma_{ion}$, where σ_{ion} is the cross-section for ionisation (Section 2.4). The decrease in flux dF over an altitude decrease dz is equal to the ionisation rate in a volume of section unity and height dz, i.e.

$$dF/dz = F \, n_n \, \sigma_{ion}. \tag{7.10}$$

The density of neutrals varies as $n_n \propto e^{-z/H}$, if it is not significantly altered by ionisation (which holds true if the density of charged particles $n \ll n_n$). Substituting this variation in (7.10) and integrating between z and infinity where $F \to F_\odot$ (the flux incident at the top of the atmosphere, assuming normal incidence), we find

$$\ln(F_\odot/F) = \int_z^\infty dz \, n_n \, \sigma_{ion} \quad \Longrightarrow \quad F/F_\odot = e^{-n_n \sigma_{ion} H}. \tag{7.11}$$

The ionisation rate is maximum at the altitude z_{max} where $n_n F$ is maximum, i.e. where

$$\frac{d}{dz}\left[n_n e^{-n_n \sigma_{ion} H}\right] = 0 \quad \Longrightarrow \quad \frac{dn_n}{dz} - n_n \sigma_{ion} H \frac{dn_n}{dz} = 0$$

which yields

$$n_n(z_{max}) = 1/(\sigma_{ion} H). \tag{7.12}$$

The altitude z_{max} is given by

$$n_n(z_{max})/n_0 = e^{-z_{max}/H} = 1/(n_0 \sigma_{ion} H)$$

where n_0 is the density of neutrals at $z = 0$ (assuming $n_0 \sigma_{ion} H > 1$). We deduce the altitude of the maximum of ionisation

$$z_{max} \sim H \times \ln(n_0 \sigma_{ion} H). \tag{7.13}$$

The ionising flux (7.11) at this altitude is from (7.12) $F(z_{max}) = F_\odot e^{-1}$. The corresponding density $n(z_{max})$ of ions or electrons is determined by equilibrium between ionisation and recombination, which occurs at the rate $\beta_{rec} n^2$, β_{rec} being a typical recombination coefficient (Section 2.4). This yields

$$F \, n_n \, \sigma_{ion} = \beta_{rec} \, n^2 \qquad (7.14)$$

in which we substitute (7.11) and (7.12) to deduce the maximum electron density

$$n(z_{max}) \sim \left(\frac{F_\odot e^{-1}}{\beta_{rec} H} \right)^{1/2}. \qquad (7.15)$$

This is a remarkable result: the maximum plasma density of the ionospheric layer does not depend on the density of the neutral atmosphere (provided its value at the bottom $n_0 > 1/\sigma_{ion} H$). This is because the more neutrals, the more particles available for ionisation, but the more absorption of the ionising radiation. Such an ionisation layer is known as a *Chapman layer*.

With the solar ionising flux (7.9), this yields the electron density

$$n(z_{max}) \sim 2 \times 10^{11} \left(H_\oplus^{1/2} d_\oplus / H^{1/2} d \right) \text{ m}^{-3} \qquad (7.16)$$

where H and d are respectively the planet's scale height and heliocentric distance, and we have estimated the dissociative recombination coefficient β_{rec} from Section 2.4 (the variation of β_{rec} with the temperature has not much consequence here since it produces a density variation $\propto T^{1/4}$).

Planetary ionospheres

How does this simple model compare with observation? To begin with, it cannot be applied to the tenuous exospheres of the Moon and Mercury, for which $n_0 \sigma_{ion} H < 1$ and which are far from equilibrium. Consider then the three inner planets having a significant atmosphere. The concentration of neutrals at the bottom is about $n_0 \simeq 2 \times 10^{25}$ m^{-3} at Earth ($T \simeq 300$ K), roughly 50 times more at Venus ($T \simeq 700$ K) and 100 times less at Mars ($T \simeq 200$ K). With the scale height and distances given in Table 7.1, we find from (7.16): $n(z_{max}) \sim 2 \times 10^{11}$ m^{-3} at Earth and Venus, and about half as much at Mars, in order of magnitude. With the above parameters, we find from (7.13) that the altitude of the ionisation layer is $z_{max} \sim 1.6 \times 10^5$ m for the Earth and Mars, and is about twice as large for Venus. In these three cases, the time for recombination $(n_{max} \beta_{rec})^{-1}$ amounts to tens to hundreds of seconds, so that these ionospheres should disappear at night.

Such an ionisation layer is indeed observed at these three planets. However, there is another ionisation layer at Earth, at a somewhat higher altitude, which is 10 times denser and does not disappear at night. It cannot be explained by this simple model because it is made essentially of ions O$^+$, which are monoatomic so that they cannot recombine dissociatively; they have thus a much greater lifetime, letting other physical processes act [16].

What about the giant planets? Their atmospheres being made mainly of hydrogen, the major ion is H^+, which cannot recombine dissociatively and has therefore a very long lifetime; hence chemical reactions and transport play a major role [3]. And what about Pluto? If the escape rate is small and (7.15) holds, we deduce with the parameters of Table 7.1 the plasma density at ionisation maximum: $n(z_{max}) \sim 2 \times 10^9$ m^{-3} in order of magnitude.

What is the temperature of the ionospheres? During ionisation, the photon energy in excess of the ionisation energy is given to the released electron,[11] which subsequently shares it with the other electrons (and, albeit more slowly, with ions and also neutrals). This heats the electrons to temperatures that increase with altitude up to $10^3 - 10^4$ K on the sunlit side.

Finally, with the above parameters, the density (7.12) of neutrals at ionisation maximum is of the order of magnitude of 10^{17} m^{-3} for the Earth, Venus and Mars; it is therefore much greater than the plasma density. Ionospheres are therefore weakly ionised plasmas near and below the altitude of ionisation maximum. However, as altitude increases, the plasma density decreases, so that the recombination time $(n\beta_{rec})^{-1}$ ultimately becomes greater than the time of hydrostatic adjustment $\sim (H/g)^{1/2} \sim H/V_S$, making the plasma density determined essentially by gravity. As a result, the plasma density falls off exponentially at high altitudes, with a scale height $\sim 2k_BT/mg$ (m being the mean ion mass and T an average plasma temperature), that is greater than the one of the neutrals. At high enough altitudes, therefore, the plasma becomes the major constituent of the environment.

7.1.5 Planetary magnetic fields and ionospheric conductivity

Each planet rotates about an axis making a small angle (less than 24°) with the normal to its orbit, with the exceptions of Venus (177°), Uranus (98°) and Pluto (~120°). All planets rotate in the same sense about themselves and about the Sun.

For planets having a suitable electrically conducting part, a dynamo may result (see Section 3.3), and indeed, all the explored planets have been found to have a significant global magnetic field, except Venus and Mars.[12] The Earth, Jupiter and Saturn have magnetic fields relatively close to dipolar, with magnetic dipole axes close to their spin axis. They are believed to have highly conducting dynamo regions, made of iron for the Earth, and of metallic hydrogen for Jupiter and Saturn. In contrast, the magnetic fields of Uranus and Neptune both have large quadrupole moments and large tilts between their spin

---

[11] The ion shares a much smaller fraction because of its large mass.
[12] Mars has been recently discovered to have strongly magnetised regions in its crust, even though there is virtually no global field at present [1].

and magnetic axes. The latter planets are not large enough to have a core of metallic hydrogen, and have poorly conducting interiors. Detailed reviews may be found in [69] and [18].

Ionospheres conduct electricity – a property that is crucial for their interaction with the solar wind. The electric conductivity depends on the mean collision frequency of the charged particles (Section 2.1). Since ionospheres are weakly ionised, we must take into account the collisions with neutrals. With the parameters found above, we may estimate the electric conductivity from (2.101), replacing the electron–ion collision frequency ν_{ei} by the electron–neutral collision frequency (2.13), which yields $\sigma_0 \sim 1 - 10$ mho m^{-1} near the maximum of ionisation of the Chapman layers. At higher altitudes, since the neutral density falls off more rapidly than the plasma density, the conductivity increases, and when the atmosphere is sufficiently ionised for the electron–ion collision frequency ν_{ei} to become larger than ν_{en}, the electric conductivity is given by (2.102), which yields $\sigma_0 \sim 10^2$ mho m^{-1} at high altitudes.

These values, however, do not account for the magnetic field. The magnetic field does not change the conductivity in the direction parallel to itself, but it reduces it considerably in the perpendicular directions (Section 2.3). As a result, the electric field is small along the magnetic field, but may be large in the perpendicular directions.

7.2 Basics of the interaction

The interaction of an object with the solar wind depends on the solar wind flow and on three major properties of the object:

- its size, compared to the basic plasma scales,

- its physical nature and the nature of its environment,

- its magnetic field.

Detailed reviews of these interactions may be found in [4], [70], [5], [49] and [12].

7.2.1 Properties and spatial scales of the flow

Flow properties

The flow properties that play an important role in the interaction are the Mach number, the plasma particle flux $n_w v_w$ and ram pressure $\rho_w v_w^2$ (with $\rho_w \simeq n_w m_p$),[13] the kinetic pressure P, the magnetic pressure P_m and the orientation of the magnetic field. With the parameters of Sections 1.3, 5.4 and 6.1, we have

---

[13] With an additional contribution from the helium ions.

Basics of the interaction

in average at heliocentric distance d

$$n_w v_w \sim [3-2] \times 10^{12} \, (d_\oplus/d)^2 \; \text{m}^{-2}\text{s}^{-1} \tag{7.17}$$

$$\rho_w v_w^2 \simeq [2.1-2.6] \times 10^{-9} \, (d_\oplus/d)^2 \; \text{Pa} \tag{7.18}$$

$$P \sim n_w k_B (T_e + T_p) \sim 2 \times 10^{-11} \, (d_\oplus/d)^{2+\alpha} \; \text{Pa} \quad \alpha \sim 0-1 \tag{7.19}$$

$$P_m \sim B_w^2/2\mu_0 \sim 6 \times 10^{-12} \, (d_\oplus/d)^2 \left[1 + d_\oplus^2/d^2\right] \; \text{Pa} \tag{7.20}$$

$$B_\phi/B_r \sim \cos\theta \, (d/d_\oplus) \quad \text{at latitude } \theta \tag{7.21}$$

where the extreme values of the brackets in (7.17) and (7.18) correspond to slow and fast wind respectively, and the index α in (7.19) accounts for the poorly known radial dependence of the temperatures (Section 5.4). In the ecliptic, where most bodies lie, the average particle flux and ram pressure are close to the first values of the brackets. The particle and magnetic pressures are similar in order of magnitude, and therefore so are the speed of compressive waves and the Alfvén speed. The Mach number M is of the order of magnitude of 10 for both Alfvén waves and sonic waves (and therefore smaller by a factor of about $1/\sqrt{2}$ for magnetosonic waves).

Close to the ecliptic ($\theta = 0$), the average solar wind magnetic field is relatively close to the velocity direction inward of the Earth distance $d_\oplus = 1$ AU, makes an angle of about 45° at $d_\oplus$ in the ecliptic, and becomes more and more perpendicular to the velocity farther away in the ecliptic plane. On average in the ecliptic plane, where most of the bodies lie, its component perpendicular to the ecliptic is zero. However, we have seen that there are permanent fluctuations, in addition to large perturbations produced by interactions between slow and fast wind and by coronal mass ejections – which we have seen to occur frequently near the ecliptic plane.

The 'thermal'[14] speed of electrons is

$$v_{the} = (2k_B T_e/m_e)^{1/2} \sim 2 \times 10^6 \, (d_\oplus/d)^{\alpha/2} \quad \alpha \sim 0-1 \tag{7.22}$$

$$\Rightarrow n_w v_{the} \sim 10^{13} \, (d_\oplus/d)^{2+\alpha/2}. \tag{7.23}$$

The electron thermal speed is greater than the solar wind speed, itself much greater than the proton thermal speed, i.e.

$$v_{thp} \ll v_w \ll v_{the}.$$

Hence an object lying in the solar wind sees a typical electron moving at a speed $\sim v_{the}$, and a typical proton moving at $\sim v_w$.

Since most of the solar wind energy lies in bulk motion, the major parameter governing the interaction with bodies is the ram pressure $\rho_w v_w^2$ given in (7.18). As other parameters, it exhibits large fluctuations. This is illustrated in Fig. 7.3, which shows measurements made aboard the spacecraft Ulysses during the fast latitude scans performed around solar activity minimum and maximum, along

---

[14] We use this usual notation, even though the velocity distribution is not Maxwellian; the temperature is then the kinetic temperature, defined in (5.89).

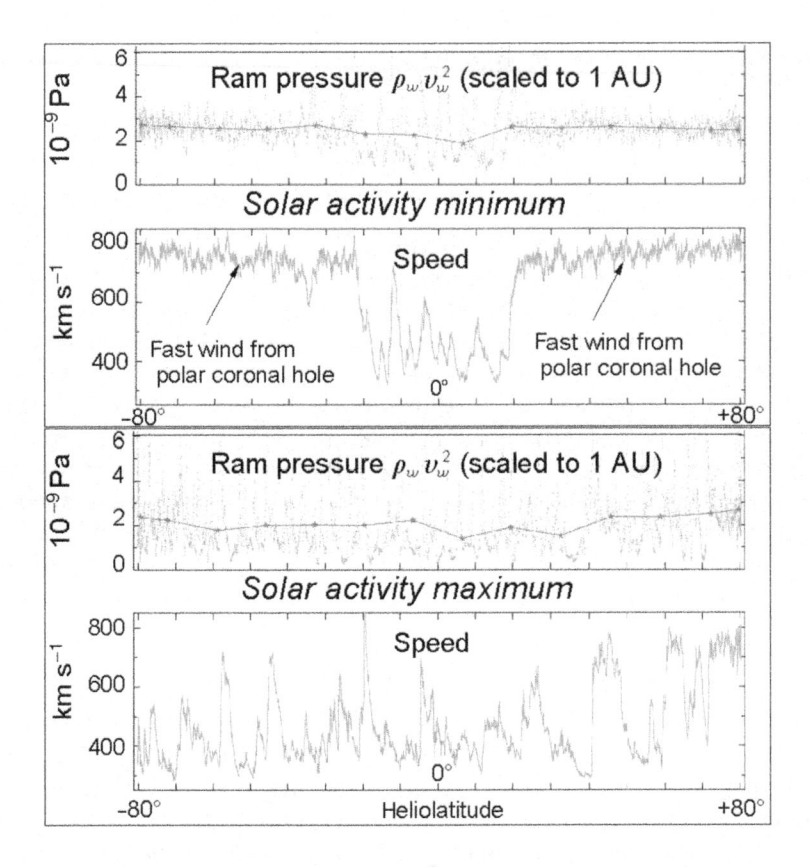

Figure 7.3 Solar wind ram pressure $\rho_w v_w^2$ (scaled to 1 AU by multiplying the data by $(d/d_\oplus)^2$) measured on Ulysses during the fast latitude scans performed near solar activity minimum and maximum. The speed is shown to identify the fast and slow winds. (K. Issautier, personal communication, based on data from the instruments SWOOPS and URAP).

the trajectory shown in Fig. 1.20. The main cause of variation are the fast/slow wind transitions and the coronal mass ejections (Section 6.3), which take place essentially at low heliolatitudes near solar activity minimum, and at virtually all latitudes at activity maximum. Since most bodies lie near the ecliptic plane (i.e. at small heliolatitudes), they see relatively large fluctuations of $\rho_w v_w^2$ during all the solar cycle, with a variance that is barely less than the mean value. However, one sees in Fig. 7.3 that despite these large fluctuations, the average value is extremely stable, varying only from about 2.1×10^{-9} Pa in the slow wind to 2.6×10^{-9} Pa in the fast wind.

Spatial scales

The main plasma scales determining the interaction are the mean distance between particles $n_w^{-1/3}$, the Debye length L_D, the particle gyroradii in the solar

Basics of the interaction

wind magnetic field and their mean free path for collisions l_f. The mean electron gyroradius is $r_{ge} \sim v_{the}/\omega_{ge}$; for the protons, we have indicated the gyroradius of a proton moving at the solar wind speed, v_w/ω_{gp}, and at the Alfvén speed, $V_A/\omega_{gp} \equiv c/\omega_{pp}$ (also called the proton inertial length) respectively.[15] With the parameters given in Sections 5.4 and 6.1, we have at heliocentric distance d

$$n_w^{-1/3} \sim 5 \times 10^{-3} (d/d_\oplus)^{2/3} \text{ m} \tag{7.24}$$

$$L_D \sim 10 \times (d/d_\oplus)^{1-\alpha/2} \text{ m} \quad \alpha \sim 0-1 \tag{7.25}$$

$$r_{ge} \sim 3 \times 10^3 \frac{(d/d_\oplus)^{1-\alpha/2}}{\left(1 + d_\oplus^2/d^2\right)^{1/2}} \text{ m} \tag{7.26}$$

$$V_A/\omega_{gp} \equiv c/\omega_{pp} \sim 10^5 \times d/d_\oplus \text{ m} \tag{7.27}$$

$$v_w/\omega_{gp} \sim 10^6 \frac{d/d_\oplus}{\left(1 + d_\oplus^2/d^2\right)^{1/2}} \text{ m} \tag{7.28}$$

$$l_f \sim d_\oplus \times (d/d_\oplus)^{2(1-\alpha)} \text{ m} \tag{7.29}$$

$$n_w^{-1/3} \ll L_D \ll r_{ge} \ll c/\omega_{pp} \ll v_w/\omega_{gp} \ll l_f.$$

Consider first the objects smaller than the mean electron gyroradius; this includes dust particles, space probes and small asteroids. These objects are also much smaller than the collision-free paths. At these small scales, the interaction with the wind involves the trajectories of individual particles, so that we must adopt a kinetic point of view. Space probes and larger objects, whose size is larger than the interparticle distance, trail (in the solar wind frame) a wake depleted of particles. Their size compared to the Debye length (~ 10 m at 1 AU) plays an important role in the building of the charge on the surface of the object and around it. We address these questions in the next section.

Next, consider objects larger than the ion gyroradius, as our Moon and large asteroids. At this scale, some MHD concepts may be useful, although not necessarily sufficient. Assume first that the body has no atmosphere and no magnetic field, and is made of insulating matter, as our Moon and most asteroids. The surface of the body absorbs the solar wind particles, creating a wake on the rear side, with minor magnetic field perturbations. If, however, the object is either electrically conducting, or has an ionosphere or a magnetic field of its own, and the size involved is large enough that the magnetic Reynolds number is large, then the solar wind magnetic field and the plasma frozen to it are stopped by the obstacle. A bow shock forms ahead, whereas the plasma sweeps around the object, stirring the magnetic field into a long tail. This happens for planets and comets, with differences that depend on the kind of atmosphere and on the magnetic field. We shall examine these interactions in more detail in the following sections.

---

[15] As we already noted, since we have in order of magnitude in the solar wind $V_A \sim v_{thp}$ and $T_e \sim T_p$, the electron gyroradius is of the same order of magnitude of the electron inertial length c/ω_p. Likewise, the gyroradius of a proton moving at the thermal speed, v_{thp}/ω_{gp}, is roughly equal to its inertial length c/ω_{pp}, where $\omega_{gp} = \omega_{ge}(m_e/m_p)$ and $\omega_{pp} = \omega_p (m_e/m_p)^{1/2}$.

The intermediate scale, between the electron and ion gyroradii, requires some care, since the complicated geometry makes kinetic calculations difficult, and usual MHD cannot be used.[16]

7.2.2 Being small: electrostatic charging and wakes

The surface of an object immersed in a plasma collects and emits charged particles. In general, the incoming and outgoing electric currents do not balance each other, making the net surface electrostatic charge (and potential) vary, which in turn changes the currents. This proceeds until the charge and potential set up so as to ensure the balance of the currents. How can this charge be estimated? And what is the electric field around the object? These questions are important for understanding the behaviour of dust grains, and for the design of instruments carried by space probes. We give below simple estimates; detailed reviews may be found in [86] and [36].

Consider an object of size much smaller than the gyroradii – and therefore than the free paths (but still greater than the atomic scales) lying in the solar wind. At this scale, the plasma particles do not feel the magnetic field and have no collisions. To keep the problem simple, we approximate the body by a sphere of radius R having a conducting surface. Let us estimate the electrostatic charge Q on the surface and its electrostatic potential Φ with respect to the distant unperturbed plasma.

Charging on the back of an envelope

We can get an order-of-magnitude estimate in the following way (Fig. 7.4). A sunlit surface is subjected to the solar ionising radiation. The surface thus ejects a flux of photoelectrons, that is somewhat smaller than the flux of ionising photons (7.9), but not by much more than one order of magnitude. On the other hand, the surface collects solar wind electrons and ions. Since the thermal speeds satisfy $v_{thp} \ll v_w \ll v_{the}$, the electron flux is of the order of magnitude of the random flux $n_w v_{the}$ given in (7.23), whereas the ion flux is about $n_w v_w$ (on the projected surface) – much smaller than the electron flux. Comparing the photoelectron and the solar wind electron fluxes, we see that with typical parameters, an uncharged surface ejects many more photoelectrons than it collects electrons from the ambient plasma. Hence it charges positively, until this positive charge binds sufficiently the photoelectrons to make their flux balance the one of solar wind electrons. For doing so, the electric potential Φ of the surface must provide the photoelectrons with a potential energy $-e\Phi$ that outweighs their typical kinetic energy of a few electronvolts. We deduce that a *sunlit surface* must charge to a *positive potential of several volts* in the solar wind.

---

[16] The interaction of asteroids with the solar wind has been modelled using the so-called Hall MHD, which assumes that electrons are frozen to the magnetic field, whereas ions are not (Baumgärtel, K. *et al.* 1994, *Science* **263** 653).

Basics of the interaction 353

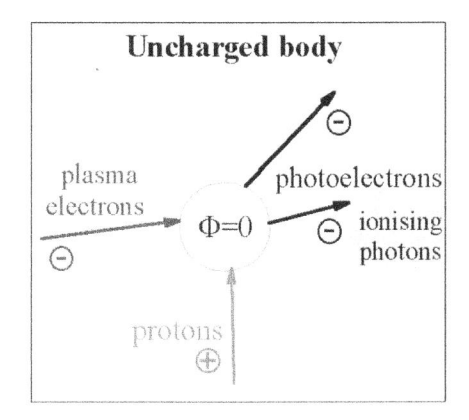

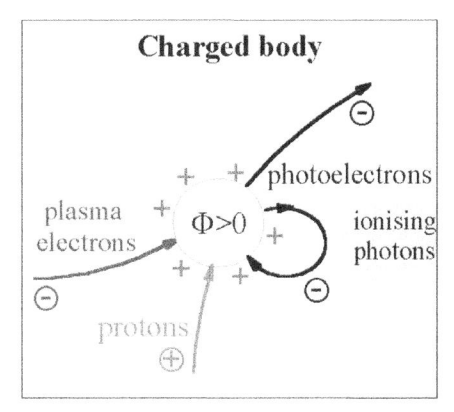

Figure 7.4 Charging of a sunlit (conductive) body in the solar wind. If the surface is uncharged, the flux of ejected photoelectrons far outweighs the flux of incoming plasma electrons (itself much larger than the flux of plasma ions). This charges the body positively, until the electrostatic potential traps enough photoelectrons to make the net current vanish.

What about a shadowed surface? It does not emit photoelectrons, but it collects solar wind electrons and ions, with fluxes about $n_w v_{the}$ and $n_w v_w$ respectively. Because $v_{the} \gg v_w$, an (uncharged) shadowed surface collects many more electrons than protons, and therefore charges negatively. This proceeds until the negative charge repels sufficiently the incoming electrons to make their flux balance the proton one. For doing so, the electrostatic potential must ensure that the potential energy $-e\Phi$ outweighs the typical kinetic energy $k_B T_e$ of the solar wind electrons, which is about 10 eV. We deduce that a *shadowed surface* must charge to a *negative potential of several tens of volts* in the solar wind.

Charging currents and electrostatic potential

Let us put these estimates on a more quantitative footing, and address some complications that have been swept under the rug. Consider first the photoelectrons ejected from an uncharged sunlit surface. The ejection rate is smaller than the flux $F_\odot$ given in (7.9) by an amount that depends on the physical and chemical structure of the surface; with typical sunlit materials, the photoemission rate is

$$N_{ph0} \sim \delta \, F_\odot \, S_\perp \quad \text{s}^{-1} \qquad \delta \sim 0.07 - 0.7 \qquad (7.30)$$

where $S_\perp$ is the projected sunlit area perpendicular to the solar direction, and δ the integrated yield that depends on the surface material; the smaller figure is relevant for low-yield materials such as graphite, the larger for high-yield photoemitters such as insulators, including silicates (see [32] and references therein). Photoelectrons are ejected with a typical energy of 1–2 eV, which may be approximated by a temperature

$$T_{ph} \sim 1 - 2 \times 10^4 \text{ K}. \tag{7.31}$$

This holds for a surface that is not electrically charged. For a charged surface whose potential Φ with respect to the distant undisturbed plasma is positive, the escaping photoelectrons are those emitted with a kinetic energy greater than $e\Phi$. Integrating over the velocities, this yields (Problem 7.6.1) the photoemission rate

$$N_{ph} = N_{ph0} \left(1 + e\Phi/k_B T_{ph}\right) \exp(-e\Phi/k_B T_{ph}) \qquad \Phi > 0 \tag{7.32}$$

in spherical symmetry, which holds for a small (roughly spherical) object; just how small the object has to be is derived below. In plane geometry (which is relevant for large bodies), the limitation in energy for the escaping photoelectrons concerns only the ones perpendicular to the plane, so that the escaping photoelectron flux is in this case $N_{ph} = N_{ph0} e^{-e\Phi/k_B T_{ph}}$ (Problem 7.6.1).

Consider now the fluxes of incoming solar wind electrons and protons. Suppose that the surface is not charged and that it absorbs all the charged particles impinging on its surface. The electron impact rate is then roughly equal to the electron random flux

$$N_{e0} \sim n_w \langle v \rangle_e S/4 \sim 1.5 \times 10^3 n\, T_e^{1/2}\, S \quad \text{s}^{-1} \tag{7.33}$$

where $\langle v \rangle_e = (8 k_B T_e / \pi m_e)^{1/2} \simeq 2 v_{the}/\sqrt{\pi}$ is the average electron speed and S is the surface of the object (as already noted in another context, the factor $1/4$ in (7.33) arises because for a given infinitesimal surface element, $n_w/2$ electrons per unit volume are approaching from one side with an average perpendicular velocity $\langle v \rangle_e / 2$). Note that electrons that are energetic enough do penetrate into the object and excite electrons within it, producing secondary emission of electrons; to keep the problem simple, we neglect this effect, which is important only for impinging electrons of energy greater than several hundreds of electronvolts.[17]

This holds for an uncharged surface. A positively charged surface attracts electrons, so that it collects more of them than the above value. If there are no intervening barriers of effective potential, the potential of the body increases its effective radius for collecting particles to a value $r_{\it eff}$ that is given from conservation of energy and angular momentum by $r_{\it eff}/R = \left(1 + 2e\Phi/m_e v^2\right)^{1/2}$ for an electron of speed v. Integrating over the velocities with a Maxwellian distribution yields (Problem 7.6.1)

$$N_e = N_{e0} \left(1 + e\Phi/k_B T_e\right) \qquad \Phi > 0. \tag{7.34}$$

Note again that this holds for a three-dimensional geometry. The same reasoning yields $N_e = N_{e0}$ for a plane geometry (relevant for large bodies, as we shall see below), and $N_e = N_{e0} \left(1 + e\Phi/k_B T_e\right)^{1/2}$ for a long cylinder (Problem 7.6.1).

---

[17] However, it should be kept in mind that under some circumstances, secondary electron emission may be important even from a qualitative point of view, in producing multiple charge equilibrium states and bifurcations (see Meyer-Vernet, N. 1982, *Astron. Astrophys.* **105** 98 and references therein).

Basics of the interaction 355

We neglect for simplicity the solar wind proton contribution, which is smaller than the electron one.

Compare the outgoing and ingoing electron fluxes. From (7.23)–(7.30) and (7.9), we have

$$N_{ph0}/N_{e0} \simeq (2-20)\,(d/d_\oplus)^{\alpha/2} \qquad \alpha \sim 0-1 \qquad (7.35)$$

so that the balance of the photoelectron current (7.32) and the solar wind electron current (7.34) yields the electric potential at equilibrium

$$\Phi \simeq (k_B T_{ph}/e)\,\ln\left[(2-20)\,\frac{1+e\Phi/k_B T_{ph}}{1+e\Phi/k_B T_e}\right] \qquad (7.36)$$
$$\sim 1-10\ \mathrm{V}$$

where we have assumed a three-dimensional geometry, and dropped the weak radial variation in (7.35). In a planar geometry (relevant for large bodies), the bracket in (7.36) reduces to its value for $\Phi = 0$.

There are two main complications to this simple picture. First, if the surface is not conductive, it should not be necessarily equipotential,[18] so that the balance of currents must be made separately for the sunlit and the shadowed parts. Second, a shadowed surface is generally in the wake of the object, where protons are depleted; this reduces the proton flux on the surface, making the potential even more negative, as we shall see below.

We can see in these estimates an echo of the kinetic calculations made in Section 5.5 for studying the solar wind acceleration. In the latter case, the electrostatic potential had to be greater than in a static atmosphere, because otherwise, the escaping electron flux would have been too large compared to the proton one because of the differences in masses. In the present case, the gravitational force is absent, and the scale is vastly smaller so that the plasma is not quasi-neutral, but the basic principle remains the same.

Deducing the electric charge

Having estimated the electric potential of the surface, we can now calculate its electric charge. In vacuum, the (radial) electrostatic field around a body of charge Q is given from Poisson's equation as $Q/4\pi\epsilon_0 r^2$ at distance r. Since, however, the solar wind is a plasma, the body is surrounded by a Debye sheath which shields the field farther than a distance of about L_D. Hence for a small object of radius $R \ll L_D$, the field is nearly equal to the Coulomb field out to a fraction of L_D – which amounts to several radii R, so that the capacitance has roughly the same value as in vacuum, i.e. for a sphere

$$R \ll L_D: \qquad C = Q/\Phi \simeq 4\pi\epsilon_0 R. \qquad (7.37)$$

---

[18]Note that among the various causes of potential difference, the electric field $\boldsymbol{v}_w \times \boldsymbol{B}$ amounts to a difference of potential of about 3×10^{-3} V m^{-1} at 1 AU, which is negligible, except for very large bodies.

With the typical potential found above, we deduce that a small object of radius R in the solar wind carries a number of elementary charges

$$Q/e \sim \text{a few } 10^9 \times R.$$

This amounts to a hundred electrons for a dust grain of size 0.1 µm.

Consider now a large object, of radius $R \gg L_D$. Since the potential decreases with distance at the scale L_D, the electric field at the surface is roughly $E \sim \Phi/L_D$; on the other hand, from Gauss's law, it is equal to $E = Q/4\pi\epsilon_0 R^2$, and so

$$R \gg L_D: \quad C = Q/\Phi \sim 4\pi\epsilon_0 R^2/L_D. \tag{7.38}$$

For these large objects, the geometry is roughly plane, so that the charging is one-dimensional.

Note that the photoelectrons contribute to the shielding with their proper Debye length. The photoelectron density near the object is about $n_{ph} \sim N_{ph0}/(\langle v \rangle_{ph} S/4)$ (with the mean speed $\langle v \rangle_{ph} = (8k_B T_{ph}/m_e)^{1/2} \simeq 7 \times 10^5$ m s^{-1}); substituting the photoemission rate (7.30) and temperature (7.31), this gives a photoelectron Debye length of about $(\epsilon_0 k_B T_{ph}/n_{ph} e^2) \sim d/d_\oplus$ m. Dust grains are very small at this scale, but spacecraft are not.

Charging timescale

The above estimates are equilibrium values. How much time does the charging process take? This may be estimated by viewing the object in the plasma as an electric circuit of capacitance C and resistance $R \sim 1/|dI/d\Phi|$, where I is the current flowing between the object and the plasma in absence of equilibrium. Since each charging process of rate N contributes to a current eN, and, from (7.32) and (7.34), has a derivative of the order of magnitude of $eN/k_B T$, we have roughly $|dI/d\Phi| \sim |edN/d\Phi| \sim e^2 N/k_B T$, where N and T are respectively the rate and temperature of the charging process that yields the greatest derivative. For a sunlit object in the solar wind, the charging timescale is thus in order of magnitude

$$\tau \sim RC \sim C k_B T_{ph}/(N_{ph0} e^2). \tag{7.39}$$

Substituting the values found above, we find

$$\tau \sim \frac{5}{R_{\mu m}} \frac{d^2}{d_\oplus^2} \text{ s} \tag{7.40}$$

for a dust grain of radius $R_{\mu m}$. On the other hand, for a large object we find by substituting (7.38) and (7.25) a timescale of a few 10^{-6} s at distance $d_\oplus$, independent of size, since both C and N then vary as R^2.

Basics of the interaction

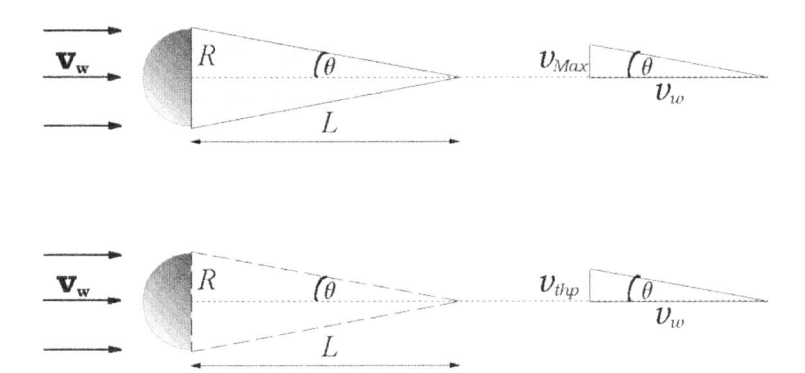

Figure 7.5 The wake of a body in the solar wind. Top: if all the particles had speeds smaller than a maximum value v_{max} (in the solar wind frame) their velocities as seen by the body would lie in a cone of half-angle θ, producing a conic cavity devoid of particles (hatched). Bottom: with a Maxwellian velocity distribution of thermal speed $v_{thp} \ll v_w$, the cavity is only partially depleted.

Wakes

Let us consider in more detail what happens downstream of an object like a spacecraft in the solar wind. The incoming solar wind particles are absorbed on the front side, which tends to create a depleted wake behind. Since the thermal electrons move much faster than the bulk solar wind, they tend to fill the wake, but the protons cannot easily do so. What is the length of this wake?

For simplicity, let us neglect the speed of the object with respect to the Sun, which is generally much smaller than v_w. Imagine that all the particles have speeds smaller than some value $v_{max} \ll v_w$; in this case, their speeds in the frame of reference of the object lie in a cone of half-angle v_{max}/v_w, so that the object trails a cone of this half-angle devoid of particles (Fig. 7.5, top). With a Maxwellian distribution for protons, the cone is not perfectly defined nor empty, but the outline of the picture remains true, producing a cone-like region devoid of protons, of length $\sim Rv_w/v_{thp}$ (Fig. 7.5, bottom). This simple picture neglects the electric field which has two effects. First, it curves the ion paths if the energy $e\Phi$ is not small compared to their kinetic energy $m_p v_w^2/2$; second, the ion depletion produces a negative space charge which repels the electrons, so that they are also somewhat depleted in the wake, though less so than the ions. This picture holds mostly for large bodies since the Coulomb field prevails near small bodies.

The wake has important effects on the potential distribution on large objects if their surface is not conductive. To avoid this, space probes are generally covered with a conductive coating (Fig. 7.6).

This simplified picture neglects the magnetic field of the solar wind, which introduces complications, as does the intrinsic magnetic field of the body if it has one (see for example [63]).

Figure 7.6 The spacecraft Ulysses before launch, covered with a conductive coating. (Photograph by Dornier GmbH.)

7.2.3 Being large: the importance of conductivity

Consider now scales that are not smaller than the particle gyroradii, so that the magnetic field comes into play. This has several consequences, that depend in a major way on the electric conductivity of the body.

To begin with, consider a solid body having no atmosphere nor a large-scale magnetic field of its own.[19]

Insulating body: the Moon example

Assume first that the body has a negligible electrical conductivity. In this case, no current can flow in the body, so that it does not react to the solar wind magnetic field which therefore penetrates it with few perturbations.

Let us put this rough argument on a quantitative footing. The time for the magnetic field to diffuse through a body of radius R and conductivity σ is $\tau_{diff} \sim \mu_0 \sigma R^2$; if it is smaller than the timescale for changes in the ambient magnetic field, or for the solar wind to flow past the body – which is roughly R/v_w – the solar wind magnetic field penetrates the body without being much perturbed. This holds if

$$\mu_0 \sigma R v_w < 1 \qquad (7.41)$$

which is the condition for the magnetic field not to be frozen into the body.

For an object of the size of the Moon (Table 7.1) and the typical solar wind speed v_w, (7.41) reads: $\sigma < 10^6$. This inequality should hold in much of

---

[19]We neglect here for simplicity the small-scale crustal magnetic fields, as do exist for example on the Moon and Mars.

Basics of the interaction 359

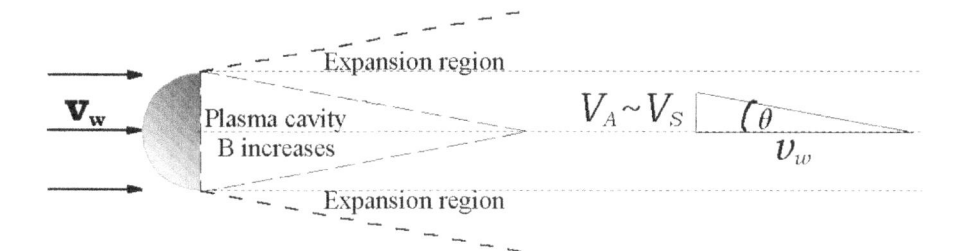

[Figure: sketch showing $\mathbf{V}_w$ arrows impinging on a Plasma cavity (B increases), with Expansion regions above and below, and angle θ with $V_A \sim V_S$ and v_w labeled]

Figure 7.7 Sketch of the interaction of a large insulating body with the solar wind. Since $V_A \sim V_S$, the magnetosonic wave speeds are all equal in order of magnitude, and so are the angles $\theta \sim 1/M$ that the depleted cone and the expansion region make to the wake axis.

the interior of the Moon, since the solar wind magnetic field has indeed been observed to be relatively unperturbed in its vicinity. This seems to agree with other inferences on the Moon's internal structure [76].

So the solar wind impinges on the insulating body, largely unperturbed. Incidentally, the absorption of solar wind particles by the surface of the Moon enables one to infer the past solar wind composition from the analysis of lunar rock samples.

The absorption of particles on the upstream surface creates downstream a cavity depleted of plasma, somewhat as outlined in the previous section. We have seen that when the scales are smaller than the particle gyroradii so that the magnetic field does not play a role, the plasma moves sideways to fill the cavity at the typical speed $v_{thp} \sim V_S$, whereas the body moves at $-v_w$ in the plasma frame; this produces a conical cavity of half-angle $\theta \sim 1/M \sim v_{thp}/v_w$, whose edge is diffuse because of the thermal spread of the particle speeds (Fig. 7.7). Furthermore, since the electrons tend to fill the cavity faster than the (slower-moving) protons, the cavity tends to charge negatively, producing an electric field that accelerates protons from the edge of the cavity towards the axis.

The magnetic field changes this picture, because at scales larger than the gyroradii it affects the penetration of particles into the cavity, by preventing them diffusing across it. If $\mathbf{B}$ is perpendicular to $\mathbf{v}_w$ (which coincides approximately with the wake axis),[20] the particles can still diffuse along field lines to fill the cavity, which is therefore not much modified. But if the magnetic field is parallel to $\mathbf{v}_w$, i.e. to the cavity axis, it does not let the particles diffuse across itself to fill the cavity, whose size is therefore significantly increased (Fig. 7.8).

In that case, a further effect arises. Since the plasma pressure is severely reduced within the cavity, the magnetic field must increase with respect to the ambient solar wind value for the total (plasma + magnetic) pressure in the cavity to balance the ambient value sideways. In the extreme case of a void

---
[20] We neglect the heliocentric speed of the body compared to the solar wind speed.

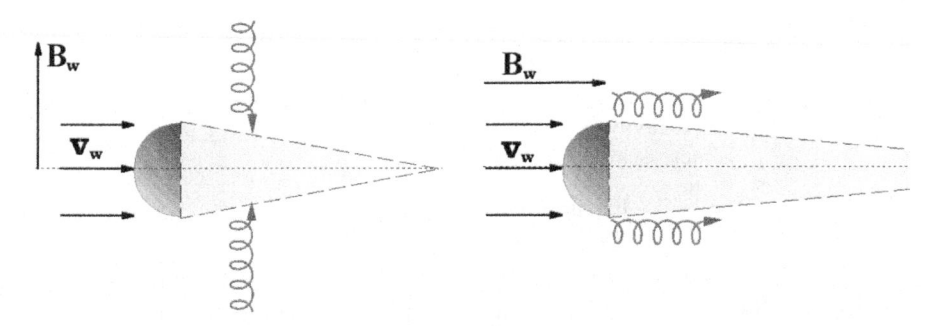

Figure 7.8 How the magnetic field modifies the cavity downstream of an insulating body. If $\mathbf{B} \perp \mathbf{v}_w$, the particles can fill the cavity by moving along the magnetic field lines, but they cannot do so if $\mathbf{B} \parallel \mathbf{v}_w$; in the latter case, the plasma-depleted cavity is of larger size.

cavity, the total pressure there is $\sim B_{cavity}^2/2\mu_0$; in order to balance the total solar wind pressure, which is roughly twice the magnetic pressure $\sim B_w^2/2\mu_0$ since the plasma and magnetic pressures are roughly equal in the solar wind, the magnetic field in the cavity must increase to the value $B_{cavity} \sim \sqrt{2} B_w$.

In addition, the plasma expands on the flanks of the cavity, making the particle density and the magnetic field decrease there, in a way that depends on the relative orientation of the solar wind magnetic field and speed, since Alfvén waves carry magnetic perturbations along $\mathbf{B}$ (in the plasma frame), whereas compressional modes may propagate in different directions at different speeds. Because of this anisotropy, the perturbations do not have a circular symmetry around the wake axis, but since all these waves propagate in the solar wind at speeds having the same order of magnitude, the angles made by the depleted conic cavity and the expansion region (Fig. 7.7), which are determined by the ratio of the distances travelled by the waves and by the solar wind, as $\theta \sim \arctan 1/M \sim 1/M$, are not very different, being all in the ballpark of 6°. The actual interaction is more complex than this rough picture, as shown by early observations of the lunar wake, and by recent ones performed with modern instruments on the spacecraft WIND (see for example [61]).

Note, finally, that since the body absorbs solar wind particles and emits photoelectrons, it acquires an electrostatic charge, somewhat as outlined in the previous section for scales greater than L_D (with minor changes due to the preferential collection of particles along the magnetic field). Because the body does not conduct electricity, the balance of currents must be made on each point of the surface, producing large differences of potential between the sunlit and shadowed sides (the latter being furthermore within the cavity). The electrostatic charging of the Moon has interesting consequences (Problem 7.6.1.).[21]

---

[21]The electrostatic charging of the Moon might be responsible for the transient dust clouds observed lying above the lunar surface (see for example Horanyi, M. et al. 1995, Geophys. Res. Lett. **22** 2079 and references therein).

Basics of the interaction 361

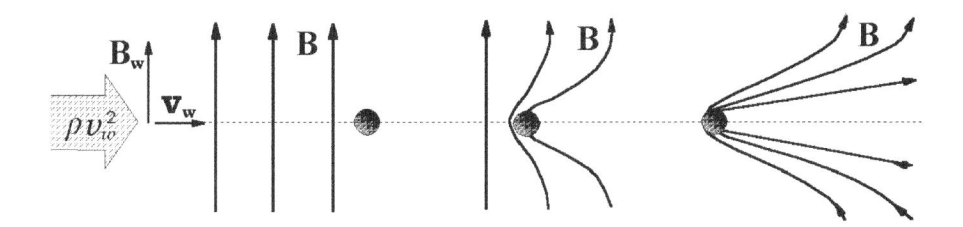

Figure 7.9 Sketch of the interaction of a large conducting body with the solar wind. The solar wind magnetic field cannot penetrate the body, forcing the incoming plasma to slow down ahead and be diverted sideways, whereas the magnetic field lines pile up ahead and a magnetic tail forms downstream. Since the solar wind flow is supersonic far upstream, a shock (not shown) is generally produced upstream. (Adapted from a sketch of a comet plasma tail by Alfvén [2].)

Large conducting body

Consider now a body of radius R (still larger than the gyroradii) with a conductivity σ large enough that

$$\mu_0 \sigma R v_w \gg 1. \tag{7.42}$$

In this case the magnetic diffusion time through the body is much longer than the time for the solar wind to sweep past it, so that the solar wind magnetic field cannot penetrate it. Since the solar wind plasma is frozen to the magnetic field, it must slow down ahead of the body and stop just before it. As the plasma slows down, the magnetic field lines pile up, just as do cars in a traffic jam. Because the magnetic field continues to be convected by the unperturbed flow sideways, the magnetic field lines bend around the object to form a magnetic tail downstream. Figure 7.9 is adapted from a sketch by Alfvén of the formation of comets' tails; we shall see later that in that case, the conducting body is the ionosphere of the comet, and that the wind is slowed down at large distances by picking up freshly ionised cometary molecules.

The magnetic field increase as the plasma slows down in approaching the object is related to the frozen-in induction equation. In a stationary state, this equation yields $\nabla \times (\mathbf{v} \times \mathbf{B}) = 0$. The problem is most easily visualised in the simple case when the solar wind speed and magnetic field are roughly perpendicular to each other. In this case, the above induction equation reads (in a one-dimensional approximation)

$$v(x) B(x) = \text{constant} \tag{7.43}$$

where x is the distance from the front of the body.

Hence, as the body is approached, the amplitude of the magnetic field increases – an increase that stops close to the body, because the frozen-in approximation breaks down at small scales (as does the one-dimensional approximation). What is the strength of the peak magnetic field B_{max}? Basically, it must be large enough that the Lorentz force stops the incoming plasma at the front of the body and deflects it on the flanks. For this to be so, the magnetic energy must be roughly equal to the ram pressure of the incoming flow, i.e.

$$B_{max}^2/2\mu_0 \sim \rho_w v_w^2 \qquad (7.44)$$
$$\Rightarrow B_{max} \sim 7 \times 10^{-8} \, d_\oplus/d \quad \text{T} \qquad (7.45)$$

where we have used (7.18). This means that the magnetic pile-up does increase the ambient solar wind magnetic field by a factor of

$$B_{max}/B_w \sim \sqrt{2} \, v_w/V_A \qquad (7.46)$$

i.e. by one order of magnitude. The vanishing of the magnetic field inside the object, in spite of the large magnetic field ahead of it, is ensured by a sheet of electric current **J** flowing perpendicular to the plane of the figure; this current is driven by the electric field induced in the frame of the body by the motion of the magnetised solar wind: $\mathbf{E} = -\mathbf{v}_w \times \mathbf{B}_w$ (Problem 7.6.2).

Finally, the above picture misses an important point: since the solar wind plasma is supersonic, it generally cannot stop without forming firstly a shock, so that the sketch shown in Fig. 7.9 holds downstream of a bow shock, where the solar wind plasma has become subsonic. Since the solar wind ram pressure is not lost in the shock (see Section 6.3), but somehow converted into plasma and magnetic pressures, the balance equation (7.44) still holds, and so does the value (7.46) of the pile-up magnetic field. We shall consider planetary bow shocks in more detail in Section 7.2.6.

How does this picture apply to actual solar system bodies? We have seen that most planets have a ionosphere and/or a magnetic field. In that case, the environment may provide the required conductivity instead of the body itself. In the case of comets (and some planets, too, albeit more weakly), the picture is modified by the ejection of neutrals from the body, which are subsequently ionised and picked up by the solar wind (Section 6.5), making it slow down, but the basic picture remains true. For planets having a magnetic field of their own, the magnetic field deflects the solar wind particles and is itself modified by the interaction. We examine these cases below.

7.2.4 Large objects with a conducting atmosphere

We have seen that most planets have ionospheres, whose conductivity is very large at high altitudes. The solar wind magnetic field therefore cannot penetrate it, and wraps around it as illustrated in Fig. 7.9, except that now the obstacle to the flow is the ionosphere rather than the body itself. Just as we saw above, the solar wind slows down ahead of the ionosphere and stops, being diverted sideways, while the magnetic field lines pile up ahead and are pulled downward to form a magnetic tail. The vanishing of the magnetic field is ensured by

Basics of the interaction

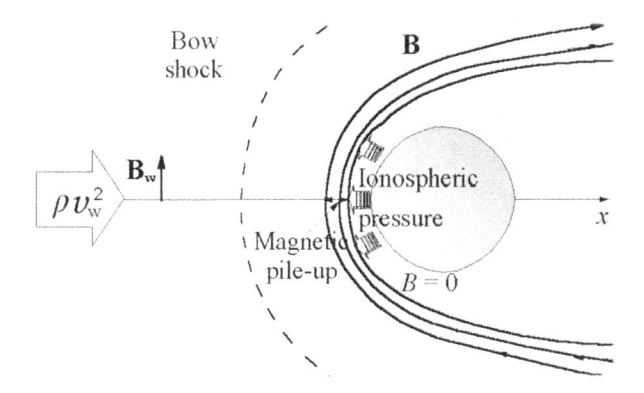

[Figure: sketch showing Bow shock, B field lines, Ionospheric pressure, Magnetic pile-up, B=0 region, with incoming ρv_w^2 and B_w labels along x axis]

Figure 7.10 Sketch of the interaction of a non-magnetised planet having an ionosphere. The flow (not shown) slows down ahead of the object and is diverted sideways, while the magnetic field lines pile up against the ionosphere and are stirred downstream into a magnetic tail rather as in Fig. 7.9. The slowing down of the supersonic solar wind flow is mediated by a bow shock ahead (Section 7.2.6).

currents flowing at the top of the ionosphere, forming a complex boundary – the *ionopause* (Fig. 7.10) [59].

In the sunward direction, the ionopause is located at the distance from the planet where the particle pressure in the ionosphere balances the pressure $\rho_w v_w^2$. The interaction proceeds somewhat as outlined above, with the pile-up magnetic field serving as a 'cushion' whose magnetic pressure $B^2/2\mu_0$ on the sunward side – roughly equal to $\rho_w v_w^2$ – balances the ionospheric pressure. As a result, the magnetic field and the plasma of solar wind origin tend to be excluded from the region bounded by the ionopause, i.e. the part of the ionosphere whose pressure is sufficient to stand off the flow.

This kind of interaction takes place, in a first-order approximation, at Venus and Mars, which both have no large-scale magnetic field.

In order to be able to stand off the solar wind, the ionospheric plasma pressure must be equal to at least $\rho_w v_w^2$. We have seen that when the ionosphere is a Chapman layer, the electron density at the maximum of ionisation is given in order of magnitude by (7.16), so that the plasma pressure $P \sim 2nk_BT$ is

$$P_{max} \sim 10^{-11} T \left(H_\oplus^{1/2} d_\oplus / H^{1/2} d \right). \qquad (7.47)$$

Comparing with the solar wind ram pressure (7.18), we see that if the ionospheric temperature is greater than a few times 100 K, the ionosphere of an Earth-like (but non-magnetised) planet can stand off the solar wind in normal conditions. Since the ionospheric pressure decreases with altitude above the maximum of ionisation, the smaller the solar wind ram pressure, the higher the altitude of the ionopause. This is illustrated in Fig. 7.11, which shows how the interaction changes with the solar wind ram pressure at Venus. When $\rho_w v_w^2$

Electron density n (cm^{-3})

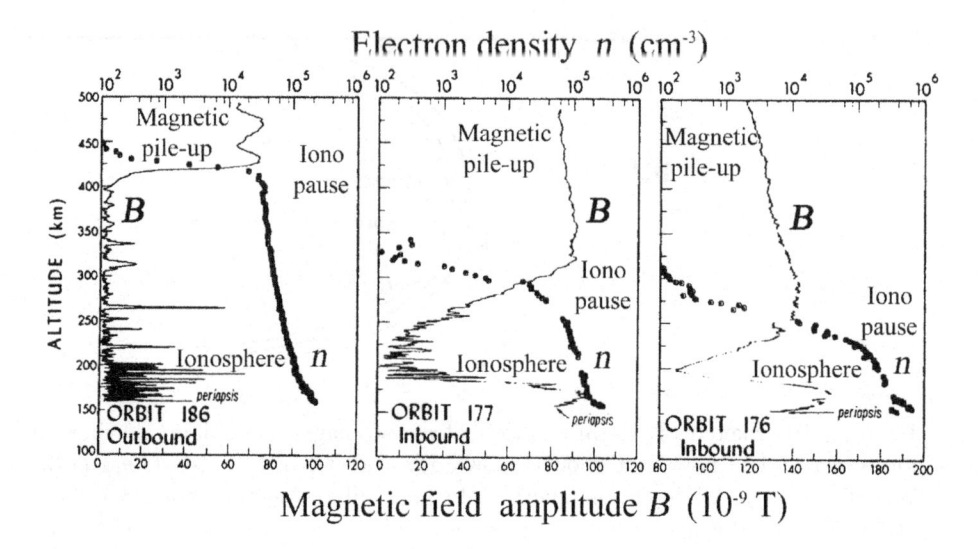

Figure 7.11 Interaction of the solar wind with Venus. Electron density and magnetic field strength versus altitude measured on the Pioneer Venus Orbiter spacecraft, for three values of the solar wind ram pressure (increasing from left to right). As the solar wind ram pressure increases, so does the magnetic pile-up, pushing the ionopause downwards. For too large a ram pressure, the ionosphere becomes partly magnetised. (Adapted from [68].)

increases (from left to right), the magnetic field strength increases according to (7.44), and the ionopause goes downward where the plasma pressure is greater (Problem 7.6.4). When the solar wind ram pressure is so large that the ionospheric pressure alone cannot stand off it, the magnetic field penetrates into the ionosphere.[22]

Figure 7.12 shows the magnetic strength near Mars, as the spacecraft Mars Global Surveyor moves from the solar wind to the ionosphere, and returns to the solar wind again. One sees that the magnetic field starts to pile up at some distance above the ionopause, forming an abrupt boundary where the solar wind plasma is replaced by plasma of planetary origin [10]. This feature is outside the scope of the traditional fluid description, for two basic reasons; first, the solar wind and planetary particles have properties that are too different for them to be treated as a single fluid; second, each kind of particle itself does not behave as a fluid; in particular, the planetary heavy ions have a gyroradius larger than the size of the planet.

For various reasons (Problem 7.6.4), the Mars environment has more difficulty in standing off the solar wind, so that the interaction at Mars resembles

---

[22] In that case, the additional pressure needed is supplied by the magnetic field that has penetrated and by the drag of ion–neutral collisions associated with some downward motion of the ionospheric plasma. If these contributions are not enough, the solar wind may impinge on the planet's surface.

Basics of the interaction

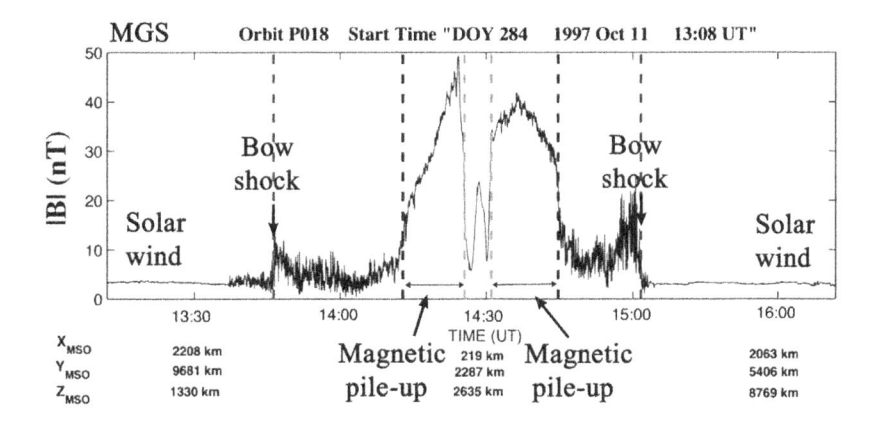

Figure 7.12 Interaction of the solar wind with Mars. Magnetic field strength measured on the Mars Global Surveyor spacecraft during part of the orbit, as it approaches the planet and goes away, showing the bow shock, the magnetic pile-up and the region of reduced magnetic field. (Adapted from [10].)

the one at Venus when the solar wind ram pressure is large, with the magnetic field penetrating somewhat into the ionosphere.

The solar wind has still another effect on the planetary environment. Dissociative recombination taking place in the ionosphere ejects atoms, that become ionised outside the ionosphere – mainly by photoionisation and by charge exchange with solar wind ions – and are then picked up by the flow (Section 6.5.3). This has several consequences. First, this contributes to slowing down the flow – a process that is most important in comets. Second, since these particles are lost from the planet, this process makes it lose part of its atmosphere. Third, the pick-up process produces plasma instabilities and waves, especially at the proton cyclotron frequency. And finally, the charge exchange (Section 2.4.4) produces X-ray emission – a process we shall study in detail for comets (Section 7.5.6).

7.2.5 Large magnetised objects

Many planets have magnetic fields of their own. We consider in this section the basic points of the interaction of the solar wind with a body (again of radius larger than the gyroradii) carrying a magnetic moment μ. We assume here for simplicity that the solar wind velocity is roughly perpendicular to μ (Fig. 7.13), and to begin with we forget the solar wind magnetic field, concentrating on the magnetic field of the planet; we shall consider the role of the solar wind magnetic field later.

The solar wind particles tend to be reflected by the planetary magnetic field, making the wind slow down and stop. This is easily seen by considering the simplified case of a non-magnetised region on the sunward side of some boundary, and a large magnetic field on the planet's side. The incident particles entering

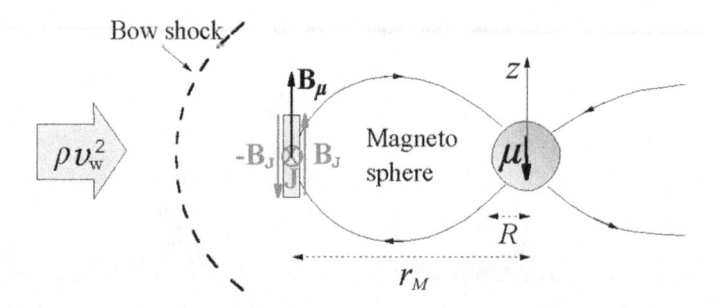

Figure 7.13 Sketch of the (sunward side) interaction of a magnetised planet with the solar wind. In order to confine inwards the planet's magnetic field, the boundary (the *magnetopause*) carries a current sheet producing a magnetic field $\mathbf{B}_J$ that roughly cancels the planet's dipole field outside the magnetopause, and therefore adds to it just inside, doubling it. The inward confined magnetic field stands off the incoming flow. The slowing down of the supersonic solar wind flow is mediated by a shock ahead (Section 7.2.6). This picture will be completed in Section 7.3.

the magnetised region make half a gyro turn and exit back in the field-free region (this requires that the gyroradii be smaller than the distance between the planet and the boundary). The difference in the gyroradii of electrons and protons produces a current perpendicular to both the initial velocity and the magnetic field, which makes for the discontinuity in magnetic field (Problem 7.6.3).

Since the stagnation of the solar wind is produced by the planet's magnetic field, it takes place at the distance from the planet where the pressure of the planet's magnetic field (which is generally much greater than the plasma pressure of the planetary environment there) balances the solar wind ram pressure; this yields $\rho_w v_w^2 \sim B_\mu^2 / 2\mu_0$, where B_μ is the magnetic field inward of the stagnation region.[23] This boundary, which particles do not cross,[24] is typically a few ion gyroradii thick; it is known as the *magnetopause*. It separates the plasma of solar wind origin from the planetary environment – made of both the planet's magnetic field and its plasma – called the *magnetosphere*.

Let us estimate the distance r_M of the magnetopause to the planet, in the magnetic equatorial plane (Fig. 7.13). A first approximation may be obtained by simply substituting into the balance equation the planet's magnetic field strength at r_M:

$$B_\mu = (\mu_0/4\pi)\,\mu/r_M^3. \tag{7.48}$$

A little reflection, however, shows that the magnetic field strength just inside the magnetopause is greater than the dipole value B_μ. This is because in order

---

[23] As we already noted, the equivalent pressure just outside the boundary is actually made of plasma and magnetic pressure – resulting from the conversion of the solar wind ram pressure at a shock and downstream of it.

[24] With some notable exceptions, discussed in Section 7.3.

Basics of the interaction

to confine inwards the planet's dipole magnetic field, the magnetopause must carry a current sheet producing a magnetic field that cancels $\mathbf{B}_\mu$ on the outward side (Problem 7.6.3). It must therefore produce, by Ampère's law, an equal but oppositely directed field on the inward side, which adds to $\mathbf{B}_\mu$, doubling it (Fig. 7.13). Hence the total magnetic field just *inside* the boundary is roughly twice larger than B_μ, so that the balance equation becomes

$$\rho_w v_w^2 \simeq (2B_\mu)^2 / 2\mu_0. \tag{7.49}$$

Substituting (7.48), we deduce the stand-off distance

$$r_M \simeq \left(\frac{\mu}{4\pi} \sqrt{\frac{2\mu_0}{\rho_w v_w^2}} \right)^{1/3}. \tag{7.50}$$

Introducing the planet's radius R and its magnetic field at equator $B_0 = (\mu_0/4\pi)\,\mu/R^3$, we get the relative distance of the magnetopause on the sunward side

$$\frac{r_M}{R} \simeq \left(\frac{2B_0^2}{\mu_0 \rho_w v_w^2} \right)^{1/6} \simeq 9 \left(\frac{\mu}{\mu_\oplus} \frac{d}{d_\oplus} \right)^{1/3} \frac{R_\oplus}{R} \tag{7.51}$$

(which may be rewritten as $r_M/R \simeq 2^{1/6} (B_0 V_A / B_w v_w)^{1/3}$ as a function of the solar wind Alfvén speed V_A and its magnetic field B_w).

The magnetopause separates the flow of solar wind origin and the magnetospheric plasma, which are slipping past each other via a tangential discontinuity.[25]

How does this picture compare with observation? Table 7.2 shows the distances of the nose of the magnetopause for the magnetised planets, from observation and from (7.51) (using the solar wind ram pressure (7.18) and the planets' properties listed in Table 7.1). One sees that the agreement is quite good, except for Jupiter, whose magnetosphere is observed to be much larger than the estimate.

Mercury has by far the smallest magnetosphere, due to the combination of two factors: the greatest solar wind dynamic pressure (because of the proximity to the Sun), and the smallest magnetic moment. At the other extreme lies Jupiter's magnetosphere, which is greater by three orders of magnitude than the volume of the Sun; if it were visible, its angular size as seen from Earth would be twice that of the Sun even though it is more than four times farther away. This huge size is due to three factors: a small solar wind dynamic pressure (because it is far from the Sun), the largest magnetic moment and a relatively dense plasma, supplied by the volcanoes of the satellite Io, which rotates with the planet and further inflates the magnetosphere in the equatorial plane. This latter effect, not included in our simple estimate, is responsible for most of the discrepancy between the observed and calculated stand-off distances. Note, finally, that the

---

[25] Just as a wind blowing over a lake induces ripples on its surface, the solar wind produces ripples in the magnetopause, by the *Kelvin–Helmholtz instability*.

Table 7.2 *The sunward distance of the magnetopause, calculated from (7.50) and observed (or inferred)*

	$d/d_\oplus$	$R/R_\oplus$	r_M/R (calc.)	r_M/R (obs.)
Mercury	0.39	0.38	1.2-1.5	1.4
Earth	1.0	1	9	8–12
Jupiter	5.2	11.2	38	50–100
Saturn	9.5	9.4	17	16–22
Uranus	19	4	22	18–25
Neptune	30	3.9	22	23–26

large tilts of the axes of Uranus and Neptune make their magnetospheres vary considerably during their rotation.

Despite the order-of-magnitude agreement of the magnetopause location given by (7.50)–(7.51) and observation, Table 7.2 shows a systematic difference: the calculated values are about 10% too small. Four factors contribute to this slight discrepancy: first, the solar wind does not necessarily arrive at a normal angle to the magnetopause; second, even if it arrives normal, the solar wind ram pressure to be put in the balance equation (7.49) is slightly smaller than ρv_w^2, because of the diversion of the flow around the magnetopause (as we shall see below); third, the dipole magnetic field is compressed by the interaction. And finally, when the solar wind magnetic field has a component directed opposite to the dipole's field, the planet's field is weakened, so that the magnetopause must move inwards to keep standing off the solar wind, and further effects arise, as we shall see in Section 7.3.

The Earth's magnetopause is by far the most extensively studied, and a concise recent review may be found in [72], while early research is nicely reviewed in [30].

7.2.6 Bow shocks

As we noted, a large body that is either conducting, or surrounded by a conducting atmosphere, or has an intrinsic magnetic field, does stop the solar wind. This generally requires a shock, where the flow becomes subsonic and changes direction to divert around the body. At the shock, the solar wind bulk kinetic energy is converted into particle and magnetic pressure, with some dissipation.

As the shocks we encountered in the solar wind proper (see Section 6.3), the planetary bow shocks are collisionless, and the dissipation involves complex microscopic plasma processes. When the shock is quasi-perpendicular and of large Mach number, the width of the shock is roughly equal to the proton gyroradius (calculated with the plasma speed in the shock frame), which is close to v_w/ω_{gp} (where ω_{gp} is the proton angular gyrofrequency downstream of the shock) [7]. As for interplanetary shocks, the compressive wave involved is

Basics of the interaction

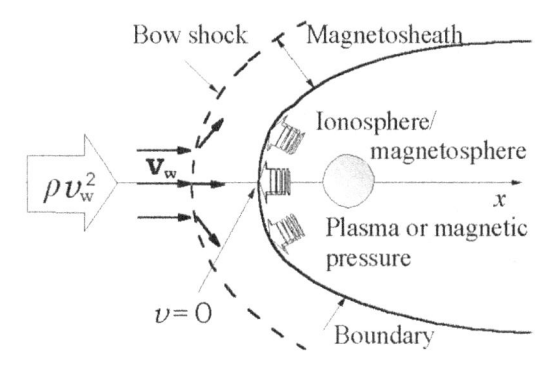

Figure 7.14 Sketch summarising the generic interaction between the solar wind and a body having either a conducting atmosphere or a magnetic field of its own. A bow shock is formed ahead, where the solar wind becomes subsonic, delimiting a magnetosheath where the subsonic flow slows down smoothly, stopping and being diverted sideways at a boundary. This boundary separates from the incoming wind the planetary plasma and its magnetic field (if it has one), either of which provides the required pressure at the boundary.

the fast magnetosonic wave. The bow shock is nearly stationary with respect to the planet (except for a small velocity mainly due to changes in the solar wind dynamic pressure), and defines a transition region – the *magnetosheath* – containing heated solar wind plasma (Fig. 7.14). It is in this region that the – now subsonic – flow slows down smoothly.

Figure 7.15 shows the flow speed and magnetic field measured on one of the Cluster spacecraft as the spacecraft moves from the magnetosheath to the solar wind. The speed and magnetic field jumps are close to a factor of four: the value expected at high Mach numbers when $\gamma = 5/3$ (see Section 2.3).

From the point of view of the interaction of the solar wind with the planet's magnetosphere or ionosphere, two important questions arise. First, we have assumed in the previous sections that the pressure in the stagnation region just before the obstacle (where the flow speed vanishes) is roughly equal to the solar wind ram pressure. Is this assumption correct? Second, how far from the obstacle (i.e. the magnetopause or the ionopause) does the shock lie?

Consider the first question, in the simple case of a quasi-perpendicular shock of large Mach number, with $\gamma = 5/3$. Let the subscripts 1 and 2 denote respectively the upstream (solar wind) and downstream (magnetosheath) values of the plasma parameters, and P the total (plasma + magnetic) pressure. We have from the Rankine–Hugoniot relations (Section 2.3)

$$\rho_2 v_2^2 \simeq \rho_1 v_1^2/4 \tag{7.52}$$

$$\rho_1 v_1^2 \simeq P_2 + \rho_2 v_2^2 \tag{7.53}$$

where (7.52) comes from $\rho_1 v_1 = \rho_2 v_2$ with $v_2 \simeq v_1/4$ (since $M_1 \gg 1$), and (7.53) uses the fact that since $M_1 \gg 1$, $\rho_1 v_1^2 + P_1 \simeq \rho_1 v_1^2$. The upstream momentum

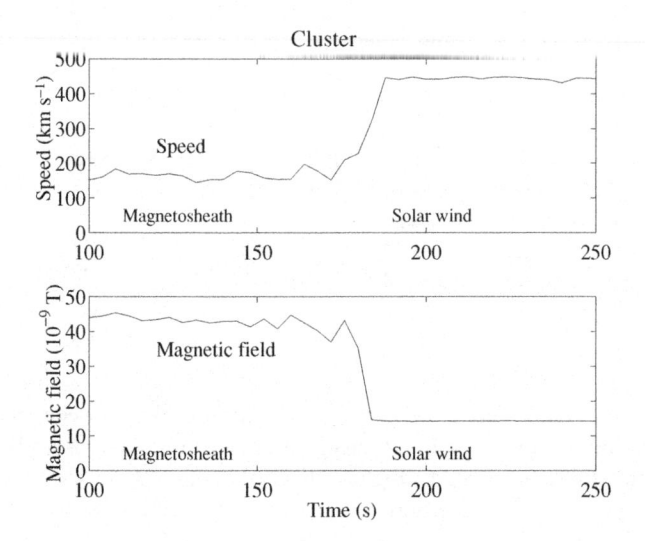

Figure 7.15 Speed and magnetic field measured on one of the Cluster spacecraft near the Earth's bow shock on 31 March 2001 (the time origin is 22.13.00). Cluster prime parameter data: speed from the ion spectrometer [67], and magnetic field from the flux gate magnetometer [8]. (Courtesy of M. Maksimovic.)

flux $\rho_1 v_1^2$ is just the solar wind ram pressure. Consider how the flow manages to slow down between the downstream side of the shock (where the parameters are ρ_2, v_2, P_2) and the stagnation point, where the speed vanishes and the pressure is P_0 (Fig. 7.14). Let us assume for simplicity that the problem is nearly one-dimensional, depending mainly on the distance x along the Sun–obstacle line. Momentum conservation yields

$$\rho v \frac{dv}{dx} = -\frac{dP}{dx}. \tag{7.54}$$

Since the fluid is subsonic in this region ($M_2 \simeq 1/\sqrt{5}$ for $M_1 \gg 1$ and $\gamma = 5/3$), we may neglect its compressibility without making a large error, so that

$$\rho v \frac{dv}{dx} \simeq \frac{d}{dx} \left(\rho v^2 / 2 \right)$$

and (7.54) may be approximated by

$$\rho v^2 / 2 + P \simeq \text{constant}. \tag{7.55}$$

Applying this equality between the downstream side of the shock (labelled by the subscript 2) and the stagnation point (where the flow speed is $v_0 = 0$), we have

$$\rho_2 v_2^2 / 2 + P_2 \simeq \rho_0 v_0^2 / 2 + P_0 = P_0. \tag{7.56}$$

Basics of the interaction

We deduce the total pressure at the stagnation point

$$P_0 \simeq \rho_2 v_2^2/2 + P_2 = -\rho_2 v_2^2/2 + \left(\rho_2 v_2^2 + P_2\right) \simeq -\rho_1 v_1^2/8 + \rho_1 v_1^2$$
$$\simeq (7/8)\,\rho_1 v_1^2 \equiv (7/8)\,\rho_w v_w^2 \qquad (7.57)$$

where we have used (7.52) and (7.53). This shows that we made indeed a small error in approximating the total pressure at the stagnation point by the solar wind ram pressure.

Now, let us examine the second question: what is the distance of the shock from the obstacle? The answer depends on the shape of the obstacle. Indeed, if the obstacle had a sharp-pointed nose, the shock would not be detached from it. The magnetosphere has a blunt nose, with a radius larger in the transverse direction than towards the Sun. In practice, its sunward shape can be modelled by a conic function whose focus is the planet, so that the distance r from the focus to a point on the curve is given as a function of the distance x along the Sun–planet line by

$$r \simeq r_M + \epsilon\,(r_M - x) \qquad (7.58)$$

where r_M is the already determined distance of the nose to the planet, and the eccentricity ϵ determines the terminator distance as $r \simeq (1+\epsilon)\,r_M$ at $x = 0$. Typically, $\epsilon \simeq 0.5$, so that the magnetosphere has a transverse radius 50% larger than the stand-off distance r_M.

The shape of the shock may be approximated by a similar curve, whose nose is located at some distance from the magnetosphere's nose. This distance is some fraction of r_M: the larger the obstacle, the larger the structure, and the farther it lies from it. Typically, the distance of the nose of the shock to the planet is roughly $r_M\,(1 + 1.1\rho_1/\rho_2) \simeq 1.3 r_M$ for $M_1 \gg 1$ and $\gamma = 5/3$ (see [65]).[26]

In the same way as the heliospheric shocks, planetary shocks are at the origin of a large variety of plasma waves. In particular, particles reflected by the shock can reach large distances upstream of the shock along magnetic field lines, producing instabilities in the region of the solar wind that is connected to the shock by magnetic field lines – a region known as the *foreshock*.

7.2.7 Not being constant: sputtering and evaporation

Up to now, we have pictured the bodies as being virtually constant, merely emitting photoelectrons (as a result of solar irradiation) and a small quantity of secondary electrons (as a result of solar wind plasma bombardment if any). Likewise, we considered the atmosphere as virtually constant, too, except for atoms that are picked up by the solar wind.

---

[26] When the upstream Mach number is not much greater than unity, the shock is weaker, so that it needs to stand further upstream of the obstacle to be able to divert the flow. On the other hand, when the adiabatic index γ is smaller (i.e. closer to 1), the plasma is more compressible, so that the shock can stand closer to the obstacle and still be able to divert the flow.

In some cases, however, the body loses so much matter that the nature of its interaction with the solar wind and its evolution are considerably modified. This happens in two main situations:

- when solar wind particles hitting the body (or its atmosphere if any) produce the ejection of a large quantity of atoms and molecules; this is referred as *sputtering*;

- when the body loses large quantities of matter by sublimation; this is the basis of cometary physics.

The latter case will be addressed in Section 7.5, and here we only comment briefly on the former case: sputtering [11]. Sputtering is a process by which an energetic particle deposits its energy in a material, leading to the ejection of particles. For this to happen, two conditions must be met:

- particles must reach the object (or its atmosphere) and deposit enough energy;

- the sputtered particles must be able to escape, i.e. their energy must exceed the binding energy of the body's surface (for sputtering from the body) or of the body's gravitational attraction (for sputtering from the atmosphere).

Sputtering of various bodies by solar wind particles has a number of consequences. For example, sputtering of the dust grains contributes to producing solar wind particles. Sputtering of the surfaces of the Moon and Mercury contributes to their environments. Sputtering of the atmosphere of Mars contributes to its atmospheric loss.

The impact of solar wind particles also affects the surface properties of bodies, producing a *space weathering* (see for example [84] and references therein).

7.3 The magnetospheric engine

'What are fireworks like?' ... asked the little Princess in an Oscar Wilde fairy tale. 'They are like the *Aurora Borealis*,' said the King ... 'only much more natural.'

For the layman and the expert alike, the aurora (Fig. 1.2) remains one of the most mysterious of Nature's displays. It is now quite well understood how energetic electrons, precipitating into the Earth's atmosphere, excite its constituents and produce the majestic draperies of light that have fascinated humans for ages and fostered a variety of myths. Likewise, the source of the huge energy involved – the solar wind – is well identified. Nevertheless, the complex chain of events that produces the precipitating electrons is still a source of debate.

In a sense, the polar atmosphere can be thought of as a giant cathode television tube on which is displayed the aurora; we understand how electrons impinging on this screen produce images; we know that the television set is

The magnetospheric engine 373

Figure 7.16 Looking after the draperies of the sky ... (Jean Effel, *Turelune le Cornepipeux*, 1975, Copyright Adagp, Paris, 2007.)

plugged into the solar wind. But we neither understand nor master the basic electronics driving the machine.

And yet, for all its mystery, this phenomenon remains insignificant on a cosmic scale. Auroras are only one of the numerous effects that solar wind perturbations produce on Earth. Basically, when the planet is hit by an interplanetary mass ejection, or encounters a transition between slow and fast wind or – more frequently – a gentler perturbation, the magnetosphere is distorted and electric currents are induced, producing changes in the magnetic field (the *geomagnetic activity*), radio emissions, and a number of nuisances, such as breakdowns of power stations and loss of orbiting satellites. Other planets having an intrinsic magnetic field and a conspicuous atmosphere, such as Jupiter, Saturn, Uranus and Neptune, are subjected to similar processes,[27] which are also expected to occur in extrasolar planetary systems. And much the same physics acts at incomparably larger scales in the universe.

Figure 7.17 illustrates such perturbations, on a rare occasion when the Sun and the planets Earth, Jupiter and Saturn were nearly aligned. According to

---

[27] Auroras have also been observed on Venus, and, recently, on Mars – controlled by the crustal magnetic field [9].

374   *Dust, asteroids, planets and comets*

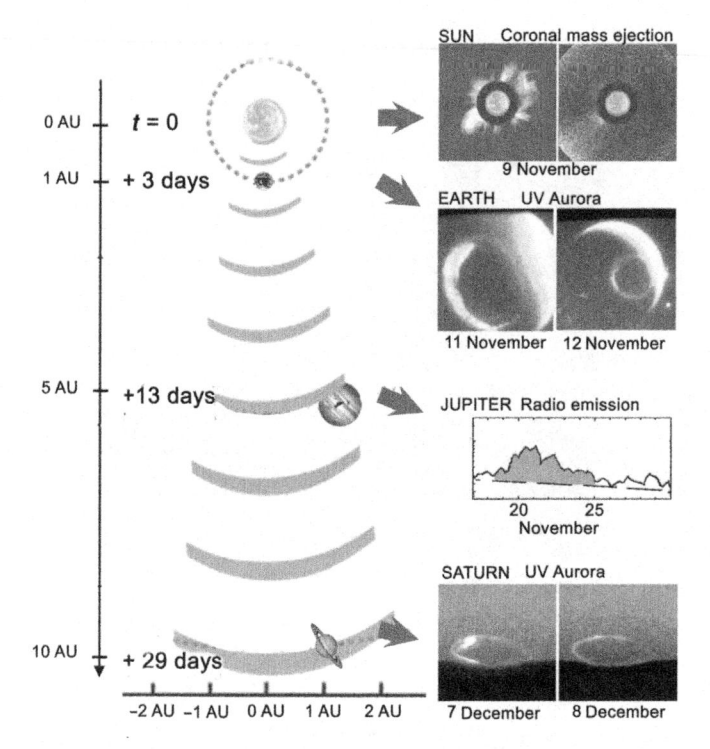

Figure 7.17 How an interplanetary shock associated with a coronal mass ejection from the Sun may have triggered auroral disturbances on Earth, Jupiter and Saturn. The three planets were nearly aligned with the Sun, and the temporal succession of events is consistent with a disturbance travelling at a speed of 600–700 km s^{-1}. The auroral oval at Earth was recorded on the spacecraft Polar, the radioemission at Jupiter was measured on Cassini, and the auroral oval at Saturn by the Hubble Space Telescope. (Adapted from [66].)

the authors [66], a coronal mass ejection at the Sun produced auroral storms at Earth three days later, enhanced radio emission at Jupiter about 13 days later and enhanced auroral emission at Saturn about 29 days later – i.e. time intervals consistent with the propagation of an interplanetary disturbance at a speed of 600–700 km s^{-1}.

How are these perturbations produced? Today's scientists have arguably improved the ingenious explanation of auroras based on the firing of dry gases proposed by Aristotle; they have built beautiful instruments and implemented them on a horde of space vehicles for performing *in situ* measurements – arguably improving on Birkeland's *terella* laboratory experiments. And yet, they still do not agree on the mechanism responsible for accelerating the particles producing auroras.

In this section, we shall derive some order-of-magnitude estimates on the structure of magnetospheres of magnetised planets, the energy provided by the solar wind, the coupling process and finally the products of the interaction.

The magnetospheric engine 375

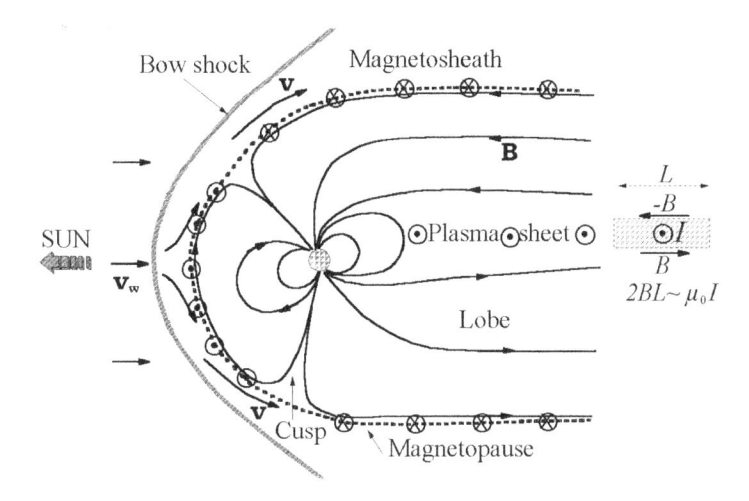

Figure 7.18 Sketch of the magnetosphere of a magnetised planet (cross-section in the noon–midnight meridian). The tail has two lobes of opposite magnetic field, of cross-section sketched in Fig. 7.21, separated by a plasma sheet that carries a current connecting with the currents flowing on the flanks of the tail, in the sense indicated by the symbols $\odot$ and $\otimes$. (The orientation corresponds to the present Earth's magnetic field.)

Among the huge (and often self-contradictory) literature on this difficult subject, accessible reviews aimed at physical processes may be found for example in [38], [39] and [19]; a tutorial on the difficult art of mapping the Earth's magnetosphere may be found in [79]; and a recent detailed review of auroral phenomena may be found in [64].

7.3.1 Basic structure

Simple picture

Figure 7.18 is a sketch of the magnetosphere in a plane containing the Sun and the planet's magnetic moment, when the latter is roughly perpendicular to the solar wind velocity. The picture assumes that two conditions are met: the scales are larger than the solar wind particle gyroradii, and the distance of the nose of the magnetopause (estimated in (7.50)) is comfortably larger than the planet's radius. This holds for the six planets listed in Table 7.2, although marginally so for Mercury.[28]

As we saw in the previous section, the solar wind becomes subsonic at the bow shock and is further decelerated and diverted in the magnetosheath, to pass round the magnetopause (dotted line) which confines the planet's magnetic field

---
[28] Due to the large dipole tilt angles of Uranus and Neptune, the orientation of their magnetic fields with respect to the solar wind varies considerably over a planetary rotation period, making their magnetospheres highly asymmetric and time variable.

and its plasma. The magnetopause has two neutral points on the sunward side, where the magnetic field turns and vanishes, forming two cusps. The solar wind drags the planet's magnetic field lines in the anti-solar direction, forming an elongated tail which has two lobes, the top one pointing to the planet, the bottom one pointing away.

The tail lobes of oppositely directed magnetic fields require a current sheet between them, just as the confinement of the planet's magnetic field requires currents flowing on the magnetopause. The sketch on the right-hand side of Fig. 7.18 shows how the current flowing along some length L of a sheet is related, by Ampère's law, to the magnetic field strength B produced on both sides, as $I \sim 2BL/\mu_0$. Other currents flow, but for the moment we spare them from the reader.

Opening the magnetosphere

Figure 7.18 does not tell the whole story. If it would, the planet's magnetic field would be nearly perfectly confined by the current sheets flowing along the magnetopause, and there would be virtually no connection between the magnetosphere and the solar wind.[29] For such a 'closed' magnetosphere, the formation of the magnetic tail requires, from a fluid point of view, some friction between the solar wind and the flanks of the magnetosphere, since, contrary to the case sketched in Fig. 7.9, the field lines of the tail do not belong to the solar wind flow.

The additional point in the story comes from the fact that the solar wind magnetic field has often a significant component antiparallel to the planet's magnetic field at the nose of the magnetopause (Fig. 7.19). In that case, the flow pushes together oppositely directed magnetic fields, favouring field line reconnection, so that the planet's magnetic field may become connected to the solar wind one (at X_1 on Fig. 7.19).

In this picture [26], reconnection produces field lines having one end attached to a polar region while the other end is connected to the solar wind, so that, because of flux freezing and magnetic tension, they are convected from the day side to the night side magnetosphere. The magnetic field lines therefore open up, forming a tail attached to the planet. No anomalous viscosity at the magnetopause is required to stir them, since they are simply drawn by the solar wind at their 'ends' belonging to it. Note the basic difference from the geometry of Fig. 7.9, where the magnetic tail entirely belongs to the solar wind, being made of solar wind field lines folded up as an umbrella around the obstacle. Beware that the terminology of 'moving magnetic field lines' should not be over-interpreted other than a convenient shorthand notation for picturing the plasma particles frozen into the magnetic flux tubes, and that the above picture is open to much criticism [37]. Indeed, the velocity of a magnetic field line is not a clear-cut concept and cannot be measured, contrary to the one of particles, and the MHD description on which it is based is not properly justified.

---

[29] Except at the cusps, and through diffusion of particles, and transfer processes associated in particular with the Kelvin–Helmholtz instability.

The magnetospheric engine 377

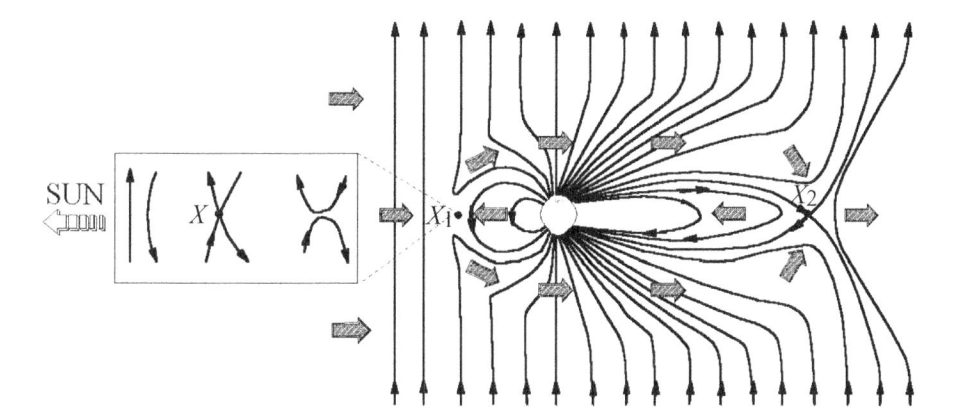

Figure 7.19 Fluid picture of an open magnetosphere. When the solar wind magnetic field has a large component along the planet's magnetic moment (and in the same sense), the planet's magnetic field becomes connected with the solar wind one, opening the magnetosphere through field lines that cross the planet's surface on polar caps (dark grey). Merging of magnetic field lines occurs on the day side (at X_1 – of which an enlarged view as a function of time is shown in the left-hand panel), whereas a disconnection occurs on the night side (at X_2). This drives a circulation pattern (big arrows.) To avoid excessive 'geobias', the planet's magnetic moment is here opposite to the Earth's present one. (Adapted from [26].)

This produces the steady state magnetic pattern sketched in Fig. 7.19, (where only a part of the tail's length is shown.) The magnetic field lines connected to the solar wind cross the planet in oval-shaped regions surrounding the magnetic poles – the *polar caps* (shown in dark grey). The anti-sunward flow driven by the solar wind over the polar caps requires a compensating return flow at lower latitudes, so that a disconnection of field lines must occur somewhere in the tail (X_2 in Fig. 7.19). The corresponding large-scale convection pattern postulated in this picture is shown by big arrows.

This gives a hint as to the location of the auroras. Convection towards the planet drives particles from the tail along the field lines towards the demarcation between closed and open lines, i.e. the ovals bounding the polar caps. It is therefore not surprising that particles precipitate there to produce auroras (Fig. 7.17). The question of their acceleration is another matter, which we shall examine later.

This picture must be taken with a large pinch of salt. Indeed, the figure might suggest that the process is steady; this is not so (see for example [47]), as the opening of the magnetosphere depends critically on the component of the solar wind magnetic field along the planet's magnetic moment. As we have seen, many planets have their magnetic moment not far from the normal to the ecliptic, and we saw in Section 6.1 that the average solar wind magnetic field is zero along this direction, whereas the variance is not. Hence the pattern shown represents some time average of a process that is highly sporadic as it depends critically on solar wind disturbances. When the solar wind magnetic field has a large enough component along the planet's magnetic moment (i.e. a southward component at Earth, or a northward one at Mercury,[30] Jupiter and Saturn[31]), magnetic energy builds up in the tail, to be released later, producing magnetic substorms, auroras and associated phenomena, as we shall see below.

Magnetospheres being complex engines powered by the solar wind, their behaviour depends not only on the energy available, but on the efficiency of the mechanics (compare the fuel consumption of a French Citroën 2CV and of a modern car) and on the relative timescales of the driving process and of the engine itself (think of Earth-based technicians driving a vehicle on a remote planet through a radio link).

7.3.2 Energy, coupling and timescales

Energy from the solar wind

The solar wind provides the kinetic energy flux $v_w \left(\rho_w v_w^2 / 2 \right)$, to which the magnetosphere presents the cross-sectional area $S_M \sim \pi \left(1.5 r_M \right)^2$ (from (7.58) with $\epsilon \sim 0.5$). The maximum energy available is therefore

$$P_K \sim \left(\rho_w v_w^3 / 2 \right) S_M. \tag{7.59}$$

For a planet of magnetic moment μ at heliocentric distance d, this yields from (7.18) and (7.50)

$$P_K \sim 10^{13} \left(\frac{\mu}{\mu_\oplus} \right)^{2/3} \left(\frac{d_\oplus}{d} \right)^{4/3} \text{ W}. \tag{7.60}$$

This energy serves mainly to produce the magnetosphere, drawing the tail, and only a small fraction is dissipated within, producing the various perturbations observed. The latter occur via friction with neutrals near the planet, and electric currents, so that one expects them to be driven by the electric field $\mathbf{E}_w = -\mathbf{v}_w \times \mathbf{B}_w$ induced by the flow of the magnetised solar wind, of magnitude

$$E_w \sim 1.6 \times 10^{-3} d_\oplus / d \quad \text{V m}^{-1} \tag{7.61}$$

from Section 6.1. The corresponding flux of electromagnetic energy is the Poynting vector $\mathbf{E}_w \times \mathbf{B}_w / \mu_0$. Over the magnetosphere cross-section, this yields the electromagnetic power

$$P_{EM} \sim \frac{B_w^2}{\mu_0} v_w S_M \sim 1.5 \times 10^{11} \left(\frac{\mu}{\mu_\oplus} \right)^{2/3} \left(\frac{d_\oplus}{d} \right)^{4/3} \text{ W} \tag{7.62}$$

---

[30] The case of Mercury requires some care, because of the small scales involved, and of the scarcity of the exosphere.

[31] As already noted, the cases of Uranus and Neptune depend on the phase along a planetary rotation.

The magnetospheric engine 379

(hereafter, B_w is the component of the solar wind magnetic field perpendicular to the velocity, and magnetic compression is not taken into account). This electromagnetic power P_{EM} is smaller than the kinetic value P_K by roughly the factor $(V_A/v_w)^2$, which amounts to at most a few per cent.

It is interesting to compare these figures to the solar radiative energy impinging on the planet

$$P_\odot \simeq L_\odot \frac{\pi R^2}{4\pi d^2} \sim 2 \times 10^{17} \left(\frac{R}{R_\oplus} \frac{d_\oplus}{d} \right)^2 \text{ W}. \tag{7.63}$$

This radiative input $P_\odot$ is far greater than the kinetic input P_K, itself much greater than P_{EM}. This puts into perspective any possible effects on the Earth's climate due to solar wind magnetosphere interactions, but does not exclude them since the nature of the system makes it extremely sensitive to small changes, as we already noted in Section 6.4.2.

Another interesting figure is the total magnetic energy of the planet's dipolar magnetic field, i.e. the magnetic energy density $B^2/2\mu_0$ of the dipole field (see Appendix) integrated above the planet's surface

$$W_{mag} = \frac{B_0^2}{\mu_0} \frac{4\pi R^3}{3} \simeq 10^{18} \left(\frac{\mu}{\mu_\oplus} \right)^2 \left(\frac{R_\oplus}{R} \right)^3 \text{ J}. \tag{7.64}$$

This is roughly equal to the power P_K available during one day.

Coupling

How much of the incident energy available is used by the magnetosphere, and how does it manage to use it?

The coupling is thought to act through electric currents, driven by the electric field $\mathbf{E}_w$ induced by the solar wind (which is approximately conserved at the shock and in the magnetosheath), producing some kind of dynamo. Let us consider the case when the coupling is expected to be large, i.e. when the solar wind magnetic field at the planet has a large component parallel to its magnetic moment (and in the same sense), so that the magnetosphere is open. The electric field $\mathbf{E}_w$ drives large-scale currents, as illustrated in Fig. 7.20, which close in the ionosphere (J_{ion}) and along the planet's magnetic field ($J_\parallel$),[32] in addition to the currents on the magnetopause and across the plasma sheet shown in Fig. 7.18, and also circuits closing the ionosphere to the tail (not shown).

Because the electric conductivity is very large along $\mathbf{B}$, the currents $J_\parallel$ flow with almost no resistance, so that the difference of potential on the solar wind side (more precisely, on the magnetosheath side) maps to the one on the ionosphere side. Most of the dissipation occurs in the latter region, where the dense atmosphere brakes the ion motion by collisional coupling, producing a greater electric resistance. In this picture, the power dissipated by the current J_{ion} is provided by the 'generator' current J that draws energy from the solar wind flow acting on the open magnetic field lines.

---
[32] The latter are called *Birkeland currents*, as K. Birkeland first inferred their existence.

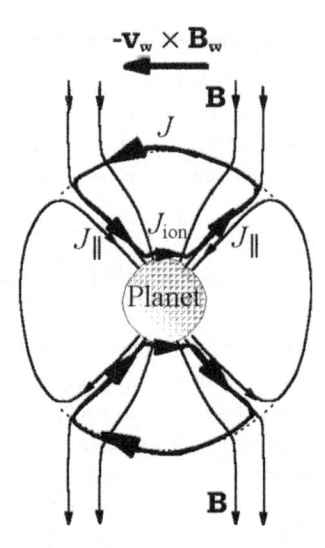

Figure 7.20 Driving current system (heavy lines) that couples the solar wind to the polar-cap ionosphere. The sketch is in a plane perpendicular to the Sun–planet axis (as viewed from the Sun, if the planet is the Earth). (Adapted from [38].)

Therefore the oval-shaped caps surrounding the poles, where these open field lines cross the planet's surface, play an important role. These caps are driven by the solar wind, whose perturbations modify them, and, as we already noted, it is near their boundary that most auroral phenomena are observed to occur. Indeed, one sees in Fig. 7.19 that, on the night side, particles coming along magnetic field lines from the centre of the tail to the planet tend to impact the ionosphere near the (night side) boundary of each polar cap, whereas on the day side, particles may precipitate along newly reconnected field lines, near the (day side) boundary of each cap.

To estimate the size of these ovals, we shall approximate them by circles – a drastic approximation which is justified for order-of-magnitude estimates only; indeed, the polar caps should be smaller on the sunward side and larger in the anti-solar direction because the field lines are stretched out towards the tail. Figure 7.19 sketches the geometry in a meridian plane, showing that the closed field lines cross the tail axis between the points X_1 and X_2; we therefore see that the magnetic flux through a polar cap (i.e. passing along the open field lines) is close in order of magnitude to the magnetic flux of the dipole magnetic field along field lines that cross the equatorial plane farther than r_M (the distance of the nose of the magnetopause). If θ_{PC} is the colatitude of the equatorward edge of a polar cap, the magnetic flux through a polar cap is

$$\Phi_{PC} \sim 2\pi R^2 \theta_{PC}^2 B_0 \qquad (7.65)$$

where R is the planet's radius, B_0 is the surface magnetic field at equator (half

The magnetospheric engine 381

the polar value) and the angle θ_{PC} is assumed to be small. On the other hand, the flux along magnetic field lines closing farther than r_M is

$$\int_{r_M}^{\infty} B_0 (R/r)^3 \, 2\pi r \, dr = 2\pi B_0 R^3 / r_M. \tag{7.66}$$

Equating both fluxes, we deduce

$$\theta_{PC} \sim \sqrt{R/r_M}. \tag{7.67}$$

For the Earth, this yields $\theta_{PC} \sim 18°$.[33]

We may use this result to estimate the difference of potential available for driving the ionospheric currents across the polar caps, producing energy dissipation through atmospheric friction. From Fig. 7.20 (and because there is almost no resistance along the path of $J_\parallel$), this is roughly the difference of potential available on the solar wind side of the electric circuit. In order of magnitude, this is equal to the electric field $\mathbf{E}_w$ integrated along the arc length subtended by the angle $2\theta_{PC}$ at the distance r_M, i.e. $\varphi \sim 2 E_w \theta_{PC} r_M$. This yields, using (7.67),

$$\varphi \sim 2 E_w \sqrt{R r_M} \sim 1.6 \times 10^4 \frac{R}{R_\oplus} \frac{d_\oplus}{d} \left(\frac{r_M}{R} \right)^{1/2} \tag{7.68}$$

$$\sim 5 \times 10^4 \left(\frac{\mu}{\mu_\oplus} \right)^{1/6} \left(\frac{d_\oplus}{d} \right)^{5/6} \left(\frac{R}{R_\oplus} \right)^{1/2} \text{ V} \tag{7.69}$$

where we have substituted (7.61) and (7.50). For the Earth, this yields about 50 kV, in agreement with observation. The estimate (7.69), however, should not be taken too seriously, as it builds on a coupling process that is not fully understood, and on several unproven assumptions.[34] Recall that we have assumed that the solar wind magnetic field is oriented adequately for the magnetosphere to be 'open' to the solar wind.

Let us make some order-of-magnitude estimates about the magnetic tail. We approximate it by a cylinder of radius R_T where the magnetic field strength is B_T, in opposite directions in the north and south lobes (Fig. 7.21); because of the variation along the tail, this approximation is acceptable for the near part of the tail only, up to lengths not too much greater than the diameter.

In order of magnitude, the magnetic flux through a tail lobe $\Phi_T \sim B_T \pi R_T^2 / 2$ is equal to the flux Φ_{PC} through a polar cap. Using (7.65) and (7.67), we deduce $(B_T/B_0)(R_T/R)^2 \sim 4R/r_M$. Because of the flaring, the radius of the tail is typically $R_T \sim 2 r_M$, which yields $B_T/B_0 \sim (R/r_M)^3$. With the value (7.51) of r_M, we deduce $B_T^2/\mu_0 \sim \rho_w v_w^2/2$, whence, substituting (7.18),

$$B_T \sim 2 \times 10^{-8} d_\oplus / d \text{ T}. \tag{7.70}$$

---
[33] A more correct but still accessible estimate may be found in [40].
[34] We have also implicitly assumed that the total ionospheric current corresponding to J_{ion} is not strong enough to perturb significantly the magnetic field at the magnetopause, i.e. that $B_{ion}^2/\mu_0 < \rho_w v_w^2$ with $B_{ion} \sim \mu_0 I_{ion}/r_M$ and $I_{ion} = \Sigma_{ion} \varphi$ where Σ_{ion} is the electric conductance across the magnetic field integrated over the height of the ionosphere where the current density j_{ion} is significant (see [41]). See also [83].

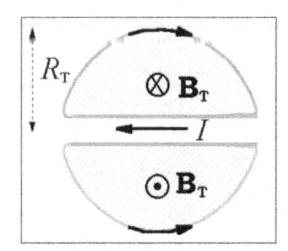

Figure 7.21 Sketch of the cross-section of the near part of the tail (looking towards the Sun, if the planet is the Earth).

This enables one to get an estimate of the cross-tail current flowing in the plasma sheet separating the lobes

$$I \sim \frac{2}{\mu_0} B_T L \sim 4 \times 10^6 \left(\frac{\mu}{\mu_\oplus} \frac{d_\oplus^2}{d^2} \right)^{1/3} \frac{L}{R_T} \quad \text{A} \tag{7.71}$$

per length L along the tail, where we have substituted (7.70), and $R_T \sim 2 r_M$ with r_M given by (7.51).

With these values of R_T and B_T, the magnetic flux across a tail lobe is

$$\Phi_T \sim 5 \times 10^6 \frac{d_\oplus}{d} \left(\frac{R}{R_\oplus} \right)^2 \left(\frac{r_M}{R} \right)^2 \quad \text{Wb} \tag{7.72}$$

and the magnetic energy in a length L of the tail is

$$W_{Ltail} \sim \frac{B_T^2}{2\mu_0} \pi R_T^2 L \sim 10^{15} \frac{d_\oplus}{d} \frac{\mu}{\mu_\oplus} \frac{L}{R_T} \quad \text{J.} \tag{7.73}$$

This means that at Earth for example, the magnetic energy is about 10^{15} J in a length of the tail equal to its radius; this is roughly equal to the solar wind electromagnetic power P_{EM} tapped during two hours.

Let us estimate the power furnished by the solar wind to draw the magnetic tail. The magnetic tension $B_T^2/2\mu_0$ applied to the tail cross-sectional area πR_T^2, with the driving velocity $\sim v_w$, produces the work $P_T \sim \pi R_T^2 \left(B_T^2/2\mu_0 \right) v_w$. Substituting R_T and B_T, this yields $P_T \sim \pi \rho_w v_w^3 r_M^2 \sim P_K$. This suggests that most of the kinetic power incident on the magnetosphere is used to draw the tail, producing a flux of magnetic energy along the tail. A fraction of this power is dissipated in driving the flow towards the planet and in its vicinity (against atmospheric friction), and is converted through tail reconnection at the tail current sheet (X_2 in Fig. 7.19) – accelerating particles in various processes – and the remainder is lost to the downstream solar wind wake.

Note that, since an appreciable part of the ionospheric currents connect to the tail, the potential drop ϕ is expected to hold also (in order of magnitude) across the tail's diameter ($2R_T$). Since the induced electric field E_w across the same distance would produce a potential drop $\sim E_w 2R_T$, (7.68) implies that only about 10% of the available electric field maps onto the magnetosphere.

Timescales

How long does the magnetosphere take to react to solar wind perturbations? Since the magnetosphere's behaviour in the changing solar wind seems to involve phases of energy build-up until the stress reaches some critical level, upon which energy is released, an order-of-magnitude estimate may be obtained by calculating the time t_T required for the induced potential ϕ to build up a magnetic flux equal to Φ_T through the cross-section of a tail's lobe. From Faraday's law, we have $\phi \sim \partial \Phi_T/\partial t \sim \Phi_T/t_T$, whence

$$t_T \sim \frac{\Phi_T}{\phi} \sim 250 \frac{R}{R_\oplus} \left(\frac{r_M}{R}\right)^{3/2} \sim 7 \times 10^3 \left(\frac{\mu}{\mu_\oplus} \frac{R_\oplus}{R} \frac{d}{d_\oplus}\right)^{1/2} \text{ s} \qquad (7.74)$$

using the expressions (7.72) for Φ_T, (7.69) for ϕ and (7.51) for r_M. For the Earth, this amounts to about two hours. Since the magnetic tail probably cannot tolerate a doubling of the magnetic flux because its dynamics is constrained by the solar wind, one expects it to reconfigure itself after a perturbation on a timescale smaller than the above value.

During this time, the solar wind – and the extremity of the field lines connected to it – travels the distance $L_T \sim v_w t_T$, i.e., in units of the planet's radius,

$$L_T/R \sim 16 \, (r_M/R)^{3/2}. \qquad (7.75)$$

This may tentatively be taken as an estimate of the length of that part of the tail that is connected to the planetary magnetic field. For the Earth, this comes to $L_T \sim 500 R_\oplus$. For Jupiter, this yields a length of several astronomical units, and indeed, on some occasions, the planet Saturn happens to be engulfed in Jupiter's magnetic tail.

Note that at the distance L_T, the magnetic field in the lobes is no longer equal to B_T, having decreased to match the solar wind value, because the linkage of the tail's field lines to the solar wind makes the magnetic flux decrease along the tail, while the flaring of the tail increases the tail's diameter.

Rotation and drift

Additional energy is furnished to magnetospheres by satellites and by planetary rotation, a process that is especially important for rapidly rotating magnetised planets. Both sources of energy play a major role on Jupiter, which spins rapidly, and has satellites that are conducting (or having a conducting ionosphere) and ejecting copious amounts of matter. The same holds to a lesser extent on Saturn. The tapping of the planetary rotational energy slows down the planet's spin; the amount, however, is imperceptible, contrary to what occurs in some other astrophysical systems, as for example pulsars.

The atmosphere of a rotating planet is put into corotation with the planet by frictional forces. So is the ionosphere, which is collisionally coupled to the atmosphere. Because the ionosphere is magnetised and the plasma cannot move in bulk across the field lines, the magnetosphere tends to corotate, too. To what distance corotation holds depends on the other factors controlling the dynamics, in particular the mechanical stresses produced by the electromagnetic fields and by matter ejected along field lines, in relation to the ability of the ionosphere to conduct sufficient current for the Lorentz forces to ensure mechanical equilibrium (see [82] and references therein).

Relatively close to the planet, the gradient and curvature of the magnetic field are important. The planetary dipolar magnetic field provides a geometry adequate for confining charged particles on shells around the planet, while they gyrate around field lines, oscillate along them and drift azimuthally around the planet (Section 2.2). Since ions and electrons drift in opposite senses, this produces a *ring current*, flowing as an annulus circling the planet, whose most energetic particles constitute the *radiation belts* (the so-called *Van Allen belts* at Earth),[35] whose relativistic electrons produce synchrotron radiation.

Whether the dynamics is governed primarily by the planet's rotation, the drifts, or the solar wind, depends on the distance to the planet and on the particle energy. At small distances (where the rotation speed and the magnetic gradients are greater), the planet's rotation and drifts dominate, the latter being dominant for high-energy particles, whereas at large distances the solar wind driven circulation dominates.

A rough estimate of the distance r at which the solar wind driven circulation dominates may be obtained by comparing the corotation speed Ωr to the circulation speed $\sim E_w/B(r)$. The solar wind circulation is expected to win against corotation when $E_w/B(r) > \Omega r$. With $E_w \sim v_w B_w$ and $B(r) \sim B_0 (R/r)^3$, this happens at a distance

$$r > \left(\frac{\mu_0 \mu \Omega}{4\pi v_w B_w}\right)^{1/2} \quad \text{or} \quad \frac{r}{r_M} > \left(\frac{\Omega r_M}{V_A}\right)^{1/2} \qquad (7.76)$$

where we have substituted the expression (7.50) of r_M, V_A is the solar wind Alfvén speed, and we have neglected a numerical factor close to unity. With the parameters of Tables 7.1 and 7.2, one finds that at Earth, the solar wind wins against corotation outward of about $0.4 r_M \sim 4 R_\oplus$. For Jupiter, on the other hand, the factor Ωr_M is greater by more than two orders of magnitude, so that (7.76) predicts that the solar wind effect is too weak to dominate corotation within the magnetosphere. In that case, however, corotation is limited instead by outward transport of matter which yields too large a torque at large distances to be supported by Lorentz forces, which are limited ultimately by the current that the ionosphere is able to carry.

---

[35] The most energetic particles (energies > 1 MeV for protons and $>$ several tens of keV for electrons) are created as a by-product of the interaction of cosmic rays with the atmosphere, via β decay of neutrons produced by the interaction of cosmic protons with the nuclei of atmosphere molecules.

7.3.3 Storms, substorms and auroras

Several factors make planetary magnetospheres powerful accelerators of particles. First of all, they are very dilute media, since relatively few plasma sources are available, whereas the volume is large. Some solar wind plasma does penetrate through the magnetopause, but not much, and the ionosphere is not a large source either, given its small size compared to the magnetosphere.[36] Hence particles undergo few collisions and a large part of the energy available serves to accelerate a small fraction of the particles to high energies.

The particle drift towards the planet in the tail, due to the global circulation, may produce Fermi and betatron acceleration (Section 8.2.2). Consider for example a particle initially in the tail at distance r from the planet; inward diffusion with conservation of the first adiabatic invariant up to the distance $\sim R$ (where the magnetic field strength B is about $(r/R)^3$ greater) increases the energy of its motion perpendicular to $\mathbf{B}$ by the factor $(r/R)^3$. An electron having initially an energy of 10 eV (typical in the solar wind) at a distance of $r \sim 10R$ in the tail may acquire in this way an energy of 10 keV, whereas a proton having initially an energy of 1 keV may acquire 1 MeV.[37] The larger the size of the magnetosphere, the greater the energy acquired in this way. This process requires, however, the particles' magnetic moment to be conserved, which requires their gyroradius to be smaller than the curvature radius of the magnetic field lines – a condition that is not met by energetic particles.

A great variety of other, not yet fully understood, acceleration processes are inferred to occur, making magnetospheres powerful laboratories for understanding particle acceleration in the Universe. In particular, magnetic field reconnection transforms magnetic energy into particle kinetic energy. And a part of the energy of the electrons producing the auroras is thought to be produced in the region of upward field aligned currents $J_\parallel$ (Fig. 7.20), by an electric field parallel to the magnetic field lines (and directed outward).

As we already noted, magnetospheres undergo frequent perturbations, due to solar wind variations. There are basically two kinds of perturbations: magnetic *storms*, which follow large perturbations and energy inputs from the solar wind, and *substorms*, which are produced by gentler variations, and whose mechanism is still under debate. In both cases, the orientation of the solar wind magnetic field plays an important role (see for example [48]).

Storms

Consider first what happens when a large solar wind perturbation, such as an interplanetary mass ejection, hits a magnetosphere. The incident ram pressure increases, which compresses the magnetosphere; according to (7.50), an

---

[36] Some planets have satellites that are a copious source of particles, as for example Jupiter's satellite Io.

[37] This holds for particles moving roughly across $\mathbf{B}$. Conservation of the second adiabatic invariant during the motion along $\mathbf{B}$ from small latitudes (distance r) to high latitudes (distance R) would increase the parallel speed by the factor r/R, whence the parallel energy by $(r/R)^2$.

increase of the ram pressure by a factor of 10 in a large storm decreases the distance of the nose by a factor of 2/3; this increases the current at the nose of the magnetopause, so that the magnetic field at the planet's surface *increases*. These phenomena, however, are only preliminaries, for two reasons. First, because of the increase in solar wind speed and magnetic field, the electric field E_w increases, as does the incident energy flux. Second, because of the curved shape of the field lines at an interplanetary mass ejection (Fig. 6.12), the magnetic field encountering the magnetopause has generally a geometry adequate for opening the magnetosphere. The magnetosphere is thus energised, and particles are driven inward from the tail by the anti-sunward convection. This populates the ring current with particles from the tail that are accelerated in the process.

As we saw in Section 2.2, the ring current *decreases* the planet's magnetic field. The amount of magnetic field change at the planet's equator is related to the total kinetic energy in the ring particles W_{ring} by $\Delta B = 2 W_{ring}/\mu$, i.e.

$$\frac{\Delta B}{B_0} = -\frac{2}{3}\frac{W_{ring}}{W_{mag}} \tag{7.77}$$

where B_0 is the initial planet's equatorial magnetic field, W_{mag} its magnetic energy (7.64) and the effect of magnetopause and tail currents is neglected; this is a straightforward consequence of Virial theorem (Problem 7.6.5) ([24], [62]). During a geomagnetic storm, the energy in the ring current may typically increase to several times 10^{15} J. From (7.77), using (7.64), this produces a magnetic field dip of $\Delta B \sim 3 \times 10^{-3} B_0 \sim 10^{-7}$ T. Note that if the incident electromagnetic power P_{EM} given in (7.62) is increased by a factor of 10 during a storm, and the magnetosphere taps it, the time required to put several times 10^{15} J into the ring current is of the order of magnitude of the timescale t_T given by (7.74).

Energetic electrons coming from the tail not only populate the ring current; they also come towards the planet along field lines, impacting the atmosphere in the auroral ovals (Fig. 7.19), producing powerful auroras. These electrons are observed to have typical energies of a few keV, and to produce a total auroral power of about 10^{11} W; this is roughly 10% of the increased power P_{EM}. Furthermore, the auroral ovals widen because the increased ring current stretches the magnetic field so as to increase the magnetic flux threading the tail magnetic field lines. This makes the auroral perturbations observable at smaller latitudes.[38]

Figure 7.22 illustrates an extreme case of magnetic storm observed on Earth [80]. It shows the magnetic field recorded at Bombay (according to a recent calibration), during the giant magnetic storm of September 1859 that followed the solar event observed by Carrington (Section 1.1). This is the most intense geomagnetic storm recorded in history, and the widening of the auroral ovals is illustrated by the fact that the auroras produced were visible over much of the Earth's surface. One sees in the figure that the main phase lasted for a little more than one hour, which is of the order of magnitude of the time t_T, while the

---

[38] A simplified estimate may be found in [71].

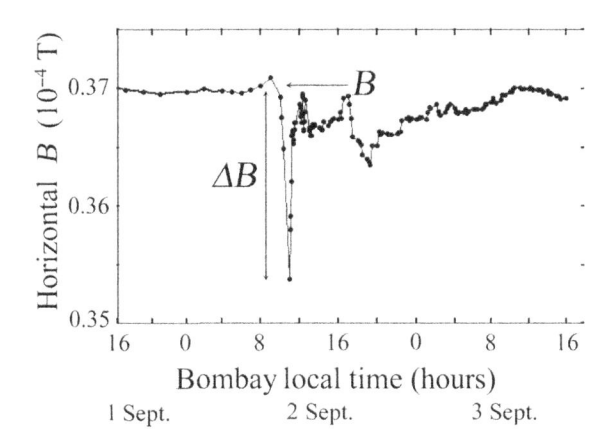

Figure 7.22 A giant magnetic storm: magnetic field recorded at Bombay after the solar event observed by Carrington in 1859. (Adapted from [80].)

maximum negative variation in magnetic field strength was $\Delta B \sim -0.016 \times 10^{-4}$ T. This corresponds, from (7.64) and (7.77), to an energy in the ring current of $W_{ring} \sim 6 \times 10^{16}$ J. This suggests that if the energy was furnished by the electromagnetic flux P_{EM}, it was at this time 100 times greater than the normal value given by (7.62).

For comparison, the strongest storm in the last century, which produced a breakdown of the power station of the Canadian province Quebec during 9 hours in March 1989, is reported to have caused a magnetic field decrease three times smaller than the value shown in Fig. 7.22.

Substorms

Magnetic storms are relatively rare. Much more frequent are gentler perturbations in which the incident solar wind power does not change much, but the direction of the solar wind magnetic field changes so as to open the magnetosphere; at Earth, this occurs typically a few times per day. In this case, the magnetic tail reorganises itself, producing a chain of events known as a *substorm*, that lasts typically about one hour ($\sim t_T$). The magnetic pressure in the lobes increases, the tail is stretched and the plasma sheet gets thinner; when the structure becomes unstable, a part of the tail collapses, ejecting downward a piece of the plasma sheet (a *plasmoid*), and releasing energy. Figure 7.23 is a sketch of what might possibly occur.

Let us assume that a length L of the tail delivers its magnetic energy W_{Ltail} during the substorm, and that for the tail to rebuild itself, this energy must be replaced by tapping the solar wind electromagnetic energy P_{EM} during the timescale t_T estimated above. In this case, the length L must ensure $W_{Ltail} \sim$

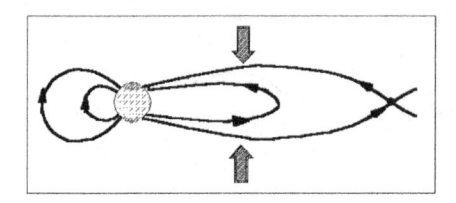

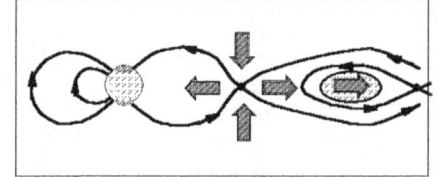

Figure 7.23 Sketch of magnetic field lines in the tail just before (left) and during (right) a magnetospheric substorm. A new reconnection region is formed in the near tail, which leads to a disconnection of the plasma sheet, which is ejected as a plasmoid.

$P_{EM} \times t_T$. Substituting (7.73), (7.62) and (7.74), we deduce

$$\frac{L}{R_T} \sim \left(\frac{\mu}{\mu_\oplus} \frac{d}{d_\oplus}\right)^{1/6} \left(\frac{R_\oplus}{R}\right)^{1/2} \tag{7.78}$$

where $\mu/\mu_\oplus$, $d/d_\oplus$ and $R/R_\oplus$ are respectively the planet's magnetic moment, heliocentric distance and radius normalised to the Earth's values. This suggests that a length of the tail equal in order of magnitude to its radius may deliver its magnetic energy and be rebuilt by the 'normal' solar wind during the tail's timescale. For the Earth, this corresponds to a length of about $20R_\oplus$, an energy of about 10^{15} J (from (7.73)), and a time of about two hours.

This magnetic energy serves to accelerate particles that populate the ring current and precipitate in the auroral ionosphere near the auroral ovals, powering auroras, while a part is dissipated in the ionospheric currents which increase dramatically, and another part returns to the solar wind as ejected bullets of plasmas (Fig. 7.23) or plasma flow.

A major difference with storms is that the solar wind perturbation mainly serves as a trigger of the reconfiguration of the tail producing the event, whose energy was already present as magnetic energy in the tail. Solar wind energy is used afterwards to reconfigure the tail which has been partly destroyed. In some sense, the building of magnetic energy and its sudden release, producing particle acceleration and heating, reminds one of solar flares. Both phenomena involve power law distributions of events (see for example [81]) and are still not fully understood; they are the subject of much debate (see for example [50]).

Substorms and their associated auroral perturbations occur frequently, and all planets having an internal magnetic field and an atmosphere exhibit auroral phenomena. However, the rotation of the planet and the effects of satellites makes the problem (and in particular the structure of auroral ovals) more complex than outlined above.

The energetic particles in magnetospheres are sources of radiation in various frequency ranges. Particles trapped in the radiation belts remain there for long times and therefore respond with a long timescale to perturbations. In contrast, the precipitation of particles in the auroral ovals closely follows solar

The magnetospheric engine

wind perturbations.[39] They not only produce the visible and ultraviolet light observed in the auroras; they also produce powerful radioemissions, at frequencies corresponding to the electron gyrofrequency at the planet. This power has been shown to be closely proportional to the solar wind power incident on the magnetosphere. For the five planets having an internal magnetic field and a conspicuous atmosphere (Earth, Jupiter, Saturn, Uranus and Neptune), the power emitted in radio is

$$P_{radio} \sim 10^{-5} P_K \sim 10^{-3} P_{EM} \qquad (7.79)$$

in order of magnitude (see [23], [87]), with P_K and P_{EM} respectively given in (7.60) and (7.62). An interesting consequence is that when Saturn happens to find itself inside the magnetic tail of Jupiter, therefore becoming isolated from the solar wind, its radioemission drops out.

Space weather

Geomagnetic storms and substorms, in addition to various particles coming from space – from energetic nuclei to large boulders – in the Earth's environment, produce numerous effects that may be very harmful to our modern technology. This is known as *space weather*. In Carrington's times, a solar perturbation disturbed telegraph systems. Nearly a century later, during the Second World War, emission of radio waves from the Sun was discovered through the perturbations created on radar signals, a technology that was in rapid development during the war. More than half a century later, technology has harnessed space, and its extreme sophistication has made it extremely sensitive to space weather perturbations (see [35]). Figure 7.24 summarises some of these effects [46].

The electric currents induced in the Earth and in the ionosphere during a geomagnetic storm affect technological systems involving long conductors. Solar radio noise affects communications. Energetic particles and radiation harm sensitive electronics on satellites and put severe constraints on human space flight. In particular, as we already noted, if good electrical connections are not established between the various materials on a spacecraft, differential charging may produce lightning-like breakdown discharges. Variations in magnetic field may perturb satellites' attitude control. Impacts of meteoroids and larger objects may have harmful consequences. Changes in the Earth's atmosphere produced by solar perturbations may increase the atmospheric drag on satellites. Hopefully, progress in understanding these phenomena will result in capability of predicting and perhaps preventing them, and space weather forecasting is a great challenge for twenty-first-century science.

---

[39] Additional auroral phenomena are driven by the motion of conductive (or surrounded by a conductive ionosphere) satellites through the planet's magnetic field. This occurs near the footprint in the planet's ionosphere of the magnetic field lines passing near the satellite orbits.

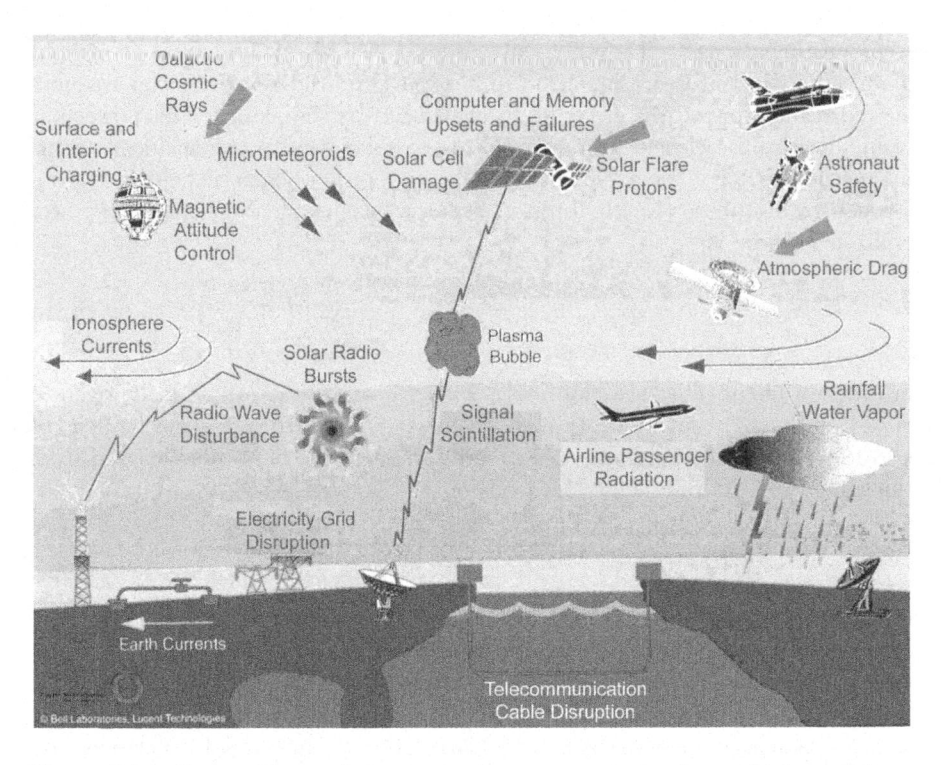

Figure 7.24 Some effects of space weather on our technology. (Adapted from [46].)

7.4 Physics of heliospheric dust grains

Dust grains are intermediate between plasma particles – essentially subjected to electromagnetic forces, and large bodies – essentially subjected to gravitation. They interact with the plasma via exchange of particles, which provides them with a net electric charge (in addition to a change in momentum and energy) and provides the plasma with a source or loss of particles (and momentum and energy). Their electric charge also makes them a plasma component, so that in some cases they contribute to plasma waves and instabilities. This has produced the emergence of a new discipline: physics of dusty plasmas [31], [56].

7.4.1 Forces

Because of their smallness, dust grains have large size-to-mass and area-to-mass ratios. They are thus affected by forces that, for larger bodies, are negligible compared to gravity: radiation forces (due to the momentum carried by photons), and plasma drag (due to the momentum carried by plasma particles), both proportional to area, and the Lorentz force (due to the grain electric charge in the magnetic field), proportional to size.

Physics of heliospheric dust grains

As other bodies, dust grains are subjected to the gravitational force. Except in the vicinity of planets and other massive objects, the gravitational attraction on a dust grain of mass m is due to the Sun and thus directed radially towards the Sun and equal to

$$F_G = mM_\odot G/d^2 \qquad (7.80)$$

at heliocentric distance d.

Radiation forces

Absorption and/or scattering and reflection of solar radiation by a dust particle transfers momentum, producing a radiation pressure force directed radially outwards. This is easily calculated for a perfectly absorbing object of cross-section S in circular orbit. It absorbs per unit time the energy $P_\odot = L_\odot S/4\pi d^2$, and therefore the momentum $P_\odot/c$, directed radially outwards, which yields the *radiation pressure* force

$$F_{rad} = P_\odot/c \simeq 4.5 \times 10^{-6} \left(d_\oplus/d\right)^2 S \quad \text{N.} \qquad (7.81)$$

Since it is antiparallel to the gravitational attraction F_G, its magnitude is usually expressed as the ratio

$$\beta = \frac{F_{rad}}{F_G} = \frac{L_\odot S/\left(4\pi d^2\right)}{c\, F_G} \simeq \frac{0.2}{r_{\mu m}} \quad \text{(perfectly absorbing)} \qquad (7.82)$$

where we have substituted $S = \pi r^2$, the expression (7.80) of F_G with $m \simeq 10^4 \times r^3$ kg (for a spherical grain of radius r and mass density 2.5×10^3 kg m^{-3}), and the solar parameters.

Radiation pressure is not the only radiative force, because in the particle's frame, sunlight is viewed as coming at a small angle v/c to the radial. This aberration produces a drag force on the orbital motion equal to v/c times the radiation pressure force. This is known as the *Poynting–Robertson* drag. This drag force, antiparallel to the orbital speed, superimposed to the outward radiation pressure force, may also be viewed in the heliocentric frame as coming from the mass–energy equivalence. Absorption of radiation increases the mass at the rate $dm/dt = P_\odot/c^2$, so that, in the heliocentric frame, conservation of angular momentum around the Sun produces to first order a decrease in the orbital speed v given by $dv/v = -dm/m$ (for a particle in circular orbit), whence $mdv/dt = -vP_\odot/c^2$, which yields the Poynting–Robertson drag.

Although it is much smaller than the radiation pressure, this drag has important consequences because it causes secular changes in the orbital energy and angular momentum, which cumulate over long periods of time.

In practice, dust particles are not perfectly absorbing and they scatter radiation, so that the radiation forces are reduced by an amount which depends on the grain material and size (a detailed review may be found in [14].) Scattering plays an important role for sizes smaller than the wavelength. Since the solar spectrum peaks at a wavelength $\lambda_0 \simeq 0.5$ μm and is relatively narrow, scattering acts, with typical materials, for grain sizes shortward of this value. In practice, the radiation forces decrease sharply for grain sizes smaller than about $\lambda_0/2\pi \sim 0.1$ μm. We therefore have

$$\beta = \frac{F_{rad}}{F_G} \simeq \beta_{rad} \frac{0.2}{r_{\mu m}} \qquad (7.83)$$

with $\beta_{rad} \sim 1$ in order of magnitude for $r > 0.1$ μm, and decreasing strongly for smaller sizes.

Radiation pressure is therefore generally smaller than gravity, except for graphite and metallic grains which strongly backscatter sunlight. These particles can therefore be ejected from the heliosphere by the radiation pressure. In fact, ejection may even occur when $\beta < 1$, because for a particle to be ejected, all that is necessary is that its kinetic plus gravitational energy be positive. For a grain of mass m and speed v at heliocentric distance d, this specific energy is equal to $W = v^2/2 - F_G d/m$ (independent on m) per unit mass, in the absence of radiation pressure. Consider the simple case of a grain in circular orbit (i.e. $v^2/d = F_G/m$), that is suddenly released from a massive parent body. It suddenly feels the radiation pressure force, which is equivalent to a sudden decrease of the gravity force by the factor $1 - \beta$, so that the equivalent energy becomes $W' = v^2/2 - d(1-\beta) F_G/m$. Substituting $v^2/d = F_G/m$, this yields $W' = v^2 (\beta - 1/2)$, which is positive if $\beta > 1/2$. For elliptic orbits, still smaller values of β may produce ejection if the grain is released at perihelion.

In contrast, the Poynting–Robertson drag decreases the total energy, making the orbit shrink. Let us estimate the time to spiral into the Sun, for a particle initially in circular orbit at speed v. The drag force $F_{rad} v/c$ makes the grain energy decrease at the rate $dW/dt = -F_{rad} v^2/c$. The grain falls into the Sun when its energy has decreased by an amount of the order of magnitude of $mv^2/2$, which occurs after a time of the order of magnitude of

$$\tau_{PR} \sim \frac{mv^2/2}{-dW/dt} \sim \frac{mc}{2F_{rad}} \sim 3 \times 10^3 \, r_{\mu m} \, \frac{d^2}{d_\oplus^2} \quad \text{years.} \qquad (7.84)$$

Since $F_{rad} c$ is just the rate of radiative energy absorption by the grain, the time $\tau \sim mc/2F_{rad}$ is about the time for the grain to absorb the equivalent of its own mass in radiation.

We shall see in Section 7.5 another consequence of the radiation pressure on dust grains: the cometary dust tails.

Plasma drag

The impact of solar wind ions also exerts an outward pressure on dust particles, of the order of magnitude of the ram pressure $\rho_w v_w^2$ per unit cross-section.[40]

---

[40]This force is not significantly affected by the temperature of the particles and by the grain electric charge because both the particle thermal energy and the electrostatic energy due to the grain electrostatic potential with respect to the plasma are much smaller than the plasma bulk kinetic energy.

Physics of heliospheric dust grains 393

Comparing (7.18) and (7.81), we see that this force is weaker than the radiation pressure by more than three orders of magnitude. However, it is not negligible because a dust grain moving at speed v in circular orbit sees the solar wind coming at an angle to the radial equal to v/v_w, an aberration angle greater by more than three orders of magnitude than the aberration angle v/c relevant for incoming sunlight. As a result, even though the solar wind ram pressure is negligible compared to the radiation pressure, the corresponding drag on the orbital motion is comparable to Poynting–Robertson drag, and adds to it.

Lorentz force

We have seen in Section 7.2.2 that sunlit objects in the solar wind are charged to a positive electric potential $\Phi \sim 1 - 10$ V, and that a dust grain of radius r carries an electric charge $q \simeq 4\pi\epsilon_0 r\Phi$. It is thus subjected to the Lorentz force $q\mathbf{v}_{rel} \times \mathbf{B}_w$ where $\mathbf{B}_w$ is the solar wind magnetic field and $\mathbf{v}_{rel}$ the grain's speed relative to the solar wind. Since the grain speed with respect to the Sun is small compared to the solar wind speed, the relative speed is roughly equal to the solar wind speed $\mathbf{v}_w$, so that the Lorentz force is roughly $\mathbf{F}_L \sim q\mathbf{E}_w$ with E_w given in (7.61). This yields

$$\frac{F_L}{F_G} \sim \frac{10^{-2}}{r_{\mu m}^2} \frac{d}{d_\oplus}. \tag{7.85}$$

The Lorentz force is therefore important for grains of radius 0.1 μm and smaller. The consequences of the dynamics are indicated by the value of the gyroradius

$$r_g \sim \frac{v_w}{qB_w/m} \sim 2 \times 10^4 \, r_{\mu m}^2 \, d. \tag{7.86}$$

Only for very small grains is the gyroradius smaller than the relevant scales. In that case, the grain behaves as a plasma particle, gyrating around field lines and being subjected to magnetic drifts. Even though the Lorentz force is very small for larger grains, it may have long-term effects, depending on the inclination of their orbit and on the phase in the solar cycle. At low inclinations, the grains see alternating signs due to passages on either side of the heliospheric current sheet, whereas at high inclinations, they see a more constant magnetic field during a large fraction of their orbit (near solar activity minimum).

Finally, subtle effects are produced by the changes in the charge carried by the grains. The fluctuations in charge can lead to radial diffusion. In addition, when the grain's electric potential varies, the electric charge takes some delay to reach its equilibrium value. When this delay, given by (7.40), is comparable to the dynamic timescales, significant energy and momentum may be exchanged between the grain and the plasma, producing important orbital changes. This may happen near planets, and may be responsible for the capture of interplanetary grains by planetary magnetospheres [42].

7.4.2 Evaporation

Dust grains are heated by solar radiation. A grain of radius r at heliocentric distance d receives per unit time the energy $P_\odot \sim L_\odot \times \pi r^2/4\pi d^2$ on its sunlit cross-section. If it radiates as a blackbody at temperature T, it loses energy at the rate $\sigma_S T^4 \times 4\pi r^2$, so that its blackbody equilibrium temperature is

$$T \sim \left(\frac{L_\odot}{16\pi\sigma_S d^2}\right)^{1/4} \sim 280 \left(\frac{d_\oplus}{d}\right)^{1/2} \text{ K}. \qquad (7.87)$$

Particles smaller than about 0.1 μm deviate from this value, as do particles whose material significantly reflects or scatters sunlight.

With melting temperatures in the range 1000–1500 K for many materials, most dust grains are expected to evaporate inward of 0.05 AU $\sim 10\,R_\odot$. Sublimation, however, begins at a much lower temperature; in particular, water ice begins to sublimate at 1–3 AU.

This releases particles, which are quickly photoionised and picked up by the solar wind (Section 6.5). Dust particles also recycle solar wind particles into pick-up ions by adsorption and desorption. Both effects are revealed by measurements of the velocity distribution of pick-up ions (a recent review may be found in [54]).

7.5 Comets

Comets are fascinating objects. Left to their own devices, they are mere aggregates of dust and snow about as big as a terrestrial mountain [43]. There are billions of such objects moving around the Sun in the space beyond the planets. Whenever some perturbation sends one of these chunks of ice (Fig. 7.25) closer to the Sun, its surface boils up and the interaction with the solar radiation and corpuscular emission produces a transient activity, involving a huge variety of

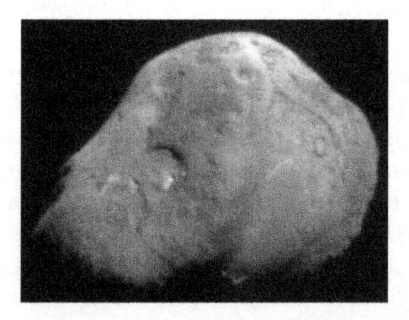

Figure 7.25 The nucleus of comet Tempel 1 (largest dimension about 7 km), photographed in July 2005 during the Deep Impact mission, just before a projectile was fired against the nucleus in order to study its structure. (Image by NASA/JPL-Caltech.)

Comets

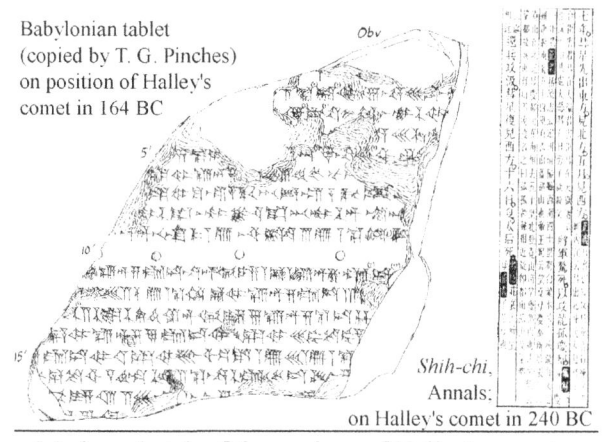

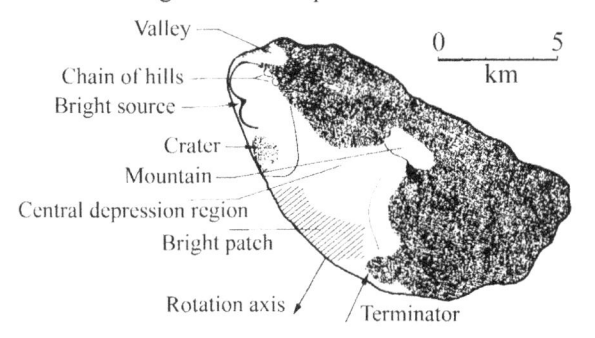

Figure 7.26 Comments on Halley's comet. Top: in 164 BC from a Babylonian tablet, and in 240 BC from a Chinese text (adapted from [78]). Bottom: interpretation of a photograph of the nucleus, taken on the Giotto spacecraft during a close fly-by in 1986 AC (adapted from [45]).

physical and chemical processes and offering sky displays that have captured human imagination and skills throughout the ages (Fig. 7.26).

The surface layers of the body sublimate, liberating molecules and dust. Molecules blow away, unimpeded by the small gravity of the object, producing a huge expanding atmosphere in which complex chemistry acts in the presence of solar ionising radiation. The expanding gas drags out the dust. Dust grains, subjected to solar radiation pressure, follow curved trajectories – as do tennis balls submitted to Earth's gravity – generating a curved dust tail. Ionisation produces an ionosphere, of a size much larger than the nucleus itself, whose electric conductivity excludes the solar wind magnetic field, making the wind stop ahead of the comet and sweep around it. So do the magnetic field lines frozen in the wind, which bend round and produce a long tail antiparallel to the solar direction, which guides the cometary ions (Fig. 7.9). The solar wind is affected by the comet far ahead, via the cometary ions that it picks up, which slow it down.

M. Leverrier, pour étudier plus commodément le nouvel astre, prend le parti de le faire entrer à l'Institut.

Figure 7.27 Mr Leverrier, in order to better study the comet, introduces it to the Institute. (Drawing by Cham, 1848.)

All this makes comets exciting challenges for physicists (Fig. 7.27). One finds there solids, gases, plasmas and radiation, playing together in surprising ways, while the nucleus itself, because it is small and spends most of its time in the distant heliosphere, may represent the most pristine material in the solar system.

The solar wind interaction with a comet has much in common with that of a non-magnetised planet we studied in Section 7.2.4. However, the small mass and large emission rate of the comet produce two main differences. First, the freshly ionised cometary ions are picked up by the solar wind at so great a rate that they slow it down considerably, even very far from the comet, producing an interaction region that is larger than the body itself by many orders of magnitude. Second, the cometary ions populate the cometary environment, including the magnetic tail, in such a large concentration that they are visible from large distances, sometimes with the naked eye (Fig. 1.4).

Comets played a major role in the discovery of the solar wind, thanks to Ludwig Biermann's pioneering work on plasma tails (Section 1.1), which behave as distant solar wind probes. Nearly half a century later, comets surprised astronomers by revealing a different aspect of their interaction with the solar wind, producing X-rays that are footprints of minor wind constituents; this X-ray emission releases a tiny part of the energy of the solar corona stored into highly ionised atoms, until their journey with the solar wind makes them encounter a comet. And more surprises are still to come, as, among other

Comets   397

spacecraft, an ambitious mission – Rosetta – is en route to catch up with a comet during its journey towards the Sun, and to attempt a soft landing on the nucleus.

We make below simplified estimates of the main physical processes, focusing on the interaction with the solar wind. Among the huge literature on the subject, a classical (though somewhat outdated) account is [55]; a general review may be found in [28] and [29], while a concise tutorial on plasma processes may be found in [21].

7.5.1 Producing an atmosphere

Comets are on highly elliptical orbits. They spend most of their life far from the Sun, according to Kepler's second law,[41] where they are mere boulders difficult to detect. As they approach the Sun and catch more radiation, their temperature increases and they melt, or rather directly sublimate (because of the small vapour pressure) at an increasing rate, which produces the various aspects of their activity.

The evaporation rate may be easily estimated from basic physics. Let us approximate a comet's nucleus by a sphere of radius R made of water ice,[42] at heliocentric distance d. It receives from the Sun the power $L_\odot/4\pi d^2$ per unit (sunlit) cross-section. When it is close enough to the Sun that evaporation is strong, most of this energy serves to evaporate molecules. In this case, since the energy required to evaporate an H_2O molecule is W_{H_2O} (the latent heat of evaporation), the evaporation rate is $L_\odot/4\pi d^2 W_{H_2O}$ (per unit sunlit cross-section). With the cross-section πR^2, the emission rate is therefore (close enough to the Sun)

$$Q \sim \frac{\pi R^2 L_\odot}{4\pi d^2 W_{H_2O}} \sim 5 \times 10^{22} \, R^2 \times \left(\frac{d_\oplus}{d}\right)^2 \quad \text{molecules s}^{-1} \quad (7.88)$$

where we have substituted the sublimation energy of an H_2O molecule $W_{H_2O} \sim 0.5$ eV $\sim 8 \times 10^{-20}$ J. This holds provided the evaporation is strong enough that the energy radiated by the nucleus ($\sim 4\pi R^2 \times \sigma_S T^4$) may be neglected, which occurs typically closer than 2 AU from the Sun. For a body of equivalent radius $R \sim 5 \times 10^3$ m – as Halley's comet – at heliocentric distance $d \simeq d_\oplus$ (1 AU), (7.88) yields the evaporation rate

$$Q \sim 10^{30} \quad \text{molecules s}^{-1}. \quad (7.89)$$

---

[41] The equation of motion (in reduced co-ordinates) yields the angular velocity $d\theta/dt = L/r^2$, with $a(1-e^2) = L^2/M_\odot G$ (a is the semi-major axis of the ellipse and e the eccentricity), and L (the angular momentum per unit mass) is a constant. Hence the area swept by the radius vector joining the Sun and the orbiting body is the same in equal time intervals (Kepler's second law). Therefore, the farther the body, the smaller the angular velocity about the Sun.

[42] Beware that an actual comet's nucleus rather looks as the sketch of Fig. 7.26. Not only is the shape far from spherical, because the gravitational force is insufficient to bring it into hydrostatic equilibrium against the forces ensuring the rigidity, but it is far from homogeneous and contains numerous chemical species in addition to water ice.

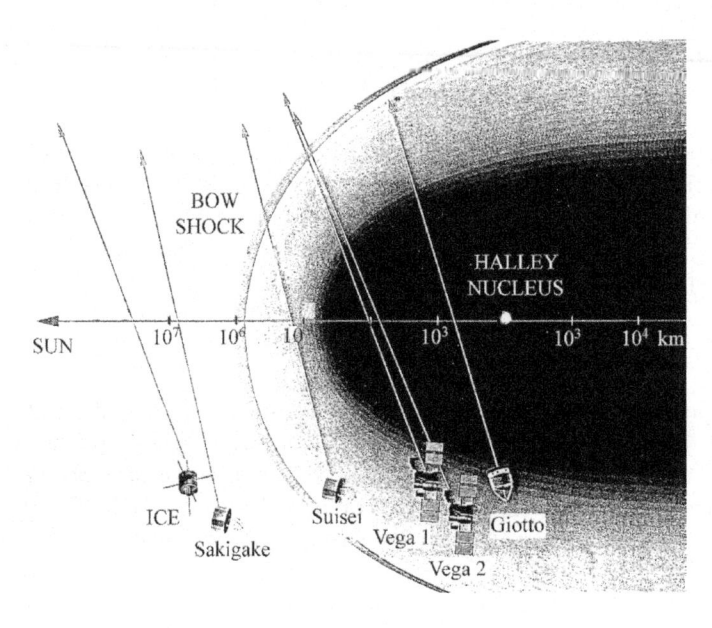

Figure 7.28 The fleet of spacecraft that encountered Halley's comet in March 1986. All encounters took place on the sunward side of the comet, at a relative speed in the range 60–80 km s^{-1}, and a heliocentric distance in the range 0.8–0.9 AU.

A fleet of spacecraft encountered Halley's comet in 1986 when it was at 0.8–0.9 AU from the Sun (Fig. 7.28). They found an emission rate of this order of magnitude.

Consider now what happens to the evaporated molecules. In the spirit of these simple estimates, we ignore the complex chemistry and assume a single chemical species, H_2O, of molecular mass

$$m_0 \simeq 18 \times m_p. \tag{7.90}$$

At 1 AU from the Sun, the temperature of the nucleus's surface is typically $T_0 \sim 200$ K, lower than the blackbody value given in (7.87), mainly because the nucleus uses most of the absorbed solar energy to evaporate molecules rather than to radiate. Let us then compare:

- the thermal speed of an H_2O molecule $v_{th0} = (2k_B T_0/m_0)^{1/2} \sim 430$ m s^{-1},

- the speed of liberation from the comet's nucleus (of mass M), $v_L = (2MG/R)^{1/2} \sim 2$ m s^{-1}, for $R \sim 5 \times 10^3$ m and $M \sim 200 \times 4\pi R^3/3$ kg (using the estimated density of such a loose aggregate [43]).

Since $v_{th0} \gg v_L$, the molecules do escape, producing an atmosphere that is accelerated as it expands. This comes about because the expansion cools the gas, so that the thermal energy stored in the degrees of freedom of the molecules is transformed into bulk flow energy (while a part is used to drag the dust grains).

The maximum flow speed may be estimated from Bernoulli's theorem. In contrast to the fully ionised hot solar atmosphere at the origin of the solar wind (Section 5.1), the cometary flow transports a negligible heat flux, and expands in a nearly adiabatic way.[43] Another difference from the solar wind acceleration is that the gravitational energy is negligible, so that Bernoulli's equation (5.1) reduces to

$$\frac{v^2}{2} + \frac{\gamma}{\gamma-1}\frac{k_B T}{m_0} = \text{constant}. \qquad (7.91)$$

Applying (7.91) between the nucleus surface (where the temperature is T_0 and the mean radial speed is $v \sim v_{th0}/\sqrt{\pi}$),[44] and a distance where the temperature has decreased significantly and the flow speed is v_n, we find

$$\frac{v_n}{v_{th0}} \simeq \left(\frac{1}{\pi} + \frac{\gamma}{\gamma-1}\right)^{1/2}. \qquad (7.92)$$

This 'terminal' speed v_n is reached when the adiabatic radial expansion has significantly cooled the gas, which becomes true at a distance from the nucleus of several radii R.

The adiabatic index depends on the number N of degrees of freedom of the molecules as $\gamma = 1 + 2/N$. An H$_2$O molecule has many more degrees of freedom than a point charge: 3 for translation, plus 3 for rotation (the temperature is too low for the vibration levels to be populated [22][45]). This yields $N = 6$, whence $\gamma = 4/3$. The terminal speed (7.92) is therefore

$$v_n \simeq (9k_B T_0/m_0)^{1/2} \simeq 0.9 \times 10^3 \text{ m s}^{-1}. \qquad (7.93)$$

Indeed, measurements of the expansion velocity of the atmosphere of Halley's comet made aboard the spacecraft Giotto gave just 0.9 km s^{-1}. Note that this acceleration requires the flow to experience a transition from subsonic to supersonic speed, and it is amusing to find here another echo of the solar wind acceleration problem.[46]

Therefore, farther than the distance where the terminal speed v_n is reached, the gas streams outwards at a constant speed v_n of about 1 km s^{-1}. For isotropic expansion, the density of neutral molecules at distance r from the nucleus is therefore

$$n_n = \frac{Q}{4\pi v_n r^2} \qquad (7.94)$$

provided it is not significantly reduced by ionisation.

---

[43] A realistic estimate should take into account the energy loss due to the friction by dust grains (Problem 7.6.7).

[44] Because of the averaging over direction, the mean radial velocity at the surface is half the average speed $(8k_B T_0/\pi m_0)^{1/2}$, i.e. $v_{th0}/\sqrt{\pi}$ (Section 2.3).

[45] Otherwise one should add 6 more degrees of freedom since vibration counts double because the energy includes kinetic and potential energies.

[46] This transition from subsonic to supersonic speed may be viewed as a nozzle problem. Here the converging part of the nozzle is produced by the gas–dust drag, whereas for the solar wind the converging part of the nozzle was produced by gravity (Problem 7.6.7).

7.5.2 Ionising the atmosphere

The molecules are subjected to the solar ionisation flux $F_\odot$, which dissociates and ionises them. Far from the comet, this flux is given by (7.9). Upon arriving close to the nucleus, however, the solar ionising flux is smaller, because a large part of it has been used in ionising the atmosphere along its path, as we saw for planetary ionospheres. The mean free path of photons is $l_{ph} \sim 1/n_n \sigma_{ion}$, where $\sigma_{ion} \sim 10^{-21}$ m² is the cross-section for ionisation (Section 2.4). The solar ionising flux is reduced significantly when l_{ph} becomes smaller than the distance r. Substituting the neutral density (7.94) in the above expression of l_{ph}, we find that $l_{ph} < r$ at distances $r < Q\sigma_{ion}/4\pi v_n$. With the above parameters, this yields $r < 10^2$ km (in order of magnitude) for Halley's comet near 1 AU.

Farther than this distance, the ionising flux is nearly $F_\odot$, so that the photoionisation rate of a neutral is $F_\odot \sigma_{ion}$. The solar ionising flux, however, is not the only cause of ionisation; charge exchange (Section 2.4) with solar wind particles contribute significantly, with a cross-section σ_{ex}. With a flux $n_w v_w$ of solar wind protons, the charge exchange ionisation rate of a neutral is $n_w v_w \sigma_{ex}$, whence the total ionisation rate $F_\odot \sigma_{ion} + n_w v_w \sigma_{ex}$. With the charge exchange cross-section $\sigma_{ex} \sim 2 \times 10^{-19}$ m² given in (2.147) and the solar wind flux (7.17), we find the average timescale for ionisation at 1 AU

$$\tau_{ion} = (F_\odot \sigma_{ion} + n_w v_w \sigma_{ex})^{-1} \sim 10^6 \text{ s} \qquad (7.95)$$

varying as the square of the heliocentric distance since both $F_\odot$ and $n_w v_w$ vary as d^{-2}. Since neutrals are moving radially at constant velocity v_n and are ionised at the rate τ_{ion}, their number density n_n is given by the continuity equation as

$$\frac{1}{r^2} \frac{d}{dr} (n_n v_n r^2) = -\frac{n_n}{\tau_{ion}} \qquad (7.96)$$

when the plasma concentration is sufficiently small that recombination is negligible. Making the change of variable $n_n r^2 = x$ and putting v_n out of the derivative, we find $n_n r^2 \propto e^{-r/r_{ion}}$, with

$$r_{ion} \simeq v_n \tau_{ion} \sim 10^6 \text{ km.} \qquad (7.97)$$

Since (7.94) holds at short distances, this yields

$$n_n = \frac{Q}{4\pi v_n r^2} e^{-r/r_{ion}}. \qquad (7.98)$$

The neutral density (7.94) is thus significantly reduced by ionisation outwards of the distance r_{ion}, which represents the size of the neutral atmosphere – the *coma* – surrounding the comet.

What happens to the ions? First of all, they collide with neutrals, so that they tend to travel at roughly the same speed v_n, and they collide with electrons, with which they tend to recombine. Farther away, they are picked up by the solar wind, a point we shall consider later.

Comets

Consider first the collisions between ions and neutrals. The exchange of momentum couples them, so that the corresponding free path of ions is $l \simeq 1/n_n \sigma_{col}$, where the cross-section for collisions is (Section 2.1)

$$\sigma_{col} \sim 10^{-18} \text{ m}^2 \tag{7.99}$$

for water-group chemical species (at the temperature there). The ions and neutrals are coupled collisionally when the free path $l \ll r$, i.e. when $n_n \sigma_{col} \times r \gg 1$. Substituting the expression (7.94) of n_n (which holds for $r < r_{ion}$), we find that ions and neutrals are collisionally coupled within the radial distance $Q\sigma_{col}/4\pi v_n$. With the above parameters relevant for Halley's comet at 1 AU, this yields a distance of about 10^5 km.

Consider now recombination. The density n of ions is determined by the balance between ionisation and recombination, as in planetary ionospheres, if recombination has enough time to operate, i.e. if the timescale for recombination $1/n\beta_{rec}$ is much smaller than the dynamic timescale r/v, where v is the ion speed. In this case, the ion density is given by (7.14), whence

$$n \sim \left(\frac{n_n}{\beta_{rec} \tau_{ion}} \right)^{1/2} \sim \left(\frac{Q}{4\pi \beta_{rec} r_{ion}} \right)^{1/2} \frac{1}{r} \tag{7.100}$$

where we have substituted (7.94) and (7.95). This holds provided $1/n\beta_{rec} \ll r/v$, whence, with (7.100), (7.94) and the approximation $v \sim v_n$:

$$\frac{\beta_{rec} Q}{4\pi v_n^3 \tau_{ion}} \gg 1. \tag{7.101}$$

With the above parameters, the approximate production rate (7.89) of Halley's comet at 1 AU, and the recombination coefficient (from Section 2.5 with $T_e \sim 10^3$ K)

$$\beta_{rec} \sim 4 \times 10^{-13} \text{ m}^3 \text{ s}^{-1} \tag{7.102}$$

the non-dimensional number in (7.101) is greater than 10, so that the inequality holds true, and the order of magnitude of the ion density in the comet's ionosphere is given by (7.100) as $n \sim 10^{16}/r$. The spacecraft Giotto found indeed a density of ions varying as $1/r$ at small distances, roughly equal to 3×10^9 m^{-3} at $r \simeq 4000$ km, which agrees with this simplified estimate.

7.5.3 Pick-up of cometary ions

The above estimates hold sufficiently close to the nucleus that the solar wind has not much effect. Consider now what happens far from the comet, at distances where the solar wind dominates the problem. The freshly ionised cometary ions are picked up by the solar wind (Section 6.5.3). This loads the solar wind with fresh mass, slowing it down, and – as a by-product – produces waves at the frequency of gyration of the cometary ions.

To make an order-of-magnitude estimate, let us use a one-dimensional approximation, assuming that the parameters change essentially along the comet–Sun axis x. Let ρ and v be respectively the mass density and speed of the wind at distance x ahead of the comet. The mass added to the solar wind per unit time in a slice of width dx and section unity is equal to m_0 (the ion mass) times the number of ions produced per unit time in this volume, $(n_n/\tau_{ion})\,dx$. Substituting the neutral density (7.98), and (7.97), this yields

$$-\frac{d}{dx}(\rho v) = \frac{m_0 Q}{4\pi r_{ion} x^2} e^{-|x|/r_{ion}}. \tag{7.103}$$

Integrating between a large distance, where the solar wind mass flux has the undisturbed value $\rho_w v_w$, and the distance r, we find

$$\rho(r)\,v(r) \sim \rho_w v_w + \frac{m_0 Q}{4\pi r_{ion} r} \tag{7.104}$$

where we have neglected the exponential term, so that the result cannot be applied farther than r_{ion}. Not only does the mass flux of the loaded solar wind increase as the comet is approached, but the pressure increases, too, because the density increases and the pick-up ions had their bulk speed in the solar wind frame transformed into kinetic energy, as we saw in Section 6.5.3. This pressure gradient produces a force that slows down the solar wind. The flow is significantly modified when the second term on the right-hand side of (7.104) is roughly equal to the first, i.e. at a distance

$$r_{pick\text{-}up} \sim \frac{m_0 Q}{4\pi r_{ion} \rho_w v_w} \sim \frac{Q}{10^{30}} \times 5 \times 10^5 \text{ km} \tag{7.105}$$

where we have substituted the above numerical values $m_0 \sim 18 \times m_p$, $r_{ion} \sim 10^9$ m, and the typical solar wind mass flux (from (7.17) with $\rho_w \simeq m_p n_w$).

Since these pick-up ions make the supersonic solar wind aware of the presence of the comet and make it slow down, one may wonder whether a shock is necessary to accommodate the obstacle. Detailed analysis shows that a shock may form at a distance of the order of magnitude of $r_{pick\text{-}up}$, when the plasma is not sufficiently compressible to put up with an addition of momentum density of the order of magnitude of the solar wind undisturbed value.

For Halley's comet, a weak bow shock was indeed observed at about this distance; since the solar wind had already been significantly slowed down, the shock was weak. The major role played by cometary ions in the dynamics makes the physics of this shock differ from the ones near planets, not only because of mass loading, but also because the ion gyroradius – which sets the scale of the structure – is much larger for cometary ions than for protons (since it is proportional to the square root of the mass), so that for comets having a small production rate, it is barely smaller than the relevant cometary scales.

Comets

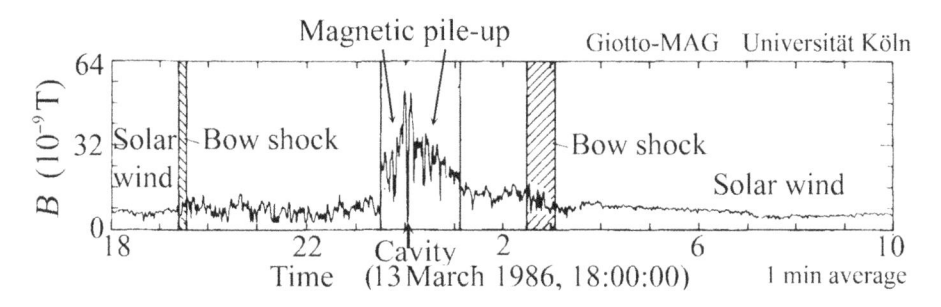

Figure 7.29 Magnetic field strength measured on the spacecraft Giotto along the trajectory shown in Fig. 7.28, as it crossed the bow shock, the magnetic pile-up region and the cavity of vanishing magnetic field (indicated by the arrow at the bottom). (Adapted from [60].)

7.5.4 Magnetic pile-up

The cometary plasma density is large enough for the electric conductivity (Section 7.1) to produce a large magnetic Reynolds number in the plasma–neutral coupling region, so that the magnetic field is frozen to the plasma at the relevant speeds and scales. Therefore, as the comet is approached, the solar wind magnetic field is expected to behave somewhat as for non-magnetised planets, piling up ahead while the speed decreases, vanishing in its vicinity while the flow slows down, and draping around it, producing a tail in the anti-sunward direction. This behaviour has been observed by the spacecraft that encountered Halley's comet on the sunward side (Fig. 7.28), and on the anti-sunward side by the ICE spacecraft, which crossed the tail of comet Giacobini–Zinner (Fig. 7.30 below).

Figure 7.29 shows the magnetic field strength measured by the magnetometer aboard Giotto [60] along the trajectory sketched in Fig. 7.28. The similarity to Fig. 7.12 showing the magnetic field at Mars is striking. The field magnitude increases at the bow shock, behind which the plasma is extremely turbulent, up to a transition where the field begins suddenly to pile up, to reach a maximum of $B_{max} \simeq 60 \times 10^{-9}$ T. This strength is close to the value (7.45), whose magnetic pressure balances the solar wind ram pressure. The magnetic field vanishes in a cavity of radius $r_{cavity} \sim 5000$ km.

This distance marks a complex transition since on the nucleus side, the cometary plasma is flowing outwards with the neutrals, whereas on the solar side the plasma is made of the loaded solar wind flowing anti-sunwards. This produces a stagnation region where the plasma speed vanishes and changes in direction.

What provides the force required to stand off the solar wind, or rather to balance the magnetic force associated with the 'magnetic cushion' produced by the pile-up? For non-magnetised planets, this is the pressure of the ionosphere. This is not so for comets, because the ion pressure is far smaller than the magnetic pressure $B_{max}^2/2\mu_0$. Since, however, this interaction takes place in the region where ions and neutrals are collisionally coupled, the outward moving neutrals come to the rescue in providing a friction force to the stagnating ions, which can balance the magnetic force.

Let us estimate the distance where this takes place. The magnetic force is of the order of magnitude of the magnetic curvature force F_B; since at the cavity boundary the curvature of the field lines is $\sim 1/r_{cavity}$, we have $F_B \sim B_{max}^2/(\mu_0 r_{cavity})$ per unit volume. It must be balanced by the ion–neutral friction force, which is of the order of magnitude of the neutral ram pressure $m_0 n_n v_n^2$ times the cross-section σ_{col} per ion (assumed to be at rest), i.e. $m_0 n_n v_n^2 \times \sigma_{col} \times n$ per unit volume. Substituting n_n from (7.94), the ion density n from (7.100), and rearranging, we deduce the radius of the cavity

$$r_{cavity} \sim \left(\frac{Q}{4\pi}\right)^{3/4} \left(\frac{m_0 \sigma_{col} v_n}{B_{max}^2/\mu_0}\right)^{1/2} \frac{1}{(r_{ion} \beta_{rec})^{1/4}} \quad (7.106)$$

$$\sim \left(\frac{Q}{10^{30}}\right)^{3/4} \times 4 \times 10^3 \text{ km} \quad (7.107)$$

where we have substituted the numerical values determined above. This order of magnitude estimate is close to the value observed at Halley's comet.

7.5.5 The plasma tail

The draping of the magnetic field around the nucleus as a windsock channels the cometary ions within a flat ribbon squeezed between two flux ropes of opposite magnetic polarity, producing a tail in the anti-solar direction, as first suggested by Ludwig Biermann (Section 1.1), although the actual agent of the interaction – the solar wind magnetic field – was identified later by Hannes Alfvén [2] (Fig. 7.9).

The spacecraft ICE crossed the tail of comet Giacobini–Zinner at about 8000 km from the nucleus, at a relative speed of 21 km s^{-1}, when it was at 1 AU from the Sun and emitting $Q \sim 3 \times 10^{28}$ molecules s^{-1}. Figure 7.30 shows the magnetic field measured by the inboard magnetometer, superimposed onto an optical image [74]. The magnetic field is draped around the comet, producing a magnetic tail 10^4 km wide (dotted line), where the magnetic field peaks at about 60×10^{-9} T in each lobe, and vanishes in the 1200 km wide plasma sheet where the electron density was $n \simeq 7 \times 10^8$ m^{-3},[47] and the electron temperature $T_e \simeq 1.5 \times 10^4$ K [57].[48] Note that if the temperature of the ions – which was not measured – was similar, the total plasma pressure would be $P = 2nk_B T_e \simeq 3 \times 10^{-10}$ Pa, smaller by a factor of five than the magnetic pressure $B^2/2\mu_0$ in the adjacent lobes. This suggests that either the tail was not in dynamic equilibrium, or the plasma sheet ions were much hotter than the electrons.[49]

---

[47]With thin structures, which may be dynamic phenomena.

[48]Too cold for the inboard particle analyser to be able to measure the density and temperature.

[49]The neutrals cannot come to the rescue to balance the forces as they do at the magnetic cavity, because this takes place outside the plasma–neutral coupling region, which for comet Giacobini–Zinner is smaller than Halley's by a factor roughly equal to the ratio of their emission rates, i.e. about 1/30.

Comets

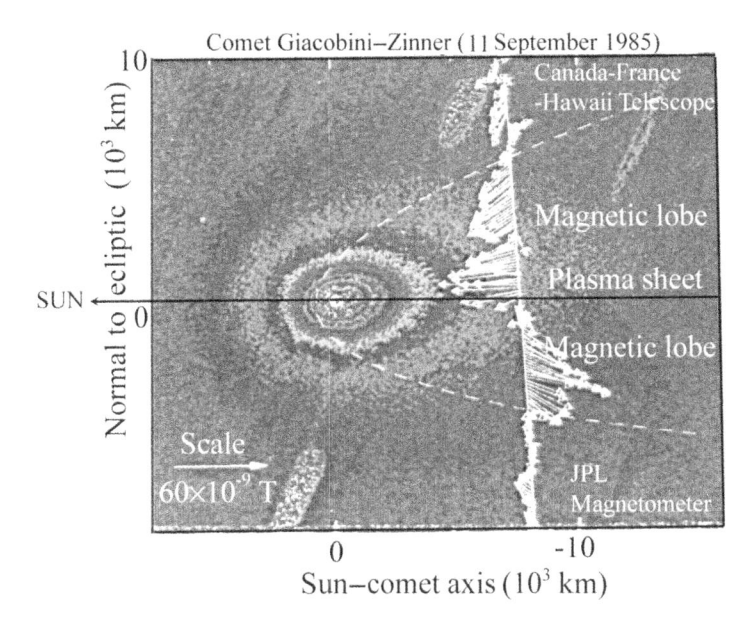

Figure 7.30 Magnetic field measured on the spacecraft ICE when it crossed the tail of comet Giacobini–Zinner at about 1 AU from the Sun, projected onto an image of the comet taken with the Canada–France–Hawaii Telescope. This illustrates the draping of the magnetic field, producing two lobes of opposite polarity. (Adapted from [74].)

The plasma tails of comets exhibit a number of structures. When a comet encounters a change in solar wind properties, essentially the direction of the magnetic field and/or the ram pressure, the tail may disconnect from the nucleus, sometimes launching bullets, and reform itself to adapt to new solar wind conditions. This occurs at slow/fast wind transitions or solar mass ejections, and may produce multiple tail structures as may be seen in Fig. 1.4. A study of such tail disconnections may be found for example in [13]. This behaviour is reminiscent of substorms occurring at magnetised planets, although the comet's magnetic tail is made of bona fide solar wind magnetic field lines, contrary to the one of magnetised planets, whose structure depends on badly understood processes connecting the magnetic field of the planet to the solar wind one.

How long are comets' plasma tails? The cometary ions survive to recombination for a long time and may be detected very far away. The magnetic tail, however, relies on the draping of solar wind magnetic field lines, which itself relies on plasma speed gradients. As the draped magnetic field transfers momentum to the cometary ions via magnetic curvature forces in the tail, these ions are accelerated; as they catch up with the solar wind, the draping decreases and vanishes, which sets the end of the magnetic tail.

To estimate the length of the magnetic tail, consider the electric current I flowing in the plasma sheet that separates the lobes of opposite magnetic fields

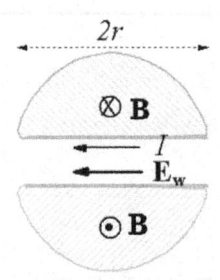

Figure 7.31 Sketch of the cross-section of the magnetic tail. The current in the plasma sheet separating the lobes of opposite polarity is $2B/\mu_0$ per unit length perpendicular to the figure.

(of amplitude B). From Ampère's law, this current is $2B/\mu_0$ per unit length of the tail (Fig. 7.31), whence $I \sim 2BL_T/\mu_0$ for a tail of length L_T. If the solar wind electric field $E_w \sim v_w B_w$ acts over the distance $2r$, the power dissipated is $P \sim E_w \times 2rI \sim 4v_w B_w BL_T r/\mu_0$ (B_w denotes the component of the solar wind magnetic field perpendicular to the velocity).

This electric power cannot exceed the flux of kinetic energy of the solar wind $P_K \sim v_w \left(\rho_w v_w^2/2\right) \pi r^2$ across the magnetic tail of radius r. Writing $P < P_K$, with $r \leq r_{pick\text{-}up}$ and $B > B_w$, we deduce an upper limit to the length of the magnetic tail

$$L_T < \frac{\pi}{8} r_{pick\text{-}up} \times \frac{\rho_w v_w^2}{B_w^2/\mu_0} \sim \frac{m_0 Q v_w / r_{ion}}{32 B_w^2/\mu_0} \qquad (7.108)$$

$$\sim \frac{Q}{10^{30}} \times 3 \times 10^7 \text{ km}$$

where we have substituted $r_{pick\text{-}up}$ from (7.105) and the solar wind parameters (see Section 7.2.1). The length of the magnetic tail is therefore limited to a distance that is greater than $r_{pick\text{-}up}$ by the ratio of the solar wind kinetic and (perpendicular) magnetic energies, i.e. roughly the square of the magnetic compression. This ratio is nearly two orders of magnitude at 1 AU from the Sun. Model calculations yield a similar result, albeit from somewhat different arguments [85].

7.5.6 X-ray emission

Given the low temperature and density of the cometary plasma, the discovery that comets emit X-rays came as a surprise. This radiation consists of emission lines in 'soft' X-rays (energies less than 1 keV) and extreme ultraviolet. It is not produced by scattering and fluorescence of solar X-rays, which occur at Venus and Mars, but by a subtle interaction between highly ionised solar wind ions and cometary neutrals ([20] and references therein).

Comets

We have seen that the solar corona is so hot that ions can be stripped of most of their electrons, and are then carried out with the solar wind.[50] When these ions happen to encounter a neutral molecule, they take one of its electrons, producing a reaction such as

$$O^{n+} + H_2O \longrightarrow \left(O^{(n-1)+}\right)^* + H_2O^+ \qquad (7.109)$$

for oxygen. Similar reactions take place with other elements and various levels of ionisation. In the above example, the ion O^{n+} (an oxygen atom stripped of all its electrons but $8-n$) acquires an electron, and is left in a highly excited state, symbolised by $()^*$. Because of the large unscreened nuclear charge, the energy available is large, and de-excitation down to the ground state results in emission of X-ray photons. The maximum energy available may be estimated by noting that the upper electron of $O^{(n-1)+}$ 'sees' an effective charge $q = ne$ and has therefore a potential energy $\sim -W_{Bohr} \times n^2$ in its ground state (Section 2.5). Since the energy in the excited state has a much smaller modulus because the electron is nearly free, the transition involves an energy of nearly $n^2 \times W_{Bohr}$, which comes to about 900 eV for $n = 8$ (a fully ionised oxygen atom). The larger the atomic number and the ionisation level, the greater the energy available. In practice, the transitions taking place are subtly controlled by the parameters and by the selection rules of atomic states.

How much power may be emitted in this way? Let α be the (cumulated) proportion relative to protons of the most effective solar wind ions ($\alpha \sim 10^{-3}$), and let us approximate their speed by the solar wind speed v_w. The effective ion flux is therefore the solar wind flux $n_w v_w$ times α. If each encounter with a neutral produces an average energy W_{ph} with an average cross-section σ_{ex}, the power emitted per unit volume is

$$P_X \sim (\alpha n_w v_w) \sigma_{ex} W_{ph} n_n. \qquad (7.110)$$

Substituting the neutral density (7.98) and the parameters $W_{ph} \sim 20 \times W_{Bohr}$ (~ 0.3 keV), $\sigma_{ex} \sim 2 \times 10^{-19}$ m^2 from (2.147), and integrating over volume, we get the total X-luminosity

$$L_X = \int_0^\infty P_X \times 4\pi r^2 dr \sim (\alpha n_w v_w) \sigma_{ex} \times 20 W_{Bohr} Q \tau_{ion} \qquad (7.111)$$

$$\sim \frac{\alpha}{10^{-4}} \frac{Q}{10^{30}} \times 2 \times 10^9 \text{ W} \qquad (7.112)$$

where we have substituted the solar wind flux, the cross-section, and the ionisation radius (7.97) and timescale (7.95) relevant at 1 AU. This order-of-magnitude estimate agrees with the values observed. Temporal variations of the flux of heavy ions in the solar wind translate directly into variations of X-ray emission. Note that this exchange of charge cannot take place when the highly ionised ion

---

[50] They are unable to recombine because the timescale for recombination is much greater than the dynamic timescale (see Section 2.5).

encounters another ion because of the Coulomb electrostatic repulsion between charges of equal sign.

We find here an echo of the heating of the solar corona. Some of the energy available in the corona is used to strip heavy ions of their electrons, which remain in this highly ionised state during their journey in the solar wind, until they encounter a neutral particle and deliver the energy. Analysis of the emission spectra may provide a novel tool for probing the solar wind minor ions in regions difficult to access.

This effect not only takes place at comets, but also near planets, which are a source of neutrals, and within the heliosphere, which contains a significant quantity of interstellar neutrals (Section 8.1.2).

7.5.7 The dust tail

Consider finally what happens to the dust. Close to the nucleus, the moving gas pushes the grains outwards. When the gas density has sufficiently decreased, the drag becomes negligible, and the motion of the dust grains is driven by the solar gravity and radiation pressure. We have seen that the radiation pressure is equivalent to reducing the effective gravitational attraction. Since the comet's nucleus itself follows a trajectory determined mainly by solar gravity, dust grains follow orbits that are more open than that of the nucleus, the more so as the grain is smaller, by virtue of (7.83).

It is amusing to find here an echo of a problem we encountered in Section 6.1, when calculating the shape of solar magnetic field lines produced by the Sun's rotation. In the solar case, magnetic field lines follow the trajectories of plasma parcels 'launched' from a rotating point at distances where gravity is negligible, so that they follow straight lines at the constant solar wind speed. The problem is similar for the grains for which $\beta \simeq 1$, so that the radiation pressure annuls the gravitational attraction. In this case, once launched at about $1 \, \text{km s}^{-1}$ near the nucleus, the grains continue on straight lines, just as do the plasma parcels in the solar wind problem. They therefore follow trajectories as shown in Fig. 6.1, with two main differences: first, the comet's nucleus does not move on a circular orbit (contrary to a point on the solar surface), second, the scale with respect to the object is different because the grain's 'launch' speed is much smaller than that of the comet's nucleus (whereas the solar wind speed is much greater than the rotation speed at the solar surface).

For most grains, however, $\beta \neq 1$, so that they are subjected to a net force that curves the trajectory somewhat as does gravity for objects launched at the surface of a planet. Grains emitted at different times follow different orbits, as do those of different size, which are subjected to a different force. This produces the diffuse curved tail of comets, whose visible radiation is essentially sunlight reflected from the dust grains.

The small grains for which the radiative force is the largest ultimately fill the whole heliosphere and are ultimately lost. The large ones, which are subjected to a small radiative force, follow orbits close to that of their comet progenitor; they give rise to the meteor showers observed at regular intervals when the Earth happens to cross these orbits, producing the so-called shooting stars.

7.6 Problems

7.6.1 Electrostatic charging in space

Calculate the current of photoelectrons and of solar wind electrons for an object with different geometries, as explained in the text. Show that the current of protons is in general negligible. What happens in the wake for a body having an insulated surface? Try to estimate the electrostatic charging of the Moon on both its sunlit and shadowed parts. How does the surface potential of the Moon vary when it passes from the solar wind environment to the Earth's magnetosphere?

Hints

When calculating the currents, look at Problem 5.7.7. For an insulating surface, note that the net current must vanish locally on each point. The rear side of the object is in the wake, which is depleted of protons, but much less so of electrons. Since the rear side is generally in shadow, the incoming electron and proton fluxes must balance each other. The potential of the surface must therefore be sufficiently negative to curve the proton paths sufficiently to make them enter the wake; this requires $e\Phi \sim -m_p v_w^2/2$, which yields a potential of nearly -10^3 V; in fact, electrons are also depleted, and barriers of potential are produced, making the calculation far more complex than this rough estimate. Beware that when the Moon is immersed in the Earth's magnetosphere, the presence of high-energy electrons may result in electron secondary emission playing an important role in the charging process.

7.6.2 Magnetic pile-up

Imagine a body of size R made up of a material of conductivity σ immersed in the solar wind flow (Fig. 7.32). Calculate the current density induced in the cube by the relative motion. What is the current necessary to cancel the magnetic field in the cube? Show that the magnetic field produced by this current adds to the ambient magnetic field, doubling it. How do the currents close?

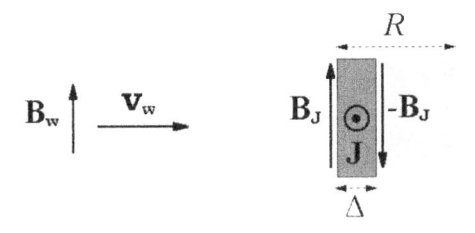

Figure 7.32 Current sheet in a conducting body immersed in the solar wind flow, and corresponding magnetic field.

Hints

The current density is $J = -\sigma v_w B_w$ (where v_w and B_w are the unperturbed solar wind values), since the product $v_w B_w$ is conserved as the solar wind slows down and the magnetic field lines pile up. From Maxwell's equation $\nabla \times \mathbf{B} = \mu_0 \mathbf{J}$, a sheet of current density J and width Δ produces a magnetic field $B_J \sim \mu_0 J \Delta$, whence

$$B_J \sim \mu_0 \sigma v_w B_w \Delta. \tag{7.113}$$

Let B_{max} be the total magnetic field in front of the cube. To cancel it behind the sheet requires $B_J = B_{max}$, i.e. $\sigma \Delta \sim B_{max}/(\mu_0 v_w B_w)$. Substituting (7.46), this yields $\mu_0 \sigma V_A \Delta \sim 1$. Since the size of the cube must be $R > \Delta$, we have

$$\mu_0 \sigma V_A R > 1. \tag{7.114}$$

The lines of current in the body must diverge towards its extremities, to make the current density decrease sufficiently to enable current closure.

7.6.3 Chapman–Ferraro layer

Consider a very simplified model of the magnetopause, in which solar wind particles are coming from the left-hand side of the boundary, where the magnetic field is assumed to be negligible. On the right of the boundary, they suddenly see a magnetic field B, which makes them turn and reverse their direction of motion. Because the protons have a much greater gyroradius than the electrons, they penetrate much farther, producing a current sheet that flows downwards in a slab perpendicular to Fig. 7.33.

Show that any proton entering the boundary over a region $2r_p$ wide (r_p is the gyroradius in the field $\mathbf{B}$) will cross the plane $y = 0$.

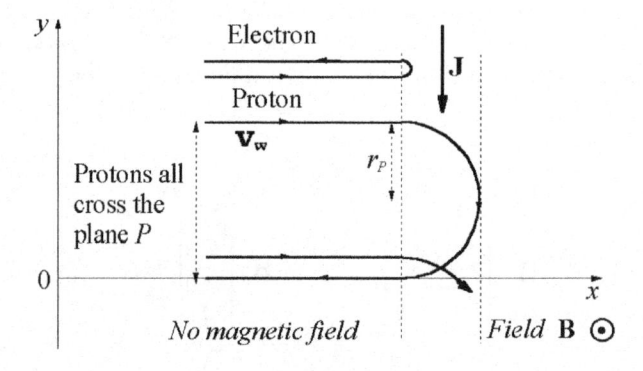

Figure 7.33 A sketch of the particle trajectories at the magnetopause, with a simplified model.

Deduce that with a density n of protons arriving at speed v_w, this yields a current $I \sim 2\rho v_w^2/B$ flowing downward per unit thickness perpendicular to the figure. Since the current sheet serves to cancel the dipolar field B on the left, while it doubles it on the right-hand side, we have $I = 2B/\mu_0$ which yields (7.49) to a numerical factor of order unity.

Show that with electrons and protons coming both at the solar wind speed, the geometric mean of the electron and proton radii is equal to the electron skin depth c/ω_p.

How does the actual situation differ from this simplified model?

Hints

What about charge neutrality? What happens when two counterstreaming plasmas are in presence? Is the magnetic field in the magnetosheath negligible?

7.6.4 Interaction of the solar wind with Venus and Mars

Deduce from Fig. 7.11 the solar wind dynamic pressure during the measurements, and the approximate temperature of the plasma at the ionopause, for the three cases shown. Compare the latter with the measured values [58], and comment. Try to find reasons why the ionosphere of Mars has more difficulty than that of Venus in standing off the solar wind.

7.6.5 Ring current

Show that the magnetic field produced at the surface of a planet of radius R by the gradient drift of a charge of energy W in the equatorial plane is (counted along the axis antiparallel to the planet's magnetic moment)

$$\Delta B_D = \frac{-3\mu_0 W}{4\pi R^3 B_0}. \qquad (7.115)$$

Show that the particle (circling at distance r from the planet) has a magnetic moment $\mu = -W/B$ along that same axis, where $B = B_0 (R/r)^3$ and produces at the planet's equator a magnetic field (along that same axis)

$$\Delta B_\mu = \frac{\mu_0 W}{4\pi R^3 B_0}. \qquad (7.116)$$

Deduce that the magnetic field decrease produced by a ring of drifting particles of total kinetic energy W_{ring} is given by (7.77) (the Dessler–Parker–Sckopke relationship).

Derive this relation from the Virial theorem.

Assume that the energy in the ring current is 10^{15} J and approximate it by an annulus circling the Earth at a distance of $5\,R_\oplus$. Show that the corresponding current is a few 10^6 A.

Hints

To derive (7.77) from the Virial theorem, apply it both to the planet alone and to the planet + magnetosphere system, neglecting the electrostatic energy, the mutual gravitational energy of the planet + magnetosphere system, and the magnetic self-energy of the magnetosphere. Note that the mutual energy of the planet's magnetic dipole and the ring current is $\mu\Delta B$, whereas the magnetic energy of the planet's dipole above the surface is $W_{mag} = \mu B_0/3$.

7.6.6 Does Vesta have a magnetosphere?

The asteroid Vesta, of diameter $\simeq$ 530 km, is one of the three largest bodies of the main asteroid belt, and studies of meteorites suggest that it may have a conductive core and once had an intrinsic magnetic field. Its distance to the Sun is about 2.4 AU. Measurements of the solar reflectivity of its surface suggest that it is not subjected to direct impact of the solar wind particles [84]. Propose possible explanations.

Hints

Compare the size of Vesta to the basic solar wind scales. Vesta has no atmosphere, but might have a significant electric conductivity and/or a significant intrinsic magnetic field. In the latter case, what is the magnetic moment required to possibly stand off the solar wind? Alternatively, calculate the minimum value of the conductivity in a region of given size for the magnetic Reynolds number to be greater than unity, so that an induced magnetosphere may possibly exist. Given the size of Vesta, can the interaction be described by MHD? See [84] and [63].

7.6.7 Gas–dust drag in a comet: another nozzle problem

Consider the outflow of a dusty gas from a cometary nucleus at small distances. The gas starts at a subsonic speed at distance $r = R$ (the radius of the nucleus), and acquires a supersonic speed at some distance where the temperature has decreased considerably. To estimate how the flow speed varies with distance, make three (unrealistic) assumptions: first, the flow has a spherical symmetry; second, the gas (of radial velocity v) is slowed down by collisions with dust, which produces a drag force equal to $\nu_{gd}v$ per unit mass (where ν_{gd} is the collision frequency of the neutrals with the dust, approximated by $\nu_{gd} \sim V_S/l_{gd}$ with $l_{gd} \sim 1/n_d\sigma_{gd}$ the free path of molecules for collisions with dust grains); and, third, the flow is adiabatic.

Show that the problem is similar to the de Laval nozzle studies in Problem 5.7.1, with the nozzle cross-section variation dS/S being replaced by the factor $dr\,(2/r - v/V_S l_{gd})$, so that the constriction of the equivalent nozzle is provided by the drag force.

Deduce the critical distance where the gas becomes supersonic, as a function of l_{gd}.

Estimate this distance, compare with other scales, and comment.

References

[1] Acuna, M. H. *et al.* 2001, Magnetic field of Mars: summary of results from the aerobraking and mapping orbits, *J. Geophys. Res.* **106** 23403.

[2] Alfvén, H. 1957, On the theory of comet tails, *Tellus* **9** 92M.

[3] Atreya, S. K. 1986, *Atmospheres and Ionospheres of the Outer Planets and their Satellites*, New York, Springer.

[4] Bagenal, F. 1985, in *Solar System Magnetic Fields*, ed. E. R. Priest, Dordrecht, The Netherlands, D. Reidel, p. 224.

[5] Bagenal, F. 1992, Giant planet magnetospheres, *Annu. Rev. Earth Planet. Sci.* **20** 289.

[6] Bakich, M. E. 2000, *The Cambridge Planetary Handbook*, Cambridge University Press.

[7] Bale, S. D. *et al.* 2003, Density-transition scale at quasi-perpendicular collisionless shocks, *Phys. Rev. Lett.* **91** 265004.

[8] Balogh, A. *et al.* 1997, in *The Cluster and Phoenix Missions*, ed. C. P. Escoubet *et al.*, New York, Kluwer, p. 65.

[9] Bertaux, J. L. *et al.* 2005, Discovery of an aurora on Mars, *Nature* **435** 790.

[10] Bertucci, C., C. Mazelle and M. H. Acuna 2005, The interaction of the solar wind with Mars from Mars Global Surveyor: MAG/ER observations, *J. Atmos. Sol.-Terrestr. Phys.* **67** 1797.

[11] Betz, G. and K. Wien 1994, Energy and angular distribution of sputtered particles, *Int. J. Mass Spectr. Ion Process.* **140** 1.

[12] Blanc, M. *et al.* 2005, Solar system magnetospheres, *Space Sci. Rev.* **116** 227.

[13] Brosius, J. W. *et al.* 1987, The cause of two plasma-tail disconnection events in comet P/Halley during the ICE-Halley radial period, *Astron. Astrophys.* **187** 267.

[14] Burns, J. A. *et al.* 1979, Radiation forces on small particles in the solar system, *Icarus* **40** 1.

[15] Ceplecha, Z. *et al.* 1998, Meteor phenomena and bodies, *Space Sci. Rev.* **84** 326.

[16] Chamberlain, J. W. and D. M. Hunten 1987, *Theory of Planetary Atmospheres: An Introduction to their Physics and Chemistry*, London, Academic Press.

[17] Cole, G. H. A. and M. M. Woolfson 2002, *Planetary Science*, Bristol UK, Institute of Physics.

[18] Connerney, J. E. P. 1993, Magnetic fields of the outer planets, *J. Geophys. Res.* **98** 18659.

[19] Cowley, S. W. H. 1991, The plasma environment of the Earth, *Contemp. Phys.* **32** 235.

[20] Cravens, T. E. 2002, X-ray emission from comets, *Science* **296** 1042.

[21] Cravens, T. E. and T. I. Gombosi 2004, Cometary magnetospheres: a tutorial, *Adv. Space Res.* **33** 1968.

[22] Crovisier, J. 1984, The water molecule in comets: fluorescence mechanisms and thermodynamics of the inner corona, *Astron. Astrophys.* **130** 361.

[23] Desch, M. D. and M. L. Kaiser 1984, Predictions for Uranus from a radiometric Bode's law, *Nature* **310** 755.

[24] Dessler, A. and E. N. Parker 1959, Hydromagnetic theory of geomagnetic storms, *J. Geophys. Res.* **64** 2239.

[25] Dohnanyi, J. S. 1969, Collisional model of asteroids and their debris, *J. Geophys. Res.* **74** 2531.

[26] Dungey, J. W. 1961, Interplanetary magnetic field and the auroral zones, *Phys. Rev. Lett.* **6** 47.

[27] Elliott, J. L. *et al.* 2003, The recent expansion of Pluto's atmosphere, *Nature* **424** 165.

[28] Festou, M. C. *et al.* 1993, Comets. I. Concepts and observations, *Astron. Astrophys. Rev.* **4** 363.

[29] Festou, M. C. *et al.* 1994, Comets. II. Models, evolution, origin, and outlook, *Astron. Astrophys. Rev.* **5** 37.

[30] Feynman, J. 2002, The good old days: John Spreiter in the 1960s, *Planet. Space Sci.* **50** 417.

[31] Goertz, C. K. 1989, Dusty plasmas in the solar system, *Rev. Geophys.* **27** 271.

[32] Grard, R. J. L. 1973, Properties of the satellite photoelectron sheath derived from photoemission laboratory measurements, *J. Geophys. Res.* **78** 2885.

[33] Grün, E. *et al.* 1985, Collisional balance of the meteoric complex, *Icarus* **62** 244.

[34] Grün, E. *el al.* eds. 2001, *Interplanetary Dust*, New York, Springer.

[35] Harris, R. A. ed. 2001, SP-476 *7th Spacecraft Charging Technology Conf.*, Noordwijk, The Netherlands, ESA/ESTEC.

[36] Hastings, D. and H. B. Garrett 1996, *Spacecraft–Environment Interactions*, Cambridge University Press.

[37] Heikkila, W. J. 1973, Critique of fluid theory of magnetospheric phenomena, *Astrophys. Space Sci.* **23** 261.

[38] Hill, T. W. 1983, in *Solar–Terrestrial Physics*, ed. R. L. Carovillano and J. M. Forbes, Dordrecht, The Netherlands, D. Reidel, p. 261.

[39] Hill, T. W. and A. J. Dessler 1991, Plasma motions in planetary magnetospheres, *Science* **252** 410.

[40] Hill, T. W. and M. E. Rassbach 1975, Interplanetary magnetic field direction and the configuration of the day side magnetosphere, *J. Geophys. Res.* **80** 1.

[41] Hill, T. W., A. J. Dessler and R. A. Wolf 1976, Mercury and Mars: the role of ionospheric conductivity in the acceleration of magnetospheric particles, *Geophys. Res. Lett.* **3** 429.

[42] Horanyi, M. 1996, Charged dust dynamics, *Annu. Rev. Astron. Astrophys.* **34** 383.

[43] Hughes, D. W. 1994, Comets and asteroids, *Contemp. Phys.* **35** 75.

[44] Hunten, D. M., T. H. Morgan and D. E. Shemansky 1988. In *Mercury*, ed. F. Vilas *et al.*, Tucson AZ, University of Arizona Press, p. 562.

[45] Keller, H. U. *et al.* 1987, Comet P/Halley's nucleus and its activity, *Astron. Astrophys.* **187** 807.

[46] Lanzerotti, L. J. 2001, in *Space Weather*, *Geophys. Monogr. Ser.* **125**, ed. P. Song *et al.*, Washington DC, American Geophysical Union, p. 11.

[47] Lemaire, J. 1985, Plasmoid motion across a tangential discontinuity (with application to the magnetopause), *J. Plasma Phys.* **33** 425.

[48] Lemaire, J. *et al.* 2005, The influence of a southward and northward turning of the interplanetary magnetic field on the geomagnetic cut-off of cosmic rays, on the mirror point positions of geomagnetically trapped particles and on their rate of precipitation in the atmosphere, *J.A.S.T.P.* **67** 719.

[49] Luhmann, J. G. *et al.* eds. 1992, *Venus and Mars: Atmospheres, Ionospheres, and Solar Wind Interactions*, *Geophys. Monogr. Ser.* **66**, Washington DC, American Geophysical Union.

[50] Lui, A. T. Y. 1996, Current disruption in the Earth's magnetosphere: observations and models, *J. Geophys. Res.* **101** 13067.

[51] Luu, J. X. and D. C. Jewitt 2002, Kuiper belt objects: relics from the accretion disk of the sun, *Annu. Rev. Astron. Astrophys.* **40** 63.

[52] McDonnell, J. A. M. et al. eds. 1978, *Cosmic Dust*, New York, Wiley.

[53] Mann, I. and H. Kimura 2000, Interstellar dust properties derived from mass density, mass distribution, and flux rates in the heliosphere, *J. Geophys. Res.* **105** 10317.

[54] Mann, I. et al. 2004, Dust near the sun, *Space Sci. Rev.* **110** 269.

[55] Mendis, D. A. et al. 1985, The physics of comets, *Fund. Cosmic Phys.* **10** 1.

[56] Mendis, D. A. and M. Rosenberg 1994, Cosmic dusty plasma, *Annu. Rev. Astron. Astrophys.* **32** 419.

[57] Meyer-Vernet, N. et al. 1986, Physical parameters for hot and cold electron populations in comet Giacobini-Zinner with the ICE radio experiment, *Geophys. Res. Lett.* **13** 279.

[58] Nagy, A. F. and T. E. Cravens 1997, in *Venus II*, ed. S. W. Bougher et al., Tucson AZ, University of Arizona Press, p. 189.

[59] Nagy, A. F. et al. 2004, The plasma environment of Mars, *Space Sci. Rev.* **111** 33.

[60] Neubauer, F. M. 1987, Giotto magnetic field results on the boundaries of the pile-up region and the magnetic cavity, *Astron. Astrophys.* **187** 73.

[61] Ogilvie, K. W. et al. 1996, Observations of the lunar plasma wake from the WIND spacecraft on December 27, 1994, *Geophys. Res. Lett.* **23** 1255.

[62] Olbert, S. et al. 1968, A simple derivation of the Dessler–Parker–Sckopke relation, *J. Geophys. Res.* **73** 1115.

[63] Omidi, N. et al. 2002, Hybrid simulations of solar wind interaction with magnetised asteroids: general characteristics, *J. Geophys. Res.* **107** 1487.

[64] Paschmann, G. et al. 2002, Auroral plasma physics, *Space Sci. Rev.* **103** 1.

[65] Petrinec, S. M. 2002, The location of the Earth's bow shock, *Planet. Space Sci.* **50** 541.

[66] Prangé, R. et al. 2004, An interplanetary shock traced by planetary auroral storms from the Sun to Saturn, *Nature* **432** 78.

[67] Reme, H. et al. 1997, in *The Cluster and Phoenix Missions*, ed. C. P. Escoubet et al., New York, Kluwer, p. 303.

[68] Russell, C. T. 1992, in *Venus and Mars: Atmospheres, Ionospheres, and Solar Wind Interactions*, Geophys. Monogr. Ser. **66**, ed. J. G. Luhmann et al., Washington DC, American Geophysical Union, p. 225.

[69] Russell, C. T. 1993, Magnetic fields of the terrestrial planets, *J. Geophys. Res.* **18** 18681.

[70] Russell, C. T. 1993, Planetary magnetospheres, *Rep. Prog. Phys.* **56** 687.

[71] Schulz, M. 1997, Direct influence of ring current on auroral oval diameter, *J. Geophys. Res.* **102** 14149.

[72] Shue, J. H. and P. Song 2002, The location and shape of the magnetopause, *Planet. Space Sci.* **50** 549.

[73] Sicardy, B. et al. 2003, Large changes in Pluto's atmosphere as revealed by recent stellar occultations, *Nature* **424** 168.

[74] Slavin, J. A. et al. 1986, The structure of a cometary type I tail: ground-based and ICE observations of P/Giacobini–Zinner, *Geophys. Res. Lett.* **13** 1085.

[75] Smyth, W. H. and M. L. Marconi 1995, Theoretical overview and modeling of the sodium and potassium atmospheres of the Moon, *Ap. J.* **443** 371.

[76] Sonett, C. P. 1982, Electromagnetic induction in the Moon, *Rev. Geophys. Space Phys.* **20** 411.

[77] Srama, R. et al. 2004, The Cassini cosmic dust analyser, *Space Sci. Rev.* **114** 465.

[78] Stephenson, F. R. and C. B. F. Walker eds. 1985, *Halley's Comet in History*, London, British Museum.

[79] Stern, D. P. 1994, The art of mapping the magnetosphere, *J. Geophys. Res.* **99** 17169.

[80] Tsurutani, B. T. et al. 2003, The extreme magnetic storm of 1–2 September 1859, *J. Geophys. Res.* **108** 1268.

[81] Uritsky, V. M. et al. 2002, Scale-free statistics of spatiotemporal auroral emissions as depicted by Polar UVI images: dynamic magnetosphere is an avalanching system, *J. Geophys. Res.* **107** 1426.

[82] Vasyliunas, V. M. 1983, in *Solar–Terrestrial Physics*, ed. R. L. Carovillano and J. M. Forbes, Dordrecht, The Netherlands, D. Reidel, p. 479.

[83] Vasyliunas, V. M. 2004, Comparative magnetospheres: lessons for Earth, *Adv. Space Res.* **33** 2113.

[84] Vernazza, P. et al. 2006, The color of asteroids: a novel tool for magnetic field detection? The case of Vesta, *Astron. Astrophys.* **451** L43.

[85] Wegmann, R. 2002, Large-scale disturbance of the solar wind by a comet, *Astron. Astrophys.* **389** 1039.

[86] Whipple, E. C. 1981, Potentials of surfaces in space, *Rep. Prog. Phys.* **44** 1197.

[87] Zarka, P. *et al.* 2001, Magnetically-driven planetary radioemissions and application to extrasolar planets, *Astrophys. Space Sci.* **277** 293.

8

The solar wind in the Universe

> 'To an astronomer, man is nothing more than an insignificant dot in an infinite universe,' someone once said to Einstein. To which Einstein replied: 'But I realise that the insignificant dot who is man is also the astronomer.'

What is the solar wind to an astronomer? I shall not address this question in detail, but only briefly examine how the solar wind interacts with its cosmic environment, and how it compares with some other cosmic winds. I also address briefly the physics of cosmic rays, their acceleration and their interaction with the solar wind.

8.1 The frontier of the heliosphere

In studying the interaction of the solar wind with objects in Chapter 7, we omitted one major obstacle: the interstellar medium.

In the long term, the interstellar medium is the direct partner of the Sun in the galactic system; it served as a source of matter for the formation of the Sun and the Solar System; in turn the Sun returns matter to the interstellar medium via the solar wind at a rate of about 10^6 tonnes per second, a figure that will change dramatically in the final phase of its life.

In the short term, we saw in Section 5.2 that the interstellar medium plays a major role in the solar mass ejection, by providing the low-pressure exit of the solar wind nozzle. We address in this section this aspect of the interaction, studying the large-scale structure of the cavity – the heliosphere – that the solar wind carves in the interstellar medium. In the spirit of this book, we focus on basic physics, and refrain from doing things in full detail. More may be found in the reviews [17], [14] and [6] and in the older but still excellent account [19].

Table 8.1 *Main physical properties of the local interstellar medium as inferred in 2005.*

Number density of neutral hydrogen, n_n (m^{-3})	Number density of protons, n_I (m^{-3})	Temperature, T_I (K)	Speed relative to the Sun, v_I, making an angle of about 6° to the ecliptic (m s^{-1})	Magnetic field strength, B_I (10^{-10} T)
0.2×10^6	$(0.04 - 0.07) \times 10^6$	$(6 - 7) \times 10^3$	2.6×10^4	2–3

The Sun lies in the remote part of our galaxy, and its motion carries it in various regions of the interstellar medium, the inhomogeneous and unsteady medium separating the stars [7]. The Sun is presently travelling through a small interstellar cloud: our *Local Cloud*, in which it is expected to remain embedded for several thousand years more.

8.1.1 The Local Cloud

We penetrate here into unsafe ground, since no spacecraft has ever explored this medium. Furthermore, being inside the structure, our view is highly biased, and our knowledge is based on ingenious – but unsecured – inferences from various observations. As happens in the exploration of new territories, discoveries are accumulating faster than the digestion rate of textbook writers.[1]

To a first-order approximation, the Local Cloud is believed to be made of partially ionised hydrogen of temperature a little under 1 eV, moving at a few tens of km s^{-1} with respect to the Sun in a direction making a small angle to the ecliptic plane, and permeated with a magnetic field of a few 10^{-10} T (Table 8.1); nothing being simple, this field is neither roughly parallel nor roughly perpendicular to the velocity. The medium also contains dust grains and relativistic charged particles – *cosmic rays*; the latter are negligible in number density, but represent an appreciable part of the total energy density.

The neutral particles (whose mean free path is comparable to the size of the heliosphere) and the dust grains interact relatively weakly with the solar wind plasma; the main agents of interaction are the interstellar charged particles and magnetic field, which control the large scale structure of the heliosphere via their pressure, confining it into an elongated bubble with a tail, somewhat as a planetary magnetosphere. Table 8.2 lists the energy densities of these plasma constituents. This shows that:

- the kinetic energy of bulk motion, $n_I m_p v_I^2 / 2$,
- the thermal energy of protons and electrons, $2 \times 3 n_I k_B T_I / 2$,
- and the magnetic energy $B_I^2 / 2\mu_0$,

are each in the range of $(1 - 4) \times 10^{-14}$ J m^{-3}.

---

[1]The Local Cloud in which the heliosphere is presently embedded belongs to a family of interstellar clouds known as the *local fluff*, or *CLIC* (*Complex of Local Interstellar Clouds*).

The frontier of the heliosphere 421

Table 8.2 *Energy density of the interstellar plasma near the Sun*

Bulk plasma motion, $\rho_I v_I^2/2$, with $\rho_I \simeq n_I m_p$ (J m^{-3})	Proton + electron thermal motion, $3 n_I k_B T_I$ (J m^{-3})	Magnetic field, $B_I^2/2\mu_0$ (J m^{-3})
$(2-4) \times 10^{-14}$	$(1-2) \times 10^{-14}$	$(2-4) \times 10^{-14}$

The energy density of high-energy particles, the cosmic rays, is estimated to be about $w_{CR} \simeq 3 \times 10^{-13}$ J m^{-3} ([7] and Problem 8.4.1), i.e. of the same order of magnitude as the total of the above contributions. However, cosmic rays contribute relatively weakly to controlling the heliosphere because of their large radius of gyration in the heliospheric magnetic field. We shall see below that only the lower energy cosmic rays have a sufficiently small gyroradius to interact significantly with the bulk plasma, and they represent a badly known fraction of the total cosmic ray energy. Furthermore, for relativistic particles, the pressure is one-third of the energy density, compared to 2/3 for non-relativistic particles (Section 2.1). As a consequence, cosmic rays may be estimated to contribute an equivalent pressure of roughly a tenth of their energy density w_{CR} or less.

As a consequence of the similar energy densities of the plasma bulk motion, the thermal motion and the magnetic field, the corresponding speeds are comparable, too. More precisely, the bulk speed v_I of the interstellar plasma is somewhat smaller than the proton thermal speed $(2k_B T_I/m_p)^{1/2} \simeq 10^4$ m s^{-1} and the sound speed, but it is close to the Alfvén speed $B_I/\sqrt{\mu_0 \rho_I} \simeq (2.1-2.7) \times 10^4$ m s^{-1} and to the magnetosonic speed.

8.1.2 Basics of the interaction

The plasma

The interaction of the solar wind with the interstellar plasma has much in common with its interaction with Solar System bodies (Section 7.2). From the point of view of the solar wind, the flowing magnetised interstellar plasma is an obstacle that forces it to stop. Since the solar wind is supersonic, its slowing down is mediated by a shock, where it becomes abruptly subsonic and can subsequently slow down smoothly. It stops farther away at a contact surface (the *heliopause*), that separates solar from interstellar material. The region between the solar wind terminal shock and the heliopause contains solar wind plasma that has been slowed down, heated and compressed at the shock, and continues to slow down until it encounters the heliopause; this so-called *heliosheath* is akin to planetary magnetosheaths (Section 7.2). Since the wind supplies in permanence new matter towards the frontier, the only way for the flow to exit is to turn down at the stagnation point and be ejected downward, forming a heliotail somewhat as the tail of a planetary magnetosphere (Fig. 8.1). The geometry of the solar wind magnetic field (which is highly wound up at large distances (Fig. 6.2, right panel)) makes the structure relatively complex.

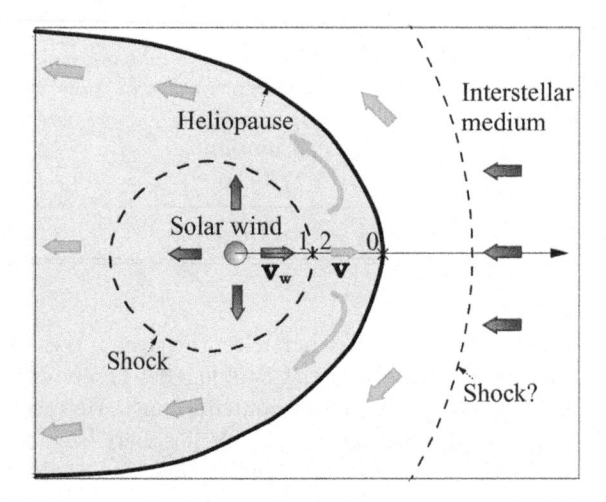

Figure 8.1 Simplified sketch of the interaction of the solar wind with the interstellar wind due to the relative motion of the Sun and the local interstellar medium. Both winds (dark grey arrows) slow down upon encountering each other, and stagnate at the heliopause (black line). The solar wind becomes subsonic (1–2) at a shock (dashed) and then slows down to stop at the heliopause (0) and turn sideways to form a tail (light grey arrows). The interstellar wind slowdown is mediated by a shock if the bulk speed is greater than the compressive wave speeds. The interstellar magnetic field is expected to distort the structure as shown in Fig. 8.2.

Consider now the point of view of the interstellar medium. It sees the magnetised heliosphere as an obstacle on which it arrives at the speed v_I (the relative speed between the local interstellar medium and the Sun). It must stop ahead of it and sweep around, somewhat as the solar wind encountering a large body of the solar system, with, however, two main differences. First, since it is presently not clear whether the bulk speed is larger than the speed of compressive plasma waves in the interstellar medium (and if it is so, this is by a very small amount), a plasma bow shock does not necessarily form. Second, the interstellar medium contains a huge proportion of neutrals.

The neutrals

The interstellar neutrals interact relatively weakly with the solar wind plasma. The main interaction is by charge exchange between proton and hydrogen atoms, and the corresponding free path is large enough for the neutrals to flow relatively freely through the solar wind, to a first approximation.

To a second approximation, however, the charge exchange between protons and neutrals produces significant effects. This exchange takes place both outside and inside the heliosphere. Outside the heliosphere, the exchange of one electron between interstellar H atoms and interstellar protons produces fresh interstellar

The frontier of the heliosphere 423

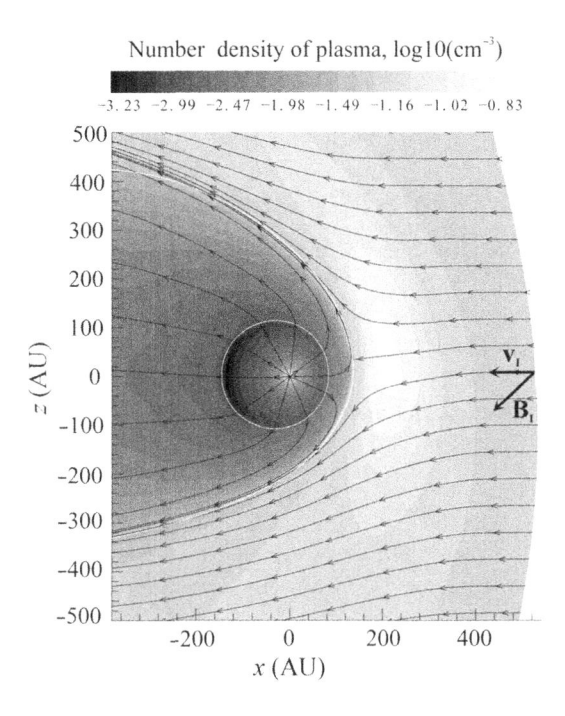

Figure 8.2 A model of the distortion of the heliosphere produced by the interstellar magnetic field (B_I). Isolines of the plasma density, and plasma stream lines, showing the asymmetry of the interstellar pile-up region, of the heliospheric shock and of the heliopause. (Courtesy of V. Izmodenov; adapted from [12].)

neutrals having the properties of the interstellar plasma. Since this plasma is compressed and heated in slowing down before the heliopause, the charge exchange produces a 'wall' of interstellar hydrogen of relatively large density and temperature just before the heliopause, on the interstellar side. We shall see later that similar *hydrogen walls* have been detected from the absorption spectra of several other stars, providing an indirect detection of distant stellar winds. Inside the heliosphere, when an interstellar H atom gives its bound electron to a solar wind proton, this produces a neutral having the solar wind speed – which leaves the system – and a proton which is quickly picked up by the solar wind, making it slow down. Furthermore, when an interstellar H atom gives its electron to a highly ionised heavy ion of the solar wind, this produces X-ray emission, just as occurs for planetary and cometary neutrals (Section 7.5).

Since in approaching the Sun, both the solar ionising radiation and the solar wind density are increasing, the question arises as to how close to the Sun interstellar atoms can penetrate before being ionised by photoionisation and by charge exchange. This distance d_{ion} may be estimated by noting that the ionisation timescale τ_{ion} of neutrals at this distance is of the order of magnitude of the time they take to travel this distance, i.e. $\tau_{ion} \sim d_{ion}/v_I$. The ionisation timescale τ_{ion} may be estimated as in the case of comets (Section 7.5). Ionisation is produced by:

- the solar ionising flux $F_\odot$ with the cross-section σ_{ion},
- the solar wind flux $n_w v_w$ with the charge exchange cross-section σ_{ex}.

The total rate of ionisation per atom is therefore $\tau_{ion}^{-1} = F_\odot \sigma_{ion} + n_w v_w \sigma_{ex}$. We found in Section 7.5 that $\tau_{ion} \simeq 10^6$ s at 1 AU ($d_\oplus$) from the Sun, on average over the solar cycle, varying as the square of the heliocentric distance r. If the parameters for hydrogen ionisation are of the same order of magnitude as those for cometary atoms, we have $\tau_{ion} \sim 10^6 \times (r/d_\oplus)^2$ at the heliocentric distance r, so that the heliocentric distance d_{ion} is given by $10^6 \times (d_{ion}/d_\oplus)^2 \sim d_{ion}/v_I$, whence[2]

$$d_{ion}/d_\oplus \sim 10^{-6} d_\oplus/v_I \sim 6. \tag{8.1}$$

Hence, ionisation near the Sun is sufficiently weak that an interstellar atom can penetrate down to a few AU from the Sun before being ionised. We shall return to the fate of these atoms later.

8.1.3 The size of the solar wind bubble

The size of the heliospheric bubble has been the source of speculations for half a century, mainly because of our ignorance of the state of the interstellar medium. This fostered a long suspense as to when the distant space probes would encounter the boundary, a suspense that ended in December 2004, when Voyager 1, the most distant spacecraft of the remote fleet (Section 6.3), heading towards the nose of the heliopause, finally encountered the heliospheric shock.

A simple estimate of the heliocentric distance of the shock, where the solar wind becomes subsonic (labels 1–2 in Fig. 8.1) before slowing down smoothly and stopping at a stagnation point (label 0 in Fig. 8.1) may be obtained by using a similar reasoning to that in Section 7.2.6. The total pressure at the solar wind stagnation point (label 0) is $P_0 \simeq 7 \rho_w v_w^2/8$, ρ_w and v_w being the solar wind mass density and speed just inside the shock (label 1 in Fig. 8.1). Neglecting the solar wind deceleration due to charge exchange, the solar wind ram pressure given in (7.18) yields at the shock distance d_{TS}: $\rho_w v_w^2 \simeq 2.1 \times 10^{-9} (d_\oplus/d_{TS})^2$, whence

$$P_0 \simeq 1.8 \times 10^{-9} (d_\oplus/d_{TS})^2. \tag{8.2}$$

Contrary to the situation in the solar wind, where the bulk kinetic energy dominates the thermal and magnetic energies, in the interstellar wind, all sources

---

[2]This order-of-magnitude estimate gives the same result as a more correct calculation. Indeed, the probability of ionisation upon travelling the distance dr (which takes the time dr/v_I) is $(dr/v_I)/\tau_{ion}$, so that the probability of ionisation of a neutral travelling from infinity to the distance d_{ion} is $\int_{d_{ion}}^{\infty} dr/(v_I \tau_{ion})$. Substituting the value of τ_{ion} at distance r, integrating and equating the result to unity yields (8.1).

of energy contribute significantly, so that the interstellar equivalent pressure is made of:

- the ram pressure $n_I m_p v_I^2$,
- the ion + electron thermal pressure $2 n_I k_B T_I$,
- the magnetic pressure $B_I^2/2\mu_0$,
- the cosmic ray effective pressure P_{CR}.

As we noted above, we approximate the effective cosmic ray pressure by $P_{CR} \sim 3 \times 10^{-14}$ Pa, and deduce from the values of Table 8.2 the total equivalent interstellar pressure $P_I \sim 1.4 \times 10^{-13}$ Pa, which must balance the pressure P_0. We deduce from (8.2) the heliocentric distance of the solar wind terminal shock

$$\frac{d_{TS}}{d_\oplus} \sim \left(\frac{1.8 \times 10^{-9}}{1.4 \times 10^{-13}} \right)^{1/2} \sim 10^2. \quad (8.3)$$

This puts the distance of the nose of the solar wind terminal shock at about 100 AU in order of magnitude, a value that varies with the solar cycle and the solar wind perturbations.

The magnetometer of Voyager 1 (whose plasma instrument is no longer operational) found a sudden abrupt magnetic field increase at 94 AU from the Sun (and a heliocentric latitude of 34°), a signature of the shock traversal that was backed up by a simultaneous increase of both the intensity and isotropy of galactic cosmic rays.

8.2 Cosmic rays

Cosmic rays are charged particles of high energy, which are ubiquitous in the Universe. The bulk of them are revealed by their electromagnetic radiation, but those reaching the Solar System play a unique role since we can observe them directly. Early studies of cosmic rays referred to those arriving in the Earth's atmosphere, but they are now measured in various regions, outside the ecliptic plane (aboard the Ulysses spacecraft), and even at the frontier of the heliosphere (aboard the Voyager spacecraft). The term *high-energy particles* is rather loose. In a wide sense, it might denote the particles of energy greater than average whose concentration exceeds their normal thermodynamic share, that is the Maxwellian level. The term, however, is generally used in a stricter sense, and rather denotes particles of energy greater than about 1 MeV per nucleon or electron.[3] Since for a proton the rest mass is $m_p c^2 \simeq 0.9$ GeV and for an electron about 0.5 MeV, the above energy range includes but is not limited to relativistic velocities.

---

[3] For example on Ulysses, the range of the detector of 'low-energy particles' is 50 keV to 5 MeV for ions and 30–300 keV for electrons, whereas the range of the detector of 'cosmic ray particles' (including those of solar origin) is 0.3–600 MeV per nucleon for ions and 4–2000 MeV for electrons.

These particles share a significant part of the energy in the Universe. We saw in Section 8.1 that their total energy density in the interstellar medium is estimated to be of the order of magnitude of 3×10^{-13} J m^{-3} (see Problem 8.4.1). If this is dominated by particles of energy around 1 GeV, i.e. $10^9 \times e \sim 10^{-10}$ J, the corresponding number density is of the order of 10^{-3} m^{-3}, smaller by many orders of magnitude than the average density of thermal particles in the interstellar medium.

Most of these particles come from outside the heliosphere, but the heliosphere produces them, too, since we have seen that the solar atmosphere, the planetary magnetospheres and the various shock structures in the heliosphere are powerful accelerators of particles.

We examine in this section the basic observed properties of cosmic rays, how they may be accelerated, and how the solar wind may affect them, focusing on basics. One must keep in mind that the question of the origin of cosmic rays is still an unsolved problem, despite nearly a century of research since the announcement of their discovery by V. Hess in 1912. Likewise, there is still no complete theory of how the solar wind affects them, and the available literature is merely an attempt to decode the as yet mysterious cosmic ray language. Reviews may be found in the books [15] and [3], and for example in [2] and [17] and the volumes [9] and [6].

8.2.1 Cosmic rays observed near Earth

Figure 8.3 shows the energy distribution of cosmic rays observed near Earth, by various methods using space and Earth-based detectors [4].[4] The distribution includes particles of various types, of which about 98% are protons and heavier nuclei, and the rest are electrons. It is plotted as differential flux (per steradian) versus energy, and shows two major properties. First, the spectrum is well represented by a power law over a large range of energies (11 orders of magnitude), although a comparison with a single power law (dotted line) shows breaks at a 'knee' and, to a lesser extent, at an 'ankle'. Overall, the differential flux may be expressed as

$$dF/dW \propto W^{-p} \qquad (8.4)$$

with the exponent p lying in the range $2.7 - 3.1$ for $W > 10^{10}$ eV. We shall see below that a relatively straightforward acceleration mechanism predicts indeed a power law spectrum, which is modified, however, by the particle propagation and losses throughout the Galaxy.

Second, the curve exhibits a pronounced attenuation at energies less than about 10^9 eV. This part of the spectrum is shown in Fig. 8.4 separately for hydrogen, helium and heavier nuclei, at different phases of the solar cycle; in each

---

[4]Because the Earth's atmosphere shields the primary cosmic rays, their direct observation must be made from space. However, at energies greater than about 10^{14} eV, the flux is so small (see Fig. 8.3) that space-bound detectors run out of statistics, and one must rely on Earth-based detectors, which yield indirect measurements, via the ionisation produced by cosmic rays in the atmosphere.

Cosmic rays

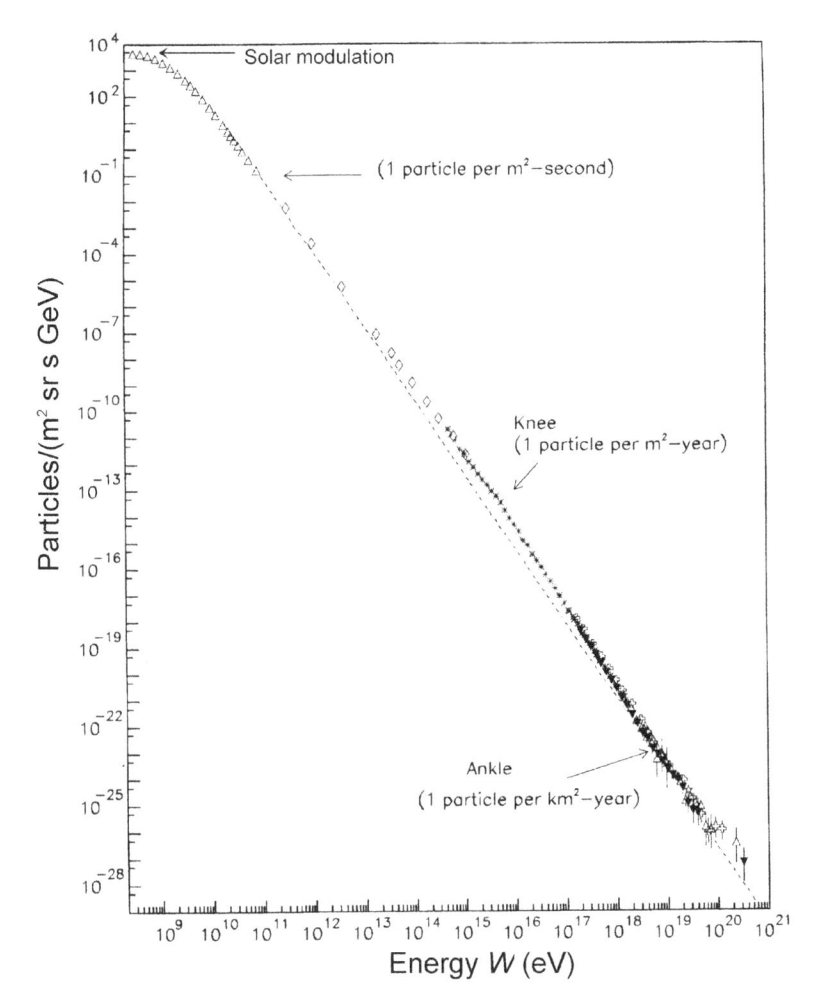

Figure 8.3 Measured energy spectrum of cosmic rays above 0.1 GeV (plotted as the differential flux per steradian dF/dW versus energy W), compared to a power law of index about -3 (dotted line). (Adapted from [4].)

set of spectra, the top (bottom) curve is relative to solar minimum (maximum) respectively [21]. One sees that the fluxes are strongly attenuated for energies per nucleon below about 10^9 eV, and that the attenuation increases as solar activity increases.

The modulation with solar activity is illustrated in more detail in Fig. 8.5, which shows the flux of protons of energy around 2×10^8 eV as a function of time from 1972 to 1997 (top), together with the mean sunspot number (bottom). This behaviour is attributed to cosmic ray propagation through the heliospheric magnetic field, whose structure complicates considerably as solar activity increases. One sees in Fig. 8.5 that the modulation depends not only on the sunspot number, but also on the parity of the solar cycle, through the direction of the solar

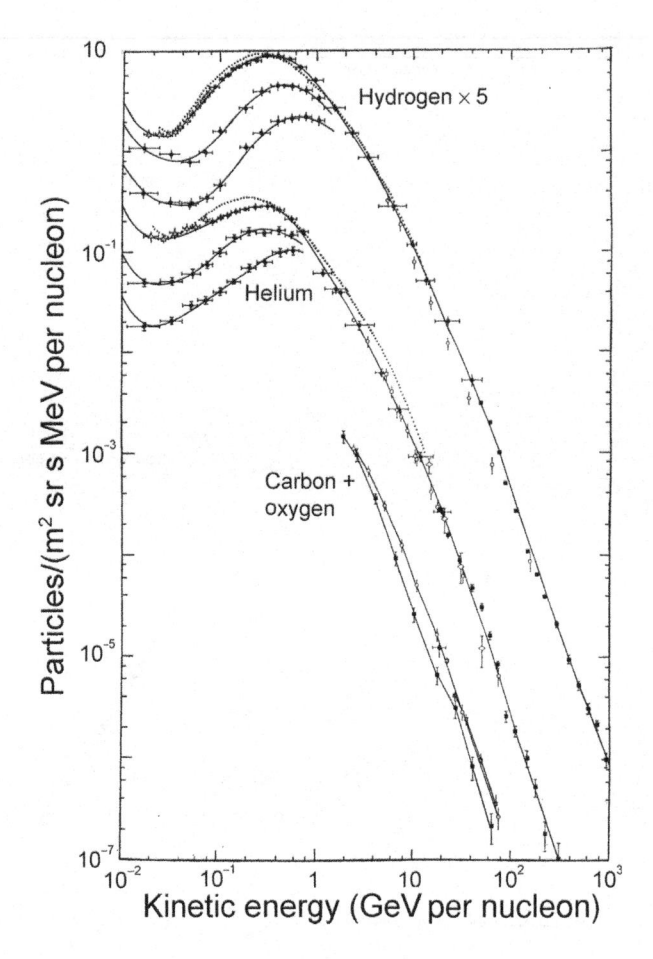

Figure 8.4 Measured energy spectrum of cosmic rays around 1 GeV (10^9 eV) per nucleon, plotted as differential flux versus energy per nucleon, separately for hydrogen, helium and heavier nuclei, at three epochs of the solar cycle. In each set of spectra, the top (bottom) curve is relative to solar minimum (maximum). Solid lines are smoothed fits to the data, dotted lines are sunspot minimum spectra measured in earlier studies. (Adapted from [21].)

magnetic dipole. This direction is indicated by the sign of the parameter A, which (by convention) is positive when the magnetic field points away from the Sun in the north hemisphere (the case shown in the figures of Sections 1.3 and 6.1), a configuration that we have seen to reverse every 11 years. This variation with the sense of the solar magnetic dipole holds because most cosmic rays are charged with the same sign (positive), and the Lorentz force reverses with the heliospheric magnetic field; the modulation then depends on the sign of A. In particular, the variation with the sunspot number is faster at solar minima having $A < 0$ than at those having $A > 0$. We shall find an explanation later.

Cosmic rays 429

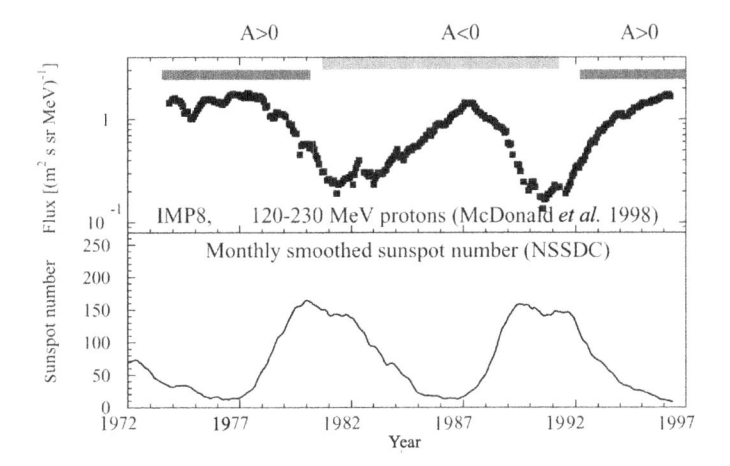

Figure 8.5 Modulation of cosmic rays with solar activity. Flux of protons of energy 120–239 MeV as a function of time from 1972 to 1997 (top), with the corresponding sunspot number (bottom). $A > 0$ ($A < 0$) denotes time periods when the solar magnetic field points outwards in the north (south) solar polar region respectively. (Adapted from [10].)

Figure 8.4 shows another intriguing point: there is an upturn in the cosmic ray spectra at low energies, and detailed measurements show that the spectra exhibit a bump at energies around 10–100 MeV (see Fig. 8.10 below). At such energies, the particles' properties are found to be different from those of galactic cosmic rays, whence their name *anomalous cosmic rays*. They are singly ionised, as are most pick-up ions, in contrast to the highly stripped galactic cosmic rays, their composition is similar to that of the interstellar medium, and their flux increases with heliocentric distance.

At still lower energies lie the particles that have been accelerated by transient events in the solar atmosphere and in other regions of the heliosphere, up to energies of the order of a few MeV per nucleon, above the background of solar wind ions and picked up particles (Section 6.5) whose energy is several keV per nucleon [16].

We shall examine these points in more detail in the following sections.

At the other extreme of the energy spectrum plotted in Fig. 8.3, near 10^{20} eV, the measurements are still inconclusive because of the rarity of the particles. This motivates a strong observational and theoretical activity because particles of energy per nucleon greater than 10^{20} eV pose a serious challenge for conventional theories.

Not only are those particles so energetic that even the boldest theoreticians run out of ideas to find a plausible scenario for accelerating them, but anyway, they would lose most of their energy before reaching the Earth, by colliding with photons of the cosmic microwave background pervading the Universe. Worse still, their huge energies make their gyroradii so large in the galactic and intergalactic magnetic fields that these fields should not perturb them. Their arrival directions should thus point back to their sources in the sky, which does not appear consistent with the available observations. All these difficulties are so serious that they pose a challenge to standard particle physics and cosmology [4].

8.2.2 Rudiments of the acceleration of particles

Let us examine how the particles may be accelerated. A basic requirement for particles to be accelerated to energies much above thermal is that the energy be not shared between all constituents of the medium. Namely, collisions and other equilibration mechanisms must be rare. This usually requires the medium to be dilute. Plasmas, however, share two properties that facilitate particle acceleration:

- the particles carry an electric charge and are thus subject to the electromagnetic field,

- the collisional free path is proportional to the energy squared and thus extremely large for energetic particles.

Since the subject of particle acceleration in plasmas may fill several volumes, we only consider a few basic acceleration mechanisms.

First of all, because the Lorentz force (Section 2.2) does not change the energy of a particle if the electric field vanishes, any acceleration mechanism relies ultimately on the electric field.

Consider the simple case of a static electric field. If the electric field amplitude is such that the energy gained by a particle of charge e over a mean collisional free path $eE \times l_f$ is greater than the thermal energy, then since the free path increases rapidly with speed, all the suprathermal particles of the velocity distribution (whose free path is still greater) will be accelerated without any impediment, and will thus run away from the bulk of the distribution. Because large-scale electric fields of large amplitude are rare in plasmas because of their large electrical conductivity, this acceleration process works generally only at small scales, for example at sites of magnetic field reconnection.

In practice, particles are generally accelerated by varying magnetic fields, which produce electric fields via Maxwell's equation $\nabla \times \mathbf{E} = -\partial \mathbf{B}/\partial t$. They are ubiquitous in the Universe on large scales, and on small scales too, due to MHD waves and turbulence.

Rigidity and radius of gyration

We saw in Section 2.2 that the dynamics of charged particles in a magnetic field is governed by two basic parameters:

- the rigidity $R = pc/q = (v/c)(W/q) = (v/c)[(W/e)/Z]$,

- the radius of gyration $r_g \sim p/qB = R/Bc$,

where v is the particle speed, W is the energy, $p = vW/c^2$ the relativistic momentum, $Z = q/e$ the number of elementary charges and B the magnetic field strength. In practical units, for a particle made of A nucleons, this may be written

$$R \simeq \frac{v}{c} \times \frac{A}{Z} \times \frac{W/e}{A} \equiv \frac{v}{c} \times \frac{A}{Z} \times \text{energy per nucleon (in eV)} \qquad (8.5)$$

$$r_g \sim \frac{R}{Bc} \sim \text{energy per nucleon (in eV)} \times \frac{v}{c} \times \frac{A}{Z} \times \frac{3 \times 10^{-9}}{B(\text{T})} \text{ m}. \qquad (8.6)$$

Particles of different charges and masses but with the same rigidity have the same dynamics in a given magnetic field configuration (Section 2.2). If two particles have the same velocity, and therefore the same Lorentz factor γ, they have the same energy per nucleon ($W/A \simeq \gamma m_p c^2$), and their rigidity depends only on their mass to charge ratio A/Z; whereas $A/Z = 1$ for protons, we have seen that for most heavy elements stripped of all their electrons, $A/Z \simeq 2$, so that all completely ionised elements have the same value of A/Z to a factor of two.

The gyroradius depends on the ratio of the rigidity to the magnetic field strength. It determines the minimum scale of a magnetic field structure capable of affecting the dynamics of the particle. From this argument, the galactic magnetic field $B \sim 10^{-10}$ T of spatial scale about 10^{19} m is expected to confine the bulk of cosmic rays of energy at least up to the 'knee' of Fig. 8.3; some observations are consistent with the production of these cosmic rays by shocks produced by the remnants of explosions of massive stars (the so-called *supernovae*; see Fig. 8.11 below) in the Galaxy, by a mechanism we shall examine later.

Betatron acceleration

Consider a particle in a uniform magnetic field, and let the magnetic field increase slowly compared to the particle gyration, so that the magnetic moment of the particle (the first adiabatic invariant) is conserved (Section 2.2). The momentum perpendicular to $\mathbf{B}$ thus increases as $p_\perp^2 \propto B$, whereas the parallel momentum $p_\parallel$ remains constant, so that the total particle energy increases.

This *betatron acceleration*, however, is a reversible process. Sooner or later the magnetic field should decrease, and when it returns to the initial value, the particle loses its energy gain. Assume, however, that before this happens, some irregularities scatter the particle faster than the gyration so that the magnetic moment is not conserved, whereas energy is conserved. This tends to isotropise the velocity distribution, making the parallel energy increase at the expense of the perpendicular one. If the magnetic field now decreases and returns to its original value, the perpendicular energy decreases accordingly, but the particle keeps the parallel energy gained during the stochastic part of the cycle, producing a net energy gain during the cycle.

This basic acceleration process appears under many guises, and is known as *magnetic pumping*. It is an instructive example of the interplay between a reversible effect and stochasticity.

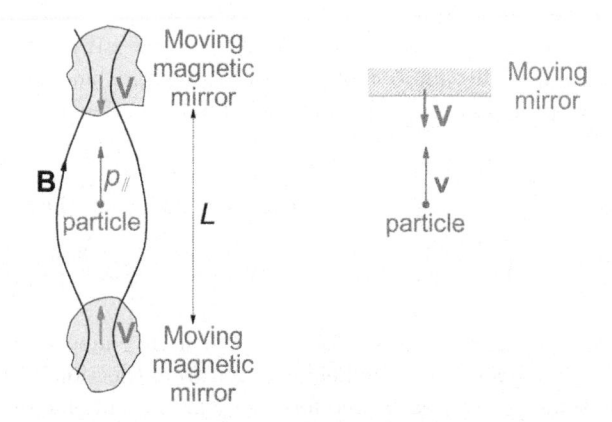

Figure 8.6 Fermi acceleration of a particle trapped between two magnetic mirrors (left), and the equivalent acceleration of a particle upon head-on reflection by a moving mirror (right).

A simple form of betatron acceleration occurs in planetary magnetospheres, when time variations enable particles to move from a large distance where the planetary magnetic field is small, to a small distance where it is large. This may increase the particle perpendicular energy by a factor of the order of the ratio of the magnetic fields, i.e. the cube of the size of the magnetosphere expressed in planetary radius.

Magnetic pumping is expected to take place at small scales, in low-frequency waves.

Fermi acceleration

Fermi acceleration, originally proposed by Fermi in 1949 to explain the acceleration of cosmic rays, is at the basis of most acceleration mechanisms thought to act in astrophysics; as the betatron mechanism, it requires a stochastic process to act. Fermi acceleration is based on the scattering of particles on large clumps of plasma that distort the magnetic field, producing magnetic mirrors which reflect the particles. The particle's energy is not changed in the mirror's rest frame, but if the mirror is moving towards the incident particle, the particle gains energy upon reflection, just as does a tennis ball pushed by a racket. Repeated scattering of the particles by randomly moving 'mirrors' (irregularities of the turbulent magnetic field) produces a net transfer of energy from the moving irregularities to the individual charged particles.

Let us study this in more detail. Just as betatron acceleration, Fermi acceleration may be understood from the conservation of an adiabatic invariant (Section 2.2), in this case the second invariant. Consider a particle trapped between two *magnetic mirrors*, which may be formed by plasma clouds of large magnetic field (Fig. 8.6, left), and of mass much greater than the one of the particle itself. Let the distance L between the mirrors decrease slowly compared to the particle parallel motion, so that the second adiabatic invariant is conserved. The parallel momentum thus increases as $p_\| \propto 1/L$, making the total energy increase. Note that, as in the case of magnetic pumping, this acceleration cannot proceed indefinitely unless some isotropisation process acts, because the increase in $p_\|$ decreases the pitch angle of the particle, so that sooner or later it will no longer be reflected by the magnetic mirrors. Another difficulty is that the mirrors cannot approach each other indefinitely. If the mirrors are moving away instead of approaching each other, L increases, making $p_\| \propto 1/L$ decrease and return to its initial value after a cycle. A gain in energy thus requires the intervention of a stochastic process.

The acceleration by a magnetic mirror moving towards the particle is similar to that of a tennis ball by a racket. Consider the simple case of a particle of speed v impinging normally on a mirror (assumed infinitely massive) moving at speed $V \ll c$ towards the particle in the same direction (Fig. 8.6, right). Assume first the particle to be non-relativistic, so that in the frame of the mirror its speed is simply $v + V$. Upon elastic reflection, the particle's speed just changes of sign in this frame, becoming $-(v+V)$. Transforming back to the observer's frame, the particle's new speed is $-(v+V) - V = -(v+2V)$. The particle has thus gained the speed $\Delta v = 2V$ upon reflection. Generalising the calculation to relativistic particle velocities, we find that upon head-on reflection, a particle of momentum p (at normal incidence) gains the energy $\Delta W \simeq 2Vp$ for $V \ll c$. Using $p = vW/c^2$, and generalising to different velocity directions, we find that the relative energy gain is

$$\Delta W/W = -2\left(\mathbf{v} \cdot \mathbf{V}\right)/c^2 \qquad (8.7)$$

since only the projection of $\mathbf{v}$ on $\mathbf{V}$ plays a role. This energy variation is positive for a head-on collision, and negative for a following collision, just as for a tennis ball.

This would be great, were it not for a severe problem: the particle sees mirrors that are moving towards it and away from it, which at first sight should cancel the effect. This is not exactly so, however, because, just as you get more rain on the front windscreen of your car than on the rear one, there are slightly more head-on reflections than tail-on ones. To calculate the balance, we note that the probabilities are proportional to the relative velocities of approach of the mirror and the particle, which are greater for head-on than for tail-on reflections. For $V \ll v$, the relative excess of head-on collisions over tail-on collisions is $2V/v$, so that the net relative energy gain per reflection is in average (considering reflection at normal incidence only)

$$\frac{\langle \Delta W \rangle}{W} \simeq \frac{2V}{v} \times \frac{2vV}{c^2} = 4\frac{V^2}{c^2}. \qquad (8.8)$$

This energy gain is of second order in the small parameter V/c, whence its name: *second-order Fermi acceleration,* so that the acceleration rate is in practice very small. It is generally much smaller than the escape rate of the particles from the scattering region, making this acceleration process rather ineffective.

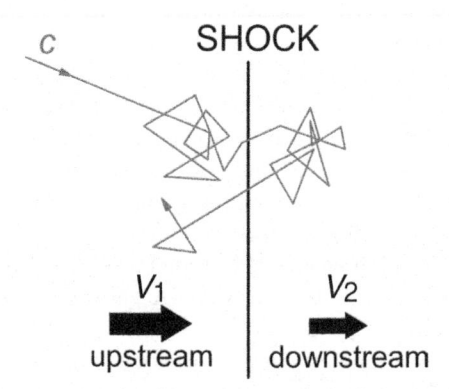

Figure 8.7 First-order Fermi acceleration of a particle near a (quasi-parallel) shock.

First-order Fermi acceleration at shocks

The original Fermi acceleration process is very slow because of the averaging between head-on and tail-on reflections, with only a small excess of head-on reflections, so that it is of second order in the small parameter V/c. The process would be much more effective if there were only head-on reflections. This is exactly what happens at shocks under adequate conditions [1]. The trick is that the medium undergoes a speed decrease from upstream to downstream at the shock, so that from the point of view of the medium on each side of the shock, the other side is moving towards it, whereas both the upstream and downstream regions are full of irregularities that scatter the particles and isotropise the velocity distributions in each frame.

To understand this, consider a plane shock, in which the velocity decreases from V_1 upstream, to $V_2 = V_1/n$ downstream, in the frame where the shock is at rest (Fig. 8.7). For a strong shock of adiabatic index $\gamma = 5/3$, we have seen in Section 2.3 that $n = 4$. In the frame of reference of the upstream medium, in which the velocity distribution of the particles is isotropic, the downstream medium (and its scattering irregularities) are coming head-on at speed $V_1 - V_2$ (the velocity difference between the scattering centres upstream and downstream); likewise, from the point of view of the downstream medium, the upstream medium (and its scattering irregularities) are coming head-on at the same speed. Therefore, each time an average particle (of velocity randomised by scattering on the irregularities) traverses the shock, it sees the plasma irregularities on the other side coming head-on at speed $V_1 - V_2$ in average. This is the basis of the diffusive acceleration at shocks, which is expected to have a great importance in astrophysics given the ubiquity of shocks in the Universe.

As an average particle of speed v traverses the shock downstream and back upstream, the total relative energy gain is given by twice the value given in (8.7), with the velocity $V = V_1 - V_2$, and $(\mathbf{v} \cdot \mathbf{V}) = -vV\cos\theta$, θ being the angle of incidence of the particle to the shock, which satisfies $0 < \theta < \pi/2$. For

Cosmic rays

ultra-relativistic particles ($v \simeq c$), this yields

$$\frac{\Delta W}{W} \simeq 4\frac{V_1 - V_2}{c}\cos\theta. \qquad (8.9)$$

We must average this expression over the directions of incidence. Since the rate of arrival of the particles to the shock is proportional to the normal component of the velocity ($\propto \cos\theta$), and the energy gain (8.9) is itself proportional to $\cos\theta$, we must average $\cos^2\theta$, which yields a factor $\langle\cos^2\theta\rangle_0^{\pi/2} = 1/3$. Hence, for a round trip across the shock and back again, the fractional energy increase of the particles is finally, on average

$$\frac{\langle\Delta W\rangle}{W} \simeq \frac{4}{3}\frac{V_1 - V_2}{c} \simeq \frac{4V_1}{3c}\times\frac{n-1}{n} \qquad (8.10)$$

where $n = V_1/V_2$ is the shock compression ratio. To work out the resulting energy distribution of the accelerated particles, we must estimate their probability of escape from the shock. Ultra-relativistic particles of number density n_{CR} arrive on the shock at the rate $n_{CR}\times c/4$ (the factor $1/4$ comes from averaging over the directions of arrival, as found in Section 7.2.2). On the other hand, the rate at which they are swept away from the shock (without returning) is, since this occurs downstream, $n_{CR}V_2 = n_{CR}V_1/n$. The escape probability of particles is thus $(V_1/n)/(c/4) = 4V_1/nc$. The ratio of the timescales of energy release and acceleration is therefore

$$\frac{4V_1/nc}{\langle\Delta W\rangle/W} \simeq \frac{3}{n-1} \qquad (8.11)$$

where we have substituted (8.10). We deduce (Problem 8.4.2, or using the reasoning of Section 4.5.3) that the differential energy distribution is given by (4.35), at high energies, i.e. $dN/dW \propto W^{-(\kappa+1)}$ with $\kappa = 3/(n-1)$, i.e.

$$dN/dW \propto W^{-(n+2)/(n-1)}. \qquad (8.12)$$

The particles thus emerge from the acceleration site with a power law spectrum, whose index depends on the shock compression ratio, and not on the shock speed, nor on the detailed geometry or the scattering process, as long as the shock may be considered as plane.[5] This makes this acceleration process universal.[6] For a strong shock with adiabatic index $\gamma = 5/3$, we have $n = V_1/V_2 \simeq 4$, so that the high-energy spectrum has the form $dN/dW \propto W^{-2}$, which agrees well with the observed cosmic ray spectrum (above 1 GeV and below the 'knee' of Fig. 8.3), when due account is made of particle propagation in the interstellar medium.

Three important comments are in order. First, the accelerated particles are expected to be confined within some distance from the shock, with a concentration that decreases farther away. Second, this mechanism requires the particles

---
[5] Namely, the radius of curvature of the shock must be much greater than the other scales.
[6] However, the timescale of acceleration does depend on the diffusion process.

to make multiple shock traversals. The scale of diffusion (and the particle gyroradius) must therefore be greater than the shock thickness. Since the shock thickness is of the order of magnitude of the gyroradius of the particles of the ambient medium, this condition means that the particles must already move faster than average, in order to pass freely between the upstream and downstream sides and be significantly accelerated.[7] Third, the average magnetic field has been implicitly assumed to be quasi-parallel to the shock normal (a so-called quasi-parallel shock; see Fig. 6.14, left).

Shock drift acceleration

In the opposite case, when the magnetic field makes an appreciable angle to the shock normal (a quasi-perpendicular shock; see Fig. 6.14, right), the motion of the plasma across the magnetic field produces an electric field $-\mathbf{V} \times \mathbf{B}$ (conserved upon shock traversal) in the frame of the shock. Furthermore, particles drift along the shock surface along $\mathbf{B} \times \nabla \mathbf{B}$ for positive charges (and the opposite for negative charges), perpendicular to both the magnetic field and its gradient (Section 2.2). Since we have seen that B increases upon traversing the shock from upstream to downstream (Section 6.3), the drift is in the same sense as the shock electric field for positive charges (and the opposite for negative charges), so that the particles are accelerated by the shock electric field whatever the sign of their charge. We shall see an application of this *shock drift acceleration* below.

A review of the physics of particle acceleration at shocks may be found in [13]. These mechanisms are responsible of a large part of particle acceleration in the Universe, from solar, heliospheric and planetary shocks, to distant astrophysical objects.

8.2.3 Modulation of galactic cosmic rays by solar activity

Galactic cosmic rays entering the heliosphere are subjected to:

- scattering by magnetic field irregularities due to turbulence, and by larger transient structures as the solar mass ejections,

- drifts due to the magnetic field gradient and the curvature of the field lines, determined by the three-dimensional structure of the heliospheric magnetic field, including the current sheet separating opposite magnetic polarities,

- outward convection and adiabatic deceleration as they follow the large-scale magnetic field of the expanding solar wind.

As a result, theoretical models have to consider the detailed structure of the magnetic turbulence, which we have seen to be far from understood, the

---

[7]These particles, however, may simply be the fast-speed particles of an ambient near-equilibrium velocity distribution.

Cosmic rays

full three-dimensional structure of the heliosphere, including the heliospheric current sheet, the sporadic structures, the terminal shock and the heliopause. Another difficulty is that the problem is not static; cosmic rays arrive at Earth after a long journey in the heliosphere where they interact with plasma emitted from the Sun at different times in the past, so that there is a time lag between observed cosmic ray properties and solar activity.

Scattering by turbulence

Scattering by magnetic field irregularities produces a diffusion that is maximum for particles whose gyroradius is of the order of magnitude of the scale of inhomogeneity. If the gyroradii are smaller, the particles simply conserve their adiabatic invariants, while in the opposite limit the particles only feel the average magnetic field. We have seen in Section 6.4 that magnetic turbulence produces irregularities at scales up to $l_{out} \sim 0.02$ AU at 1 AU. We thus expect that for the particles to be affected by the magnetic irregularities, their gyroradius (8.6) should satisfy $r_g \leq l_{out}$. With the average magnetic field $B \sim 5 \times 10^{-9}$ T (Section 6.1) at 1 AU, the energy per nucleon should therefore satisfy

$$\text{energy per nucleon} < \frac{B}{3 \times 10^{-9}} \times l_{out} \qquad (8.13)$$

$$\sim \frac{5 \times 10^{-9} \times 3 \times 10^9}{3 \times 10^{-9}} \sim 5 \times 10^9 \text{ eV} \qquad (8.14)$$

for the cosmic rays observed near Earth to be affected by the heliospheric magnetic field. Since at larger heliospheric distances, the average magnetic field strength and the outer scale of the irregularities vary roughly in opposite ways, this result should not be very sensitive to the heliocentric distance.

We thus expect turbulence to produce a break in the energy spectrum at energies below a few GeV, which agrees with the spectra plotted in Figs. 8.3 and 8.4. At these energies, a particle encountering a magnetic irregularity of amplitude δB and size l changes the inclination of its trajectory to the average magnetic field (the pitch angle) by about $\delta B/B$, so that its guiding centre is displaced by a distance $r_g \delta B/B$ if $l \sim r_g$. For the pitch angle to change by 1 radian (so that the particle has lost memory of its initial value), the particle must encounter a number N of such random deviations, producing a total deviation $\sqrt{N} \delta B/B \sim 1$ over a path $N r_g$. The free path is therefore in this simple case $r_g \times (B/\delta B)^2$.

Drift in the large-scale heliospheric magnetic field

Cosmic rays are also affected by the large-scale heliospheric magnetic field. The gradient and curvature of the magnetic field of Parker's spiral structure (Section 6.1) produces a drift velocity of order of magnitude $W/(qBr)$ at distance r, oriented along $\mathbf{B} \times \nabla \mathbf{B}$ for positive charges (Section 2.2.3). The resulting trajectories are shown in Fig. 8.8 in a meridional plane for a 2 GeV proton, when the heliospheric magnetic field $\mathbf{B}$ points outwards in the northern hemisphere

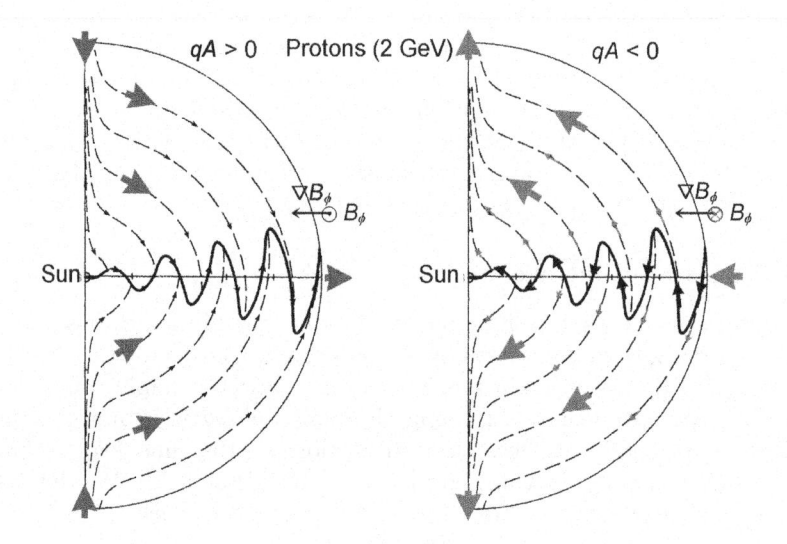

Figure 8.8 Trajectories (in meridional projection) of 2 GeV protons drifting in the heliospheric magnetic field, when **B** points outwards ($A > 0$, left) or inwards ($A < 0$, right) in the northern hemisphere. The wavy curve is the heliospheric current sheet along which the particles exit ($qA > 0$) or enter ($qA < 0$) the heliosphere; the parameters are the same as in Fig. 6.6, where $A > 0$. (Adapted from reference [16] of Chapter 6.)

($A > 0$, left), or inwards ($A < 0$, right). The bold line is the meridional cross section of the current sheet, which is given by (6.20) when the angle α between the solar dipole and rotation axes is small, a configuration that is relevant near solar activity minimum. One sees that in this case, the cosmic rays circulate in a quasi-static flow, entering over the poles and exiting along the sheet when $qA > 0$ (left), and the reverse when $qA < 0$ (right). Odd and even solar cycles are therefore expected to exhibit a different behaviour, as observed. In particular, when $qA > 0$, cosmic rays are not very sensitive to the shape and inclination of the current sheet, since they exit along it; in contrast, in cycles with $qA < 0$, cosmic rays are more sensitive to the shape of the current sheet, since they enter the heliosphere along it, and their observed flux should decrease as solar activity increases because the increasing complexity of the current sheet leads to additional scattering. The consequences may be seen on Fig. 8.5.

Figure 8.9 shows the effect of the current sheet that separates magnetic fields of opposite polarity. Near solar activity minimum, it has the simple wavy structure sketched in Fig. 6.6, whose tilt (wavy amplitude) and complexity increases as activity increases. Figure 8.9 sketches the trajectory of a particle of positive electric charge located at less than a radius of gyration of the current sheet. The magnetic field changes of sign at the current sheet, and so does the sense of gyration of the particle, making it drift outward (inward) if the magnetic field points outwards (inwards) in the northern hemisphere.

Cosmic rays

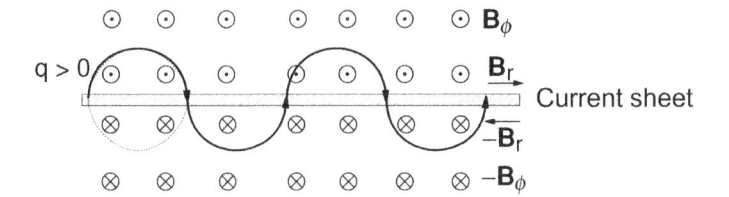

Figure 8.9 Sketch of the drift of cosmic rays along the current sheet. The dots and crosses show the direction of the azimuthal magnetic field, whose radial component points outwards (inwards) above (below) the current sheet. The heavy line sketches the trajectory of a positively charged particle, which gyrates clockwise around the field above the current sheet; as it crosses the current sheet, the sense of gyration reverses, so that the particle continues to drift outwards along the sheet. If either the sense of **B** or the sign of the charge q reverses, the drift proceeds inwards rather than outwards.

8.2.4 'Anomalous cosmic rays'

Figure 8.10 shows the energy spectrum (plotted as differential energy flux) of *anomalous cosmic rays* measured aboard the spacecraft Voyager 1 (at about 57 AU from the Sun and 32° heliolatitude). These particles are believed to be interstellar atoms that entered the heliosphere, and have been ionised either by charge exchange with the solar wind protons or by photoionisation. They are then immediately picked up by the solar wind and are carried out to the termination shock. A small percentage of these may be accelerated in this vicinity to cosmic ray energies and propagate back into the inner heliosphere, where they are observed as anomalous cosmic rays.

The maximum energy that may be acquired by the particles in this way may be estimated by simple arguments. The termination shock (Fig. 8.1) is approximately spherical so that it lies roughly along the heliospheric magnetic field, which is highly wrapped at this large distance (Fig. 6.2, right). In this quasi-perpendicular shock, the particles drift along the shock face, being accelerated by the shock electric field. They may therefore gain an energy equal to their charge times the total electric potential between the equator and the pole. With the electric field given in (6.12) this voltage is given by

$$\Phi_{EP} = \int_{\pi/2}^{0} E_\theta \times r d\theta = 1.5 \times 10^{11} \int_{\pi/2}^{0} E_\theta \times r_{AU} d\theta$$
$$\simeq 2 \times 10^8 \text{ V}. \tag{8.15}$$

Since the particles are singly ionised, they can gain a maximum energy of about 200 MeV, which is of the order of magnitude of the maximum observed energy of these particles (cf. Fig. 8.10). Since we have seen in Section 6.1 that the electric field points north (south) if $B_r > 0$ (< 0) respectively, the particles are expected to gain energy by drifting along the terminal shock from the equator to the poles during the $A > 0$ phase of the solar cycle (when **B** points outwards

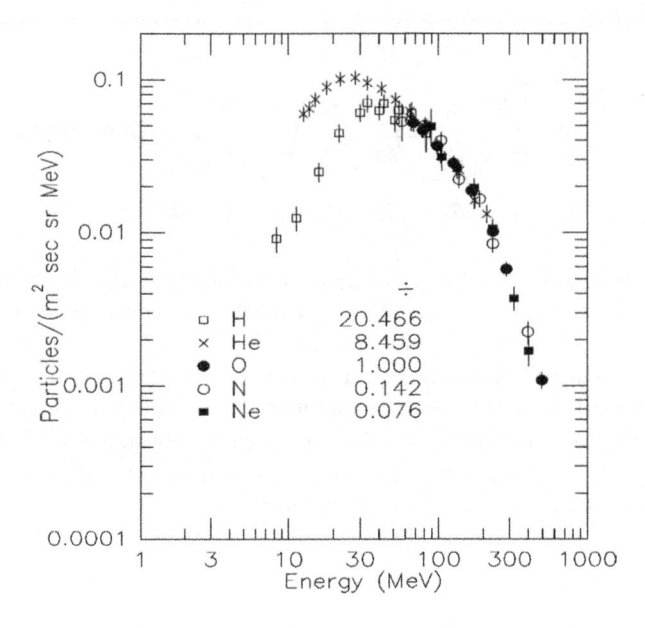

Figure 8.10 Energy spectrum of anomalous cosmic rays measured aboard the spacecraft Voyager 1 in 1994 (from days 157 to 365), plotted as differential fluxes versus energy. The fluxes of different ions are normalised to the one of oxygen by the factors indicated. (Adapted from [18].)

in the northern hemisphere), and from poles to equator during the $A < 0$ phase.

8.3 Examples of winds in the Universe

Virtually any object in the Universe is ejecting matter. Depending on the object, the ejection varies from steady to chaotic, from symmetrical to jet-like, and involves different physical effects, ranging from thermal evaporation to explosive events, radiation pressure or centrifugal ejection.

Solar mass ejections are examples of explosive events, which drive huge shocked balls of magnetised plasma, producing particle acceleration and other perturbations along their trajectory. We saw in Section 7.3 the consequences of such perturbations hitting the Earth's environment.

Farther from us, and rather frequent, are explosions of massive stars at the end of their life, known as *supernovae*. When fusion in the star's core ceases, the star can no longer resist gravitational attraction and collapses, and then recoils when the repulsion between nuclei begins to overcome gravity. The energy is transferred to the envelope of the star, which explodes, producing a shock where fusion processes act. This yields huge shocks propagating through the interstellar medium, which are revealed by the intense production of energetic particles

Examples of winds in the Universe 441

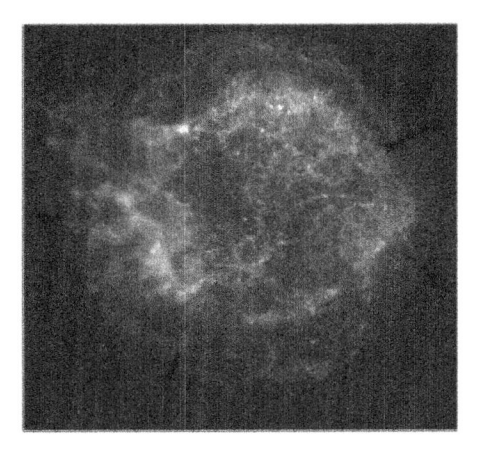

Figure 8.11 Image of the supernova remnant *Cassiopeia A* by the Chandra X-ray Observatory. (Adapted from [11].)

and electromagnetic waves (Fig. 8.11).[8] Not only may they be responsible for the acceleration of a part of the observed cosmic rays, as we already noted, but they supply heavy elements to the interstellar medium, providing the stuff from which new stars are born.

On the other hand, a host of objects eject matter in a steady way. The fast solar wind is the best-studied example of relatively steady ejection of stellar plasma. We have seen that planets and comets also eject matter in a quasi-steady way. Most stars eject quasi-steady winds, as do galaxies, too [5].

Figure 8.12 shows the cavity carved by the wind of a young bright star in the surrounding medium. This *astrosphere* is hugely larger than the heliosphere, by more than four orders of magnitude. The wind of this young star is flowing faster than the solar wind by a factor of four, and is inferred to have a mass flux greater by more than eight orders of magnitude than the solar wind, according to the scaling of the astrosphere's size as the square root of the wind ram pressure (cf. Eq.(8.2) and Problem 8.4.3).

Unfortunately, most of those winds are so distant that observation is not sufficient to fully constrain the proposed theories of their formation. Astronomers construct hypothetical scenarios, and estimate their validity by comparison with the observed electromagnetic emission. However, even though modern observation has considerably widened the spectral extent of the observed radiation, which ranges from radio waves to gamma rays, the available data leave much freedom to theoreticians and modellers.

8.3.1 Some basic physical processes in mass outflows

Most steady mass outflows are expected to involve a coronal plasma at the 'surface' of the parent object, and a process that drives this gas to supersonic

---
[8]In the jargon of astronomy, such an object is known as a *Type II supernova*.

Figure 8.12 Image of an astrosphere, the cavity – about 3×10^{17} m across – carved in the surrounding medium by the wind of a young bright star. (Image from the Hubble Space Telescope, credit ESA/NASA, Yaël Nazé (University of Liège, Belgium) and You-Hua Chu (University of Illinois, Urbana, USA).)

speeds and to escape from gravitational confinement. The restraining force is generally the gravitational attraction of the parent object, MG/r^2 (per unit mass) at radial distance r. The outward push, however, may involve various agents:

- the thermal pressure gradient producing a force, $-\nabla p$ per unit volume, as occurs at the Sun but we saw is not sufficient to produce the solar wind,

- non-thermal contributions, in the form of various plasma waves or suprathermal particles or both, probably produced by (turbulent) magnetic fields, as is expected to occur at the Sun, in a form still to be determined,

- radiation pressure, generally acting on atoms, via absorption of photons producing atomic transitions, and/or on dust grains coupled to the atmosphere,

- pulsations of the star, which produce shocks propagating in the atmosphere, heating and compressing the gas,

- the centrifugal force, acting via the magnetic field lines which entrain the plasma frozen to those lines by a 'sling-shot' effect produced by the magnetic tension force.

The three latter mechanisms require conditions far from those occurring at the Sun.

For radiation pressure on the gas to play an important role, the star's radiation temperature must be large enough (see Problem 8.4.4), which means that the star emits most of its radiation in the ultraviolet spectral range, where the atoms of the atmosphere have many spectral lines. The radiation pressure is mainly produced by absorption in these lines. The process is effective because there is a velocity gradient in the wind, so that atoms see the star's radiation Doppler-shifted to a frequency different from the one at which the atoms closer to the star saw (and absorbed) it.

For radiation pressure on dust grains to play an important role, the grains must be present at a place where their (collisional) coupling to the gas is sufficient to drive the wind. This requires the temperature to be sufficiently low to enable grain formation (instead of evaporation) close to the star, and a large density and mass loss rate. Thus, the star's effective temperature must be small enough, whereas its luminosity and mass loss must be large enough. In practice, this requires stars of very large size.

On the other hand, for the centrifugal force to play an important role in the wind acceleration, the object must be rapidly rotating and magnetised (unless the centrifugal force exceeds the gravitational force). This process, which results in an appreciable loss of angular momentum, may have played an important role in the young Sun, and explain how it lost most of its initial rotation speed at the beginning of its life.

In some galaxies, stars in formation and double stars, the geometry is far from spherical, and the collimation of the flow plays a major role, producing jets along two opposite directions, which often lie along the rotation axis. An extreme case of radiation pressure driven winds occurs in high-energy objects like pulsars and galactic nuclei, where a highly magnetised rotating object produces a large-amplitude electromagnetic wave at the rotation frequency, behaving as a giant (non-linear) magnetic dipole antenna.

8.3.2 Some empirical results on stellar winds

We have focused in this book on the fast solar wind, for two main reasons. It occupies most of the heliosphere, at least near solar activity minimum, and its relative stationarity and simple geometry should make it easier to model. And yet, we still do not have a proper theory of its origin. As we saw in Chapter 5, most theories stick to a (multi-)fluid description, adorned with more and more ad hoc refinements to accommodate more and more refined observations, whereas others claim that the issue is instead in a kinetic description able to address properly the heat transport, a problem still not fully solved, and whose solution may ultimately involve both points of view.

The fact that it is not stationary and its complex geometry make the slow wind's modelisation more complex, even though its lower speed is easier to reach. However, despite the differences between the slow and fast winds, and their different origins on the Sun, we have seen that their basic properties are similar:

- their particle fluxes differ by only 50%, providing a total mass loss $M'_\odot \simeq [1.-1.5] \times 10^9$ kg s^{-1}, which amounts to about $2 \times 10^{-14} M_\odot$ per year,

- their momentum fluxes differ still less, providing a ram pressure $\rho_w v_w^2 \simeq [2.1-2.6] \times 10^{-9}$ Pa at 1 AU from the Sun (proportional to the inverse squared distance),

- their total energy fluxes – the sum (per unit surface on the Sun) of the kinetic energy of the wind $M'_\odot v_w^2/2$ and the energy $M'_\odot M_\odot G/R_\odot$ needed to lift it out of the Sun's gravitational potential – are nearly equal to

$$\frac{M'_\odot}{4\pi R_\odot^2} \times \left(\frac{v_w^2}{2} + \frac{M_\odot G}{R_\odot}\right) \simeq 70 \text{ W m}^{-2} \qquad (8.16)$$

$$\simeq 10^{-6} \times \frac{L_\odot}{4\pi R_\odot^2}, \qquad (8.17)$$

showing that the engine producing the solar wind has a very small efficiency of about 10^{-6}, whatever the type of wind.

The equality of the energy fluxes is all the more interesting in that since the enthalpy at the base of the solar wind is much smaller than the other contributions (Section 5.3.1), the above expression is close to the heat flux required at the base to power the wind in the absence of other energy sources, if the expansion is spherical.

Still more interesting, the total energy flux to drive the wind is found to be similar within a factor of ten for a sample of cool stars covering nine orders of magnitude in mass loss when the Sun is included.[9] Namely, for each of these stars of mass loss M', luminosity L, mass M and radius R, one finds

$$\frac{M'}{4\pi R^2} \times \left(\frac{v_w^2}{2} + \frac{MG}{R}\right) \sim 10^2 \text{ W m}^{-2}. \qquad (8.18)$$

The same sample of stars also follow a widely quoted empirical relation which is usually written in the form (Fig. 8.13)[10]

$$M' \sim 5 \times 10^{-13} \times \frac{(L/L_\odot)(R/R_\odot)}{(M/M_\odot)} M_\odot \text{ per year.} \qquad (8.19)$$

It is probably significant that the latter empirical relation is also met by other kinds of cool stars, albeit with a different numerical factor, and is met by the Sun, too, within a factor of 20. Such a relation is suggested by dimensional arguments, and is not unexpected since the wind is ultimately powered by the star's luminosity and restrained by the gravitational attraction. Its fundamental significance, however, is not a trivial matter. A hint may be obtained by noting that since for these winds, the specific kinetic energy $(v_w^2/2)$ is smaller or of the same order of magnitude as the specific gravitational energy MG/R, the empirical law (8.19) may be rearranged in physical (SI) units as

$$M' \times \left(\frac{v_w^2}{2} + \frac{MG}{R}\right) \sim 2 \times 10^{-5} \times L. \qquad (8.20)$$

This means that the ratio between the stellar luminosity and the energy required to power the wind is the same for the stars obeying this relation. This suggests that the engines powering these stellar winds have a similar efficiency, of order

---
[9] From Reimers (1988) in [5], p. 25.
[10] Details on this relation also proposed by Reimers (1988) may be found in reference [27] of Chapter 5.

Examples of winds in the Universe

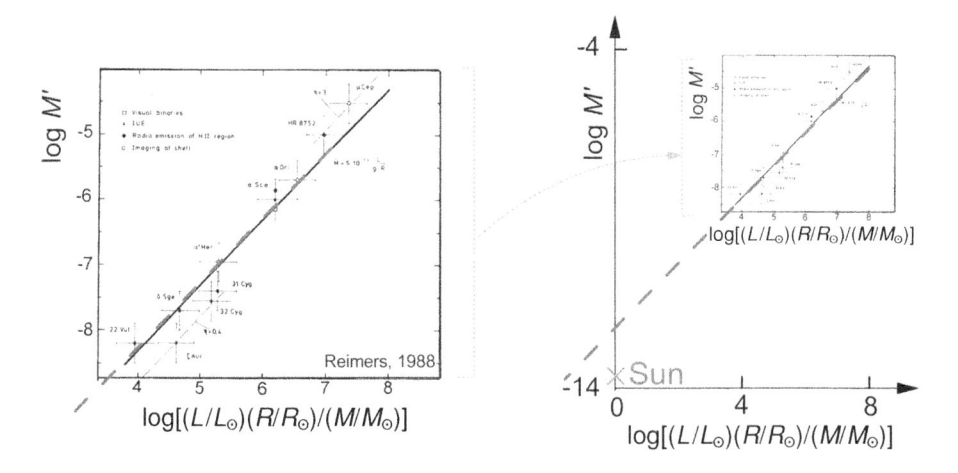

Figure 8.13 An empirical relation between the mass loss of stars (in units of $M_\odot$ per year) and the ratio of their luminosity to their gravitational potential MG/R (normalised to the solar values). Right: comparison with the Sun.

of magnitude 10^{-5}, i.e. of the same order of magnitude as the solar value (cf. (8.17)).

The data, however, are vastly insufficient. Figure 8.13 illustrates a major observational problem. Stars whose mass loss is measured relatively easily (Fig. 8.13, left) have mass losses greater than the solar one by more than five orders of magnitude, so that there is a large gap between the solar case and the other measured stellar winds. This is due to the difficulty of measuring distant winds of small mass loss; indeed, traditional measurements of winds apply down to a few times $10^{-10} M_\odot$ per year, still smaller by more than four orders of magnitude than the solar wind.

However, recent studies have opened new possibilities, based on the interaction of the wind with the neutrals of the surrounding interstellar medium. We have seen that interstellar neutrals exchange charge with interstellar protons and with solar wind protons and heavier atoms (Section 8.1), with two important consequences. First, the charge exchange with highly charged ions of the solar wind should produce X-rays that may be detected [20]. Second, the charge exchange with interstellar protons produces a wall of compressed hot hydrogen atoms just outside the astropause surrounding the star, which yields a detectable signature in the ultraviolet spectra.[11] This has enabled detection of a number of astrospheres, providing an indirect detection of solar-like winds [22].

8.3.3 The efficiency of the wind engine

Do the above empirical laws imply that the basic properties of winds can be deduced from the gross properties of the parent object? The answer is far from

---
[11]Through absorption of the so-called Lyman-α line – photons of energy $h\nu$ near the ionisation energy W_{Bohr} of the hydrogen atom.

trivial since we have seen that the wind also depends on the medium surrounding the object.

A familiar example may be instructive. The weather at any point on the Earth cannot be predicted from only the gross properties of the Earth and of its atmosphere, the solar luminosity and the co-ordinates of the point. So it is for the local mass loss from any point on a cosmic object. However, the typical velocity of winds on Earth can be deduced from the global properties of the atmosphere in a simple way [8].

A familiar example: the Earth's atmospheric wind engine

The kinetic energy involved for raising an atmospheric mass m_{atm} to a wind speed of v_w is $W_w \sim m_{atm} v_w^2/2$. With the atmospheric scale height H, the wind timescale is $\tau \sim H/v_w$, so that the power dissipated is $W_w/\tau \sim m_{atm} v_w^3/2H$. This power is furnished by the solar energy input, which is $L_\odot \pi R^2/4\pi d^2$ on a planet of radius R at heliocentric distance d.[12] If α_{ef} is the efficiency of the atmospheric engine, the typical wind speed in the planet's atmosphere is therefore given (in order of magnitude) by writing

$$m_{atm} v_w^3/2H \sim \alpha_{ef} L_\odot R^2/4d^2. \tag{8.21}$$

Substituting the mass of the atmosphere $m_{atm} = \rho_0 \times 4\pi R^2 H$ where ρ_0 is the mass density at the base of the atmosphere, this yields

$$v_w \sim \left(\frac{\alpha_{ef} L_\odot}{8\pi d^2 \rho_0}\right)^{1/3}. \tag{8.22}$$

With the solar luminosity $L_\odot$, the distance $d = d_\oplus$ (1 AU) and the mass density in the Earth's low atmosphere ($\rho_0 \sim n_0 \times 28 m_p$; see Section 7.1), we find a typical speed of atmospheric winds $v_w \sim 10 \times \alpha_{ef}^{1/3}$ m s^{-1}. The wind speed in the small-scale winds of the lower Earth's atmosphere is indeed about 10 m s^{-1} (with the global circulation involving greater speeds). This crude estimate suggests that the efficiency of the Earth's atmospheric engine for producing winds is not much smaller than unity. This is not too surprising since if this atmosphere were an ideal heat engine operating between temperatures T and $T - \Delta T$, the efficiency would be $\Delta T/T$; with ΔT equal to the temperature difference over a scale height in the Earth's atmosphere, this yields $\alpha_{ef}^{1/3}$ of order of magnitude unity. This estimate is little more than an improved dimensional analysis; obtaining a more precise value would require taking into account the precise mechanisms at work in the atmosphere, in particular the role of water vapour in the energy balance, and of the planet's rotation.

The efficiency of stellar wind engines

This simple example illustrates the difficulty of determining basic wind properties from first principles. The above order-of-magnitude estimate at Earth

---

[12] Approximating the planet's albedo by 1 in the spirit of this order-of-magnitude estimate.

Figure 8.14 Producing the winds of the Universe. (Jean Effel, *Turelune le Cornepipeux*, 1975, copyright Adagp, Paris, 2007.)

was reasonably successful because the efficiency of the wind engine (a nondimensional factor) did not produce a very small (difficult to estimate) factor in the final result. For the solar wind, however, (8.17) indicates that the efficiency is extremely small.[13] And we have seen (cf. (8.20)) that this is so for the stellar winds of the cool giant stars obeying the relation (8.19). Interestingly, for all their differences, the engines powering the solar wind (8.17) and the wind of these stars (8.20) have a similar efficiency of order of magnitude 10^{-6} to 10^{-5}.

Whereas it is relatively easy to estimate from first principles the total power furnished by the parent object, estimating the efficiency of the engine producing the wind is a more difficult task, that requires a subtle understanding of the wind production mechanism. For most cosmic winds, this goal is not yet attained, despite some imaginative proposals (Fig. 8.14).

---

[13]Of the same order of magnitude as the small convective temperature difference at the base of the convective zone estimated in Section 3.2.3.

8.4 Problems

8.4.1 Energy density of cosmic rays

From the flux of cosmic rays plotted in Fig. 8.3, estimate the differential energy spectrum of the number density of cosmic rays. Deduce an estimate of the total density of energy of cosmic rays, and show that the result is close to the value given in Sections 8.1–8.2.

8.4.2 Power law distribution of accelerated particles

Consider an acceleration process in which the timescale of energy increase is $1/\alpha$, so that the energy increase as a function of time may be written

$$dW/dt = \alpha W. \tag{8.23}$$

Show that the particles escape from the acceleration region with a timescale τ, so that the number density of particles of given energy varies with time as

$$dN/dt = -\alpha N - N/\tau. \tag{8.24}$$

Deduce that the differential energy spectrum of the accelerated particles has the form

$$dN/dW \propto W^{-\gamma} \quad \text{with } \gamma = 1 + \alpha^{-1}/\tau. \tag{8.25}$$

Hint

Note that $dN/dW = (dN/dt)/(dW/dt)$.

8.4.3 The size of an astrosphere

The astrosphere shown in Fig. 8.12 has a diameter of about 3×10^{17} m, and the wind of the parent star flows at about 2×10^6 m s^{-1}.

Estimate the mass flux of the star.
Compare to Fig. 8.2 and comment.

Hint

Assume that the size of the astrosphere scales as the distance to the star of the wind terminal shock.

8.4.4 Instability of a star's atmosphere produced by radiation pressure

Consider a star of mass M, radius R, surface temperature T_{eff} and luminosity L, and assume that an optically thin ionised layer of hydrogen lies just above its surface.

Show that the outward radiative force on an electron in the ionised layer is

$$F_{rad} = \frac{1}{3} a T_{eff}^4 \sigma_T \qquad (8.26)$$

$$= \frac{\sigma_T L}{3\pi c R^2} \qquad (8.27)$$

where σ_T is the Thomson cross-section (1.11)–(1.12).

Since the gravitational attraction on a proton is $F_G = m_p M G/R^2$, show that the star will be unstable and explode if

$$L/M > 3\pi c m_p G/\sigma_T. \qquad (8.28)$$

Substitute the numerical constants and show that the star is unstable if $L/M > 5$ (in SI units), or, substituting the Sun's parameters,

$$\frac{L}{L_\odot} \geq 3 \times 10^4 \frac{M}{M_\odot}. \qquad (8.29)$$

Beware that a realistic estimate should take into account the opacity and the role of line absorption by atoms, and consider the role of convection.

Look at Fig. 3.1 and comment.

Hints

From (3.28), the radiation pressure is $aT_{eff}^4/3$, with aT_{eff}^4 the radiative energy density. From (1.2), the star's luminosity is $L = \sigma_S T_{eff}^4 \times 4\pi R^2$ with the Stefan constant $\sigma_S = ac/4$.

References

[1] Bell, A. R. 1978, The acceleration of cosmic rays in shock fronts. I, *Mont. Not. Roy. Astron. Soc.* **182** 147.

[2] Belov, A. 2000, Large scale modulation: view from the Earth, *Space Sci. Rev.* **93** 79.

[3] Berezinskii, V. S. et al. 1990, *Astrophysics of Cosmic Rays*, Rotterdam, The Netherlands, North-Holland.

[4] Bhattacharjee, P. and G. Sigl 2000, Origin and propagation of extremely high-energy cosmic rays, *Phys. Rep.* **327** 109.

[5] Bianchi, L. and R. Gilmozzi eds. 1988, *Mass Outflows from Stars and Galactic Nuclei*, New York, Kluwer.

[6] Cairns, I. H. et al. eds. 2004, To the edge of the solar system and beyond, *Adv. Space Res.* **34** 1–225.

[7] Ferrière, K. 2001, The interstellar environment of our galaxy, *Rev. Mod. Phys.* **73** 1031.

[8] Golitsyn, G. S. 1970, A similarity approach to the general circulation of planetary atmospheres, *Icarus* **13** 1.

[9] Grzedzielzki, S. and D. E. Page eds. 1990, *Physics of the Outer Heliosphere*, Oxford UK, Pergamon.

[10] Heber, B. and R. A. Burger 1999, Modulation of galactic cosmic rays at solar minimum, *Space Sci. Rev.* **89** 125.

[11] Hughes, J. P. *et al.* 2000, Nucleosynthesis and mixing in Cassiopeia A, *Ap. J.* **528** L109.

[12] Izmodenov, V. *et al.* 2005, Direction of the interstellar H atom inflow in the heliosphere: role of the interstellar magnetic field, *Astron. Astrophys.* **437** L35.

[13] Jones, F. C. and D. C. Ellison 1991, The plasma physics of shock acceleration, *Space Sci. Rev.* **58** 259.

[14] Lallement, R. 2001, The interaction of the heliosphere with the interstellar medium, in *The Century of Space Science*, ed. J. A. M. Bleeker *et al.*, New York, Kluwer, p. 1191.

[15] Longair, M. S. 1997, *High-Energy Astrophysics*, Cambridge University Press.

[16] Mason, G. M. 2001, Heliospheric lessons for galactic cosmic ray acceleration, *Space Sci. Rev.* **99** 119.

[17] Smith, E. J. 2000, The heliosphere and beyond. II, *Proc. Int. School of Physics 'Enrico Fermi'*, IOS Press.

[18] Stone, E. C. *et al.* 1996, The distance to the solar wind termination shock in 1993 and 1994 from observations of anomalous cosmic rays, *J. Geophys. Res.* **101**, A5, 11017.

[19] Suess, S. T. 1990, The heliopause, *Rev. Geophys.* **28** 97.

[20] Wargelin, B. J. and J. J. Drake 2001, Observability of stellar winds from late-type dwarfs via charge exchange X-ray emission, *Ap. J.* **546** L57.

[21] Webber, W. R. and J. A. Lezniak 1974, The comparative spectra of cosmic-ray protons and helium nuclei, *Astrophys. Space Sci.* **30** 361.

[22] Wood, B. E. *et al.* 2000, Hydrogen Lyα absorption predictions by Boltzmann models of the heliosphere, *Ap. J.* **542** 493.

Appendix

Some fundamental constants and derived parameters

Speed of light	c	3×10^8 m s^{-1}
Gravitational constant	G	6.67×10^{-11} m^3 kg^{-1} s^{-2}
Boltzmann constant	k_B	1.38×10^{-23} J K^{-1}
Electron mass	m_e	0.91×10^{-30} kg
Proton mass	m_p	1.67×10^{-27} kg
Electron charge	e	1.6×10^{-19} C
Permittivity of free space	ϵ_0	8.85×10^{-12} F m^{-1}
Permeability of free space	$\mu_0 = 1/\epsilon_0 c^2$	$4\pi \times 10^{-7}$ H m^{-1}
Planck constant	h	6.63×10^{-34} m^2 kg^{-1} s^{-1}
Reduced Planck constant	$\hbar = h/2\pi$	1.05×10^{-34} m^2 kg^{-1} s^{-1}
Radiation constant	$a = \pi^2 k_B^4 / 15 \hbar^3 c^3$	7.56×10^{-16} J m^{-3} K^{-4}
Stefan–Boltzmann constant	$\sigma_S = ac/4$	5.67×10^{-8} W m^{-2} K^{-4}
Binding energy of the H atom	$W_{Bohr} = m_e e^4 / 8\epsilon_0^2 h^2$	2.17×10^{-18} J $= 13.6$ eV
Bohr radius	$r_{Bohr} = 4\pi\epsilon_0 \hbar^2 / m_e e^2$	5.3×10^{-11} m

Solar parameters

Solar radius	$R_\odot$	7×10^8 m
Solar mass	$M_\odot$	2×10^{30} kg
Solar luminosity	$L_\odot$	3.84×10^{26} W
Gravitational acceleration on the Sun	$g_\odot = M_\odot G / R_\odot^2$	272 m s^{-2}

Conversion factors

1 AU (Sun–Earth distance)	$d_\oplus$	1.5×10^{11} m
1 Ångström	1 Å	10^{-10} m
1 arc–sec on the Sun	$1.54 \times 10^{-6} \pi d_\odot$	730 km
Year	1 yr	3.16×10^7 s
Joule (SI unit of energy)	$1\text{ J} = 1\text{ kg m}^2\text{ s}^{-2}$	10^7 ergs
Electron volt	1 eV	1.6×10^{-19} J
Tesla (SI unit of magnetic field)	1 T	10^4 G

Commonly used symbols

$\mathbf{r}, r, t$	Position, radial distance, time
$f, \omega = 2\pi f$	Frequency, angular frequency
$\mathbf{k}$	Wave vector
$\mathbf{v}$	Velocity
T, P	Temperature, pressure
n, ρ	Electron (or ion) number density, mass density
w, W	Density of energy, energy
q, m	Charge, mass of a particle
μ	Mean particle mass ($\mu \simeq m_p/2$ in H plasma)
μ (associated to $\mathbf{B}$ or μ_0)	Magnetic moment
$\mathbf{E}, \mathbf{B}$	Electric field, magnetic field
$\omega_p = \left(\frac{ne^2}{\epsilon_0 m_e}\right)^{1/2}, f_p = \omega_p/2\pi \simeq 9 \times n^{1/2}$	Plasma (angular) frequency, plasma frequency
$f_{ge} = (eB/m_e)/2\pi \simeq 2.8 \times 10^{10} \times B$	Electron gyrofrequency
$L_D = \left(\frac{\epsilon_0 k_B T}{ne^2}\right)^{1/2} \simeq 69 \times (T/n)^{1/2}$	Debye length
$l_f \sim 10^8 \times T^2/n$	Mean free path of electrons, ions
$v_{the} = (2k_B T/m_e)^{1/2} \simeq 5.5 \times 10^3 T^{1/2}$	Electron thermal velocity ($\sqrt{2} L_D \omega_p$)
$V_A = \left(B^2/\mu_0 \rho\right)^{1/2}$	Alfvén speed
$\gamma = d\ln P/d\ln \rho$	Adiabatic index
$V_S = (\gamma k_B T/\mu)^{1/2}$	Sound speed
$\sim$	Equal to within an order of magnitude
$\simeq$	Equal to within a factor of two

When dealing with a large scope of subjects and conventions, the choice of symbols poses a dilemma. Rather than choosing unconventional symbols or multiply subscripts, I have used the same (conventional) symbol for different entities, when no confusion is possible. For example, μ denotes the mean mass per particle or the magnetic moment, depending on the context. The symbol $\mathbf{v}$ denotes the fluid velocity or the velocity of individual particles, depending on the context, except in Chapter 2 where the fluid velocity is noted $\mathbf{V}$, to avoid confusion.

Appendix

Characteristic scales

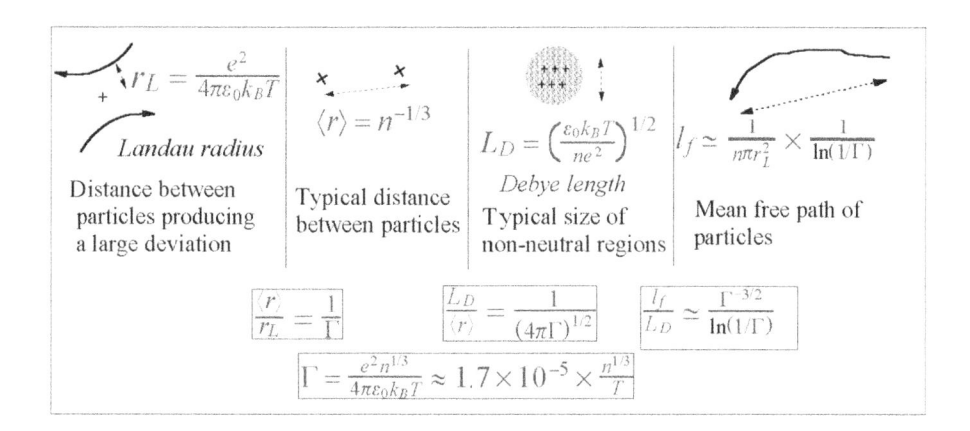

$$r_L = \frac{e^2}{4\pi\varepsilon_0 k_B T}$$

Landau radius

Distance between particles producing a large deviation

$$\langle r \rangle = n^{-1/3}$$

Typical distance between particles

$$L_D = \left(\frac{\varepsilon_0 k_B T}{ne^2}\right)^{1/2}$$

Debye length

Typical size of non-neutral regions

$$l_f \simeq \frac{1}{n\pi r_L^2} \times \frac{1}{\ln(1/\Gamma)}$$

Mean free path of particles

$$\frac{\langle r \rangle}{r_L} = \frac{1}{\Gamma}$$

$$\frac{L_D}{\langle r \rangle} = \frac{1}{(4\pi\Gamma)^{1/2}}$$

$$\frac{l_f}{L_D} \simeq \frac{\Gamma^{-3/2}}{\ln(1/\Gamma)}$$

$$\Gamma = \frac{e^2 n^{1/3}}{4\pi\varepsilon_0 k_B T} \approx 1.7 \times 10^{-5} \times \frac{n^{1/3}}{T}$$

Dynamics of a charged particle (q, m) in a magnetic field

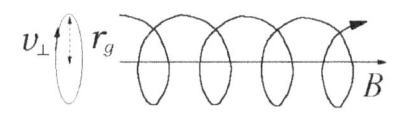

Change of reference frame with $v \ll c$	$\mathbf{B}' = \mathbf{B};\ \mathbf{E}' = \mathbf{E} + \mathbf{v} \times \mathbf{B}$
Cyclotron frequency	$\omega_g = \frac{|q|B}{m},\ f_g = \frac{1}{2\pi}\frac{|q|B}{m}$
Radius of gyration (Larmor)	$r_g = \frac{mv_\perp}{|q|B}$
Relativistic (and $\mathbf{F}_\parallel = 0;\ \partial \mathbf{B}/\partial t = 0$)	$m \to m\left(1 - v^2/c^2\right)^{-1/2}$
Electric field $\mathbf{E}$ or force $\mathbf{F}$	Drift speed: $\frac{\mathbf{E}\times\mathbf{B}}{B^2}$ or $\frac{\mathbf{F}\times\mathbf{B}}{qB^2}$
Weak B gradient ($\frac{\nabla B}{B} \ll \frac{1}{r_g}$)	Drift: $\left(\frac{mv_\perp^2}{2} + mv_\parallel^2\right)\frac{\mathbf{B}\times\nabla B}{qB^3}$
Slow variations ($\frac{\nabla B}{B} \ll \frac{1}{r_g},\ \frac{1}{B}\frac{\partial B}{\partial t} \ll \omega_g$)	$\mu = \frac{mv_\perp^2}{2B} \simeq$ constant
	(relat.: $\mu m \simeq$ constant)

Many particles (density n, temperature T, mass m)

Non-degenerate electrons	$T \gg T_F = \frac{\hbar^2}{2m_e}\frac{(3\pi^2 n)^{2/3}}{k_B} \simeq 4.2 \times 10^{-15} n^{2/3}$
Non-relativistic electrons	$T \ll m_e c^2/k_B \simeq 6 \times 10^9$ K
Three-dimensional	$f(\mathbf{v}) = \frac{n}{\left(\pi v_{th}^2\right)^{3/2}} e^{-\left(v_x^2 + v_y^2 + v_z^2\right)/v_{th}^2}$
Maxwellian distribution	
Most probable speed	$v_{th} = (2k_B T/m)^{1/2} = \sqrt{2/3}\langle v^2 \rangle$

Non-degenerate perfect gas

	Non-relativistic particles	Photons
Energy density w	$3nk_B T/2$	aT^4
Pressure P	$2w/3$	$w/3$
Adiabatic transformations	$P \propto n^{5/3}$	$P \propto n^{4/3}$

Fully degenerate electrons

	Non-relativistic	Ultra-relativistic
Pressure	$(3\pi^2)^{2/3}\hbar^2 n^{5/3}/(5m_e)$	$(3\pi^2)^{1/3}\hbar c n^{4/3}/4$

Appendix

Field of a magnetic dipole

If $\mathbf{J} = 0$, then Maxwell's equations yield: $\nabla \times \mathbf{B} = \nabla \cdot \mathbf{B} = 0 \Rightarrow \nabla^2 \mathbf{B} = 0 \Rightarrow$
$\mathbf{B}$ = magnetic field of a dipole + terms smaller by $1/r^n$ ($n \geq 1$).

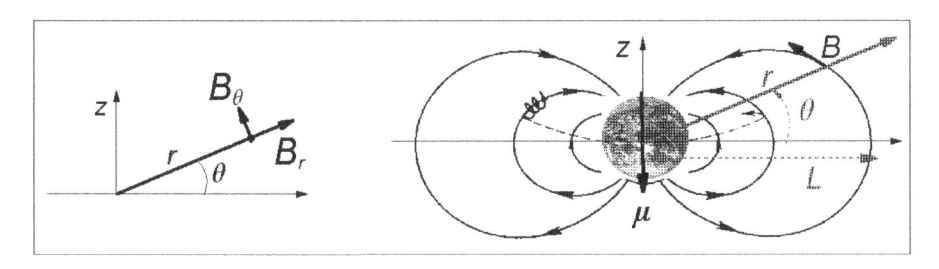

$$B_r = \frac{\mu_0}{4\pi} \frac{\mu}{r^3} \times (2\sin\theta) \quad \mu > 0 \text{ if } \boldsymbol{\mu} \text{ points to } z > 0$$

$$B_\theta = \frac{\mu_0}{4\pi} \frac{\mu}{r^3} \times (-\cos\theta)$$

$\Rightarrow$ Field lines: $r = L\cos^2\theta$

Example: the Earth

Magnetic moment $\mu_\oplus$	Radius $R_\oplus$	Magnetic field at equator B_0
-8×10^{22} Am^2	6.4×10^6 m	3×10^{-5} T

Useful vector identities

$$\mathbf{A} \cdot (\mathbf{B} \times \mathbf{C}) = \mathbf{B} \cdot (\mathbf{C} \times \mathbf{A}) = \mathbf{C} \cdot (\mathbf{A} \times \mathbf{B})$$
$$\mathbf{A} \times (\mathbf{B} \times \mathbf{C}) = (\mathbf{A} \cdot \mathbf{C})\mathbf{B} - (\mathbf{A} \cdot \mathbf{B})\mathbf{C}$$

$$\nabla \cdot (\phi \mathbf{A}) = \mathbf{A} \cdot \nabla \phi + \phi \nabla \cdot \mathbf{A}$$
$$\nabla \times (\phi \mathbf{A}) = \nabla \phi \times \mathbf{A} + \phi \nabla \times \mathbf{A}$$

$$\mathbf{A} \times (\nabla \times \mathbf{A}) = \nabla (A^2/2) - (\mathbf{A} \cdot \nabla) \mathbf{A}$$
$$\nabla^2 \mathbf{A} = \nabla (\nabla \cdot \mathbf{A}) - \nabla \times (\nabla \times \mathbf{A})$$

$$\nabla \cdot (\mathbf{A} \times \mathbf{B}) = \mathbf{B} \cdot (\nabla \times \mathbf{A}) - \mathbf{A} \cdot (\nabla \times \mathbf{B})$$

$$\nabla \times (\nabla \phi) = 0$$
$$\nabla \cdot (\nabla \times \mathbf{A}) = 0$$

$$\int_S d\mathbf{S} \cdot \mathbf{A} = \int_V dV (\nabla \cdot \mathbf{A})$$
$$\int_S d\mathbf{S} \times \mathbf{A} = \int_V dV (\nabla \times \mathbf{A})$$

Surface S encloses volume V; $d\mathbf{S}$ is a unit vector normal to the surface element and oriented outwards.

$$\int_C d\mathbf{l} \cdot \mathbf{A} = \int_S d\mathbf{S} \cdot (\nabla \times \mathbf{A})$$
$$\int_C d\mathbf{l} \times \mathbf{A} = \int_S (d\mathbf{S} \times \nabla) \times \mathbf{A}$$

Contour C encircles open surface S.

$$\nabla r = \mathbf{r}/r$$
$$\nabla \cdot \mathbf{r} = 3$$
$$\nabla \times \mathbf{r} = 0$$

$\mathbf{r}$ is radius vector from the origin.

$$(\mathbf{B} \cdot \nabla) \mathbf{A} = \sum_j B_j \partial \mathbf{A}/\partial x_j$$

Index

acceleration of particles, 430–436
    betatron, 431
    Fermi, 432–436
    first-order Fermi, 434–436
    magnetic pumping, 431
    magnetospheres, 385–389
    runaway, 430
    shock drift, 436
accretion, 232–236, 279–280
adiabatic processes, 75
    CGL, 81
    index, 75, 454
    invariants, 65–66
    temperature gradient, 127
Alfvén
    ballerina, 36, 296–299
    theorem, 89
    wave, 93
asteroid
    belt, 336
    magnetised, 357, 412
    mass distribution, 339
    Vesta, 412
astrosphere, 442, 448
aurora
    Earth display, 3, 372–389
    at giant planets, 374
    ovals, 374, 377, 386, 388
    physics, 385–389

belt
    asteroid, 336
    Kuiper, 336
    radiation, 384
Bernoulli theorem, 74–75
Biermann, 6
Birkeland, 4
    currents, 379

Bohr
    energy, 101
    radius, 101
Boltzmann equation, 70
Boojum, 213
breeze, 230, 235

Carrington, 2
CGL relations, 81
chaos, 320
charge exchange, 107
    at comets, 407
    interstellar neutrals, 422
    at planets, 365
collisions, 48–54
comet, 394–408
    atmosphere, 397–399
    coma, 400
    dust tail, 408
    Halley, 395
    ionosphere, 400–401
    magnetic cavity, 404
    magnetic pile-up, 403–404
    mass loading, 401
    nucleus, 394, 395
    pick-up, 401–402
    plasma tail, 5, 404–406
    X-ray emission, 406–408
composition
    solar, 326
    solar wind, 326
conductivity
    electric, 86–87
    thermal (collisional), 81–84
continuity equation, 69
convection, 125–135
    mixing length, 128
    Schwarzschild criterion, 127

convective
    collapse, 153
    derivative, 68
    energy flux, 130
Coriolis force, 143
corona, see solar, corona
coronal mass ejection, 176, 197–203, 309–311, 374
Corotating Interaction Region, 308–309
corotation
    magnetosphere, 383
    solar atmosphere, 300
cosmic rays, 425–440
    anomalous, 429, 439–440
    energy density, 421, 425
    energy spectrum, 427
    modulation, 427, 436–438
cross-section
    charge exchange, 107
    ionisation
        impact, 105
        radiative, 104–105
    neutrals, 48
    recombination
        dissociative, 106
        radiative, 104–105
    Thompson's, 21
current
    Birkeland, 379
    Chapman–Ferraro, 410
    electrostatic charging, 352–355
    heliospheric sheet, 36, 296–299
    in magnetosphere, 379, 382, 385
    magnetic pile-up, 409
    ring, 386, 411
cyclotron
    damping, 100
    frequency, 60
    radius, 60

de Laval nozzle, 277–279
    at comets, 412
    solar wind, 232
Debye shielding, 44–47, 107
drift velocity, 61–63
dust
    evaporation, 394
    Lorentz force, 393
    plasma drag, 393
    at comets, 408
    drag on gas, 412
    mass distribution, 339
    physics, 390–394
    Poynting–Robertson drag, 391–392
    radiation pressure, 391
dynamos, 138–145

Earth
    aurora, 372–375
    basic properties, 337
    bow shock, 370
    climate, 10, 379
    ionosphere, 346
    magnetopause, 368
    magnetosphere, 372–389
    wind speed, 446
ecliptic, 24
electric field
    Dreicer, 259
    Hall, 95
    magnetospheres, 381
    solar wind, 257–259, 268–272, 295–296
electrostatic charging, 352–356
energy
    Bohr, 101
    dipolar magnetic field, 379
    equation, 75, 82–85
    Fermi, 56–57
    interstellar medium, 421
    ionisation, 101
    kinetic
        non-relativistic, 43
        perfect, non-degenerate gas, 454
        ultra relativistic, 57
    magnetic, 91
    nuclear fusion, 118
    radiation, 116
enthalpy, 75
Euler equation, 75

Fermi energy
    non-relativistic, 56
    planet interior, 342
    stellar interior, 122
    ultra-relativistic, 57
fluid, 72–85, 100
    infinite hierarchy, 73
    versus kinetic, 84–85

Index

Fokker–Planck equation, 70
fragmentation equilibrium, 340
free path
    charged particles, 53, 84
    neutrals, 48
    photons, 117

giant planets
    aurora, 374
    ionospheres, 347
    magnetospheres, 367
guiding centre, 62
gyrofrequency, 60

heat
    conductivity (collisional), 81–84
    transport, *see* heat flux
    water sublimation, 397
heat flux, 81–85
    collisional, 81–84, 256
    kinetic, 263
    non-collisional, 272, 284–285
    observed, 257
    polytrope, 239, 256
heating of corona, 203–213
helicity, 143, 181–185, 191, 198–200
heliosphere, 419–425
    heliopause, 421
    heliosheath, 421
    hydrogen wall, 423, 445
helium, 327
hydrogen
    binding energy, 101
    wall, 423, 445

Inca quipus, 181
induction equation, 87
inertial length, 313
instability
    convective, 127
    convective collapse, 153
    fire-hose, 273
    kink, 180
    two-stream, 99, 318
intermittency, 320, 325
interstellar medium, 419
    energy, 421
    Local Cloud, 420
        solar wind interaction, 421
    local fluff, CLIC, 420
    magnetic field, 421
    neutrals, 422
ionisation, 101–107
    at comets, 400–401
    atmospheres, 344–347
    by charge exchange, 106
    by compression, 102, 343
    by impact, 105
    Chapman layer, 346
    in heliosphere, 424
    radiative, 103–105
    Saha formula, 103
    solar wind, 110
isothermal, 75

Jeans theorem, 70
Jupiter
    basic properties, 337
    magnetic tail, 383
    magnetopause, 368
    radioemission, 374

kinetic, 66–72, 100
Kuiper belt, 336

Landau
    damping, 99–100
    radius, 50
length
    Debye, *see* Debye shielding
    inertial, 95
Liouville theorem, 69
Lorentz
    factor, 59
    force, 59, 90

magnetic
    buoyancy, 151–153
    bottle, 64
    cloud, 310
    diffusion, 87–88, 139
        turbulent, 144
    dipole, 455
    energy, 91
    flux rope, 198
    flux tube, 150
    forces, 90–92
    helicity, 181–185, 198–200
    mirror, 64

magnetic (*cont.*)
   moment, 63–64
   monopoles, 58
   pressure, 91
      magnetospheres, 366
      solar atmosphere, 176
      solar wind, 348
   storm, 3, 385–387
   substorm, 387–389
magnetic field, 58
   amplification, 139, 147
   dipolar, 34, 455
   energy, 379
   force-free, 177
   freezing, 88–90
   interstellar medium, 421
   pile-up, 361, 364, 403–404, 409
   planets, 347
   potential, 177
   reconnection, 95, 376
   spiral structure, 291–299
   twist, 178–179
Magnetohydrodynamics, *see* MHD
magnetosphere
   aurora, 385–389
   basic structure, 375–385
   Chapman–Ferraro, 410
   circulation, 384
   corotation, 383
   current, 379, 382, 385
      Birkeland, 379
      ring, 384, 386, 411
   cusps, 376
   energy input, 378
   induced, 361–365
   magnetised planets, 365–389
   magnetopause, 366
   magnetosheath, 368
   particle acceleration, 385–389
   polar caps, 376
   radiation belts, 384
   storm, 385–387
   substorm, 387–389
   tail, 381–383
Mars
   aurora, 373
   basic properties, 337
   interaction with solar wind, 363
   ionosphere, 346

mass loss
   solar wind, 232, 238, 243, 248
   stars, 443–447
Maxwell equations, 58
Mercury, 337
   atmosphere, 344
   basic properties, 337
   magnetopause, 368
   magnetosphere, 367
meteors, 336–341
MHD, 85–96, 100
   discontinuities, 314–315
   non-ideal, 94–96
   shocks, 311–315
   turbulence, 323–325
   waves, 92–94
mirror asymmetry, 59, 143, 181–185
momentum equation, 73, 82
Moon
   atmosphere, 344
   basic properties, 337
   electrostatic charging, 409
   wake, 358–360

Navier–Stokes equation, 82, 320
Neptune
   basic properties, 337
   magnetopause, 368
nuclear fusion, 119

opacity
   photosphere, 133
   solar interior, 117

Parker, 6
   critical point, 228, 233, 243
   spiral, 291–299
pick-up, 328–329, 365
   at comets, 401–402
pitch angle, 60
Planck emissivity, 11
planets
   atmosphere, 344
      scale height, 337
   basic properties, 337
   bow shock, 368–371
   ionopause, 363
   ionosphere, 344–347
      Chapman layer, 344–346
      electric conductivity, 348

Index

magnetic field, 347
    dipolar, 455
magnetosheath, 368
magnetosphere, 361–389
mass density, 343
rotation, 347
plasma
    beta, 92
    collisions, 50–54
    coupling parameter, 44, 51
    degenerate, 56
    diamagnetic, 60
    drag, 392–393
    free path, 53, 85
    frequency, 47, 55
    oscillations, 54
    transport, 83–85
Pluto, 337
    atmosphere, 344, 347
    basic properties, 337
polytrope, 75
Poynting–Robertson force, 391–392
pressure, 72
    Fermi, 56–57, 454
    ionisation, 102
    kinetic, 72, 348
        ultra-relativistic, 57
    magnetic, 91, 348
    perfect, non-degenerate gas
        non-relativistic, 44, 72, 454
        ultra-relativistic, 57, 454
    ram, 348

quantum degeneracy, 56

radiation
    belts, 384
    blackbody, 11
    energy density, 116
    flux in stars, 127
    force, 391–392
    G-band, 155
    losses, 186–187
radius
    Bohr, 101
    cyclotron, 60
    of gyration, 60
    Landau, 50
    Larmor, 60

random walk, 82
ratio of specific heats, 75
recombination
    dissociative, 106
    radiative, 105
Reynolds number, 82
    magnetic, 88
rigidity, 430

Saturn
    aurora, 374
    basic properties, 337
    magnetopause, 368
scale height
    corona, 21, 177
    photosphere, 15, 132, 177
    solar atmosphere, 177
shock, 77–80
    drift acceleration, 436
    Fermi acceleration, 434–436
    magnetised collisionless, 311–315
    MHD, 311–315
    planets, 368–371
    Rankine–Hugoniot, 80
    terminal, 424–425
shower curtain, 74
solar
    activity, 142–158, 191–203
        cycle, 16
    atmosphere
        scale height, 177
    basic properties, 9
    chromosphere, 19, 166–174
        corona transition, 189
        density, 166
        temperature, 166
    constant, 10
    convection, 125–135
    corona, 19–24, 166–216
        coronal holes, 24, 167, 191, 216, 307
        density, 166
        F, K, 336
        filaments, 193
        heating, 203–213
        temperature, 166, 190
        visible radiation, 19
        X–UV emission, 22–24, 191, 194–204

solar (cont.)
  disc, 13
    limb darkening, 14
  eruptions, 195–203
  flares, 195–203
  granulation, 15, 134–135
  interior, 124–125
    convection, 125–132
    models, 124
    nuclear lifetime, 119
  ionising radiation, 345
  loops, 174, 191
  magnetism, 137–158
    surface, 15–18, 154–158
  neutrinos, 125
  opacity, 117, 133
  oscillations, 125
  photosphere, 10, 13–15, 132
    scale height, 15, 132
  probe, 275
  prominences, 175, 193–194
  rotation, 135–137
  spectrum, 10
  spicules, 174, 191
  supergranulation, 15, 154
solar wind
  angular momentum, 300
  basic properties, 348–351
  bubble, 424–425
  current sheet, 296–299
  electric field, 257–259, 268–272,
      295–296
  exospheric, 268–274
  fast, 27, 30, 253, 308
  heat flux, 256–257
  interaction with bodies, 349–412
  magnetic cloud, 310
  magnetic field, 291–305
  mass loading, 401
  mass loss, 33, 232, 238, 243, 248, 311
  minor constituents, 326–329
  observations, 24–33
  polytrope, 239–245, 281
  pressure
    kinetic, 348
    magnetic, 348
    ram, 348
  slow, 27, 30, 308
  solar connection, 305–307
  temperatures, 253
  theory, 223–277
  turbulence, 323–325
  velocity distributions
    electrons, 264
    protons, 264
  waves, 315–318
space weather, 197, 311, 389
spacecraft
  early wind observations, 7
  Pioneer, Voyager, 24, 304
  solar observations, 9
  strawman payload, 26
  Ulysses, 28–30
  wind observations, 24–33
speed
  Alfvén, 92
  escape
    planets, 344
    Sun, 228
  magnetosonic waves, 94
  mean square, 67, 267
  most probable, 67, 267, 454
  sound, 76
sputtering, 372
star
  accretion, 232–236, 279–280
  brown dwarf, 113, 123, 343
  central temperature, 116
  lifetime, 119
  luminosity, 116–118
  main-sequence, 114
  mass, 121–123
  pulsation, 120
  stability, 121
  wind, 440–447
Sun, see solar
sunspot
  Janssen's photograph, 2
  modern photograph, 13
  physics, 156–158
  properties, 15
supernova, 431, 440

temperature, 43, 283–284
  adiabatic gradient, 127
  comet plasma tail, 404
  effective, 10
  Fermi, 57
  ionosphere, 347
  kinetic, 43

Index

photosphere, 10
solar
  corona, 190
  interior, 124
  wind, 253
stellar interior, 116
Thomson scattering, 20
time
  Alfvén, 96
  free-fall, 120
  Kelvin–Helmholtz, 119
  nuclear, 119
  pulsation, 120
transport
  heat, 81
  in plasmas, 83–85
  momentum, 81
turbulence, 318–325
  Kolmogorov law, 321–323
  Kraichnan, 330
  MHD, 323–325
  solar wind, 323–325

Uranus
  basic properties, 337
  magnetopause, 368

vector
  identities, 456
  pseudo, 59
velocity
  average, 67
  drift, 61
  filtration, 72, 109, 209–213
velocity distribution
  bi-Maxwellian, 81
  kappa, 201, 265, 282
  Maxwellian, 67–68, 454

solar wind electrons, 264
solar wind protons, 264
Venus
  basic properties, 337
  interaction with solar wind, 363
  ionosphere, 346
Virial theorem, 115
viscosity, 81–84, 249, 320
Vlasov equation, 69

wake, 357–360
waves
  conventions, 98
  electromagnetic, 96–98
  fast, 94, 311
  in solar wind, 315–318
  ion–acoustic, 76–77
  Langmuir, 98–99
  MHD, 92–94
  planets, 389
  shock, 77
  slow, 94, 311
  sound, 76
  whistler, 98
wind
  astrophysics, 442
  Earth atmosphere, 446
  star, 443–447
  Sun, *see* solar wind

X-ray emission
  comets, 406–408
  heliosphere, 326, 423
  planets, 326, 365
  solar corona, 22, 191

zodiacal light, 336

Made in the USA
Monee, IL
05 August 2023